Adaptive Control for Robotic Manipulators

Adaptive Control for Robotic Manipulators

Editors

Dan Zhang

Department of Mechanical Engineering
Lassonde School of Engineering
York University
Toronto, ON
Canada

Bin Wei

Faculty of Engineering and Applied Science
University of Ontario Institute of Technology
Oshawa, ON
Canada

CRC Press
Taylor & Francis Group
Boca Raton London New York

CRC Press is an imprint of the
Taylor & Francis Group, an **informa** business
A SCIENCE PUBLISHERS BOOK

CRC Press
Taylor & Francis Group
6000 Broken Sound Parkway NW, Suite 300
Boca Raton, FL 33487-2742

First issued in paperback 2021

Version Date: 20161010

ISBN-13: 978-0-367-78261-0 (pbk)
ISBN-13: 978-1-4987-6487-2 (hbk)

Visit the Taylor & Francis Web site at
http://www.taylorandfrancis.com

and the CRC Press Web site at
http://www.crcpress.com

Preface

The robotic mechanism and its controller make a complete system. As the robotic mechanism is reconfigured, the control system has to be adapted accordingly. The need for the reconfiguration usually arises from the changing functional requirements. This book will focus on the adaptive control of robotic manipulators to address the changed conditions. The aim of the book is to introduce the state-of-the-art technologies in the field of adaptive control of robotic manipulators in order to further summarize and improve the methodologies on the adaptive control of robotic manipulators. This will be the first book that systematically and thoroughly deals with adaptive control of robotic manipulators. Advances made in the past decades are well described in this book, including adaptive control theories and design, and application of adaptive control to robotic manipulators.

We would like to thank all the authors for their contributions to the book. We are also grateful to the publisher for supporting this project. We hope the readers find this book informative and useful.

This book consists of 16 chapters. Chapter 1 discusses the role of machine learning in control from a modelling perspective and some of the most common machine learning models such as NN and mixture models were presented. Particularly, the idea of motor primitives was introduced as a biologically inspired method for motion representation and learning. Chapter 2 reviews the model reference adaptive control of robotic manipulators. Some issues of model reference adaptive control (MRAC) for robotic manipulators are covered. Very few recent papers can be found in the area of model reference adaptive control of robotic manipulators. This chapter will provide a guideline for future research in the direction of model reference adaptive control for robotic arms. Chapter 3 develops a concurrent learning based adaptation law for general Euler-Lagrange systems. Chapter 4 discusses an autonomous space robot for a truss structure assembly using some reinforcement learning. An autonomous space robot able to obtain proficient and robust skills by overcoming errors to complete a proposed task. Chapter 5 describes adaptive control for object manipulation in accordance with the contact condition and reviews the mathematical models and basic theory for stability analysis. Chapter 6 studies the networked control for a class of uncertain dynamical systems, where the control signals are computed via processors that are not attached to the dynamical systems and the feedback loops are closed over wireless networks. Chapter 7 presents the design of adaptive control of both robot manipulator and consensus-based formation of networked mobile robots in the presence of uncertain robot dynamics and with event-based feedback. Chapter 8 proposes a hybrid controller by combining a PID controller and a model reference adaptive controller (MRAC), and also compares the convergence performance of the PID controller, model reference adaptive controller, and PID+MRAC hybrid controller. This study will provide a guideline for future research in the direction of new controller designs for manipulators in terms of convergence speed and other performances. Chapter 9 presents an approach to using an impedance-controlled ankle exoskeleton ("anklebot") for task-oriented locomotor training after stroke. The objective is to determine the feasibility of using the anklebot as a gait training tool by increasing the contribution of the paretic ankle in walking function. Chapter 10 presents the open architecture high value added robot manufacturing cells. Chapter 11 discusses two basic control problems, namely set point and trajectory following control in application to rigid body manipulators with joint flexibility. Chapter 12 presents how robotic bipedal walking control can unify locomotion, manipulation, and force-based tasks into a single framework via quadratic programs utilizing control Lyapunov functions. Two common examples where the unification can be applied are introduced: ZMP-based pattern generation and locomotion, and nonlinear dynamics with push recovery. Chapter 13 develops a robust adaptive back stepping based controller considering a general uncertain dynamic model of a multi-link robotic manipulator. The designed dynamic controller only requires knowledge of the inertia matrix of the rigid links of the manipulator and is robust to uncertainties in joint stiffness, Coriolis and centrifugal terms, friction, gravity, load torques and disturbances, and actuator inertia. Chapter 14 proposes a new adaptive switching learning control approach, called adaptive switching learning PD control (ASL-PD), for trajectory tracking of robot manipulators in an iterative operation mode. Chapter 15 presents a method to design an adaptive robust control based on online estimation of the lumped time-varying model uncertainties for tracking control of the robot manipulators. Chapter 16 is concerned with evaluation of microgenetic and microimmune optimization algorithms for solving inverse kinematics of a redundant robotic arm as a prerequisite for adaptive trajectory planning in various manipulation tasks.

The editors would like to sincerely acknowledge all friends and colleagues who have contributed to this book.

Oshawa, Ontario, Canada
March 2016

Dan Zhang
Bin Wei

Contents

1

From MRAC to Learning-Based MPC

The Emerging Importance of Machine Learning for Control of Robot Manipulators

*K. Soltani Naveh** and *P. R. McAree*

ABSTRACT

The increasing importance of machine learning in manipulator control is reviewed from two main perspectives: modeling and learning control. The chapter starts with an introduction to history and theory of Model Reference Adaptive Control (MRAC) and its application to manipulator control. Least Square Minimization (LSM) regression is highlighted as the machine learning element in indirect MRAC that seeks to find unknown parameters from a number of data points. The limitations of indirect and direct MRAC are identified. Specifically, indirect MRAC is limited by the need for persistent excitation and the use of simple modeling assumptions that lead to undesirable control performance. Direct MRAC is limited by reference model mismatch, adaptation rate, choice of control law, and relying on error feedback correction with the frequent consequence of stability and adaptation problems. Moreover, neither direct nor indirect MRAC can handle state and input constraints which are important in the control of robotic manipulators. Machine learning techniques offer the promise of overcoming these limitations. Recent developments include Concurrent-MRAC which is able to alleviate the persistent excitation requirement for LTI systems and may have application to the control of robotic manipulators. The chapter covers the broader contributions of machine learning

School of Mechanical and Mining Engineering, University of Queensland, St. Lucia, Queensland, Australia.

E-mail: p.mcaree@uq.edu.au

* Corresponding author: k.soltaninaveh@uq.edu.au

to recent developments in manipulator control that combine learning and adaptation. This includes: (i) the use of advanced modeling methods such as Mixture Models and Neural Networks and (ii) learning control methods such as Iterative Learning Control and modern Reinforcement Learning and their relation to adaptive control. State and input constraints are identified to be one of the significant remaining challenges in manipulator control. Model Predictive Control (MPC) is introduced as the control method that can handle state and input constraints in its formulation. A number of recent attempts to incorporate learning capabilities in MPC are discussed.

Keywords: Machine learning, learning control, neural networks, model predictive control

INTRODUCTION

Model Reference Adaptive Control (MRAC) was first employed for robot manipulator control in 1980s as researchers explored direct drive technology to eliminate gearboxes with the aim of faster more accurate manipulation. Before that, manipulators were generally controlled using PD controllers at each joint. Although manipulator dynamics is nonlinear, this approach worked in practice due to large gear ratios that made nonlinear inertia effects negligible from the view of the control system. In this case, it has been proven that PD controllers successfully accomplish trajectory tracking (Slotine and Li 1991). Once direct drive technology was adopted and gearboxes were eliminated faster operations became possible and undesirable gear effects such as deadzone and backlash were eliminated. However, nonlinear rigid body effects became no longer negligible and the conventional joint based PD controllers were unable to deliver required levels of performance. A MIMO control design approach that took into account the coupling and nonlinear effects was required.

Early interest in the application of MRAC to manipulator control was also motivated by model mismatch or model uncertainty. The emergence of direct drive manipulators and the push for faster operation speed, led to the development of the so-called 'computed torque' control method, which uses a rigid body model of the manipulator to compute feedforward terms which are typically applied in conjunction with a feedback strategy. Modeling errors were, however, found to be significant enough to impact on trajectory tracking. This necessitated the development of control systems that were able to estimate the unknown model parameters and compute the inputs according to the estimated plant model.

As robot manipulators were increasingly applied to problems that required interaction with changing environment, manipulator controllers had to adapt to those changes in order to maintain acceptable performance. These applications involve changing loads, varying compliance and geometry.

MRAC was introduced as a control methodology in 1961 (Astrom and Wittenmark 2013, Osburn et al. 1961). It was proposed for manipulator control in 1979 by Dubowsky (Dubowsky and Deforges 1979). The adaptation mechanism in MRAC, in principle, allows the control system to maintain a desired response robust to model mismatch and changes to the plant. One way to specify the desired behavior is through defining a reference model. This is the basis for so-called ***direct MRAC***.

Alternatively, a desired behavior can be achieved by estimating the unknown or changing parameters of the system and adapting the control input based on those estimations. This approach is known as ***indirect MRAC*** and it is the main focus of this chapter.

Indirect MRAC uses data to "learn" unknown parameters and this places it as essentially a machine learning algorithm. Machine learning is a well-studied and established area that focuses on algorithms that enable machines to adapt and change without the need to be reprogrammed. The advances that have been made in machine learning over the last 30 years offer the prospect for superior modeling approaches than classic adaptive control methods. For instance modern machine learning approaches now allow modeling a particular task such as playing table tennis (Muelling et al. 2010) while classic control approaches only work with plant models.

While MRAC was able to address most of the challenges arising from nonlinear effects, model mismatch and parameter variations, some other issues remained. As gearboxes were removed and direct drive technology replaced them, another issue appeared. The actuators would saturate when high performance was desired and this would lead to poor performance. In order to avoid constraint violation in systems, it is necessary to be able to predict or know in advance the response of the system to an input. This is the fundamental idea behind Model Predictive Control (Maciejowski 2001) which is capable of handling state and input constraints while optimizing the performance with respect to a defined metric. With the increase in computational power MPC is now applied to many systems including robotic manipulators for trajectory tracking applications (Verscheure et al. 2009). The success of MPC predominantly relies on the extent to which an analytical and linear model represents the behavior of the plant and allows predictions to be made. If the response of the plant includes unknown elements or elements that are very difficult to represent analytically or are highly nonlinear then predictions will be inaccurate or even impossible.

Just as machine learning allowed indirect MRAC to model a system with unknown parameters, it seems plausible it can help obtain better models for MPC. The composition of such models based on classical methods and modern machine learning methods will dominate part of the discussion in this chapter.

The structure of this chapter is as follows. The underlying theory of MRAC and its capabilities and limitations are explained as background to the work. Then we look at adaptation from a machine learning perspective and discuss how machine learning has the potential to improve control capabilities, especially for nonlinear systems. Following that we give a number of machine learning methods and include examples of their robotic applications. Then learning control is briefly introduced and compared to adaptive control. Finally we present MPC as the method of choice for optimal control of constrained systems and how machine learning methods are being used to combine MPC with learning and adaptive control.

FOUNDATIONS OF MRAC

Adaptive control is centered around the notion that it is necessary to adjust the parameters of a control law when faced with uncertainties that change the dynamic of a process. This uncertainty could be in the form of operational variations or structural unknowns. One particularly important application for adaptive control is in manipulator robots. For example, in manipulator control end-effector load variations can be significant. The load variations arise from interacting with a changing environment or time varying dynamics.

MRAC has been one of the well-known adaptive control methods for handling such problems. Direct MRAC allows for a desired system behavior to be specified as a reference model. This is particularly useful for MIMO systems such as robotic manipulators. Direct MRAC attempts to drive the plant response to reference model response by parametrizing the controller and estimating the "controller" parameters online. Indirect MRAC has as its objective obtaining the best estimates of "plant" parameters first and then using them to produce a suitable control input. This is why the adaptation process occurs indirectly. Common parameter estimation methods are based around Recursive Least Square (RLS) (Astrom and Wittenmark 2013), which can also be interpreted as Kalman filtering and Lyapunov based estimation (P. and Sun 1996, Slotine and Li 1987).

Direct MRAC

Direct MRAC is commonly formulated using Lyapunov methods (Astrom and Wittenmark 2013) with the advantage that such formulations are based in considerations of stability. However this approach requires identification of an appropriate Lyapunov function.

By way of introduction to the approach, consider the following first order system based off an example given in (Astrom and Wittenmark 2013):

$$\frac{dy_m}{dt} = -a_m y_m + b_m u_c$$

where $a_m > 0$ and the reference signal is bounded. Assume that the plant is described by

$$\frac{dy}{dt} = -ay + bu,$$

and the control law is defined as

$$u = \theta_1 u_c - \theta_2 y.$$

The error is then

$$e = y - y_m$$

$$\frac{de}{dt} = -ay + b(\theta_1 u_c - \theta_2 y) + a_m y_m - b_m u_c$$

$$\frac{de}{dt} = (b\theta_1 - b_m)u_c + (-a - b\theta_2)y + a_m y_m$$

$$\frac{de}{dt} = -a_m e + (b\theta_1 - b_m)u_c + (-a - b\theta_2 + a_m)y$$

Knowing the error dynamics helps in creating the Lyapunov function. Since the aim is to drive the error to zero a good candidate Lyapunov function is

$$V(e,\theta_1,\theta_2) = \frac{1}{2}e^2 + \frac{1}{2b\gamma}(b\theta_1 - b_m)^2 + \frac{1}{2b\gamma}(b\theta_2 + a - a_m)^2.$$

This function adapts the control law while driving the error to zero.

Now evaluating $\frac{dV}{dt}$ and enforcing it to be negative definite would lead to the adaptation equations as demonstrated below:

$$\frac{dV}{dt} = e\dot{e} + \frac{1}{\gamma}(b\theta_1 - b_m)\dot{\theta}_1 + \frac{1}{\gamma}(b\theta_2 + a - a_m)\dot{\theta}_2,$$

$$\frac{dV}{dt} = e(-a_m e + (b\theta_1 - b_m)u_c + (-a - b\theta_2 + a_m)y) + \frac{1}{\gamma}(b\theta_1 - b_m)\dot{\theta}_1 + \frac{1}{\gamma}(b\theta_2 + a - a_m)\dot{\theta}_2,$$

$$\frac{dV}{dt} = -a_m e^2 + (b\theta_1 - b_m)(\frac{1}{\gamma}\dot{\theta}_1 + u_c e) + (b\theta_2 + a - a_m)(\frac{1}{\gamma}\dot{\theta}_2 - ye),$$

To ensure that $\dot{V} <= 0$ we need to set $\frac{1}{\gamma}\dot{\theta}_1 + u_c e$ and $\frac{1}{\gamma}\dot{\theta}_2 - ye$ to zero leading us to the following adaptation equations:

$$\dot{\theta}_1 = -\gamma u_c e$$

$$\dot{\theta}_2 = \gamma y e.$$

This approach extends to general linear dynamic systems. Choosing a notation similar to that of Chowdhary et al. (Chowdhary et al. 2013), consider a linear system with state $x \in R^n$ and input $u \in R^m$:

$$\dot{x}(t) = Ax(t) + Bu(t).$$

Assume that the reference model that represents the desired close loop response is:

$$\dot{x}_{rm}(t) = A_{rm}x_{rm} + B_{rm}\mathrm{r}(t)$$

where A_{rm} is Hurwitz. A control law can be defined as a combination of feedforward and feedback terms:

$$u(t) = K_x^T(t)x(t) + K_r^T(t)r(t).$$

Substituting the control law into the state space equation of the system gives:

$$\dot{x}(t) = (A + BK_x^T(t))x(t) + BK_r^T(t)r(t).$$

If it is assumed that K_x^* and K_r^* exist such that:

$$A + BK_x^{*T} = A_{rm}$$

$$BK_r^{*T} = B_{rm}$$

then the tracking error dynamics is:

$$\dot{e}(t) = \dot{x}(t) - \dot{x}_{rm}(t) = [(A + BK_x^T(t))x(t) + BK_r^T(t)r(t)] - [(A + BK_x^{*T})x_{rm} + BK_r^{*T}r(t)].$$

Defining $\tilde{K} = K - K^*$ the equation can be reduced to

$$\dot{e}(t) = A_{rm}\,\mathrm{e}(t) + B\tilde{K}_x^T(t)\,\mathrm{x(t)} + B\tilde{K}_r^T(t)r(t)$$

Using Lyapunov theory it is possible to find the update law for gains K_x and K_r such that error converges to zero. To do that the well known Lyapunov equation for linear systems is used to find the positive definite matrix $P \in R^{n\times n}$ such that $\dot{V}(t) \leq 0$ for Lyapunov function $V(t) = \frac{1}{2} e^T(t)Pe(t)$:

$$A_{rm}^T P + PA_{rm} + Q = 0.$$

This leads to the following update equations:

$$\dot{K}_x(t) = -\Gamma_x x(t)e^T(t)PB$$

$$\dot{K}_r(t) = -\Gamma_r r(t)e^T(t)PB.$$

In general, the procedure involves:

1. Coming up with a controller structure.
2. Deriving the error dynamics equation based on the chosen controller.
3. Finding a Lyapunov equation and then using it to find the adaptation equations necessary to drive the error to zero.

An interesting property of the Lyapunov method is that the system response converges although the parameters are not guaranteed to converge. The fact that parameters can grow unboundedly can be problematic. This will be discussed later in the chapter.

Indirect MRAC

Indirect MRAC relies on estimating unknown system parameters to adapt control input accordingly. The well-known Least Square Minimization (LSM) forms the foundations of Indirect MRAC and it is where MRAC displays its machine learning features. LSM is sometimes implemented through a Kalman filter.

LSM has appeared in many areas of mathematics and engineering. It is simply defined as:

$$\text{minimize } \|Y - \Phi\Theta\|_2^2$$

There are different ways to interpret LSM depending on how it is written. Assume that $R(\Phi)$ is the range of matrix Φ that is the vector space formed by the columns of matrix Φ. Then $\Phi\Theta$ is a point in the space defined by $R(\Phi)$. The above definition seeks the value of Θ for which $\Phi\Theta \in R(\Phi)$ is the closest point to Y. If Y is a point somewhere in the space and $\Phi\Theta$ is a hyperplane, the solution projects Y on to $R(\Phi)$.

A well-known applications of LSM is in regression and model fitting. The aim of regression is to fit a parametric model, usually termed a hypothesis in the machine learning literature to data points. Given data points $y_i \in R$ a function:

$$y(x) = \sum_{j=1}^{N} \theta_j \phi_j(x) = \Phi^T(x)\Theta$$

is sought where $\phi_i(x)$ are basis functions and θ_i are the model parameters that combine these basis functions so that the model best fits the data. This means that the model parameters have to be obtained through some optimization process and LSM is commonly used for this, i.e.,

$$\text{minimize } \sum_{i=1}^{N} (y_i - \phi^T(x_i)\Theta)^2$$

This can be expressed in matrix form as

$$J = \sum_{i=1}^{M}(y_i - \Phi^T(x_i)\Theta)^2 = \begin{bmatrix} y_1 - \phi^T(x_1)\Theta \\ y_2 - \phi^T(x_2)\Theta \\ \vdots \\ y_M - \phi^T(x_M)\Theta \end{bmatrix} = \begin{bmatrix} y_1 \\ y_2 \\ \vdots \\ y_M \end{bmatrix} - \begin{bmatrix} \phi^T(x_1) \\ \phi^T(x_2) \\ \vdots \\ \phi^T(x_M) \end{bmatrix}\Theta$$

$$= \left\| Y - \begin{bmatrix} \phi_1(x_1) & \phi_2(x_1) & \cdots & \phi_N(x_1) \\ \phi_1(x_2) & \phi_2(x_2) & \cdots & \phi_N(x_2) \\ \vdots & \vdots & \vdots & \vdots \\ \phi_1(x_M) & \phi_2(x_M) & \cdots & \phi_N(x_M) \end{bmatrix}\Theta \right\|_2^2 = \| Y - \Phi\Theta \|_2^2$$

Here the Φ matrix is the regression matrix. If $\phi_j(x) = x^{(j)}$ where $x^{(j)}$ is the j-th element of x then the above expression reduces to the linear regression problem.

The above regression equation is sometimes called parametric estimation. The use of the regression equation is common in adaptive control whenever varying components of the system are modeled as a linear function of the unknown parameters. Recursive implementations also exist that are computational efficiency.

When using least square for parameter estimation, the Φ matrix often contains the **observations** either in its rows or columns depending on the problem description. For a good estimation, one would require that the observations provide as many independent information as possible in other words one would expect Φ to have complete rank. If some of these observations are not providing new information, i.e., the columns or rows are not independent the parameter estimation would be poor and in some cases impossible. This leads to a concept called **persistent excitation** and it will be discussed later.

Applications of MRAC to Manipulator Control

Both direct and indirect MRAC have been applied to manipulator control in order to deal with the varying load or unknown parameters (Craig et al. 1987, Maliotis 1991, Ham 1993, Dubowsky and Deforges 1979, Slotine and Li 1987, Hsia 1984, Tung et al. 2000). In both approaches, the known part of the model is used in a direct torque control or feedforward framework and only the unknown or varying part is handled by an adaptive controller. In case of direct MRAC, the adaptive controller can be a PD controller with adaptive K_p and K_d gains (Maliotis 1991) while in the case of indirect MRAC it usually takes the form of a computed torque control based on estimated parameters (Craig et al. 1987, Slotine and Li 1987). In both approaches, MRAC design is predominantly based on Lyapunov method.

The common approach in applying direct MRAC to manipulators is to combine feedforward and feedback as shown by (Maliotis 1991), who shows, in this approach the feedforward component makes use of the known portion parameters of the plant such as inertia to calculate the likely input. The adaptive feedback aims to compensate for the varying nonlinear effects by using n linear decoupled second order systems as the reference model. The feedforward component could also be interpreted as the result of feedback linearization of the plant.

To make this explicit, assume that the dynamics of a robotic manipulator is expressed by

$$M(q)\ddot{q} + C(q, \dot{q})\dot{q} + G(q)q = \tau$$

where the inertia matrix M, Coriolis and gravity terms C and G are all assumed to be composed of known and unknown parts as defined below.

$$M = M_k + M_u^* = M_k - I + M_u^* + I = M_k - I + M_u$$

$$C = C_k + C_u$$

$$G = G_k + G_u$$

The subscripts k and u define the known and unknown components. Note that M has been decomposed to known and unknown parts in a way that supports the construction of invertible matrices that are positive definite. When the above equations are substituted in the dynamic equations the following is obtained:

$$(M_k - I + M_u)\ddot{q} + (C_k + C_u)\dot{q} + (G_k + G_u)q = \tau,$$

$$M_u\ddot{q} + C_u\dot{q} + G_u q = \tau - (M_k - I)\ddot{q} - C_k\dot{q} - G_k q = u,$$

$$\tau = u + (M_k - I)\ddot{q} + C_k\dot{q} + G_k q.$$

And, if all parameters are known then $M_u = I$, $G_u = 0$, $C_u = 0$ and therefore $u = \ddot{q}$ and τ will only include the feedforward terms. However if there are unknown parameters, given a reference model then u can be calculated using adaptive terms in order to make the system behave like the reference model.

The reference model is chosen as:

$$\dot{x}_d = A_m x_d + B_m v = \begin{bmatrix} 0 & I \\ -K_p & -K_v \end{bmatrix} x_d + \begin{bmatrix} 0 \\ I \end{bmatrix} v,$$

where $x_d = [q_d \;\; \dot{q}_d]^T$. This model essentially forces the manipulator to behave like a number of uncoupled second order systems. The input u is chosen to be:

$$u = u_a + u_k,$$

$$u_k = -[K_p \;\; K_v]\,x + v$$

$$u_a = -[\Delta_1 \;\; \Delta_2]\,x + \Delta_v v$$

The first term is a state feedback and feedforward and the second term is the adaptive input that deals with the unknown variations. Using the above control law, the adaptation law can be derived using Lyapunov method:

$$\dot{\Delta}_1 = -a\omega q^T$$

$$\dot{\Delta}_2 = a\omega\dot{q}^T$$

$$\dot{\Delta}_v = b\omega v^T$$

where a and b are positive scalar gains.

Computed torque control is also used in indirect MRAC where adaptation occurs through estimation of unknown parameters (Craig et al. 1987, Slotine and Li 1987).

The parameter estimation law is derived using Lyapunov method. The process is shown below:

$$\tau = M(q)\ddot{q} + C(q,\dot{q})\dot{q} + G(q)q.$$

The control law is then defined as;

$$\tau = \hat{M}\ddot{q}_d + \hat{C}(q,\dot{q})\dot{q}_d + \hat{G}(q) - K_p\tilde{q} - K_d\dot{\tilde{q}},$$

where $\widehat{(.)}$ indicates associated quantity is an estimate, and:

$$\tilde{q} = q - q_d$$
$$\dot{\tilde{q}} = \dot{q} - \dot{q}_d.$$

A Lyapunov function is defined as $V = \frac{1}{2}(\dot{\tilde{q}}^T M(q)\dot{\tilde{q}} + \tilde{a}\Gamma\tilde{a} + \tilde{q}^T K_p\tilde{q})$ where a is the vector of unknown parameters and $\tilde{a} = \hat{a} - a$ is the estimation error. In order to make sure that $\dot{V} \leq 0$ the estimation law is chosen as:

$$\dot{\hat{a}} = -\Gamma^{-1}Y^T(q,\dot{q},\dot{q}_d,\ddot{q}_d)\dot{\tilde{q}}$$

with Y coming from

$$\tilde{M}\ddot{q}_d + \tilde{C}(q,\dot{q})\dot{q}_d + \tilde{G}(q) = Y\tilde{a}.$$

This process is more or less followed in all Lyapunov-based indirect MRAC applications in manipulator control. See for example (Tsai and Tomizuka 1989, Rong 2012, Mohan and Kim 2012).

Researchers have also used indirect MRAC in combination with direct MRAC for manipulator control. This method utilizes both tracking error and prediction error for better adaptation (Yu et al. 1993, Ciliz and Cezayirli 2004).

Limitations of MRAC and the Adoption of Machine Learning

Traditional MRAC as presented in the previous section has limitations that arise from the assumptions, control architecture and the computational method. Direct and indirect MRAC have different limitations.

In traditional MRAC, it is assumed that the unknown parameters are either constant or change very slowly (Astrom and Wittenmark 2013, Slotine and Li 1987). This assumption is true in some circumstances but not in all and it is arguably predicated on mathematical convenience rather than its appropriateness for manipulators and other high performing mechanical systems. Because of the improvements in actuator technology direct drive manipulators are capable of performing high speed high torque motions which means that variations and nonlinearities can occur at a faster rate. Therefore there is a need to account for such abrupt variations that do not fit in the classic MRAC framework.

Both direct and indirect MRAC are reliant on the so-called *persistent excitation* condition in order to guarantee stability and convergence for states and parameters (Narendra and Annaswamy 1987, Chowdhary et al. 2013). This requirement ensures that the input to the system meets the conditions for parameters to be bounded. If persistent excitation is not satisfied it can lead to robustness and stability problems due to unbounded growth of parameters. The drift in parameters often occurs abruptly and, leads to the phenomenon known as bursting. This requirement is formally defined below.

Definition

A signal u is persistently exciting (PE) of order n if the following holds (Astrom and Wittenmark 2013):

$$\rho_1 I > \sum_{k=t}^{t+m} \phi(k)\phi^T(k) > \rho_2 I$$

where $\rho_1, \rho_2 > 0$ and

$$\phi(t) = \begin{bmatrix} u(t-1) \\ u(t-2) \\ \vdots \\ u(t-n) \end{bmatrix}$$

This definition shows that the essence of PE is that there is enough variance in input so that all adaptation parameters can be evaluated based on system measurements. However requiring PE can compromise performance when smooth or high precision performance is required. In order to prevent the unbounded growth of parameters, methods such as σ modification (Ioannou and Kokotovic 1983) and e modification (Narendra and Annaswamy 1986) were proposed.

In σ modification, the update equations are of the following form:

$$\dot{K}(t) = -\Gamma(x(t)e^T(t)PB + \sigma(K(t) - \bar{K}))$$

where $\sigma > 0$ is the modification gain and $\bar{K}$ is a preselected value about which the parameter K is to be bounded. If $\overline{\Delta K} = \bar{K} - K^*$, $\tilde{K} = K - K^*$, $\dot{\tilde{K}} = \dot{K}$ then the parameter dynamics can be rewritten as below.

$$\dot{\tilde{K}}(t) = -\Gamma\sigma\tilde{K}(t) - \Gamma(x(t)e^T(t)PB - \sigma\overline{\Delta K})$$

In this equation the parameter error dynamics is a linear dynamical system driven by the state error e. If the state error does not grow unbounded, this system is bounded input bounded output stable and therefore unbounded growth of parameters is prevented.

Recently it has been shown that by using data collected over a finite period of time, it is possible to achieve stability and exponential convergence in uncertain linear dynamical systems without PE as long as the system is excited over a sufficiently *finite* period of time (Chowdhary et al. 2013). In this approach the parameter update equations are augmented in the following manner. First a set of p state-input pairs (x_j, r_j) are collected. Then the following errors are defined

$$\epsilon_{K_{r_j}} = K_r^T(t)r_j - (B^T B)^{-1} B^T B_{rm} r_j$$

$$\epsilon_{K_{x_j}} = (B^T B)^{-1} B^T (\dot{x}_j - A_{rm} x_j - B_{rm} r_j - B\epsilon_{K_{r_j}}(t)).$$

$(B^T B)^{-1} B^T$ is the psudo-inverse of B and earlier we had $BK_r^{*T} = B_{rm}$ therefore the above equations can be rewritten as below.

$$\epsilon_{K_{r_j}} = (K_r^T(t) - K^{*T})r_j = \tilde{K}_r r_j$$

$$\epsilon_{K_{x_j}} = (B^T B)^{-1} B^T (\dot{x}_j - A_{rm} x_j - B_{rm} r_j - B\tilde{K}_r^T(t) r_j)$$

$$= (B^T B)^{-1} B^T B \tilde{K}_x^T(t) x_j$$

$$= \tilde{K}_x^T(t) x_j$$

These error variables are used as feedback for update equations giving:

$$\dot{K}_x(t) = -\Gamma_x(x(t)e^T(t)PB + \sum_{j=1}^{p} x_j \epsilon_{K_{x_j}}^T(t)),$$

$$\dot{K}_r(t) = -\Gamma_r(r(t)e^T(t)PB + \sum_{j=1}^{p} r_j \epsilon_{K_{r_j}}^T(t)).$$

This approach is called Concurrent-MRAC. Chowdhary et al. (2013) have demonstrated that by relying on a richer data set from past and present performance and leveraging machine learning, significant improvements on stability and convergance can be made.

Another limitation that is encountered in indirect MRAC arises from the choice of simple parametric models. Parametric models have been used in adaptive control because they facilitate the formulation of control laws and rigorous stability analysis. However these models can be too simplistic for complex robotic systems and they can fail to capture unknown nonlinear effects (Nguyen-Tuong and Peters 2011). Especially for manipulators, different tasks have different input-output profiles and features which may not be captured by a simple parametric model and using least square fitting. Parametric models are prone to underfitting and overfitting which lead to erroneous estimations. Indirect MRAC is unable to cope with the great variety of motions that may be necessary for abstract tasks such as playing table tennis. As we shall see later, the developments in machine learning have allowed a task based control design based on collected motion data. In these methods, data from motions associated with a particular task such as playing golf (Khansari-Zadeh and Billard 2011) or playing table tennis (Muelling et al. 2010) have been used to design control systems that continuously adapt to changes in environment and even improve their performance as more observations are made.

Direct MRAC also has a number of limitations. One of the fundamental limitations lies at the core of direct MRAC and that is the choice of reference model and the control law. If the reference model has significant mismatch with the plant to the extent that it becomes unrealizable by the system then robustness and stability problems can materialize. This can be particularly problematic for a nonlinear plant like an airplane, as the use of LTI reference models cannot always capture the dynamic behavior of these systems (Crespo et al. 2010). The authors in (Krause and Stein 1987) also demonstrate that for a given combination of model mismatch and parameter uncertainty, MRAC cannot provide a solution. Similarly, the choice of control law and its parametrization can limit the adaptation capability of the controller. Control policies that are more sophisticated than linear feedback-feedforward can perform better in the face of variations. We did not identify any work that addresses the choice of control policy in direct MRAC.

Direct MRAC works based on feedback, which means its adaptation is based on local information only. This is a limiting approach especially for nonlinear systems with rich dynamic behavior. It is expected that using a global approach where adaptation is not only based on local measurements but also past measurements can lead to better performance.

Adaptation rate is a structural element in direct MRAC that has been identified as a potential source of problem. While based on LTI theory high adaptation rates improve controller performance, it is not always true when the plant is nonlinear or delays are present. High adaptation rates can lead to oscillatory behavior and even instability in such scenarios. Therefore some trade-off is inevitable in order to ensure reasonable compensation of parametric uncertainties while avoiding the undesirable effects arising from nonlinearities and delays (Crespo et al. 2010).

In addition to the structural issues and limiting assumptions in MRAC, state and input constraints are also where adaptive control provides no solution. Ignoring these constraints can lead to saturation and instability. From a practical point, this is a problem that is increasingly encountered as industry moves towards high performance systems that often need to operate close to the extreme limits. While classic control approaches fail to provide a solution to this, Model Predictive Control (MPC) is not only able to accommodate complex state and input constraints but also is able to optimize system performance with regards to a given cost function. Therefore there has been an effort to integrate MPC and adaptive control through learning methods.

Machine Learning for Control

Models are an inseparable part of control design. In fact they are central to the control engineering design process. This is true for the very classical control design approaches like linear feedback control to modern approaches such as MPC. Models are needed for compensation, prediction and adaptation.

We begin with a few definitions based on (Nguyen-Tuong and Peters 2011) who distinguish the following types of models:

1. *Forward models*: Forward models are used to predict the next state of a system given the current state and input. This is the most common model that is used in control and can be generally represented as a first order ordinary differential equation $\dot{x} = f(x, u)$.
2. *Inverse models*: Inverse models calculate an input given an initial state and a desired state. These models are used generally for feedforward calculation. A well-known example of these models is the feedforward term for serial manipulators $\tau = M\ddot{\theta} + C(\theta,\dot{\theta})$. When the mapping from state to input is not unique, the inverse model can results in poor performance of a learning algorithm. Inverse kinematics is a well-know: for each pose in the workspace there may exist multiple configurations in joint space and this ambiguity can drive poor learning outcomes.
3. *Mixed models*: These are a combination of forward and inverse models. The combination of the forward and inverse models allows to compute the inverse model for systems which suffer from non-uniqueness. In this approach the forward model guides the computation of inverse model and prevents the ill-conditions arising from non-uniqueness.

Models for many systems can be obtained from the first principles and for a long time this was how models were obtained. However there are some limitations with deriving models in this manner. First, the theoretical modeling often makes simplifying assumptions such as ignoring friction, backlash and deadzones. This is often because if we were to model the system without those assumptions then unknown parameters or very complex mathematical functions would have to be included. The resulting complexities influence the capacity to handle them within allowable control design frameworks. Second, theoretical modeling does not produce models that are fit-for-control-system purpose for complex systems such as robots with flexible structure or complex mechanical characteristics. Third, there are systems or tasks for which analytical models cannot be easily or effectively produced. An example of such a task is cutting different foods using robotic manipulators (Lenz et al. 2015). It is difficult to model the dynamics of cutting for all foods which can range from soft texture like cheese to hard texture like carrot. Insisting on theoretical modeling in the face of these limitations would only limit our ability to achieve better performance. For these reasons, data driven modeling has received an increasing level of attention.

However, the idea of using data to handle uncertainties and unknown parameters is not new. These ideas can be traced back to self-tuning controllers and in general classic adaptive control. In direct MRAC, for instance, the error signal between the plant and the desired forward model allows the system to learn the inverse model of the plant. Similarly in indirect MRAC, the inverse model of the plant is learned through parameter estimation.

Data driven modeling has the benefit of being able to encapsulate complex behaviors and nonlinear effects that are often ignored in theoretical modeling or which give rise to complex analytical expressions that are too difficult to work with. Perhaps more importantly, though, data driven modeling allows the system to cope with uncertainties and adapt to changing conditions as they occur instead of manually reprogramming the control system again and again.

Machine learning has witnessed significant development in the past thirty-five years. There are many different machine learning methods but they can be classified based on their modeling, training and operation into:

1. parametric and non-parametric algorithms
2. supervised and unsupervised learning algorithms
3. local and global learning algorithms

Parametric models have a fixed number of parameters that have to be learned from data. In contrast, non-parametric models are not constained by having a fixed number of parameters and they rely directly on data for modeling, and the number of parameters can be chosen to best meet requirements in real time. The standard linear regression is an example of parametric learning algorithm while a locally weighted linear regression is an example of non-parametric learning algorithm.

Supervised learning refers to the training process where the "right" answers are provided to the machine learning algorithm and then the algorithm learns the model based on the training data. Neural Networks, logistic regression and linear regression are all examples of supervised learning. On the other hand, in unsupervised learning, only raw data is provided and the machine learning algorithm learns a model purely based on that. Clustering algorithms and Gaussian Mixture Models are among the well-known unsupervised learning algorithms.

Local learning algorithms construct models in a local neighborhood around a query point. The model then allows the prediction of the value for that query point. On the other hand, global learning algorithms fit one model to all of the data points. Global learning algorithms may involve a mixture of models in order to capture a wider range of behaviors that may be present in the data.

Earlier linear regression otherwise known as least square estimation was presented. Linear regression is one of the oldest and most common learning methods in control. Below we provide a short study of some of the other machine learning algorithms that are common in control.

NEURAL NETWORKS AND BEYOND

Neural Network is a parametric learning method which has been one of the most active research areas in machine learning. With the increase in computational power and algorithmic improvements, they are finding applications in a wide range of areas such as facial recognition, text recognition, signal processing, control and artificial intelligence.

In control, neural networks have been used for learning both forward and inverse dynamic models. Neural networks are superior to other parametric learning methods because of the greater degree of freedom in their parameters (Nguyen-Tuong and Peters 2011). However the use of neural networks is not limited to learning models. They have also been used to solve optimization problems and learn control policies (Xia and Wang 2004, Yan and Wang 2014).

Neural networks were created as an attempt to mimic the learning ability in human brain where network of neurons are able to learn a great variety of tasks. In general neural networks consist of neurons that are arranged in three stages: (i) input layer (ii) hidden layers (iii) output layer. A typical neural network is shown in Fig. 1. Each neuron is a sigmoid function that receives a linear combination of the outputs from the nodes in the previous layer. Sigmoid function has the following form.

$$g(z) = \frac{1}{1+e^{-z}}$$

The output of sigmoid functions allow the neural network to construct features from input data in the hidden layers and then at the output layer those features are

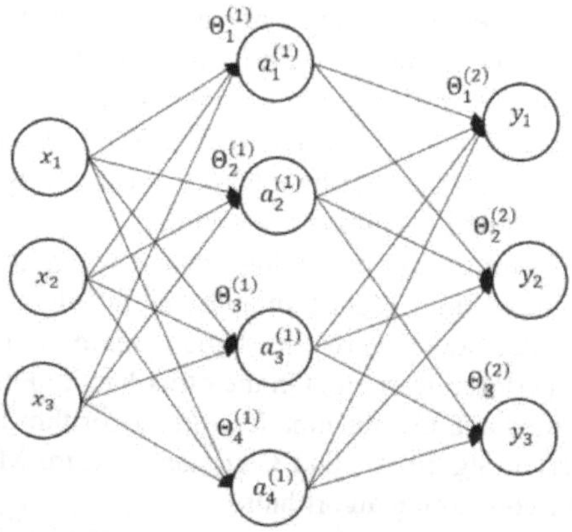

Figure 1. The representation of a typical neural network with one hidden layer.

combined to fit the data as closely as possible. Based on this description, the relation between layers is defined as

$$a_i^{(l+1)} = g(z_i^{(l)}) = g(\Theta_i^{(1)T} a^{(l)}) = \frac{1}{1+e^{-\Theta_i^{(1)T} a^{(l)}}}$$

where $a^{(l)}$ is the output from current layer, $\Theta_i^{(1)}$ is the vector of weights associated with the i th neuron in the layer $l + 1$. The weights in each layer are the parameters that most be computed in order to fit the training data. As the influence of each weight on the output, especially input layer weights, are complicated an algorithm repeatedly evaluates the gradient of the fitting error and adjusts the weights so that the fitting error is minimized. One such algorithm is called backpropagation algorithm.

One drawback in using neural networks for modeling is the long training time. A variant of neural networks called Extreme Learning Machines (ELM) have been developed that have much faster training. ELM is a neural network with one hidden layer whose input layer weights are chosen randomly while its hidden layer weights are calculated analytically using least squares. This method is much faster than the backpropagation algorithm used for ordinary neural networks.

For more complex tasks, often sequence of states and observations are important. Therefore the algorithm needs to learn features arising from data sequences which could represent temporal or spatial features. This arises in a number of applications such as speech and text recognition. In control this becomes particularly useful when we are dealing with dynamics that have temporal or structural variations such as the food cutting task in (Lenz et al. 2015). In order to introduce such learning capabilities Recurrent Neural Networks (RNN) have been developed where hidden layers not only work with the inputs coming from the previous layer but also use their previous outputs as well in order to identify features that may exist in the sequence of data. A simple representation of RNN is shown in Fig. 2.

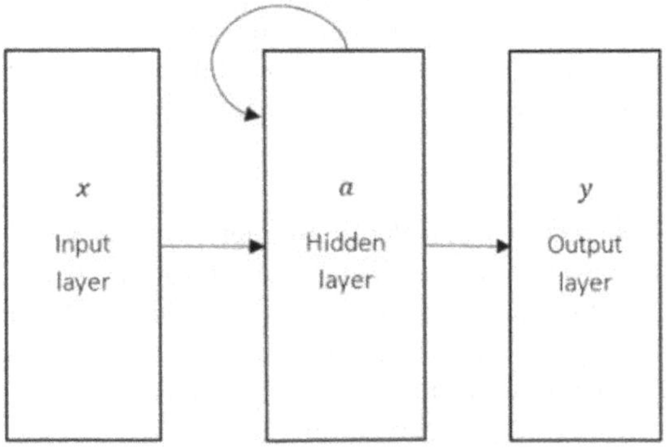

Figure 2. An abstract view of a Recurrent Neural Network.

MIXTURE MODELS

Mixture models are another group of machine learning methods that aim to model a set of data as a mixture of kernels such as Gaussians. This modeling approach has become popular in some of the recent work on manipulator control that aims to mimic the motor skills in humans. There are different theories on motor control and learning in kinesiology however among these theories Generalized Motor Programs (GMP) have been supported by many lab experiments (Schmidt and Lee 2011). Based on this theory, all motions that are performed by humans are a mixture of general motor programs. By mixing these general motor programs with different force and temporal parameterization it is possible to produce a great variety of motions. Moreover these motor programs can be executed by a different set of muscles if desired. In other words there is a decoupling between the motor programs and the execution mechanisms. Learning how to construct and improve these mixtures can serve as a source of inspiration for achieving optimized control of nonlinear systems.

Schaal et al. (Schaal et al. 2007) layed the mathematical foundation for Dynamic Motor Primitives (DMPs), which can be viewed as an attempt to formulate GMP for manipulator control. In this approach, the motion of each degree of freedom is modeled as a stable autonomous nonlinear dynamical system like:

$$\dot{x} = f(x, z)$$

The model proposed by Schaal et al. is:

$$\dot{x}_2 = \gamma\alpha_x(\beta_x(x_g - x_1) - x_2) + \gamma \boldsymbol{A}\boldsymbol{h}(z)$$
$$\dot{x}_1 = \gamma x_2$$

where $x = [q_d, \dot{q}_d]^T$ contains position and velocity for the degree of freedom, $x_g = [q_{final}, \dot{q}_{final}]^T$ contains the goal state, $\boldsymbol{A} = diag(a_1, a_2, ..., a_n)$ is the amplitude matrix used to adjust relative amplitudes between the n degrees of freedom, $\boldsymbol{h}(z)$ is a transformation function acting as a nonlinear input and z is a phase variable that acts like an internal clock so that the state does not depend on time directly. The phase variable z is defined as:

$$\dot{z} = -\gamma\alpha_z z$$

where γ and α_z are constants that are chosen through a parameter selection process. The function $\boldsymbol{h}(z)$ is the key element of this equation as it drives the nonlinear system. The values generated by $\boldsymbol{h}(z)$ dictate the sort of trajectoris that will be generated, therefore in order to produce a particular motion $\boldsymbol{h}(z)$ has to be evaluated using supervised learning. In doing so $\boldsymbol{h}(z)$ is represented as:

$$\boldsymbol{h}(z) = \sum_{i=1}^{N} \psi_i(z)\boldsymbol{w}_i z$$

where $\boldsymbol{w}_i$ is the i-th adjustable parameter of all degrees of freedom. These parameters must be learned from motion data using supervised learning in order to produce the desired motions. N is the number of parameters per degree of freedom, and $\psi_i(z)$ are the corresponding weighting functions. The weighting functions are often chosen to be Normalized Gaussian kernels with centers c_i and widths h_i:

$$\psi_i(z) = \frac{exp(-h_i(z - c_i)^2)}{\sum_{j=1}^{N} exp(-h_j(z - c_j)^2)}$$

The attractive node of this autonomous system is the desired final state, meaning that in the phase portrait of $\dot{x} = f(x, z)$, there is an attractive node at $x_g = [q_{final}, \dot{q}_{final}]^T$. The net result of this formulation is that the nonlinear autonomous system generates the desired position and velocity at each instant. If the system is perturbed the state x would follow a new path to the attractive node.

By inspecting the above system as z converges to zero with time, state x is driven to x_g, making the system behave like a nonlinear spring-damper (Schaal et al. 2007). Note that DMP acts as a reference generator only and the generated reference has to be passed to a lower level feedforward-feedback controller for execution.

The DMP equations shown above are only useful for representing motions with a zero final velocity. The formulation was further generalized by Kober (Kober 2012) to also include non-zero final velocity in order to allow modeling motions such as forehand and backhand in table tennis. A mixture of motor primitives has been used to generalize between DMPs to construct more complex motions that for example can allow a robot arm play table tennis (Muelling et al. 2010).

Another example of application of mixture models in manipulator control is the work of Khansari-Zadeh and Billard where instead of using a mixture of motor primitives as defined above, a mixture of Gaussians is used to model the nonlinear dynamical system representing a task (Khansari-Zadeh and Billard 2011). The nonlinear model is autonomous and stable and is of the following form:

$$\dot{x} = f(x),$$

where $x \in R^d$ can be reference joint positions or Cartesian positions. Similar to DMP because of the stable node the reference trajectory always reaches the desired state even in the presence of perturbation.

The machine learning algorithm has to learn $f(x)$. GMM is considered an unsupervised learning approach as the algorithm has to find out: (i) which Gaussian each sample belongs to (ii) and what are the parameters (mean, covariance, etc.) that result in maximum likelihood of the samples. Therefore GMM is often solved using Expectation-Maximization (EM) algorithm that iteratively estimates the likelihood of each sample belonging to a particular Gaussian and then uses those results to solve for the parameters (mean, covariance, etc.) that maximize sample likelihood. The algorithm eventually finds the best mixture of Gaussians that fits the data. However using EM to find GMM does not guarantee the stability of the resulting model. For this reason, Khansari-Zadeh and Billard proposed a constrained nonlinear optimization approach to obtain only the GMM that satisfies stability requirements.

Both in GMM and DMP, the demonstration data associated with the desired task is collected and the best model is fit to the data. However, the model will be only as good as the demonstration data. This can be problematic especially certain parts of the task are difficult to reproduce by an operator. In this case it is desirable that the system explore the operation space and use the resulting information to improve its performance. This leads us to the subject of learning control.

LEARNING AND ADAPTATION IN CONTROL

Learning and adaptation are closely related concepts. In order for a system to adapt it must learn about the changes in the plant through estimation or through learning the inverse model of the plant as changes occur. This is the basis of MRAC as discussed earlier. Therefore adaptation relies on learning and adaptive control has tried to use learning to adjust *controllers* in order to match the variations in the plant. However, learning can be used to make modifications at abstract levels. This allows systems to use previous performance data to find better "strategies" or better *control inputs* instead of being tied to a particular controller structure. These capabilities impart a higher level intelligence that is able to learn from mistakes or shortcomings and improve upon them. This is the subject of learning control. Learning control has the potential to address some of the challenging problems such as optimal or near optimal control of constrained nonlinear systems. This is because the learning process could allow nonlinear systems achieve optimal performance as time goes on instead of attempting to solve a nonlinear optimization problem from instant to instant.

One of the earliest attempts to implement learning control is attributed to Uchiyama (Uchiyama 1978) and Arimoto (Arimoto et al. 1984) for their work on Iterative Learning Control (ILC). There are different definitions for ILC that can be found in (Ahn et al. 2007). While each definition emphasizes on a particular feature of ILC the following definitions are important from our point of view:

1. The learning control concept stands for the **repeatability** of operating a given objective system and the possibility of improving the control input on the basis of **previous actual operation data** (Arimoto et al. 1986).
2. ILC is to utilize the system repetitions as **experience** to **improve** the system control performance even under **incomplete knowledge** of the system to be controlled (Chen and Wen 1999).

As highlighted in the definitions, ILC can be considered as a feedforward control design that uses the experience obtained from previous operation data to improve control input. Figure 3 shows the general architecture of ILC. However, it is based on the assumption that a particular motion is repeated in trials as for industrial robotic manipulators. Instead it is desirable to have a control system that can learn even if the motions are not repeated exactly and are only **similar**. Despite the limitation of repeatability assumption, ILC is an inspiring control design method for us.

Before looking at ILC formulation it is helpful to clarify how ILC and in general learning control are different from adaptive control. In adaptive control, the *controller*, which is a system, is modified while in ILC the *control input*, which is a signal, is modified. Furthermore, adaptive control methods do not take advantage of the information contained in the similarity of tasks, while ILC, and learning control in general, exploits this information for improving performance (Bristow et al. 2006).

It must be emphasized that ILC is a feedforward control and it is only able to cope with repeatable disturbances if designed well. In practice ILC is implemented

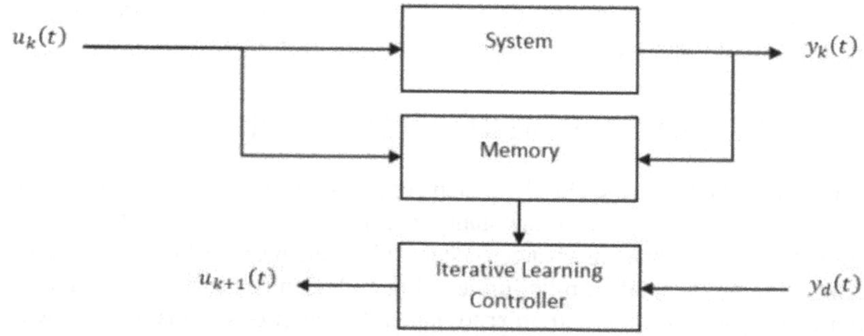

Figure 3. A representation of Iterative Learning Control architecture.

in conjunction with feedback control so that uncertain disturbances are compensated appropriately.

The formulation of ILC uses the tracking error from previous trials to adjust control input by feedforward compensation. One of the simplest ILC algorithms is the PD-type learning control. Assuming that the system needs to track a desired trajectory y_d and that $u_k(t)$ and $u_{k+1}(t)$ are the inputs at time t from trial k and $k + 1$ respectively and the tracking error is $e_k(t) = y_d(t) - y_k(t)$ the PD-type learning algorithm can be defined as:

$$u_{k+1}(t) = u_k(t) + (\Gamma_k \frac{d}{dt} + \Phi_k)e_k(t)$$

$$u_{k+1}(t) = u_k(t) + (\Psi_k \int dt + \Phi_k)e_k(t)$$

PID-type learning algorithm can be formulated in similar manner. These formulations only use the information from one previous trial. Higher Order ILC (HOILC) makes use of the information from more than one previous trial in the update equation (Ahn et al. 2007). For example if N previous trials are used in the learning equation we would have:

$$u_{k+1} = \sum_{k=1}^{N}(1-\Lambda)P_k u_k + \Lambda u_0 + \sum_{k=1}^{N}(\Phi_k e_{i-k+1} + \Gamma_k \dot{e}_{i-k+1} + \Psi_k \int e_{i-k+1})dt$$

where $\Lambda P_k u_k$ and $(1 - \Lambda)u_0$ are used to obtain the weighted average of the previous inputs. Another design approach in ILC is based on solving an optimization, usually a Quadratic Programming optimization, iteratively (Bristow et al. 2006). In this method, first state and input are written in lifted space:

$$\mathbf{u}_k = \begin{bmatrix} u_k(0) \\ u_k(1) \\ \vdots \\ u_k(N-1) \end{bmatrix}$$

$$\mathbf{x}_k = \begin{bmatrix} x_k(0) \\ x_k(1) \\ \vdots \\ x_k(N-1) \end{bmatrix}$$

$$\mathbf{e}_k = \mathbf{x}_d - \mathbf{x}_k$$

Then an appropriate cost function is defined. For example:

$$J_{k+1}(\mathbf{x}_{k+1}, \mathbf{u}_{k+1}) = \mathbf{e}_{k+1}^T Q \mathbf{e}_{k+1} + \mathbf{u}_{k+1}^T R \mathbf{u}_{k+1}$$

By solving this optimization problem the optimal update equation for control input can be obtained. If the system has any constraints then these constraints can be incorporated in the optimization. In this case, constrained optimization techniques such as active set method and interior point method have been used to find optimal control inputs by mimicking optimization steps of these algorithms at each trial (Mishra et al. 2011). The advantage of using the optimization method is that the input and state trajectory eventually converge towards their optimal values.

Learning has also been studied in other areas such as psychology, aritificial intelligence and kinesiology. Robot Reinforcement Learning (RL) is one of the learning theories that has stemmed from psychology but developed in AI and has been applied to motion control problems in the recent years. Based on reinforcement learning theory, agents learn by exploration and exploitation. Agents explore by trying different actions and observing the outcomes and then use the obtained knowledge to improve performance through exploitation. This process allows the agent to find the optimal control policy that tells what the best action is given any state.

The formulation of reinforcement learning is very similar to Markov Decision Processes (MDPs) and relies on defining a set of states and actions.

The components of a MDP problem are:

1. Set of states $s \in S$
2. Set of inputs $u \in U$
3. Transition function $T(s, a, s') = P(s'|s, a)$
4. Reward function $R(s, a, s')$
5. Initial state s_0
6. Possibly a desired final state s_{final}
7. Policy $\pi^*(s) = a^*$ that must be computed

Here states and actions are discrete. It is possible to find an optimal policy $\pi^*(s)$ by solving the MDP offline using the above information. States and actions show what states the system can be in and what actions can be taken. Transition function provides information on how a system changes its state in response to an action a taken in state s. The reward function indicates what is good and what is bad. Based on these, an optimal policy that is a mapping from states to actions can be obtained by solving the value function:

$$V(s) = \max_{a \in U} \sum_{s'} T(s,a,s')(R(s,a,s') + \gamma V(s')),$$

where γ is the discount factor and reflects the greater significance of earlier rewards.

The optimal policy indicates what is the best control input to apply in a given state if the goal is to maximize utility. While in the case of MDP this can be done by well-known value iteration and policy iteration algorithms, in RL this is not possible. This is because the transition function and reward function are unknown. In this case, the agent has to learn what the optimal policy is by taking actions and observing the outcome. This is why RL is so much harder than MDP.

Like MDPs, RL can be solved by value function methods or policy methods. In value function methods, the policy is obtained by first approximating the value function $V(s)$ or Q-values, $Q(s, a)$. Q-values are defined as:

$$Q(s,a) = \sum_{s'} T(s,a,s')(R(s,a,s') + \gamma V(s'))$$

therefore it can be seen that:

$$V(s) = \max_{a \in U} Q(s,a).$$

Q-values are very useful in that they provides a measure of utility for a state action pair (s, a) also known as q-state. This measure shows what values can be obtained if an action a is taken in state s and from then on optimal actions are taken. In other words, Q-values can indicate the quality of actions taken in a state and for this reason they form the foundation of one of the early RL methods such as Q-Learning and a variant of it known as State Action Reward State Action (SARSA) (Russell and Norvig 2010).

According to the survey in (Kober 2012), the value function methods can be divided in three categories:

1. Dynamic programming-based optimal control approaches such as policy iteration or value iteration.
2. Rollout-based Monte Carlo methods.
3. Temporal difference methods such as TD, Q-learning and SARSA.

The value function methods become intractable for high dimensional problems, however they have been successfully applied to many different robotics problems (Kober 2012).

Policy search is another approach for solving RL where an initial control policy is locally optimized to produce better controllers. Policy methods have three clear advantages in comparison to value function methods. First, policies often have fewer unknown parameters than value functions. Second, policy methods can be extended to continuous state and action space straight forwardly. Third, it is possible to incorporate expert knowledge of the system through an initial policy. This is why policy search methods have been found more suitable for high dimensional problems. There are three categories of policy methods in RL (Kober 2012):

1. Policy gradient approach based on likelihood-ratio estimation.
2. Policy update based on Expectation Maximization: a well known optimization method that consists of repeatedly estimating the expected utility (E-step) and maximizing it (M-step) until convergence.
3. Path integral methods.

According to (Kober 2012), some of the common challenges encountered in RL are:

1. *Large state space*: For high dimensional problems RL has to search in a very large state-action space for the optimal policy. This can be intractable in practice.
2. *Wear and tear*: It is possible that the system under operation experiences tear and wear due to repeated undesirable motions.
3. *Change in operating condition and erroneous data*: If the environment changes, the optimal policy also needs to change. If this change is faster than the learning rate of the system, the optimal policy will never be found.

The first two problems have been addressed mostly by using simulation environment. Using a model of the system, RL can run in simulation and come up with a policy that is optimal. However, as model mismatch is inevitable often the computed policy needs to be tweaked to work on the real system.

In order to help RL find the optimal policy in the high dimensional space demonstration data has also been used to guide policy learning. In this approach which is called imitation learning or learning from demonstration first the desired task is demonstrated by an operator and the data corresponding to that is collected to extract the initial control policy. Then the policy obtained from data can be deployed on the system and RL can be used to further improve that policy through more operations. This shows the sort of powerful control methods that can be synthesized using machine learning and learning control.

As mentioned earlier in the chapter, state and input constraints are some of the important challenges encountered in manipulator control and robotics in general. Some work on applying RL and ILC to constrained systems has been done (Andersson et al. 2015, He et al. 2007, Mishra et al. 2011). One promising approach has been to use constrained optimization formulations. This approach shares some resemblance with Nonlinear MPC (NMPC). However we believe there is still more work to be done to incorporate state and input constraints in the formulation of learning control so that effective exploration and exploitation can occur without constraint violation. In fact if the formulation incorporates the constraints, the search space for optimal actions will become smaller and the convergence rate could potentially be improved.

The needs for adaptation, learning and constraint handling has led to recent efforts to combine learning methods with MPC. This is discussed next.

LEARNING METHODS IN MPC

The presence of state and input constraints have posed a serious challenge at control design for a long time. This is particularly a problem as we aim to achieve high performance which would require operating systems close to their limits. In order to avoid constraint violation and saturation, one needs to make sure in advance that the input signal that is to be applied to the system over a period of T is not going to exceed the limits. One way is to rely on the experience obtained from operating the system. This can be captured by a learned inverse model of the system or a control policy obtained through learning control methods. Another way is to use predictions based on a model of the system.

Predictions rely on the model and for a prediction to be representative of the actual behavior of the system, the model mismatch has to be as small as possible. Any uncertainty or unknown variations not accounted for in the model would undermine the validity of predictions.

Model Predictive Control (MPC) is widely considered an effective method for controlling constrained MIMO systems. MPC has had successful applications in chemical process control for its ability to handle constraints (Morari 1989). One of the earliest industrial MPC technologies was developed by engineers in Shell Oil in early 1970s (Qin and Badgwell 2003). With increasing computational power and the rise of more efficient numerical algorithms MPC is increasingly adopted for control of many other systems such as autonomous vehicles, trajectory control, HVAC and so on. In the following section a brief introduction to MPC is presented.

MODEL PREDICTIVE CONTROL

The idea of receding horizon MPC is simple and elegant. In this approach, at time t_0 and for a prediction horizon of N_p, the controller uses the current state of the system $\mathbf{x}_0 = \mathbf{x}(\mathbf{k} \mid \mathbf{k})$ to calculate the input $\mathbf{u} = [u(k \mid k), u(k+1 \mid k), u(k+2 \mid k),..., u(k+N_p \mid k)]$

that would drive the system to the desired state $\mathbf{x}_1$ such that a cost function $J = g(x(k + N_p|k)) + \int_{i=0}^{i=N_{p-1}} l(x(k + i|k), u(k+i|k)$ is minimized and operational constraints are not violated. Then the controller applies the first element of the input to the system. In the next step, the controller takes the new state measurement/estimation and repeats the same calculations for a prediction horizon of N_p, and applies only the first element of the resulting input $\mathbf{u}$. Provided that the calculations are fast and the system models are accurate this control system will achieve optimal performance. The advantage of MPC in comparison to the classic optimal control theory methods is its ability to handle constraints by defining the control problem as a constrained optimization problem. This is invaluable in practical scenarios such as the manipulator control as actuator constraints are present. In addition, because the controller computes the input at every step based on the observed state, a feedback action is also present.

MPC computes input and state trajectories by solving an optimization problem such as:

Minimize $J_{x,u}$

subject to

$$\mathbf{x}(k + i + 1 \mid k) = f(\mathbf{x}(k + i \mid k), u(k + i \mid k)) \quad i = 0,1,2,3,\ldots, N_p-1$$

$$u_{lower} \leq u(k + i \mid k) \leq u_{upper} \quad i = 0,1,2,3,\ldots, N_p-1$$

$$\mathbf{x}_{lower} \leq \mathbf{x}(k + i \mid k) \leq \mathbf{x}_{upper} \quad i = 0,1,2,3,\ldots, N_p-1$$

$$\mathbf{x}(k \mid k) = \mathbf{x}_0$$

The forward model of the system is at the heart of MPC and is used for making predictions about the state trajectory. When the model is linear, i.e., $\mathbf{x}(k + i + 1|k) = A\mathbf{x}(k + i|k) + Bu(k + i \mid k)$ and the system is subject to linear semi-algebraic constaints. MPC can be solved easily and efficiently. The reason for this is that the optimization problem can be solved by convex mathematical programming techniques (Boyd and Vandenberghe 2004). However, if the model is nonlinear, as in the case of manipulators, solving the optimization problem is very difficult. This is somewhat expected for nonlinear systems. The complexities introduced by nonlinearities make it difficult to even simulate such systems given an input trajectory $\mathbf{u}$, let alone solving for both input and state trajectory. This can also be explained from a geometric point of view. Linear systems have a state manifold which has zero curvature and state gradient has a linear relation with the input. This means the state of the system evolves on the same hyperplane which leads to a set of linear algebraic equations. In contrast, the state manifold of nonlinear systems has varying curvature and the input does not have a linear relation with a state gradient. It is this curvature variation that prevents us from solving predictive problems using such models.

Prediction for Nonlinear Systems

Model predictions for nonlinear systems is one of the most difficult problems in MPC. For this reason, nonlinear systems are often linearized about an initial trajectory and the optimization problem is solved to obtain a new trajectory. Then the model is linearized about the new trajectory and the process is repeated until convergence is achieved. Sequential Quadratic Programming (SQP) and Sequential Convex Programming (SCP) are two optimization methods that operate in this manner and they are commonly used for Nonlinear MPC (NMPC) problems. However both methods are largely dependent on a "good" starting trajectory and may not converge to the global optimal solution.

In many problems the nonlinearities of the system are not theoretically modeled. However, their influence on model predictions cannot be ignored. In this scenario, the only option is to rely on data driven modeling to incorporate these nonlinearities into model predictions.

These issues have been recognized in the most recent research on MPC. Researchers have tried to account for the sort of nonlinear dynamics that is difficult or even impossible to model analytically. One of the first prominent works in this areas is Learning Based MPC presented by Aswani et al. (Aswani et al. 2013). The paper proposes a robust adaptive MPC that guarantees robustness and achieves performance improvement. In this approach two linear models are used in MPC. One model consists of the nominal linear model with state $\bar{x}_n$ and uncertainties d_n due to nonlinear effects. This is of the form:

$$\bar{x}_{n+1} = A\bar{x}_n + B\bar{u}_n + d_n$$

The robustness conditions are then established by introducing an invariant set for the terminal state and using tube MPC one of the common methods in robust MPC. Tube MPC uses the nominal model $\bar{x}_{n+1} = A\bar{x}_n + B\bar{u}_n$ for predictions while it accounts for the effect of disturbance d by shrinking the state constraints by some value given by set $\mathcal{R}$.

The other model is linearized with state $\tilde{x}_n$ but contains some unknown nonlinear component $\mathcal{O}_n(\tilde{x}_n, u_n)$ and is of the following form:

$$\tilde{x}_{n+1} = A\tilde{x}_n + Bu_n + \mathcal{O}_n(\tilde{x}_n, u_n).$$

$\mathcal{O}_n(\tilde{x}_n, u_n)$ can be "learned" as an "oracle" which returns the value of nonlinear component and its gradient as a function of current state and input. Using both models in the optimization problem formulated for MPC would satisfy robustness conditions while optimizing and system performance. The learning process has been shown to improve the performance of MPC by making better dynamic predictions. Indeed, it has been shown that given sufficient excitation the MPC would converge to a performance that would be achieved if the exact model of the plant was known (Aswani et al. 2013).

Yan and Wang (Yan and Wang 2014) have also carried out a similar strategy to solve robust model predictive control for nonlinear systems by adopting neural networks not only for model learning but also for solving the optimization problem. In their approach, the nonlinear system is represented as:

$$x(k+1) = f(x(k), u(k)) + w(x(x), u(k)) + v(k)$$

where $x \in \mathcal{R}^n$ is the state vector, $u \in \mathcal{R}^m$ is the input vector, $w \in \mathcal{R}^n$ represents unmodeled dynamics, $v \in \mathcal{R}^n$ is the additive bounded uncertainties and $f(.)$ is a nonlinear function representing the modeled dynamics.

This model is then linearized about the operating point (x_0, u_0) similar to the previous case:

$$\delta x(k+1) = \frac{\partial f(x_0, u_0)}{\partial x}\delta x(k) + \frac{\partial f(x_0, u_0)}{\partial u}\delta u(k) + d(k) + v(k)$$

$$\delta x(k+1) = A(k)\delta x(k) + B(k)\delta u(k) + d(k) + v(k)$$

where $d(k)$ represents the nonlinearities and unmodeled dynamics.

In order to achieve robustness the authors employ a minimax approach where the control cost is minimized for the worst uncertainty scenario:

$$\min_{\delta u} \max_{v} \boldsymbol{J}_{\delta x, \delta u, \delta v}$$

subject to:

$$\delta x(k+i+1 \mid k) = A\delta x(k+i \mid k) + B[\delta u(k+i-1 \mid k) + \Delta u(k+i \mid k)] +$$
$$d(k+i \mid k) + v(k+i \mid k)$$

$$v_{min} \leq v(k+i \mid k) \leq v_{max}$$

$$\delta x_{min} \leq \delta x(k+i \mid k) \leq \delta x_{max}$$

$$\delta u_{min} \leq \delta u(k+i-1 \mid k) \leq \delta u_{max} \quad i = 1, 2, \ldots, N$$

In order to learn the nonlinearities and unmodeled effects, an **Extreme Learning Machine (ELM)** is used. In the next step, the minimax optimization is transformed to a nonlinear convex optimization problem that is solved using a **Recurrent Neural Network (RNN)** based on the work in (Xia and Wang 2004).

Learning based MPC has been applied to trajectory tracking in robotic manipulators as well. In (Lehnert and Wyeth 2013), the authors use local weighted regression to construct m different local linear models based on observed input-output data. The learned models, which are constantly updated during operation, are then used to solve m linear MPC problems to obtain m candidate inputs to be applied to the system. These inputs are then mixed using a kernel that weighs each input according to the distance between the center of each local linear model and the current state of the system. Gaussian kernel is commonly used for calculating weights w_i.

$$w_i(z) = e^{-\frac{1}{2}(z-c_i)^T D_i (z-c_i)}$$

Where c_i is the center of the kernel and D_i controls the shape of the kernel function. Using these weights, the inputs calculated from each MPC can be mixed according to the following equations

$$u_{k+1} = \frac{1}{W} \sum_{i=1}^{m} w_i(x_k, u_k) u_{k+1}^i$$

$$W = \sum_{i=1}^{m} w_i(x_k, u_k).$$

Here u_k is the current input, u_{k+1} is the next input to be applied and u_{k+1}^i is the next input as calculated using MPC from the ith local linear model.

Although this work solves an unconstrained MPC and does not address the issues related to constraint handling, it demonstrates that manipulators can be modeled using observed input-output data. Furthermore experimental results have shown that the learning process indeed improves the tracking performance of the MPC over time.

More recently deep learning and MPC have been combined to compute the forces required for a robotic manipulator to cut a variety of foods (Lenz et al. 2015). The dynamic model of a cutting process which has position of the knife as its states and knife forces as its inputs cannot be dynamically modeled due to a great variety of food properties and complex interaction between the robot and the food. Therefore the task represents a challenging combination of time variance, nonlinearity and structural variations such as food stiffness and geometry. The deep learning architecture proposed in (Lenz et al. 2015) is able to use learned recurrent features to integrate long term information such as material properties into the model. Therefore this modeling approach is able to combine local information with global knowledge obtained from previous observations. DeepMPC is able to produce state predictions from a trajectory that is chosen based on past observations. The algorithm then evaluates MPC cost function and its gradient. After that the input is adjusted using gradient descent and the process is repeated until convergence.

CONCLUSION

In this chapter we first presented a review of MRAC and identified to the role of machine learning in traditional MRAC. We then discussed the limitations of MRAC and explained how more advanced machine learning methods have shown promise in offering more in terms of modeling more complex dynamics, removing requirements such as persistent excitation and look poised to move adaptive control methods forward significantly over what has been possible to date. The role of machine learning in control was discussed here from a modeling perspective and some of the most common machine learning models such as NN and mixture models were presented. Particularly, the idea of motor primitives was introduced as a biologically inspired method for motion representation and learning. In addition learning control and its connection with adaptive control were discussed and methods such as ILC and RL were introduced. Handling state and input constraints was identified as one of the challenges that still needs further work. We gave an introduction to MPC for its constraint handling capability and reviewed some of the recent work that have aimed to incorporate learning methods in MPC framework.

While data driven modeling and policy synthesis has demonstrated its benefits, it does not mean that analytical modeling and control approaches should be deserted altogether. Instead machine learning should play a complementary role to analytical approaches. This is an idea that has appeared in some of the literature in MRAC for manipulators where adaptation was only introduced on the unknown portion of the dynamics of the system (Maliotis 1991, Craig et al. 1987). We nevertheless see a bright future for the formulation and application of machine learning methods in control engineering.

REFERENCES

Ahn, H. -S., Y. Q. Chen and K. L. Moore. 2007. Iterative learning control: Brief survey and categorization. Systems, Man, and Cybernetics, Part C: Applications and Reviews, IEEE Transactions on, 37(6): 1099–1121.

Andersson, O., F. Heintz and P. Doherty. 2015. Model-based reinforcement learning in continuous environments using real-time constrained optimization. AAAI Conference on Artificial Intelligence.

Arimoto, S., S. Kawamura and F. Miyazaki. 1984. Bettering operation of robots by learning. Journal of Robotic Systems, 1(2): 123–140.
Arimoto, S., S. Kawamura and F. Miyazaki. 1986. Convergence, stability and robustness of learning control schemes for robot manipulators. Proceedings of the International Symposium on Robot Manipulators on Recent Trends in Robotics: Modelling, Control and Education, pp. 307–316.
Astrom, K. J. and B. Wittenmark. 2013. Adaptive Control: Second Edition, Dover Publications, Inc.
Aswani, A., H. Gonzalez, S. S. Sastry and C. Tomlin. 2013. Provably safe and robust learning-based model predictive control. Automatica, 49(5): 1216–1226.
Boyd, S. P. and L. Vandenberghe. 2004. Convex Optimization, Cambridge University Press.
Bristow, D. A., M. Tharayil and A. G. Alleyne. 2006. A survey of iterative learning control. Control Systems, IEEE, 26(3): 96–114.
Chen, Y. and C. Wen. 1999. Iterative learning control: Convergence, robustness and applications.
Chowdhary, G., T. Yucelen, M. Muhlegg and E. N. Johnson. 2013. Concurrent learning adaptive control of linear systems with exponentially convergent bounds. International Journal of Adaptive Control and Signal Processing, 27(4): 280–301.
Ciliz, K. and A. Cezayirli. 2004. Combined direct and indirect adaptive control of robot manipulators using multiple models. Robotics, Automation and Mechatronics, 2004 IEEE Conference on, 1: 525–529.
Craig, J. J., P. Hsu and S. S. Sastry. 1987. Adaptive control of mechanical manipulators. The International Journal of Robotics Research, 6(2): 16–28.
Crespo, L. G., M. Matsutani and A. M. Annaswamy. 2010. Design of a model reference adaptive controller for an unmanned air vehicle. American Institute of Aeronautics and Astronautics.
Dubowsky, S. and D. T. Deforges. 1979. The application of model reference adaptive control to robotic manipulators. International Journal of Adaptive Control and Signal Processing, 101(3): 193–200.
Ham, W. 1993. Adaptive control based on explicit model of robot manipulator. Automatic Control, IEEE Transactions on, 38(4): 654–658.
He, P., M. Rolla and S. Jagannathan. 2007. Reinforcement learning neural-network-based controller for nonlinear discrete-time systems with input constraints. Systems, Man, and Cybernetics, Part B: Cybernetics, IEEE Transactions on, 37(2): 425–436.
Hsia, T. C. 1984. Adaptive control of robot manipulators—A review. Robotics and Automation. Proceedings, 1986 IEEE International Conference on.
Ioannou, P. and P. Kokotovic. 1983. Adaptive Systems with Reduced Models, Springer Verlag: Secaucus.
Khansari-Zadeh, S. M. and A. Billard. 2011. Learning stable nonlinear dynamical systems with Gaussian mixture models. Robotics, IEEE Transactions on, 27(5): 943–957.
Kober, J. 2012. Learning Motor Skills: From Algorithms to Robot Experiments, Technische Universitat Darmstadt.
Krause, J. and G. Stein. 1987. Structural limitations of model reference adaptive controllers. American Control Conference, 1987.
Lehnert, C. and G. Wyeth. 2013. Locally weighted learning model predictive control for nonlinear and time varying dynamics. Robotics and Automation (ICRA), 2013 IEEE International Conference on.
Lenz, I., R. Knepper and A. Saxena. 2015. DeepMPC: Learning Deep Latent Features for Model Predictive Control. Robotics Science and Systems (RSS).
Maciejowski, J. M. 2001. Predictive Control with Constraints, Pearson Education (US).
Maliotis, G. 1991. A hybrid model reference adaptive control/computed torque control scheme for robotic manipulators. Proceedings of the Institution of Mechanical Engineers, Part I: Journal of Systems and Control Engineering.
Mishra, S., U. Topcu and M. Tomizuka. 2011. Optimization-based constrained iterative learning control. Control Systems Technology, IEEE Transactions on, 19(6): 1613–1621.
Mohan, S. and J. Kim. 2012. Indirect adaptive control of an autonomous underwater vehicle-manipulator system for underwater manipulation tasks. Ocean Engineering, 54: 233–243.
Morari, M. a. Z. E. 1989. Robust Process Control, Prentice-Hall.
Muelling, K., J. Kober and J. Peters. 2010. Learning Table Tennis with a Mixture of Motor Primitives. Humanoid Robots (Humanoids), (2010) 10th IEEE-RAS International Conference on.
Narendra, K. and A. Annaswamy. 1986. Robust adaptive control in the presence of bounded disturbances. IEEE Transactions on Automatic Control, 31(4): 306–315.

Narendra, K. S. and A. M. Annaswamy. 1987. Persistent excitation in adaptive systems. International Journal of Control, 45(1): 127–160.

Nguyen-Tuong, D. and J. Peters. 2011. Model learning for robot control: A survey. Cognitive Processing, 12(4): 319–340.

Osburn, P. V., H. P. Whitaker and A. Kezer. 1961. New developments in the design of adaptive control systems. Institute of Aeronautical Sciences.

P., I. and J. Sun. 1996. Robust Adaptive Control, Prentice Hall, Inc.

Qin, S. J. and T. A. Badgwell. 2003. A survey of industrial model predictive control technology. Control Engineering Practice, 11(7): 733–764.

Rong, H. -J. 2012. Indirect adaptive fuzzy-neural control of robot manipulator. High Performance Computing and Communication 2012 IEEE 9th International Conference on Embedded Software and Systems (HPCC-ICESS), 2012 IEEE 14th International Conference on, pp. 1776–1781.

Russell, S. and P. Norvig. 2010. Artificial Intelligence: A Modern Approach, Prentice Hall.

Schaal, S., P. Mohajerian and A. Ijspeert. 2007. Dynamics systems vs. Optimal control—A unifying view. *In*: P. Cisek, T. Drew and J. F. Kalaska. Computational Neuroscience: Theoretical Insights into Brain Function. Elsevier, 165: 425–445.

Schmidt, R. A. and T. D. Lee. 2011. Motor Control and Learning: A Behavioral Emphasis, Human Kinetics.

Slotine, J. -J. E. and W. Li. 1987. On the adaptive control of robot manipulators. The International Journal of Robotics Research, 6(3): 49–59.

Slotine, J. -J. E. and W. Li. 1991. Applied Nonlinear Control, Prentice-Hall International, Inc.

Tsai, M. C. and M. Tomizuka. 1989. Model reference adaptive control and repetitive control for robot manipulators. Robotics and Automation, 1989. Proceedings, 1989 IEEE International Conference on, 3: 1650–1655.

Tung, P. -C., S. -R. Wang and F. -Y. Hong. 2000. Application of MRAC theory for adaptive control of a constrained robot manipulator. International Journal of Machine Tools and Manufacture, 40(14): 2083–2097.

Uchiyama, M. 1978. Formation of High-Speed Motion Pattern of a Mechanical Arm by Trial (Japanese Title: 試行による人工の手の高速運動パターンの形成). Transactions of the Society of Instrument and Control Engineers, 14: 706–712.

Verscheure, D., B. Demeulenaere, J. Swevers, J. De Schutter and M. Diehl. 2009. Time-optimal path tracking for robots: A convex optimization approach. Automatic Control, IEEE Transactions on, 54(10): 2318–2327.

Xia, Y. and J. Wang. 2004. A recurrent neural network for nonlinear convex optimization subject to nonlinear inequality constraints. Circuits and Systems I: Regular Papers, IEEE Transactions on, 51(7): 1385–1394.

Yan, Z. and J. Wang. 2014. Robust model predictive control of nonlinear systems with unmodelled dynamics and bounded uncertainties based on neural networks. Neural Networks and Learning Systems, IEEE Transactions on, 25(3): 457–469.

Yu, H., L. D. Seneviratne and S. W. E. Earles. 1993. Combined adaptive control of constrained robot manipulators. Intelligent Robots and Systems '93, IROS '93. Proceedings of the 1993 IEEE/RSJ International Conference on, 2: 740–745.

2

Discussion on Model Reference Adaptive Control of Robotic Manipulators

Dan Zhang[1] and *Bin Wei*[2,*]

ABSTRACT

Motion control accuracy of robotic manipulators affects the overall robotic system performance. When the end-effector grabs different payloads, the joint motion of robotic manipulators will vary depending on the different payload masses. Traditional controllers have the problem of not being able to compensate the payload variation effect. Model reference adaptive control has been proposed to address the above problem. This paper discusses the model reference adaptive control of robotic manipulators initially raised by the scholar Roberto Horowitz and its associated development by other authors. Some issues of model reference adaptive control (MRAC) for robotic manipulators are presented. Very few recent papers can be found in the area of model reference adaptive control of robotic manipulators, and this paper will provide a guideline for future research in the direction of model reference adaptive control for robotic arms.

Keywords: Adaptive control, robotic manipulator, model reference approach

[1] Department of Mechanical Engineering, York University, Toronto, Ontario, M3J 1P3, Canada. E-mail: dzhang99@yorku.ca

[2] Faculty of Engineering and Applied Science, University of Ontario Institute of Technology, 2000 Simcoe Street North, Oshawa, Ontario, Canada. E-mail: Bin.Wei@uoit.ca

* Corresponding author

INTRODUCTION

The robotic manipulator control problem is generally formulated as follows, given a desired trajectory, a mathematical model of the manipulator and its interactions with the environment, find the control algorithm which sends torque commands to the actuators so that the robotic manipulator can achieve expected motion. The control design for a serial robotic manipulator involves two steps. Firstly, an end-effector motion trajectory has been given, i.e., say one wants to move the end-effector from point A to point B. From this end-effector motion trajectory, and by using the inverse kinematics, the joint motion can be determined so as to produce this desired end-effector motion trajectory. The second step is to determine the joint torque, how much torque does one have to apply to the joint so that the joint will have the desired motion? The joint torque can be determined by solving the inverse dynamic equation.

Control the robot to perform in a certain way is one of the most challenging problems because the robotic manipulator mechanism is highly nonlinear. For the robotic manipulators, the coefficients of the dynamic equations are functions of joint variables and also the function of payload mass which may be unknown or change throughout the task. When the manipulator moves, the joint variables are changing, which will cause the robotic manipulator's dynamic equation to change throughout a given task. In order to obtain a high degree of accuracy and repeatability in the manipulator performance, it is necessary to use a control system (e.g., MRAC) that will account for the changes in the dynamic characteristic of the manipulator. Conventional control methods model the manipulator as uncoupled linear subsystems, these methods can produce satisfactory performances at low speeds, but it is not efficient anymore when used for high speed and high accuracy operations. The use of the PID controller for control does not guarantee optimal control of the system or system stability (https://en.wikipedia.org/wiki/PID_controller). In order to address the above problem, adaptive control can be applied. Model reference adaptive approach is most popular and established technique.

Adaptive control adapts to a controlled system with parameters which vary, or are initially uncertain. For non-adaptive controller, the controller is designed based on the priori information of the system, i.e., one knows the system and designs the controller (e.g., PID controller) gears to that system and assume there is no change in the system. Whereas for the adaptive controller, the controller does not necessary need to depend on previous information of the system, and if there is sudden change in environment, the controller can cope with it to adapt to the changed conditions. If one considers a system that its transfer function is known, one designs a fixed classical controller, the controller will remain fixed parameters as long as it applies to the system, so one can say that this controller depends on its structure and designed on a-priori information, this is non-adaptive controller. However, if the controller is depending on posteriori information, for example, if one is changing the parameters of the controller, because of the changes of the parameters of the system or because of the disturbances coming from the environment, this controller is called adaptive. If the system is subject to unknown disturbances, or the system is expected to undergo changes in its parameters in a way which is not pre-determined from the beginning, in that case one uses adaptive control. However, in some cases one knows how the system operating condition will change, for example, for an aircraft, one knows that the aircraft controller is determined by its altitude and speed, and one expects that aircraft to fly at specific value for altitude and speed, in that case one can design a controller for each expected operating point and switch between the different controllers, this is called gain-scheduling. In other cases one knows that the parameters of the system change, but one also knows a range for the change of every parameter, in that case it is possible to design a fixed controller that can cope with different changes of the parameters, and guarantee the stability and performance, this kind of controller is robust controller.

For non-adaptive control, say fixed-gain control, according to Fig. 1, one can see that firstly when one needs to improve the performance error, the modelling accuracy will also be increased, secondly it cannot improve itself, and thirdly it is assumed that future will be much like present, ignoring environment changes, change in dynamics and structure damage.

For the adaptive control, it achieves a given system performance asymptotically, it does not trade performance for modelling accuracy, as shown in Fig. 2, and more importantly, it improves itself under unforeseen and adverse conditions.

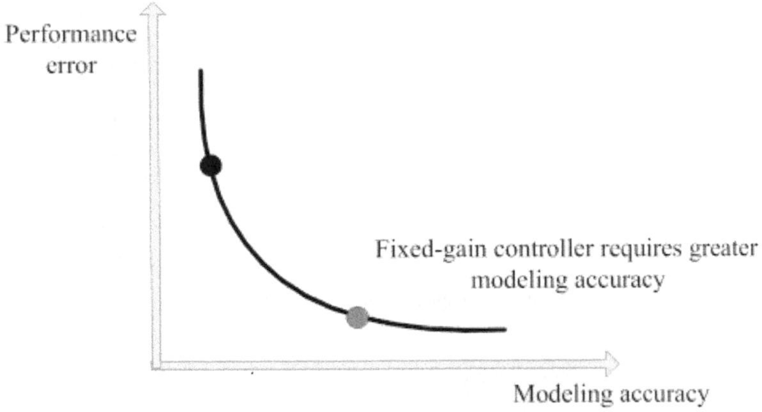

Figure 1. Non-adaptive control.

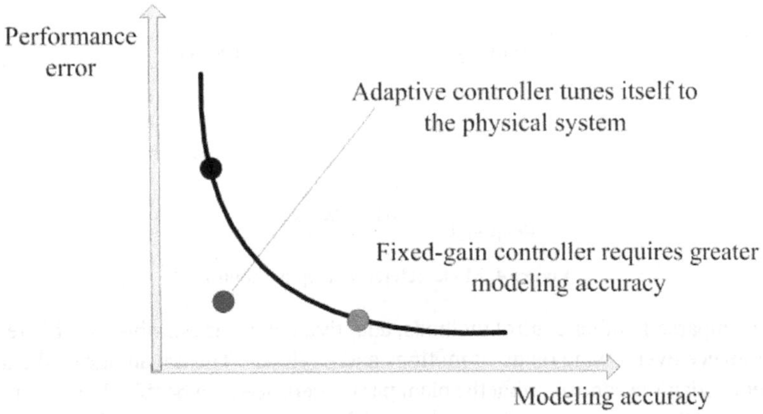

Figure 2. Adaptive control.

The adaptive control can be categorized into the following, model reference adaptive control, self-tuning adaptive control and gain-scheduled control, as shown in Fig. 3. Here we will mainly consider the model-reference approach.

For the model-reference adaptive control, the set value is used as an input to both the actual and the model systems, and difference between the actual output and the output from the model is compared. The difference in these signals is then used to adjust the parameters of the controller to minimize the difference. Figure 4 shows such control system.

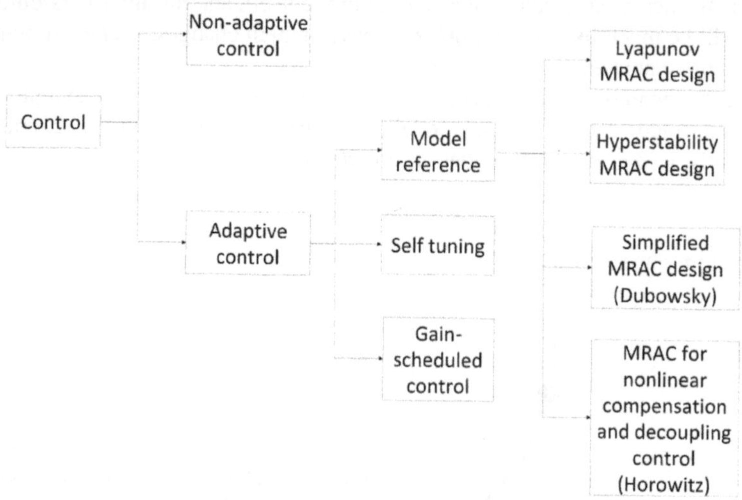

Figure 3. Adaptive control categorization.

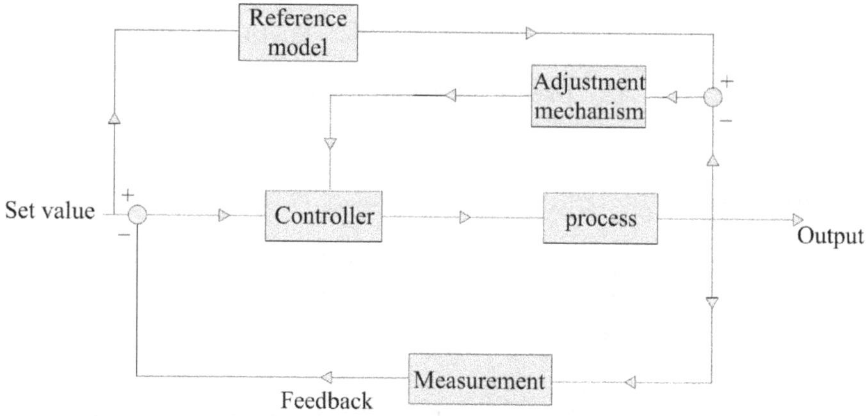

Figure 4. Model reference adaptive control.

Compared to other control methods, adaptive control is possible to achieve good performance over a wide range of motions and payloads. The advantage of the model reference adaptive control is that the plant parameters need not be fully known, instead, estimates of the plant parameters are used and the adaptive controller utilizes past input/output information to improve these estimates. However there are two drawbacks to

MRAC. Stability analysis of the system is critical as it is not easy to design a stable adaptive law. The other problem is that MRAC relies on cancellation of the non-linear terms by the reference model (Sutherland 1987). In reality, exact cancellation cannot be expected, but the non-linear terms may be made so small so as to be negligible. Model reference adaptive control method was initially introduced in (Whitaker et al. 1958), when they considered adaptive aircraft flight control systems, using a reference model to obtain error signals between the actual and desired behavior. These error signals were used to modify the controller parameters to attain ideal behavior in spite of uncertainties and varying system dynamics. The goal of an adaptive control system is to achieve and maintain an acceptable level in the performance of the control system in the presence of plant parameter variations. Whereas a conventional feedback control system is mainly dedicated to the elimination of the effect of disturbances upon the controlled variables, also known as manipulated variables. An adaptive control system is mainly dedicated to the elimination of the effect of parameter disturbances/variations upon the performance of the control system.

ADAPTIVE CONTROL OF ROBOTIC MANIPULATORS

General Adaptive Control

In traditional control system, feedback is used to reject the disturbance effect that are acting on the controlled variables in order to bring the controlled variables back to their desired value. To do that, the variables are measured and compared to the desired values and the difference is fed into the controller. In these feedback systems, the designer adjusts the parameters of the controller so that a desired control performance is achieved. This is done by having a priori knowledge of the plant dynamics. When the parameters of the plant dynamic models change with time due to disturbances, the conventional control cannot deal with it anymore as the control performance will be degraded. At this time, one needs to resort to the adaptive control. A structured approach for the design of distributed and reconfigurable control system is presented in (Valente et al. 2011). Distributed architectures are conceived as interconnected independent modules with standard interfaces which can be modified and reused without affecting the overall control structure. Whereas for the centralized control architectures, any change of the machine structure requires an extensive replacement of the control system. In RMS, modular and distributed architecture is essential to guarantee the capability of each single module or portions of the control to be adapted when a hardware reconfiguration occurs.

In (Bi et al. 2015), the sustainable manufacturing by reconfiguration of robots through using robot modules was presented. The customized modules are end-effector, suction pump and adapters, modular frame, steering guide, PLC and robot controller, sensors, power supply and indicators, and touch screen. In terms of control, there are two separate controllers to operate. One controller is used to control the robot arm and the other is programmable logic controller (PLC) that will handle user inputs and sensor data in order to inform the robot arm controller which program to run. When the robot is reconfigured, the control system needs to be reconfigured as well, i.e., the software system should be reconfigured to support the communication and interactions among system components. This is called software reconfiguration.

In (Wilson and Rock 1995), the neural networks is used for the control reconfiguration design for a space robot. The traditional controller was presented, and by using the neural networks, the traditional controller is updated to a reconfigurable controller. Two neural-network-control were developed to achieve quick adaptation controller. Firstly, a fully-connected architecture was used that has the ability to incorporate an a priori approximate linear solution instantly, this permits quick stabilization by an approximate linear controller. Secondly, a back-propagation learning method was used that allows back-propagation with discrete-valued functions. It presents a new reconfigurable neural-network-based adaptive control system for the space robot.

In (Jung et al. 2005), it studied an adaptive reconfigurable flight control system using the mode switching of multiple models. The basic idea is to use the on-line estimates of the aircraft parameters to decide which controller to choose in a particular flight condition. It is related to the multi-mode adaptive control. In (Landau et al. 2011), the basic concept of adaptive control and several kinds of categories were introduced, i.e., open-loop adaptive control, direct adaptive control, indirect adaptive control, robust control, and conventional control, etc. The adaptive control can be seen as a conventional feedback control system but where the controlled variable is the performance index. So there are two loops for the adaptive control, one is the conventional feedback loop and the other is the adaptation loop.

A control development approach is proposed in (Valentea et al. 2015) which consists of three steps: control conceptual design, application development and evaluation of solution robustness. The control system should be conceived as a set of independent and distributed control modules, capable of nesting one to each other. The structuring of control logics is the basis of the entire control development process. In order to enable the control system reconfiguration, an essential feature of the control architecture is the modularity and distribution of the control decisions across the various entities.

Adaptive Control for Robotic Manipulators

Non-adaptive controller designs often ignores the nonlinearities and dynamic couplings between joint motions, when robot motions require high speed and accelerations, it greatly deteriorate its control performance. Furthermore, non-adaptive controller designs requires the exact knowledge and explicit use of the complex system dynamics and system parameters. Uncertainties will cause dynamic performance degradation and system instability. There are many uncertainties in all robot dynamic models, model parameters such as link length, mass and inertia, variable payloads, elasticities and backlashes of gear trains are either impossible to know precisely or varying unpredictably. That is why adaptive control is needed to address the above problem.

Model reference adaptive control and its usage to robotic arms were introduced in (Neuman and Stone 1983) and (Amerongen 1981). Some design problems in adaptive robot control are briefly stated. S. Dubowsky (Dubowsky and Desforges 1979) is the first one that applies the model reference adaptive control in the robotic manipulator. The approach follows the method in (Donalson and Leondes 1963). A linear, second-order, time-invariant differential equation was used as the reference model for each degree of freedom of the manipulator arm. The manipulator was controlled by adjusting the position and velocity feedback gains to follow the model. A steepest-descent method was used for updating the feedback gains. Firstly the reference model dynamics was written, subsequently the nonlinear manipulator (plant) dynamic equation was written, but how this equation is related to the Lagrange equation is not clear, thirdly an error function was written and it follows the method of steepest descent and a set of equations was derived for the parameter adjustment mechanism, which will minimize the difference between the actual closed-loop system response and the reference model response.

An adaptive algorithm was developed in (Horowitz and Tomizuka 1986) for serial robotic arm for the purpose of compensating nonlinear term in dynamic equations and decoupling the dynamic interaction among the joints. The adaptive method proposed in this paper is different from Dubowsky's approach (Dubowsky and Desforges 1979). Three main differences are concluded as follows: firstly, in Horowitz's method, the overall control system has an inner loop model reference adaptive system controller and an outer loop PID controller, whereas the control system in Dubowsky's method is entirely based on the model reference adaptive controller; secondly, in Dubowsky's paper, the coupling among joints and nonlinear terms in the manipulator equations are ignored whereas this is considered in Horowitz's method; thirdly, in Horowitz's paper, the design method is based on the hyper-stability method whereas the adaptive algorithm design in (Dubowsky and Desforges 1979) is based on the steepest descent method. Also in (Horowitz and Tomizuka 1986), there are some issues as follows. (1) The authors stated in (Horowitz and Tomizuka 1986) that "The overall control system will have an inner loop MRAS controller and an outer loop PID action controller with fixed gains". The above statement is not consistent with the Figs. 4 and 5 in (Horowitz and Tomizuka 1986). According to Figs. 4 and 5 in (Horowitz and Tomizuka 1986), the control system has an inner loop MRAS controller, but it does not have an outer loop PID action controller. The PID controller is in the form of $K_p e + K_i \int edt + K_d \dot{e}$, but in the paper (Horowitz and Tomizuka 1986), the outer loop controller is in the form of $K_i \int e - K_p x_p - K_d x_v$, which is absolutely not a PID controller. (2) For the Figs. 4 and 5 in (Horowitz and Tomizuka 1986), the outer loop controller is in the form of $K_i \int e - K_p x_p - K_d x_v$, but in the similar paper (Horowitz et al. 1987) by the same authors, the outer loop controller is in the form of $K_i \int e + K_p x_p + K_d x_v$. It is not consistent. (3) For the Equation (13) in (Horowitz and Tomizuka 1986), the adaptive algorithm are all positive, but in (Horowitz 1983) (note that (Horowitz and Tomizuka 1986) is part of the dissertation (Horowitz 1983)), the adaptive algorithms are all negative, it is also not consistent.

Model reference adaptive control, self-tuning adaptive control and linear perturbation adaptive control are briefly reviewed in (Hsia 1986). For the model reference adaptive control, the main idea is to synthesize/design a control signal to the robot dynamic equation, which will force the robot to behave in a certain manner specified by the reference model, and the adaptive algorithm is designed based on the Lyapunov stability criterion. The MRAC methods presented in (Srinivasan 1987) is based on the theory of partitioning control, which makes them capable of compensating for non-linear terms in the dynamic equations and also to decouple the dynamic interactions between the links. It followed and used the method in (Horowitz 1983). Future research would focus on further simplification of MRAC schemes since the implementation of MRAC methods for the real time control of manipulators has proven to be a challenging task.

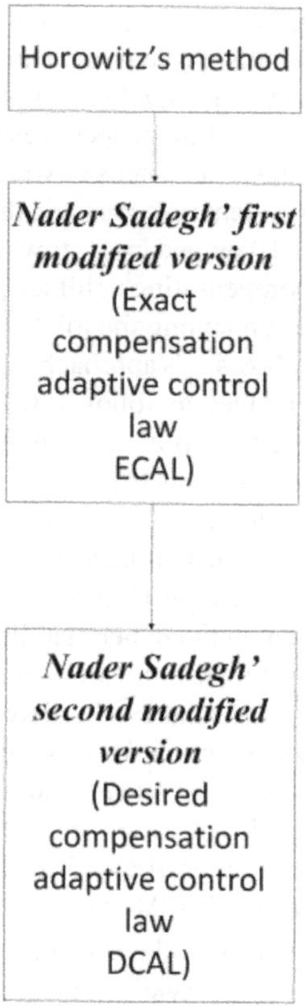

Figure 5. Modification process.

A MRAC system of 3-DOF serial robotic manipulator was presented in (Horowitz 1983). It was concerned with the application of MRAC to mechanical manipulators. Due to the dynamic equations of mechanical manipulators are highly nonlinear and complex, and also the payload sometimes varies or unknown, the MRAC was applied to the mechanical manipulators. An adaptive algorithm was developed for compensating nonlinear terms in the dynamic equations and for decoupling the dynamic interactions. Finally a 3-DOF serial manipulator was used as computer simulation and the results show that the adaptive control scheme is effective in reducing the sensitivity of the manipulator performance to configuration and payload variations. The core content of the method in (Horowitz 1983) can be concluded as four steps: first step, deterministic nonlinearity compensation and decoupling control. Because one needs to calculate the inertia matrix and nonlinear term, the second step is therefore proposed, i.e., adaptive nonlinearity compensation and decoupling control, which is to adaptively adjust the inertia matrix and nonlinear term instead of calculating them, and the adaptive algorithm was developed; final step, complete the overall control system by adding the feedback gain. In (Horowitz 1983), it did not entirely use the Landau's hyperstability design (Landau 1979), the authors used some part of it, and they themselves developed the adaptive algorithm. Because according to (Hsia 1986), Horowitz's method was separated from the Landau's hyperstability design. And also from (Sutherland 1987), it is stated that "While Landau's method replied on a pre-specified parameter matrix for a model and continuous adaptation of the plant parameters, it will be seen later that it is possible to estimate the model parameters and adapt them continuously", from this statement, it is clearly that Horowitz has his own theory to derive the adaptive algorithm, he did not use Landau's method to derive the adaptive algorithm, but how the adaptive algorithm was derived was not explicitly addressed. In (Sutherland 1987), it used the same approach with Horowitz's (Horowitz 1983) to a 2-DOF serial robotic manipulator and a flexible manipulator.

In (Tomizuka et al. 1986, Tomizuka and Horowitz 1988), the experiment on the continuous time and discrete time adaptive control on 1-DOF test stand robot arm and Toshiba TSR-500V robot were briefly conducted. The theory of Horowitz et al. (1987) is the continuation of the theory of Tomizuka et al. (1986) on a single axis direct drive robotic arm. It applies to a two axis direct drive robotic arm. In (Tomizuka et al. 1985), it presented the experiment evaluation of model reference adaptive controller and robust controller for positioning of a robotic arm under variation of payload. The results show that both method can be insensitive of the payload variation. Four adaptive control methods for the robotic arm were summarized in (Hamadi 1989), i.e., computed torque technique, variable structure systems, adaptive linear model following control, and adaptive perturbation control, and the adaptive nonlinear model following control was proposed subsequently, which combines the self-tuning regulator and the model reference adaptive control.

A modified version of the method in (Horowitz 1983) was proposed in (Sadegh and Horowitz 1987). The assumption that inertia matrix and nonlinear term are constant during adaptation can be removed by modifying the control law and parameter adaptation law. It is demonstrated that by modifying the control law (i.e., making the Coriolis and centripetal acceleration compensation controller a bilinear function of the joint and model reference velocities instead of a quadratic function of the joint velocities) and by modifying the parameter adaptation law (i.e., decomposing the nonlinear parameters in the manipulator dynamic equations into the product of two quantities: one constant unknown quantity, which includes the numerical values of the masses, moments of inertia of the links, payload and link dimensions, and the other a known nonlinear function of the manipulator structural dynamics. The nonlinear functions are assumed to be known and calculable), the assumption that the inertia matrix and nonlinear term are constant during adaptation is removed. Finally the stability of the above adaptive control law is proved. The above called "exact compensation adaptive control law (ECAL)". In the conclusion, it was found that this procedure is excessively time consuming since it involves computations of highly nonlinear functions of joint position and velocities, to overcome this difficulty, later in (Sadegh and Horowitz 1990) and (Sadegh 1987), further modified version was proposed. The modification consists in utilizing the desired joint positions and velocities in the computation of the nonlinearity compensation controller and the parameter adaptation law instead of the actual quantities, this is known as "desired compensation adaptive control law (DCAL)". The above whole modification process is shown in Fig. 5.

S. Nader applied Craig's method (Craig et al. 1986) to the Horowitz's method, so the condition inertia matrix and nonlinear term assume constant during adaptation can be removed. Craig's method is re-parametrization, i.e., decompose the manipulator dynamic equation's nonlinear parameters into the product of two quantities: one constant unknown quantity, which includes the numerical values of the masses, moments of inertia of the links, payload and link dimensions, and a known nonlinear function of the manipulator structural dynamics. The parameter adaptation law is only used to estimate the unknown constant quantities. One method of reparametrizing the manipulator's dynamic equations consists in decomposing each element of inertia matrix, nonlinear term and gravity term into products of unknown constant terms and known functions of the joint displacement vector. Or a second method consists in the re-parametrization of dynamic equation into the product of unknown constant vector, and a matrix formed by known functions of joint position.

CONCLUSION

This paper discusses the model reference adaptive control of robotic manipulators initially raised by the scholar Roberto Horowitz and its associated development by other authors. Some issues of model reference adaptive control for robotic manipulators are also presented. Very few recent papers can be found in the area of model reference adaptive control of robotic manipulators, and this paper will provide a guideline for future research in the direction of model reference adaptive control for robotic arms.

ACKNOWLEDGMENT

The authors would like to thank the financial support from the Natural Sciences and Engineering Research Council of Canada (NSERC) and Canada Research Chairs program.

REFERENCES

Amerongen, J. 1981. MRAS: Model Reference Adaptive Systems. Journal A, 22(4): 192–198.

Bi, Z. M., Y. Liu and B. Baumgartner, etc. 2015. Reusing industrial robots to achieve sustainability in small and medium-sized enterprises (SMEs). Industrial Robot: An International Journal, 42(3): 264–273.

Craig, J. J., P. Hsu and S. S. Sastry. 1986. Adaptive control of mechanical manipulators. Proceedings of the 1986 IEEE International Conference on Robotics and Automation, San Francisco, April, 1986.

Donalson, D. and T. Leondes. 1963. A model referenced parameter tracking technique for adaptive control systems. IEEE Transactions on Applications and Industry, 82(68): 241–252.

Dubowsky, S. and D. Desforges. 1979. The application of model-referenced adaptive control to robotic manipulators. Journal of Dynamic Systems Measurement and Control, 101: 193–200.

Hamadi, J. 1989. Adaptive Control Methods for Mechanical Manipulators: A Comparative Study, Master Thesis, Naval Postgraduate School.

Horowitz, R. 1983. Model Reference Adaptive Control of Mechanical Manipulators, Ph.D. Thesis, University of California.

Horowitz, R. and M. Tomizuka. 1986. An adaptive control scheme for mechanical manipulators—compensation of nonlinearity and decoupling control. Journal of Dynamic Systems, Measurement, and Control, 108(2): 1–9.

Horowitz, R., M. C. Tsai, G. Anwar and M. Tomizuka. 1987. Model reference adaptive control of a two axis direct drive manipulator arm. Proceedings of 1987 IEEE International Conference on Robotics and Automation, pp. 1216–1222.

Hsia, T. 1986. Adaptive control of robot manipulators—A review. Proceedings of 1986 IEEE International Conference on Robotics and Automation, pp. 183–189.

https://en.wikipedia.org/wiki/PID_controller.

Jung, B., S. Jeong, D. Lee and Y. Kim. 2005. Adaptive reconfigurable flight control system using multiple model mode switching. Proceedings of the 16th IFAC World Congress, Vol. 16, Part 1.

Landau, Y. 1979. Adaptive control—the model reference approach. CRC Press.

Landau, I. D. et al. 2011. Introduction to adaptive control. *In*: Adaptive Control, Communications and Control Engineering, Springer-Verlag London Limited.

Neuman, C. P. and H. W. Stone. 1983. MRAC control of robotic manipulators. pp. 203–210. *In*: K. S. Narendra (ed.). 3rd Yale Workshop on Applications of Adaptive Systems Theory. Yale Univ., New Haven, CT.

Sadegh, N. 1987. Adaptive control of mechanical manipulators: Stability and robustness analysis, Ph.D. Thesis, University of California.

Sadegh, N. and R. Horowitz. 1987. Stability analysis of an adaptive controller for robotic manipulators. Proceedings of 1987 IEEE International Conference on Robotics and Automation, pp. 1223–1229.

Sadegh, N. and R. Horowitz. 1990. Stability and robustness analysis of a class of adaptive controllers for robotic manipulators. International Journal of Robotics Research, 9(3): 74–92.

Srinivasan, R. 1987. Adaptive control for robotic manipulators, Master Thesis, Carleton University.

Sutherland, J. 1987. Model reference adaptive control of a two link manipulator, Master Thesis, Carleton University.

Tomizuka, M., R. Horowitz and G. Anwar. 1986. Adaptive techniques for motion controls of robotic manipulators. *In*: Japan—USA Symposium on Flexible Automation, July 1986, Osaka, Japan.

Tomizuka, M. and R. Horowitz. 1988. Implementation of adaptive techniques for motion control of robotic manipulators. J. Dyn. Sys., Meas., Control, 110(1): 62–69.

Tomizuka, M., R. Horowitz and G. Anwar. 1986. Adaptive techniques for motion controls of robotic manipulators. pp. 117–224. *In*: Japan—USA Symposium on Flexible Automation, Japan.

Tomizuka, M., R. Horowitz and C. L. Teo. 1985. Model reference adaptive controller and robust controller for positioning of varying inertia. pp. 191–196. *In*: Proceedings of Conference on Applied Motion Control, University of Minnesota.

Valente, A., E. Carpanzano and M. Brusaferri. 2011. Design and implementation of distributed and adaptive control solutions for reconfigurable manufacturing systems. *In*: CIRP Sponsored ICMS. International Conference on Manufacturing Systems.

Valentea, A., M. Mazzolinib and E. Carpanzanoa. 2015. An approach to design and develop reconfigurable control software for highly automated production systems. International Journal of Computer Integrated Manufacturing, 28(3): 321–336.

Whitaker, H. P., J. Yamron and A. Kezer. 1958. Design of Model Reference Adaptive Control Systems for Aircraft. Report R-164, Instrumentation Laboratory. M. I. T. Press, Cambridge, Massachusetts.

Wilson, E. and S. Rock. 1995. Reconfigurable control of a free-flying space robot using neural networks. Proceedings of the 1995 American Control Conference, 2: 1355–1359.

3

Data-Based Learning for Uncertain Robotic Systems

Anup Parikh,[1,*] *Rushikesh Kamalapurkar*[2] and *Warren E. Dixon*[1]

ABSTRACT

An adaptive controller based on the concurrent learning method is developed for uncertain Euler-Lagrange systems. Using a Lyapunov-based analysis, it is shown that this design achieves the tracking objective, as well as identifies the uncertain parameters, without requiring the well-known and restrictive persistence of excitation condition. Simulation results are provided to demonstrate faster convergence compared to gradient based adaptive controllers without concurrent learning.

Keywords: Robotic systems, concurrent learning method, adaptive controller

INTRODUCTION

Robot manipulators have been used for a variety of motion control applications, and therefore high precision control of robot manipulators has been of interest in the control community for a number of decades. The general equations of motion for robot manipulators (i.e., Euler-Lagrange system) can be used to describe the dynamics for a variety of electromechanical systems, and therefore have become a benchmark system for novel control design techniques, e.g., adaptive control, robust control, output feedback, and control with limited actuation (Lewis et al. 2003, Behal et al. 2009). Adaptive control refers to a number of techniques for achieving a tracking or regulation

[1] Department of Mechanical and Aerospace Engineering, University of Florida, Gainesville FL 32611-6250.

[2] School of Mechanical and Aerospace Engineering, Oklahoma State University, Stillwater OK 74078.

* Corresponding author

control objective while compensating for uncertainties in the model by estimating the uncertain parameters online. It is well known that least-squares or gradient based adaptation laws rely on persistence of excitation (PE) to ensure parameter convergence (Ioannou and Sun 1996, Narendra and Annaswamy 1989, Sastry and Bodson 1989), a condition which cannot be guaranteed *a priori* for nonlinear systems, and is difficult to check online, in general.

Motivated by the desire to learn the true parameters, or at least to gain the increased robustness (i.e., bounded solutions in the presence of disturbances) and improved transient performance (i.e., exponential tracking versus asymptotic tracking of many adaptive controllers) that parameter convergence provides, a new adaptive update scheme known as concurrent learning (CL) was recently developed in the pioneering work of (Chowdhary and Johnson 2011, Chowdhary 2010, Chowdhary et al. 2013). The principle idea of CL is to use recorded input and output data of the system dynamics to apply batch-like updates to the parameter estimate dynamics. These updates yield a negative definite, parameter estimation error term in the stability analysis, which allows parameter convergence to be established provided a finite excitation condition is satisfied. The finite excitation condition is a weaker condition than persistent excitation (since excitation is only required for a finite amount of time), and can be checked online by verifying the positivity of the minimum singular value of a function of the regressor matrix.

In this chapter, a concurrent learning based adaptation law for general Euler-Lagrange systems is developed. We also demonstrate faster tracking error and parameter estimation error convergence compared to a gradient based adaptation law through a simulation.

CONTROL DEVELOPMENT

Consider robot manipulator equations of motion of the form (Lewis et al. 2003, Spong and Vidyasagar 1980)

$$M(q)\,\ddot{q} + V_m(q, \dot{q})\,\dot{q} + F_d\dot{q} + G(q) = \tau \tag{1}$$

where $q(t)$, $\dot{q}(t)$, $\ddot{q}(t) \in \mathbb{R}^n$ represent the measurable link position, velocity and acceleration vectors, respectively, $M : \mathbb{R}^n \rightarrow \mathbb{R}^{n\times n}$ represents the inertial matrix, V_m : $\mathbb{R}^n \times \mathbb{R}^n \rightarrow \mathbb{R}^{n\times n}$ represents centripetal-Coriolis effects, $F_d \in \mathbb{R}^{n\times n}$ represents frictional effects, $G : \mathbb{R}^n \rightarrow \mathbb{R}^n$ represents gravitational effects and $\tau(t) \in \mathbb{R}^n$ denotes the control input. The system in (1) has the following properties (See Lewis et al. 2003).

Property 1. The system in (1) can be linearly parameterized, i.e., (1) can be rewritten as

$$Y_1(q, \dot{q}, \ddot{q})\,\theta = M(q)\,\ddot{q} + V_m(q, \dot{q})\,\dot{q} + F_d\dot{q} + G(q) = \tau$$

where $Y_1 : \mathbb{R}^n \times \mathbb{R}^n \times \mathbb{R}^n \rightarrow \mathbb{R}^{n\times m}$ denotes the regression matrix, and $\theta \in \mathbb{R}^m$ is a vector of uncertain parameters.

Property 2. The inertia matrix is symmetric and positive definite, and satisfies the following inequalities

$$m_1\,\|\xi\|^2 \leq \xi^T M(q)\,\xi \leq m_2\,\|\xi\|^2, \quad \forall \xi \in \mathbb{R}^n$$

where m_1 and m_2 are known positive scalar constants, and $\|\cdot\|$ represents the Euclidean norm.

Property 3. The inertia and centripetal-Coriolis matrices satisfy the following skew symmetric relation

$$\xi^T \left(\frac{1}{2} \dot{M}(q) - V_m(q, \dot{q}) \right) \xi = 0, \quad \forall \xi \in \mathbb{R}^n$$

where $\dot{M}(q)$ is the time derivative of the inertial matrix.

To quantify the tracking objective, the link position tracking error, $e(t) \in \mathbb{R}^n$, and the filtered tracking error, $r(t) \in \mathbb{R}^n$, are defined as

$$e = q_d - q \tag{2}$$

$$r = \dot{e} + \alpha e \tag{3}$$

where $q_d(t) \in \mathbb{R}^n$ represents the desired trajectory, whose first and second time derivatives exist and are continuous (i.e., $q_d \in \boldsymbol{C}^2$). To quantify the parameter identification objective, the parameter estimation error, $\tilde{\theta}(t) \in \mathbb{R}^m$, is defined as

$$\tilde{\theta} = \theta - \hat{\theta} \tag{4}$$

where $\hat{\theta}(t) \in \mathbb{R}^m$ represents the parameter estimate.

Taking the time derivative of (3), premultiplying by $M(q)$, substituting in from (1), and adding and subtracting $V_m(q, \dot{q})\, r$ results in the following open loop error dynamics

$$M(q)\, \dot{r} = Y_2(q, \dot{q}, q_d, \dot{q}_d, \ddot{q}_d)\, \theta - V_m(q, \dot{q})\, r - \tau \tag{5}$$

where $Y_2 : \mathbb{R}^n \times \mathbb{R}^n \times \mathbb{R}^n \times \mathbb{R}^n \times \mathbb{R}^n \to \mathbb{R}^{n \times m}$ is defined based on the relation

$$Y_2(q, \dot{q}, q_d, \dot{q}_d, \ddot{q}_d)\, \theta \triangleq M(q)\, \ddot{q}_d + V_m(q, \dot{q})\, (\dot{q}_d + \alpha e) + F_d \dot{q} + G(q) + \alpha M(q)\, \dot{e}.$$

To achieve the tracking objective, the controller is designed as

$$\tau(t) = Y_2 \hat{\theta} + e + k_1 r \tag{6}$$

where $k_1 \in \mathbb{R}$ is a positive constant. To achieve the parameter identification objective, the parameter estimate update law is designed as

$$\dot{\hat{\theta}} = \Gamma Y_2^T r + k_2 \Gamma \sum_{i=1}^{N} Y_{1i}^T (\tau_i - Y_{1i} \hat{\theta}) \tag{7}$$

where $k_2 \in \mathbb{R}$ and $\Gamma \in \mathbb{R}^{m \times m}$ are constant positive definite and symmetric control gains, $Y_{1i} \triangleq Y_1(q(t_i), \dot{q}(t_i), \ddot{q}(t_i))$, $\tau_i \triangleq \tau(t_i)$, t_i represent past time points, i.e., $t_i \in [0, t]$, and $N \in \mathbb{N}_0$. Using (1), (7) can be rewritten as

$$\begin{aligned}\dot{\hat{\theta}} &= \Gamma Y_2^T r + k_2 \Gamma \sum_{i=1}^{N} Y_{1i}^T (Y_{1i} \theta - Y_{1i} \hat{\theta}) \\ &= \Gamma Y_2^T r + k_2 \Gamma \left[\sum_{i=1}^{N} Y_{1i}^T Y_{1i} \right] \tilde{\theta}.\end{aligned} \tag{8}$$

The principal idea behind this design is to use recorded input and trajectory data to identify the uncertain parameter vector θ. The time points t_i and the corresponding τ_i and Y_{1i} used in the summation in (7) are referred to as the history stack. As shown in the subsequent stability analysis, the parameter estimate learning rate is related to the minimum eigenvalue of $\sum_{i=1}^{N} Y_{1i}^T Y_{1i}$, motivating the use of the singular value maximization algorithm in (Chowdhary 2010) for adding data to the history stack. It is also important to note that although this design uses higher state derivatives which are typically not measured (i.e., $\ddot{q}$), this data is only required for time points in the past, and therefore smoothing techniques can be utilized to minimize noise without inducing a phase shift, e.g., (Mühlegg et al. 2012).

Substituting the controller from (6) into the error dynamics in (5) results in the following closed-loop tracking error dynamics

$$M(q)\dot{r} = Y_2(q, \dot{q}, q_d, \dot{q}_d, \ddot{q}_d)\tilde{\theta} - e - V_m(q, \dot{q})r - k_1 r. \tag{9}$$

Similarly, taking the time derivative of (4) and substituting the parameter estimate update law from (8) results in the following closed-loop parameter estimation error dynamics

$$\dot{\tilde{\theta}} = -\Gamma Y_2^T r - k_2 \Gamma \left[\sum_{i=1}^{N} Y_{1i}^T Y_{1i}\right]\tilde{\theta}. \tag{10}$$

STABILITY ANALYSIS

To analyze the stability of the closed loop system, two periods of time are considered. During the initial phase, insufficient data has been collected to satisfy a richness condition on the history stack. In Theorem 1 it is shown that the designed controller and adaptive update law are still sufficient for the system to remain bounded for all time despite the lack of data. After a finite period of time, the system transitions to the second phase, where the history stack is sufficiently rich and the controller and adaptive update law are shown, in Theorem 2, to bound the system by an exponentially decaying envelope, therefore achieving the tracking and identification objectives. To guarantee that the transition to the second phase happens in finite time, and therefore the overall system trajectories can be bounded by an exponentially decaying envelope, we require the history stack be sufficiently rich after a finite period of time, i.e.,

$$\exists \underline{\lambda}, T > 0 : \forall t \geq T, \lambda_{\min}\left\{\sum_{i=1}^{N} Y_{1i}^T Y_{1i}\right\} \geq \underline{\lambda}. \tag{11}$$

The condition in (11) requires that the system be sufficiently excited, though is weaker than the persistence of excitation condition since excitation is unnecessary once $\sum_{i=1}^{N} Y_{1i}^T Y_{1i}$ is full rank.

Theorem 1. *For the system defined in (1), the controller and adaptive update law defined in (6) and (7) ensure bounded tracking and parameter estimation errors.*

Proof: Let $V : \mathbb{R}^{2n+m} \to \mathbb{R}$ be a candidate Lyapunov function defined as

$$V(\eta) = \frac{1}{2}e^T e + \frac{1}{2}r^T M(q)\, r + \frac{1}{2}\tilde{\theta}^T \Gamma^{-1}\tilde{\theta} \tag{12}$$

where $\eta \triangleq [e^T\, r^T\, \tilde{\theta}^T]^T \in \mathbb{R}^{2n+m}$ is a composite state vector. Taking the time derivative of (12) and substituting (3), (9), and (10) yields

$$\dot{V} = e^T(r - \alpha e) + \frac{1}{2}r^T \dot{M}(q)r + r^T\left(Y_2(q, \dot{q}, q_d, \dot{q}_d, \ddot{q}_d)\,\tilde{\theta} - e\right.$$
$$\left. - V_m(q, \dot{q})\, r - k_1 r\right) - \tilde{\theta}^T Y_2^T r - k_2\tilde{\theta}^T\left[\sum_{i=1}^{N} Y_{1i}^T Y_{1i}\right]\tilde{\theta}.$$

Simplifying and noting that $\sum_{i=1}^{N} Y_{1i}^T Y_{1i}$ is always positive semi-definite, $\dot{V}$ can be upper bounded as

$$\dot{V} \leq -\alpha e^T e - k_1 r^T r$$

Therefore, η is bounded based on (Khalil 2002). Furthermore, since $\dot{V} \leq 0$, $V(\eta(T)) \leq V(\eta(0))$ and therefore $\|\eta(T)\| \leq \sqrt{\frac{\beta_2}{\beta_1}}\,\|\eta(0)\|$, where $\beta_1 \triangleq \frac{1}{2}\min\{1, m_1, \lambda_{\min}\{\Gamma^{-1}\}\}$ and $\beta_2 \triangleq \frac{1}{2}\max\{1, m_2, \lambda_{\max}\{\Gamma^{-1}\}\}$. ■

Theorem 2. *For the system defined in (1), the controller and adaptive update law defined in (6) and (7) ensure globally exponential tracking in the sense that*

$$\|\eta(t)\| \leq \left(\frac{\beta_2}{\beta_1}\right)\exp(\lambda_1 T)\,\|\eta(0)\|\,\exp(-\lambda_1 t), \quad \forall t \in [0, \infty) \tag{13}$$

where $\lambda_1 \triangleq \frac{1}{2\beta_2}\min\{\alpha, k_1, k_2\underline{\lambda}\}$.

Proof: Let $V: \mathbb{R}^{2n+m} \to \mathbb{R}$ be a candidate Lyapunov function defined as in (12). Taking the time derivative of (12), substituting (3), (9), (10) and simplifying yields

$$\dot{V} = -\alpha e^T e - k_1 r^T r - k_2\tilde{\theta}^T\left[\sum_{i=1}^{N} Y_{1i}^T Y_{1i}\right]\tilde{\theta}. \tag{14}$$

From the finite excitation condition, $\lambda_{\min}\left\{\sum_{i=1}^{N} Y_{1i}^T Y_{1i}\right\} > 0, \forall t \in [T, \infty)$, which implies that $\sum_{i=1}^{N} Y_{1i}^T Y_{1i}$ is positive definite, and therefore $\dot{V}$ can be upper bounded as

$$\dot{V} \leq -\alpha e^T e - k_1 r^T r - k_2\underline{\lambda}\|\tilde{\theta}\|^2, \quad \forall t \in [T, \infty).$$

Invoking (Khalil 2002), η is globally exponentially stable, i.e., $\forall t \in [T, \infty)$,

$$\|\eta(t)\| \leq \sqrt{\frac{\beta_2}{\beta_1}}\,\|\eta(T)\|\,\exp(-\lambda_1(t - T)).$$

The composite state vector can be further upper bounded using the results of Theorem 1, yielding (13). ■

Remark 1. Although the analysis only explicitly considers two periods, i.e., before and after the history stack is sufficiently rich, additional data may be added into the history stack after T as long as the data increases the minimum eigenvalue of $\sum_{i=1}^{N} Y_{1i}^T Y_{1i}$. By using the data selection algorithm in (Chowdhary 2010), the minimum eigenvalue of $\sum_{i=1}^{N} Y_{1i}^T Y_{1i}$ is always increasing, and therefore the Lyapunov function derivative upper bound in (14), is valid for all time after T. Hence (12) is a common Lyapunov function (Liberzon 2003).

SIMULATION

A simulation of this control design applied to a two-link planar robot was performed. The dynamics of the two-link robot are given as

$$\underbrace{\begin{bmatrix} p_1 + 2p_3c_2 & p_2 + p_3c_2 \\ p_2 + p_3c_2 & p_2 \end{bmatrix}}_{M(q)} \begin{bmatrix} \ddot{q}_1 \\ \ddot{q}_2 \end{bmatrix} + \underbrace{\begin{bmatrix} -p_3s_2\dot{q}_2 & -p_3s_2(\dot{q}_1 + \dot{q}_2) \\ p_3s_2\dot{q}_1 & 0 \end{bmatrix}}_{V_m(q,\dot{q})} \begin{bmatrix} \dot{q}_1 \\ \dot{q}_2 \end{bmatrix} + \underbrace{\begin{bmatrix} f_{d1} & 0 \\ 0 & f_{d2} \end{bmatrix}}_{F_d} \begin{bmatrix} \dot{q}_1 \\ \dot{q}_2 \end{bmatrix} = \begin{bmatrix} \tau_1 \\ \tau_2 \end{bmatrix}$$

where c_2 denotes $\cos(q_2)$ and s_2 denotes $\sin(q_2)$. The nominal parameters are given by

$$p_1 = 3.473 \qquad f_{d1} = 5.3$$
$$p_2 = 0.196 \qquad f_{d2} = 1.1$$
$$p_3 = 0.242$$

and the controller gains were selected as

$$\alpha = 1.0 \qquad \Gamma = 0.1 I_5$$
$$k_1 = 0.1 \qquad k_2 = 0.1.$$

The desired trajectory was selected as

$$q_{d1} = (1 + 10\exp(-2t))\sin(t),$$
$$q_{d2} = (1 + 10\exp(-t))\cos(3t),$$

and a history stack of up to 20 data points was used in the adaptive update. The tracking and parameter estimation error trajectories are show in Figs. 1 and 2. From Fig. 2, it is clear that the system parameters have been identified. A comparison simulation was also performed without concurrent learning (i.e., setting $k_2 = 0$), representing a typical control design that asymptotically tracks the desired trajectory based on Theorem 1, with error trajectories shown in Figs. 3 and 4. In comparison to a typical gradient based adaptive controller that yields the trajectories in Figs. 3 and 4, the contribution of the concurrent learning method is evident by the exponential trajectories in Figs. 1 and 2. It is also important to note that a number techniques have been developed for improving transient performance of adaptive control architectures such as the gradient based adaptive update law simulated here (e.g. (Duarte and Narendra 1989, Krstić et al. 1993, Yucelen and Haddad 2013, Pappu et al. 2014, Yucelen et al. 2014)),

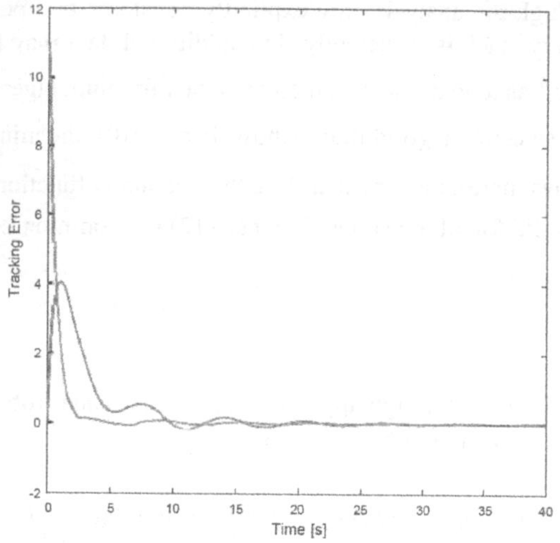

Figure 1. Tracking error trajectory using concurrent learning based control design.

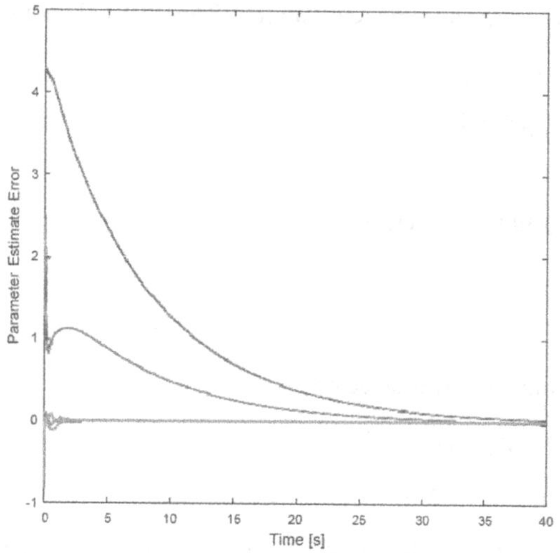

Figure 2. Parameter estimation error trajectory using concurrent learning based control design.

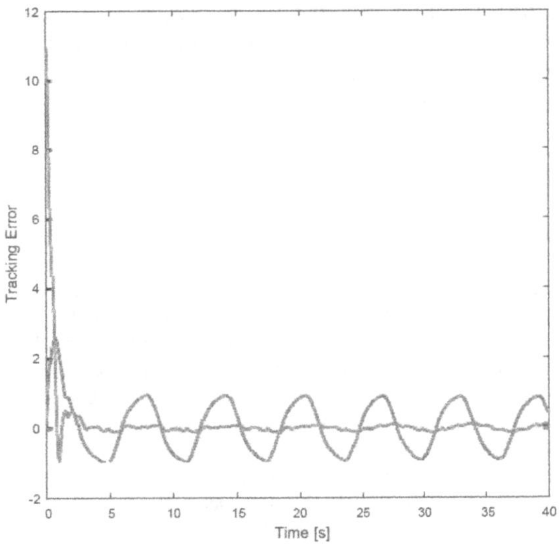

Figure 3. Tracking error trajectory using traditional gradient based control design.

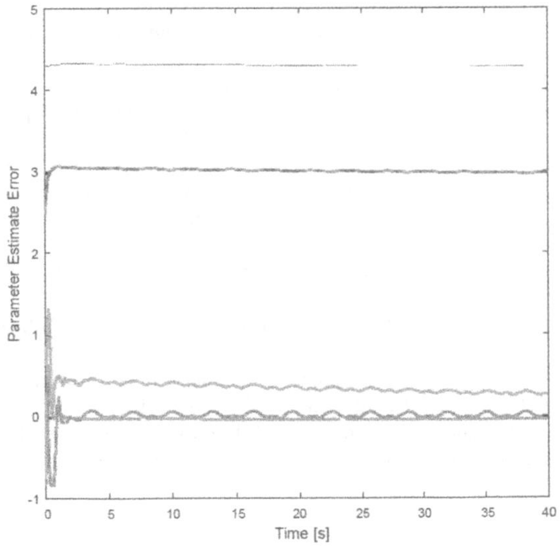

Figure 4. Parameter estimation error trajectory using traditional gradient based control design.

though cannot guarantee parameter identification, and hence exponential trajectories, without persistence of excitation.

CONCLUSION

In this chapter a novel adaptive controller is developed for Euler-Lagrange systems. The concurrent learning based design incorporates recorded data into the adaptive update law, resulting in exponential convergence of the tracking and parameter estimation errors. The excitation condition required for convergence is weaker than persistent excitation, and is easier to check online. The provided simulation results demonstrate the increased convergence rate of this design compared to traditional adaptive controllers with gradient based update laws.

REFERENCES

Behal, A., W. E. Dixon, B. Xian and D. M. Dawson. 2009. Lyapunov-Based Control of Robotic Systems. Taylor and Francis.

Chowdhary, G. V. and E. N. Johnson. 2011. Theory and flight-test validation of a concurrent-learning adaptive controller. J. Guid. Control Dynam., 34(2): 592–607, March.

Chowdhary, G. 2010. Concurrent learning for convergence in adaptive control without persistency of excitation, Ph.D. dissertation, Georgia Institute of Technology, December.

Chowdhary, G., T. Yucelen, M. Mühlegg and E. N. Johnson. 2013. Concurrent learning adaptive control of linear systems with exponentially convergent bounds. Int. J. Adapt. Control Signal Process., 27(4): 280–301.

Duarte, M. A. and K. Narendra. 1989. Combined direct and indirect approach to adaptive control. IEEE Trans. Autom. Control, 34(10): 1071–1075, October.

Ioannou, P. and J. Sun. 1996. Robust Adaptive Control. Prentice Hall.

Khalil, H. K. 2002. Nonlinear Systems, 3rd ed. Upper Saddle River, NJ: Prentice Hall.

Krstić, M., P. V. Kokotović and I. Kanellakopoulos. 1993. Transient-performance improvement with a new class of adaptive controllers. Syst. Control Lett., 21(6): 451–461. [Online]. Available: http://www.sciencedirect.com/science/article/pii/016769119390050G.

Lewis, F. L., D. M. Dawson and C. Abdallah. 2003. Robot Manipulator Control Theory and Practice. CRC.

Liberzon, D. 2003. Switching in Systems and Control. Birkhauser.

Mühlegg, M., G. Chowdhary and E. Johnson. 2012. Concurrent learning adaptive control of linear systems with noisy measurements. *In*: Proc. AIAA Guid. Navig. Control Conf.

Narendra, K. and A. Annaswamy. 1989. Stable Adaptive Systems. Prentice-Hall, Inc.

Pappu, V. S. R., J. E. Steck, K. Rajagopal and S. N. Balakrishnan. 2014. Modified state observer based adaptation of a general aviation aircraft—simulation and flight test. *In*: Proc. AIAA Guid. Navig. Control Conf., National Harbor, MD, January.

Sastry, S. and M. Bodson. 1989. Adaptive Control: Stability, Convergence, and Robustness. Upper Saddle River, NJ: Prentice-Hall.

Spong, M. and M. Vidyasagar. 1980. Robot Dynamics and Control. New York: John Wiley & Sons Inc.

Yucelen, T. and W. M. Haddad. 2013. Low-frequency learning and fast adaptation in model reference adaptive control. IEEE Trans. Autom. Control, 58(4): 1080–1085.

Yucelen, T., G. D. L. Torre and E. N. Johnson. 2014. Improving transient performance of adaptive control architectures using frequency-limited system error dynamics. Int. J. Control, 87(11): 2383–2397.

4

Reinforcement Learning of Robotic Manipulators

*Kei Senda** and *Yurika Tani*

ABSTRACT

This chapter discusses an autonomous space robot for a truss structure assembly using some reinforcement learning. It is difficult for a space robot to complete contact tasks within a real environment, e.g., a peg-in-hole task, because of the potential discrepancy between the real environment and the controller model. In order to solve problems, we propose an autonomous space robot able to obtain proficient and robust skills by overcoming error prospects in completing a task. The proposed approach develops skills by reinforcement learning that considers plant variation, i.e., modeling error. Numerical simulations and experiments show the proposed method is useful in real environments.

Keywords: Reinforcement learning, Robust skills, Plant variation/modeling error, Contact tasks

INTRODUCTION

This chapter discusses an unresolved robotics issue, i.e., how to make a robot autonomous. A robot with manipulative skills capable of flexibly achieving tasks, like a human being, is desirable. Autonomy is defined as "automation to achieve a task robustly". Skills can be considered to be solutions to achieve autonomy. Another aspect of a skill is including a solution method. Most human skill proficiency is acquired by experience. Since the way to realize autonomy is not clear, skill development must

Department of Aeronautics and Astronautics, Kyoto University, Japan.
* Corresponding author

include solution methods for unknown situations. Our problem is how to acquire skills autonomously, i.e., how to robustly and automatically complete a task when the solution is unknown.

Reinforcement learning (Sutton and Barto 1998) is a promising solution, whereas direct applications of existing methods with reinforcement learning do not robustly complete tasks. Reinforcement learning is a framework in which a robot learns a policy or a control that optimizes an evaluation through trial and error. It is teacherless learning. By means of reinforcement learning, a robot develops an appropriate policy as mapping from state to action when an evaluation is given. The task objective is prescribed, but no specific action is taught. Reinforcement learning often needs many samples. The large number of samples is due to the large number of states and actions. Online learning in a real environment is usually impractical. Most learning algorithms consist of two processes (Bertsekas and Tsitsiklis 1996): (1) online identification by trial and error sampling and (2) finding the optimal policy for the identified model. These two processes are not separated in typical learning algorithms such as Q-learning (Fujii et al. 2006). Reinforcement learning is said to be adaptive because it uses online identification and on-site optimal control design. Robustness attained using this adaptability is often impractical. It takes a very long time for online identification by means of trial and error. Real-world samples are expensive in terms of time, labor, and finances, and is referred to as a curse of real-world samples (Kober and Peters 2012). They also state that it is often more important to limit the real-world interaction time instead of limiting memory consumption or computational complexity.

In our approach, by following a robust policy rather than by online identification, reinforcement learning is used to achieve a solution to an unknown task. Such reinforcement learning to obtain a robust policy against model uncertainty is a recent research topic.

This chapter addresses an autonomous space robot for a truss structure assembly. For a space robot, it is difficult to achieve a task contacting with a real environment, e.g., peg-in-hole task, because there is the error between the real environment and the controller model. A space robot must autonomously obtain proficiency and robust skills in order to counter the error in the model. Using the proposed approach, reinforcement learning can achieve a policy that is robust in the face of plant variation, i.e., the modeling error. Numerical simulations and experiments show that a robust policy is effective in a real environment and the proposed method is used.

NEED FOR AUTONOMY AND APPROACH

Need for Autonomy

Autonomy is needed wherever robots work. Below, we discuss why and what kind of autonomy is required (Senda 2001, Senda et al. 2002) for space robots.

Space robots are required to complete tasks in the place of extra vehicular activity by an astronaut. Studies of autonomous systems are needed to realize space robots that can achieve missions under human-operator command. There are many applications for the autonomous space robots (Skaar and Ruoff 1995).

We developed ground experiments to simulate a free-flying space robot under orbital microgravity conditions (Fig. 1). Using this apparatus, we have studied robot autonomy. Targeting control-based autonomy, we developed an automatic truss structure assembly, etc. However, it has been hard to achieve perfect robot automation because of various factors. To make a robot autonomous in the face of the following challenges, we must:

(a) solve problems in the actual robot-environment;
(b) operate robustly in the face of uncertainties and variations;
(c) overcome the difficulty of comprehensively predicting a wide variety of states;
(d) identify tasks and find unknown solutions to realize robust robot autonomy.

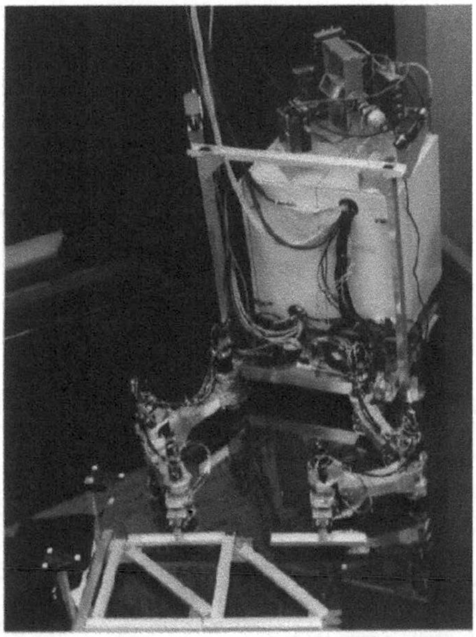

Figure 1. Photograph of space robot model and truss.

Approach to Autonomy Based on Reinforcement Learning

Human beings achieve tasks regardless of the above complicating factors, but robots cannot. Many discussions have rationalized the difficulties as originating in the nature of human skills. We have not established a means by which we can realize such skills. This chapter takes the following approach to this problem.

The fourth section approaches to factors (a) and (b) in part by way of control-based automation taking account of the robustness of controls. The autonomous level of this approach is low because only small variations are allowable.

The fifth section considers how to surmount factors (a), (b), and (c) by learning. Using a predicted model and transferring the learned results to the actual system (Asada 1999) is studied. This model is identified online and relearned. This procedure is applied where adaptability or policy modification is needed to thwart variations in the real environment. In some cases, the robot completes the targeted tasks autonomously. Learning is completed within an acceptable calculation time.

The sixth section considers factors (a), (b), (c), and (d). A peg-in-hole task has higher failure rates using the same approach as used in the fifth section. Because of small differences between the model and the real environment, failure rates are excessive.

The robot thus must gain skills autonomously. Skills similar to those of human beings must be generated autonomously. These skills are developed by reinforcement learning and additional procedures.

The seventh evaluates approach robustness and control performance of the newly acquired skills. The skills obtained in the sixth section are better than those in the fifth section.

EXPERIMENTAL SYSTEM AND TASKS

Outline of Experimental System

Our experimental system (Figs. 1 and 2) simulates a free-flying space robot in orbit. Robot model movement is restricted to a two-dimensional plane (Senda et al. 2002).

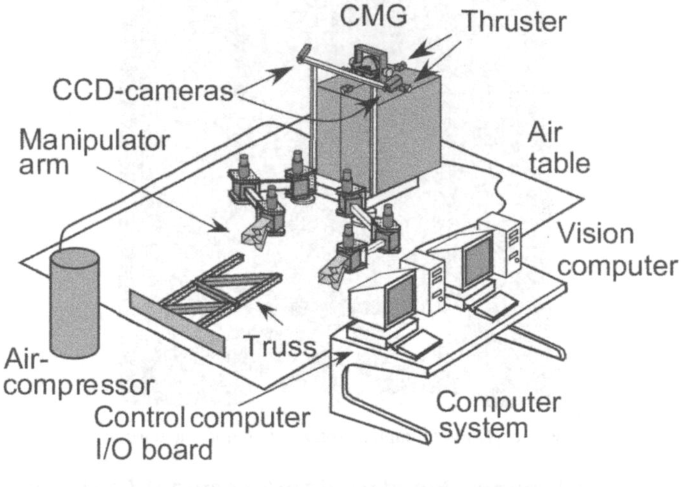

Figure 2. Schematic diagram of experimental system.

The robot model consists of a satellite vehicle and dual three-degrees-of-freedom (3-DOF) rigid SCARA manipulators. The satellite vehicle has CCD cameras for stereovision and a position/attitude control system. Each joint has a torque sensor and a servo controller for fine torque control of the output axis. Applied force and torque at the end-effector are calculated using measured joint torque. Air-pads are used to support the space robot model on a friction-less planar table and to simulate the space environment. RTLinux is installed on a control computer to control the space robot model in real time. Stereo images from the two CCD cameras are captured by a video board and sent into an image-processing computer with a Windows OS. The image-processing computer measures the position and orientation of target objects in the worksite by triangulation. Visual information is sent to the control computer via Ethernet.

The position and orientation measured by the stereo vision system involves errors caused by quantized images, lighting conditions at the worksite, etc. Time-averaged errors are almost constant in each measurement. Evaluated errors in the peg-in-hole experiments are modelled as described below. Hole position errors are modeled as a normal probability distribution, where the mean is $m = 0$ [mm], and standard deviation is $\sigma = 0.75$ [mm]. Hole orientation errors are modeled as a normal probability distribution, where the mean is $m = 0$ [rad], and standard deviation is $\sigma = 0.5\pi/180$ [rad].

We accomplish a hand-eye calibration to achieve tasks in following sections. An end-effector, a manipulator hand, grasps a marker with a light-emitting diode (LED). The arm directs the marker to various locations. The robot calculates the marker location by using sensors mounted at the joints of the manipulator arm. The vision system also measures the marker location by using the stereo image by triangulation. Measurements using these joint-angle sensors have more precise resolution and accuracy. Hence, we calibrate the visual measurements based on measurements using the joint angle sensors. We consider the joint angle sensor measurement data to be the true value.

Tasks

Truss Assembly Task

Figure 3 illustrates the truss structure assembly sequence. Manipulating a truss component and connecting it to a node proceeds the assembly process. Later this task is achieved by controls based upon mechanics understanding (Senda et al. 2002). The truss design is robot friendly for easy assembling.

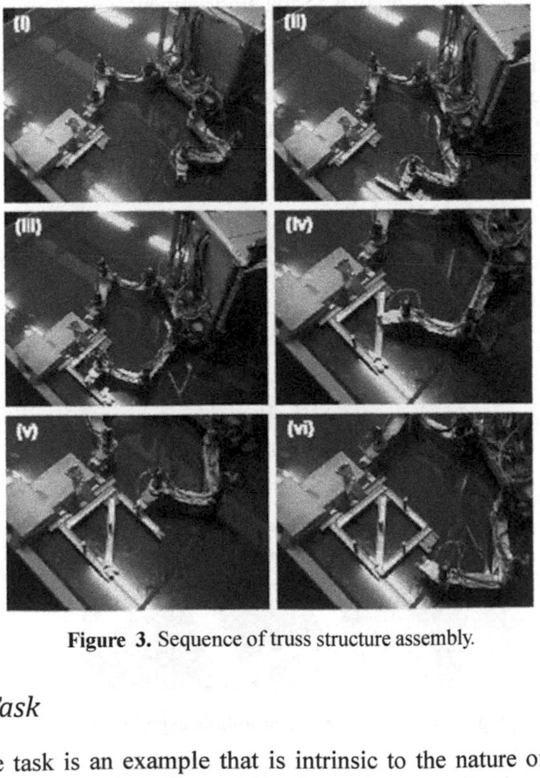

Figure 3. Sequence of truss structure assembly.

Peg-in-hole Task

The peg-in-hole task is an example that is intrinsic to the nature of assembly. The peg-in-hole task involves interaction within the environment that is easily affected by uncertainties and variations, e.g., errors in force applied by the robot, manufacturing accuracy, friction at contact points, etc. The peg easily transits to a state in which it can no longer move, e.g., wedging or jamming (Whitney 1982). Such variations cannot be modeled with required accuracy.

To complete a task in a given environment, a proposed method analyzes the human working process and applies the results to a robot (Sato and Uchiyama 2007). Even if the human skill for a task can be analyzed, the results are not guaranteed to be applicable to a robot. Another method uses parameters in a force control designed by means of a simulation (Yamanobe et al. 2006) but was not found to be effective in an environment with uncertainty. In yet another method (Fukuda et al. 2000), the task achievement ratios evaluated several predesigned paths in an environment with uncertainty. An optimal path is determined among the predesigned paths. There was the possibility a feasible solution did not exist among predesigned paths.

In the peg-in-hole experiment (Fig. 4), the position and orientation of the hole is measured using a stereo camera. The robot manipulator inserts a square peg into a similar sized hole. This experiment is a two-dimensional plane problem (Fig. 5). The space robot model coordinate system is defined as Σ_0, the end-effector coordinate system as Σ_E, and the hole coordinate system as Σ_{hl}. While the space robot completes its task, the robot grasps the structural site with other manipulator, the relative relation between Σ_0 and Σ_{hl} is fixed. State variables are defined as $[y_x, y_y, y_\theta, f_x, f_y, f_\theta]$, where (y_x, y_y) is the position of Σ_E in Σ_0, y_θ is the orientation about $\boldsymbol{k}_0$-axis, (f_x, f_y) and f_θ are the forces and torque in Σ_0 that end-effector applies to the environment.

The peg width is 74.0 [mm] and the hole width is 74.25 [mm]. The hole is only 0.25 [mm] wider than the peg. The positioning error is composed of the measurement error and the control error. The robot cannot insert the peg in the hole by position control if the positioning error is beyond ±0.125 [mm]. Just a single measurement error by stereo vision often moves the peg outside of the acceptable region.

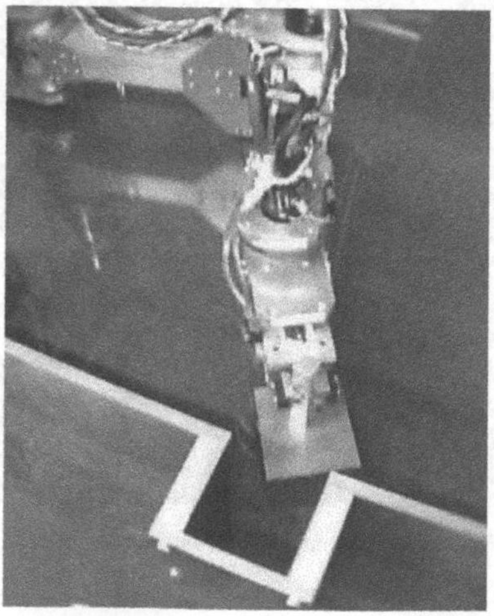

Figure 4. Photograph of peg-in-hole experiment setup.

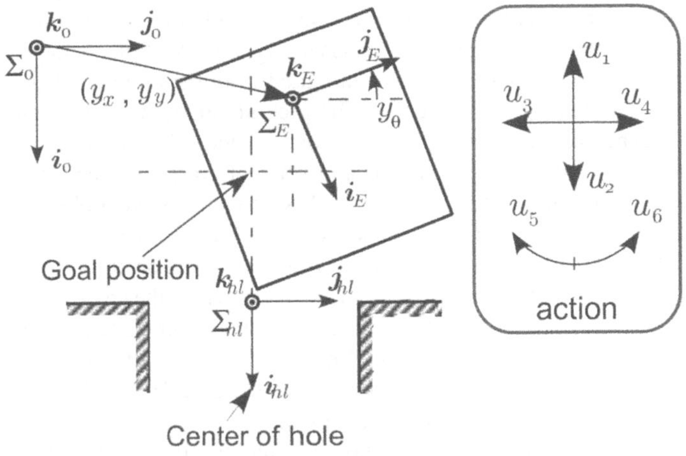

Figure 5. Definition of peg-in-hole task.

CONTROL-BASED AUTOMATION

Truss Assembly Task

Automatic truss assembly was studied via controls-based automation with mechanics understanding. The robot achieved an automatic truss structure assembly (Senda et al. 2002) by developing basic techniques and integrating them within the experimental system.

The following sensory feedback control (Miyazaki and Arimoto 1985) is used for controlling manipulators:

$$\boldsymbol{\tau} = -\boldsymbol{J}^T \boldsymbol{K}_P (\boldsymbol{y} - \boldsymbol{y}_d) - \boldsymbol{K}_D \dot{\boldsymbol{q}} \tag{1}$$

where $\boldsymbol{\tau}$ is the control input to the manipulator, $\boldsymbol{J}$ is the Jacobean matrix. The $\boldsymbol{y}$ is the manipulation variable whose elements are the hand position/orientation $[y_x, y_y, y_\theta]$, $\boldsymbol{y}_d$ is the reference value of $\boldsymbol{y}$, $\boldsymbol{q}$ is the joint angle vector, and $\boldsymbol{K}_P$ and $\boldsymbol{K}_D$ are feedback gains. When the end-effector contacts the environment and manipulation variable $\boldsymbol{y}$ is stationary under constraint, the end-effector applies force and torque to the environment:

$$\boldsymbol{f} = -\boldsymbol{K}_P (\boldsymbol{y} - \boldsymbol{y}_d) \tag{2}$$

The force and torque can be controlled by $\boldsymbol{y}_d$. This is a compliant control.

Figure 3 is a series of photographs of the experimental assembly sequence. As shown in panel (i), the robot holds on to the worksite with its right arm to compensate for any reaction force during the assembly. The robot installs the first component, member 1, during panels (ii) and (iii). The robot installs other members successively and assembles one truss unit, panels (iv)–(vi).

Control procedures for the assembly sequence are as follows. There are target markers in the experiment environment as shown in Figs. 1 and 3. Target markers are located at the base of the truss structure and at the storage site for structural parts. Each target marker has three LEDs at triangular vertices. The vision system measures the marker position and orientation simultaneously. The robot recognizes the position of a part relative to the target marker before assembly. The robot places the end-effector position at the pick-up point, which is calculated from the target marker position as measured by the stereo vision system. At the pick-up point, the end-effector grasps a handgrip attached to the targeted part. The position and orientation of the part to the end-effector is settled uniquely when the end-effector grasps the handgrip. The robot plans the path of the arm and part to avoid collision with any other object in the work environment. It controls the arm to track along the planned path. The robot plans a path from the pick-up point to the placement point, avoiding obstacles by means of an artificial potential method (Connolly et al. 1990). Objects in the environment, e.g., the truss under assembly, are regarded as obstacles. The arm is then directed along a planned trajectory by the sensory feedback control, Eq. (1). The end-effector only makes contact with the environment when it picks-up or places the part. Hence, feedback gains in Eq. (1) are chosen to make it a compliant control.

Consequently, the truss assembly task is successfully completed by controls-based automation. However, measurement error in the vision system sensors, etc. prevents assembly from being guaranteed. Irrespective of uncertainties and variations at the worksite, the space robot model requires autonomy to complete the task goal.

Peg-in-hole Task

The positioning control of Eq. (1) tries to complete the peg-in-hole task. The peg first is positioned at $y_\theta = 0$ [rad], it transits to the central axis of the hole, and it moves in a positive direction, toward $\boldsymbol{i}_0$. The peg does not contact the environment during transition from the initial state to the goal. Insertion is successful if the position control of the peg relative to the hole is free from error. Unfortunately, the peg-in-hole task is often unsuccessful because of the existing error. The robot cannot insert the peg in the hole by position control if the positioning error is greater than ± 0.125 [mm]. So, single measurement error using stereo vision is often beyond the acceptable error. The manner in which the task fails is almost the same as shown in Fig. 8 in the next section.

EXISTING METHOD FOR AUTONOMY WITH REINFORCEMENT LEARNING

Outline of Existing Method with Reinforcement Learning

Reinforcement learning (Sutton and Barto 1998) is used to generate autonomous robot action.

In "standard" learning (Fig. 6 (a)), controller $\boldsymbol{K}_Q$ is designed in advance by learning the nominal plant model $\boldsymbol{P}_N$, and it is applied to real plant $\boldsymbol{P}$. We use a policy called controller $\boldsymbol{K}_Q$, which is designed with reinforcement learning methods. When variations exist, e.g., measurement error in the vision system, unexpected obstacles appear in the environment, etc., $\boldsymbol{K}_Q$ cannot complete tasks due to poor robustness and adaptability.

As shown in Fig. 6 (b), new plant model $\boldsymbol{P}'_N$ is reconstructed using visual measurement. Controller $\boldsymbol{K}'_Q$ is designed for the reconstructed model $\boldsymbol{P}'_N$. Controller $\boldsymbol{K}'_Q$

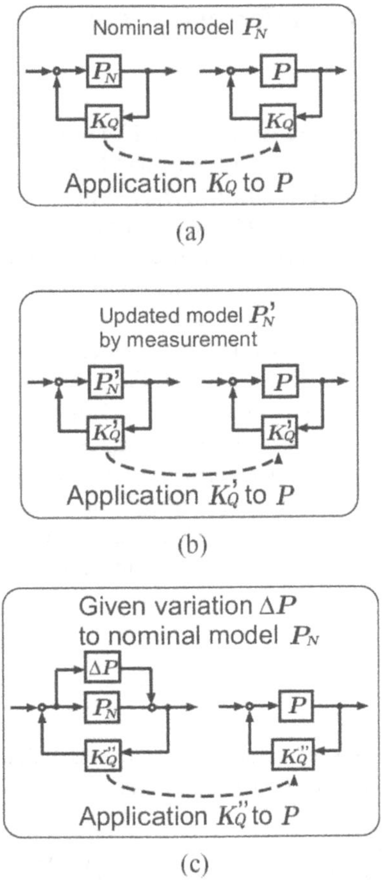

Figure 6. Learning using: (a) nominal plant model, (b) updated plant model and (c) plant model with variation.

is then applied to real plant $\boldsymbol{P}$. Learning converges within a practical calculation time and the new policy is applicable to the truss structure assembly (Senda et al. 2002). This method works well because it treats the kinematic problem without force interaction between the robot and the environment. The plant model for learning is reconstructed by visual measurement within a short time. This method has adaptability only if the model can be reconstructed accurately within a short time.

If the robot cannot complete the task with the controller due to error between the model and the real plant, the robot switches to online learning. Adaptability is realized by online identification and learning. However, this cannot be used for peg-in-hole task, because online identification requires too much time. In the next section, a reinforcement learning problem for a peg-in-hole task requires several tens of thousands of state-action pairs. It requires tens of days for online identification if a hundred samples are selected for each state-action pair and each sampling takes one second.

Existing Method with Reinforcement Learning

Problem Definition

Following general dynamic programming (DP) formulations, this chapter treats a discrete-time dynamic system in a reinforcement learning problem. A state s_i and an action u_k are the discrete variables and the elements of finite sets $\mathcal{S}$ and $\mathcal{U}$, respectively. The state set $\mathcal{S}$ is composed of N_s states denoted by $s_1, s_2, \cdots, s_{Ns}$, and an additional termination state s_0. The action set $\mathcal{U}$ is composed of K actions denoted by $u_1, u_2, \cdots, u_K$. If an agent is in state s_i and chooses action u_k, it will move to state s_j and incur a one-step cost $g(s_i, u_k, s_j)$ within state transition probability $p_{ij}(u_k)$. This transition is denoted by (s_i, u_k, s_j). There is a cost-free termination state s_0 where $p_{00}(u_k) = 1$, $g(s_0, u_k, s_0) = 0$, $Q(s_0, u_k) = 0$, $\forall u_k$. We assume that the state transition probability $p_{ij}(u_k)$ is dependent on only current state s_i and action u_k. This is called a discrete-time finite Markov Decision Process (MDP). The system does not explicitly depend on time. Stationary policy μ is a function mapping states into actions with $\mu(s_i) = u_k \in \mathcal{U}$, and μ is given by the corresponding time-independent action selection probability $\pi(s_i, u_k)$.

In this chapter, we deal with an infinite horizon problem where the cost accumulates indefinitely. The expected total cost starting from an initial state $s^0 = s_i$ at time $t = 0$, and using a stationary policy μ is

$$J^{\mu}(s_i) = \underset{s^1,s^2,\ldots}{E}\left[\sum_{t=0}^{\infty} g(s^t, \mu(s^t), s^{t+1}) \,\middle|\, s^0 = s_i\right], \tag{3}$$

where $E_x[\cdot]$ denotes an expected value, and this cost is called J-factor. Because of the Markov property, a J-factor of a policy μ satisfies

$$J^{\mu}(s_i) = \sum_{k=1}^{K} \pi(s_i, u_k) \sum_{j=0}^{N_s} p_{ij}(u_k)\{g(s_i, u_k, s_j) + J^{\mu}(s_j)\}, \ \forall s_i. \tag{4}$$

A policy μ is said to be proper if μ satisfies $J^{\mu}(s_i) < \infty$, $\forall s_i$.

We regard the J-factor of every state as an evaluation value, and the optimal policy μ^* is defined as the policy that minimizes the J-factor:

$$\mu^*(s_i) \equiv \arg\min_{\mu} \sum_{i=1}^{N_s} J^{\mu}(s_i), \ \forall s_i. \tag{5}$$

The J-factor of the optimal policy is defined as the optimal J-factor. It is denoted by $J^*(s_i)$.

The optimal policy defined by (5) satisfies Bellman's principle of optimality. Then, the optimal policy is stationary and deterministic. The optimal policy can be solved by minimizing the J-factor of each state independently. Hence, the optimal J-factors satisfy the following Bellman equation, and the optimal policy is derived from the optimal J-factors:

$$J^*(s_i) = \min_{u_k} \sum_{j=0}^{N_s} p_{ij}(u_k)\{g(s_i, u_k, s_j) + J^*(s_j)\}, \tag{6}$$

$$\mu^*(s_i) = \arg\min_{u_k} \sum_{j=0}^{N_s} p_{ij}(u_k)\{g(s_i,u_k,s_j) + J^*(s_j)\}. \tag{7}$$

Solutions

The existing type of reinforcement learning problem is solved as "standard" learning in Fig. 6 (a). It obtains the optimal policy μ^*_{nom}, which minimizes the J-factors of Eq. (7) for the nominal plant. It corresponds to controller $\boldsymbol{K}_Q$ in Fig. 6 (a). The optimal J-factor J^*_{nom} of μ^*_{nom} can be obtained by the DP-based solutions. The solutions are mentioned in reference (Sutton and Barto 1998, Bertsekas and Tsitsiklis 1996), but they are omitted here.

Learning Skill for Peg-in-hole by Existing Method

Problem Definition of Peg-in-hole

Here, the peg-in-hole task defined in subsection "Peg-in-hole task" is redefined as a reinforcement learning problem.

State variables $[y_x, y_y, y_\theta, f_x, f_y, f_\theta]$ in subsection "Peg-in-hole task" are continuous but discretized into 1.0 [mm], 1.0 [mm], $0.5\pi/180$ [rad], 2.0 [N], 1.0 [N], and 0.6 [Nm] in the model for reinforcement learning. The discrete state space has 4,500 discrete states, where the number of each state variable is [5, 5, 5, 4, 3, 3]. Robot action at the end-effector is u_1, u_2, u_3, u_4, at each of the end-effector states transiting by ± 1 in the direction of the $\boldsymbol{i}_0$-axis or $\boldsymbol{j}_0$-axis, u_5 and u_6, at each of the end-effector states transiting by ± 1 about the $\boldsymbol{k}_0$-axis of rotation. State-action space is described in the space robot model coordinate system Σ_0. The hole is 0.25 [mm] wider than the peg, and (y_x, y_y) are quantized larger than this difference.

Control in Eq. (1) is used to transit from present state s_i to the next state s_j by action u_k. The reference manipulation variable to make the transition to s_j is $\boldsymbol{y}_d^{(sj)} = \boldsymbol{y}_d^{(si)} + \delta\boldsymbol{y}_d^{(si)}$ given by $\delta\boldsymbol{y}_d^{(si)}(u_k)$ $(k = 1, 2, \cdots, 6)$, where $\delta\boldsymbol{y}_d^{(si)}$ is kept constant during transition. When the end-effector contacts the environment and manipulation variable $\boldsymbol{y}$ is stationary under constraint, the end-effector applies force and torque to the environment $\boldsymbol{f} = -\boldsymbol{K}_P(\boldsymbol{y} - \boldsymbol{y}_d)$ as Eq. (2) where $\boldsymbol{y}_d$ is the reference manipulation variable. Force and torque are controlled by $\boldsymbol{y}_d$, which is changed by an action. This is a compliant control, a force control. Tasks in which the end-effector contacts the environment, e.g., peg-in-hole, demand a control with compliance. Therefore, a compliant control is essential as a basic control.

The robot takes the next action after Eq. (1) control settles and the peg becomes stationary. Regardless of whether the peg is in contact with the environment, the robot waits for settling and proceeds to the next action. For this reason, state variables do not include velocity.

The goal is to achieve states with the largest y_x, the peg position in $\boldsymbol{i}_0$-direction, in the state space. The one step cost is $g(s_i, u_k, s_j) = 1$ for all states other than the goal state. Hence, the J-factor is the expected step number from s_i to the goal.

Learning Method

State transition probabilities for the state-action space in the previous section are calculated with sample data. Sample data are calculated with a dynamic simulator in a spatially continuous state space. The dynamic simulator is constructed with an open-source library, the Open Dynamics Engine (ODE) developed by Russell Smith. The numerical model of this simulator has continuous space, force and time in contrast to the discretized models for reinforcement learning in subsection "Problem definition". This discretized state transition model is regarded as plant P_N, and the method in Fig. 6 (a) is applied. The optimal policy μ^*_{nom}, i.e., controller $\boldsymbol{K}_Q$, is derived from the solution in subsection "Solutions". The optimal policy μ^*_{nom} is applied to the dynamic simulator with a continuous state-action space or the hardware experimental setup, the real plant $\boldsymbol{P}$ in Fig. 6 (a). This chapter does not deal with the online learning.

Learning Result

The result of a numerical simulation in which controller $\boldsymbol{K}_Q$ is applied to the environment with no position/orientation error is shown. The peg moves and arrives at the goal as shown in Fig. 7. Peg positioning is first changed to $y_\theta = 0$ [rad]. After the peg transits to the hole central axis, it is moved in a positive direction, toward $\boldsymbol{i}_0$. Then the peg is inserted into the hole. During the transition from the initial state to the goal, the peg does not make contact with the environment, and the end-effector applies force and torque, $[f_x, f_y, f_\theta] = [0, 0, 0]$.

In an environment with the hole position error of −0.5 [mm] in $\boldsymbol{j}_0$ direction, the peg does not arrived at the goal with controller $\boldsymbol{K}_Q$, see Fig. 8. The task is not completed using $\boldsymbol{K}_Q$ due to small errors caused by visual measurement, etc.

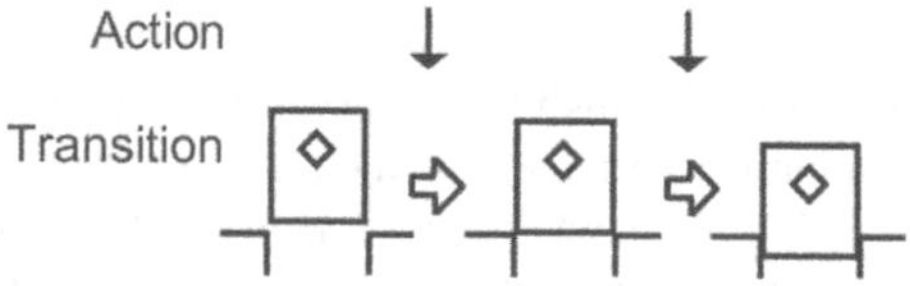

Figure 7. Trajectory of controller $\boldsymbol{K}_Q$ in a simulation without any hole position error.

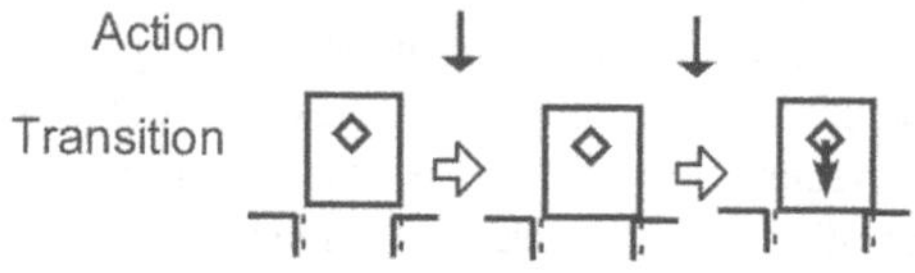

Figure 8. Trajectory of controller $\boldsymbol{K}_Q$ in a simulation with hole position error −0.5 [mm] in $\boldsymbol{j}_0$.

Discussion of Existing Reinforcement Learning Methods

If the robot cannot complete the task due to the errors, the robot switches to on-line learning. Reinforcement learning is said to be adaptive because of the online learning that uses online identification and on-site optimal control design. As mentioned before, the online learning cannot be used for peg-in-hole task, because it takes too long time for online learning by means of trial and error. Kober and Peters referred this as curse of real-world samples and mentioned it is more important to limit the real-world interaction time instead of limiting memory consumption or computational complexity (Kober and Peters 2012). Therefore, this study takes an approach different from the online learning.

On the other hand, the state and the action constitute continuous spaces in robot reinforcement learning. For such case, function approximation methods to approximate the state transition probability model, the value functions, and the policy are proposed (Bertsekas and Tsitsiklis 1996, Szepesvari 2010, Powell 2007, Lagoudakis and Parr 2003, Xu et al. 2007, Boyan 2002). The methods have some advantages, e.g., continuous action, memory saving, and so on. But, the function approximation framework also yields situations, e.g., DP solutions are no longer applicable, theoretical guarantees of convergence no longer hold, learning processes often prove unstable (Boyan and Moore 1995, Kakade and Langford 2002, Bagnell et al. 2003), and so on. Therefore, this study does not use function approximation methods.

AUTONOMOUS ACQUISITION OF SKILL BY LEARNING

Autonomous Acquisition of Skill for Peg-in-hole

A robot can achieve peg positioning or movement with contact force, and it must have basic control functions the same as a human being. Human vision measurement and positioning control are not accurate enough. However, the rate of a human failure in the same task is not as high as that of a robot. One reason for this may be skills human being brings to the task.

A human being conducting peg-in-hole task uses a typical sequence of actions (Fig. 9). First, the human being puts a corner of the peg inside the hole. The peg orientation is inclined. The peg is in contact with the environment. Two points of the peg, the bottom and a side, are in contact with the environment, as shown in the close-up, Fig. 9. The human then rotates the peg and pushes it against the hole and maintains the two contact points. The two corners are then inserted into the hole. Finally, the human inserts the peg into the hole and completes the task.

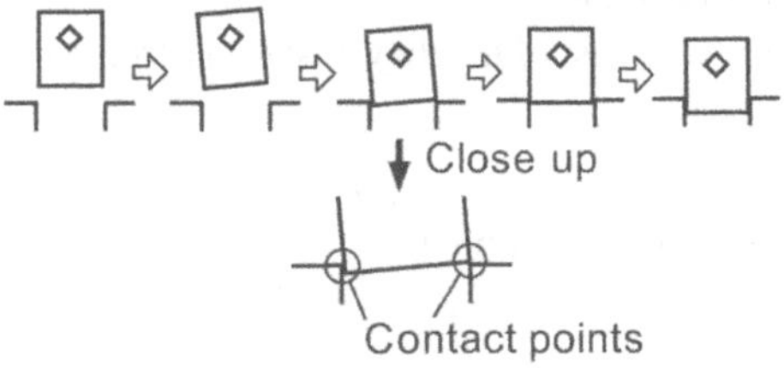

Figure 9. Human skill.

Human vision measurement accuracy and positioning control accuracy are not high. A human presumably develops skill while manipulating this situation. We conducted an experiment to check whether robot learning in the same situation could achieve the task as well as a human.

This situation conceptually corresponds to Fig. 6 (c). Plant $\boldsymbol{P}_N + \Delta\boldsymbol{P}$ denotes a variation plant with error caused by visual measurement, etc. Variation plant set $\{\boldsymbol{P}_N +$

$\Delta \boldsymbol{P}\}$ is composed of all the variation plants that can exist. Real plant $\boldsymbol{P}$ is supposed to be a member of variation plant set $\{\boldsymbol{P}_N + \Delta \boldsymbol{P}\}$. The learning robot obtains controller $\boldsymbol{K}_Q''$. The controller is able to complete the task for all of the variation plants in $\{\boldsymbol{P}_N + \Delta \boldsymbol{P}\}$.

Such reinforcement learning to obtain a policy robust to model uncertainty is a research topic currently. It was found that artificially adding a little noise would smooth model errors and avoid policy over-fitting (Jakobi et al. 1995, Atkeson 1998). Evaluations in the model were exploited to ensure that control was robust to model uncertainty (Schneider 1996, Bagnell and Schneider 2001). Then, it was established that a robust policy would be obtained by learning for a variation plant set (Senda 2008). But, its theoretical background was unclear. The following method in this study efficiently obtains a policy robust to model uncertainty and is theoretically clear.

Problem Definition for Reinforcement Learning with Variation

We assume there are N variation plants around the estimated plant (the nominal plant). We use a set composed of N variation plants for learning.

We consider difference w_l between a variation plant and the nominal plant in each state, which is a discrete variable and the element of finite set $\mathcal{W}$. Finite set $\mathcal{W}$ is composed of L differences denoted by $w_0, w_1, \cdots, w_{L-1}$. Difference w_0 indicates no difference. If an agent is in state s_i with difference w_l and chooses action u_k, it will move to s_j within a state transition probability $p_{ij}(u_k; w_l)$ and incur a one-step cost $g(s_i, u_k, s_j; w_l)$. This transition is denoted by $(s_i, u_k, s_j; w_l)$. Difference w_l can be considered as the disturbance that causes state transition probability $p_{ij}(u_k; w_0)$ to vary to $p_{ij}(u_k; w_l)$. We assume that the $p_{ij}(u_k; w_l)$ and $g(s_i, u_k, s_j; w_l)$ are given.

Difference w_l at each state is determined by a variation plant. Variation η is a function mapping states into difference with $\eta(s_i) = w_l \in \mathcal{W}$. The nominal plant is defined by $\eta_0(s_i) = w_0$ for all states s_i. The plant does not explicitly depend on time, so variation η is time-invariant. We assume that $\eta(s_i) = w_l$ is given.

A plant set composed of N plants used for learning is represented by $\mathcal{H} = \{\eta_0, \eta_1, \cdots, \eta_{N-1}\}$. Set $\mathcal{H}$ corresponds to $\{\boldsymbol{P}_N + \Delta \boldsymbol{P}\}$. Let $\rho(\eta_n)$ denote the probability that the plant variation is η_n. We call this the existing probability of variation plant η_n. We assume that $\rho(\eta)$ is given at time $t = 0$.

For set $\mathcal{H}$, the expected cost of a policy μ starting from an initial state $s\ 0 = s_i$ at $t = 0$ is

$$\overline{J}^{\mu}(s_i) = \underset{\eta, s^1, s^2, \ldots}{E} \left[\sum_{t=t_0}^{\infty} g(s^t, \mu(s^t), s^{t+1}; \eta(s^t)) \middle| \begin{matrix} s^0 = s_i, \\ \eta \in \mathcal{H} \end{matrix} \right], \tag{8}$$

which is the J-factor of this problem. This J-factor formula using the plant existing probability is

$$\overline{J}^{\mu}(s_i) = \sum_{n=0}^{N-1} \rho(\eta_n) J^{\mu,\eta_n}(s_i), \tag{9}$$

where $J^{\mu,\eta_n}(s_i)$ denotes the expected cost using the policy μ on a plant η_n starting from an initial state s_i. It satisfies

$$J^{\mu,\eta_n}(s_i)=\sum_{k=1}^{K}\pi(s_i,u_k)\sum_{j=0}^{N_s}p_{ij}(u_k;\eta_n(s_i))$$
$$\times\{g(s_i,u_k,s_j;\eta_n(s_i))+J^{\mu,\eta_n}(s_j)\}. \tag{10}$$

We define the optimal policy as

$$\mu^*(s_i)\equiv\arg\min_{\mu}\sum_{i=1}^{N_s}\overline{J}^{\mu}(s_i),\ \forall s_i \tag{11}$$

that minimizes the J-factor of every state. The J-factor of the optimal policy μ^* is defined as the optimal J-factor, represented by $\overline{J}^*(s_i)$. The objective is to obtain the optimal policy. We assume that there is at least one policy μ satisfying $\overline{J}^{\mu}(s_i) < \infty$, $\forall s_i$ in this problem. Henceforth, we will call this problem the original problem.

The variation plant in the original problem correlates with differences between any two states. Due to this correlation, the optimal policy does not satisfy Bellman's principle of optimality (Senda and Tani 2011). Therefore the optimal policy and the optimal J-factor in this problem do not satisfy Eqs. (6) and (7). In general, the optimal policy is not stationary. If policies are limited to stationary, the optimal policy is stochastic.

Therefore, another problem definition or another solution method is needed.

Solutions for a Relaxed Problem of Reinforcement Learning with Variation

We relax the original problem to recover the principle of optimality. Then, we can find the optimal J-factor efficiently by applying DP algorithms to the relaxed problem. We treat a reinforcement learning problem based on a two-player zero-sum game.

We assume that differences, w_0, w_1, $\cdots$, w_L, exist independently in each state s_i. Then the original problem is relaxed to a reinforcement learning problem (Littman 1994, Morimoto and Doya 2005, Al-Tamimi et al. 2007) based on a two-player zero-sum game (Basar and Bernhard 1995) whose objective is to obtain the optimal policy for the worst variation maximizing the expected cost. Since the correlations of differences in a variation plant are ignored, the principle of optimality is recovered.

The $\mathcal{H}_{2pzs}$ is defined as the set of variation plants consisting of all possible combinations of any differences. Since each state has L types of differences, the number of variation plants in $\mathcal{H}_{2pzs}$ is L^{N_s}. We define the optimal policy μ^*_{2pzs} as the policy minimizing the expected cost against the worst variation η^*_{2pzs} maximizing the expected cost

$$(\mu^*_{2pzs},\eta^*_{2pzs})\equiv\arg\min_{\mu}\max_{\eta\in\mathcal{H}_{2pzs}}\sum_{i=1}^{N_s}J^{\mu,\eta}(s_i), \tag{12}$$

and the optimal J-factor $J^*_{2pzs}(s_i)$ is defined as the J-factor of the optimal policy and the worst variation.

Since the principle of optimality is recovered, the optimal J-factor satisfies the following Bellman equation.

$$J^*_{2pzs}(s_i) = \min_{u_k} \max_{w_l} \sum_{j=0}^{N_s} p_{ij}(u_k; w_l)\{g(s_i, u_k, s_j; w_l) + J^*_{2pzs}(s_j)\}. \tag{13}$$

Therefore the optimal J-factor can be obtained by a DP algorithm. Using the optimal J-factor, the optimal policy and the worst variation are obtained by

$$(\mu^*_{2pzs}(s_i),\ \eta^*_{2pzs}(s_i)) = \arg\min_{u_k} \max_{w_l} \sum_{j=0}^{N_s} p_{ij}(u_k; w_l) \times \{g(s_i, u_k, s_j; w_l) + J^*_{2pzs}(s_j)\}. \tag{14}$$

The optimal policy μ^*_{2pzs} is applicable to all L^{N_s} variation plants in $\mathcal{H}_{2pzs}$. The optimal policy μ^*_{2pzs} is proper for all plants in $\mathcal{H}$ of subsection "Problem definition for reinforcement learning with variation" because $\mathcal{H} \subseteq \mathcal{H}_{2pzs}$ holds. However, the actual number of plants to which the policy should be applied is only N and $N \ll L^{N_s}$. Hence, the optimal policy of the reinforcement learning problem based on the two-player zero-sum game is often conservative and yields poor performance because the problem does not consider the existence of variation plants. We cannot solve this problem if there is no policy satisfying $J^{\mu,\eta}(s_i) < \infty$, $\forall s_i$, $\forall \eta \in \mathcal{H}_{2pzs}$, even though there policy μ exists and satisfies $\sum_i J^{\mu,\eta_n}(s_i) < \infty$, $\forall \eta_n \in \mathcal{H}$. Hence, a solution method to solve the original problem is desired.

Learning of Peg-in-hole Task with Variation

Problem Definition of Peg-in-hole Task with Variations

This section uses the same problem definition for peg-in-hole as subsection "Problem definition of peg-in-hole". The following is added to take variations into account.

The hole position and orientation is measured by the stereo vision system. These measurements involve errors caused by quantized images, lighting conditions at a worksite, etc. Time-averaged errors are almost constant while the space robot performs the task, unlike white noise whose time-averaged error is zero. Error evaluations are modeled as described below. Hole position errors are modeled as normal probability distributions, where the mean is $m = 0$ [mm], and standard deviation is $\sigma = 0.75$ [mm]. Hole orientation errors are modeled as normal probability distributions, where the mean is $m = 0$ [rad], and standard deviation is $\sigma = 0.5\pi/180$ [rad]. If the error's statistical values gradually vary, we have to estimate them on-line. The relative position and orientation between Σ_0 and Σ_{hl} are fixed during the task. The plant variations are modeled as hole position and orientation measurement errors.

Consider these errors as variations $\Delta \boldsymbol{P}$ added to nominal model $\boldsymbol{P}_N$ (Fig. 6(c)). We constructed 9 plants $\eta_0 \sim \eta_8$ for learning as listed in Table 1, where each plant has a combination of errors among [−1.0, 0.0, 1.0] [mm] in $\boldsymbol{j}_0$-axis direction and [−(0.5/180)π, 0.0, (0.5/180)π] [rad] in $\boldsymbol{k}_0$-axis rotation. Plant η_0 with no error both in $\boldsymbol{j}_0$-axis direction and $\boldsymbol{k}_0$-axis rotation is the nominal plant. The plant existing probabilities followed the above-mentioned normal probability distributions.

Table 1. Hole position of variation plant η_n from the nominal plant.

Plants	Variations	
	Position in $\boldsymbol{j}_0$ [mm]	Rotation about $\boldsymbol{k}_0$ [rad]
η_0	0.0	0.0
η_1	0.0	$-(0.5/180)\pi$
η_2	−1.0	0.0
η_3	−1.0	$-(0.5/180)\pi$
η_4	−1.0	$(0.5/180)\pi$
η_5	0.0	$(0.5/180)\pi$
η_6	1.0	0.0
η_7	1.0	$(0.5/180)\pi$
η_8	1.0	$-(0.5/180)\pi$

In the original problem, plant η_n determines the state transition probability as $p_{ij}(u_k ; \eta_n)$ for all state transitions (s_i, u_k, s_j) simultaneously. The state transition probability is represented by $p_{ij}(u_k; w_n(s_i))$ where $\mathcal{W}(s_i) = \{w_0(s_i), \ldots, w_{N-1}(s_i)\}$ and $w_n(s_i) = \eta_n(s_i)$. On the other hand, the two-player zero-sum game allows difference w_l at state s_i to be chosen arbitrary from $\mathcal{W}(s_i)$. The one step cost is $g(s_i, u_k, s_j; w_l) = 1$.

In the later simulations and experiments to evaluate the learned results, the hole position and orientation are derived from the above normal probability distribution. In the simulations, a variation in the hole position and attitude is chosen for each episode, but the variation is invariant during the episode.

Learning Method

Under the conditions of the above problem definition, a policy is obtained by the solution in subsection "Solutions for a relaxed problem of reinforcement learning with variation". It is the optimal policy μ^*_{2pzs} of the two-player zero-sum game, i.e., $\boldsymbol{K}''_Q$ in Fig. 6 (c). The optimal policy μ^*_{2pzs} is applied to the dynamic simulator with a continuous state-action space or the experimental hardware setup, which is a real plant $\boldsymbol{P}$ in Fig. 6 (c). No on-line learning is needed.

There is no proper policy for all plants in $\mathcal{H}_{2pzs}$ if the variations in Table 1 are too large. In this case, there is no policy satisfying the reinforcement learning problem based on the two-player zero-sum game. A typical approach for this situation is to make the variations smaller, to reconstruct $\mathcal{H}_{2pzs}$, and to solve the two-player zero-sum game again. This approach is repeated if we cannot obtain any solutions. This approach reduces the robustness of solutions.

Control Results

In results for numerical simulation (Fig. 10), where the peg arrives at the goal using controller $\boldsymbol{K}''_Q$ in the environment without hole position/orientation error. Peg positioning is firstly inclined, and the peg moves in the positive direction, toward $\boldsymbol{i}_0$. Then, the peg's corner is inserted in the hole. The peg makes contact with a corner of the hole. The peg transits in a positive direction, toward $\boldsymbol{j}_0$, while maintaining contact. Another corner of the peg is put inside the hole when the action in the direction of $\boldsymbol{i}_0$ and $\boldsymbol{j}_0$ is repeated. Peg positioning is changed to $y_\theta = 0$ [rad], and the peg slips into the hole. The task is completed. The learned result is similar to human skill for the peg-in-hole task in Fig. 9.

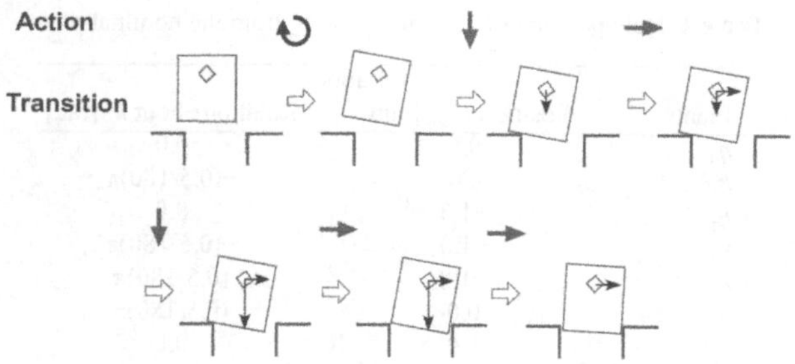

Figure 10. Trajectory of controller K_Q'' in a simulation using nominal plant η_0 without any hole position error.

The peg has arrived at the goal using controller K_Q'' for variation plants $\eta_1 - \eta_8$. The numerical results for $\eta_1 - \eta_4$ are shown in Figs. 11–14. Each transition is similar to the case of η_0, and the peg is inserted into the hole.

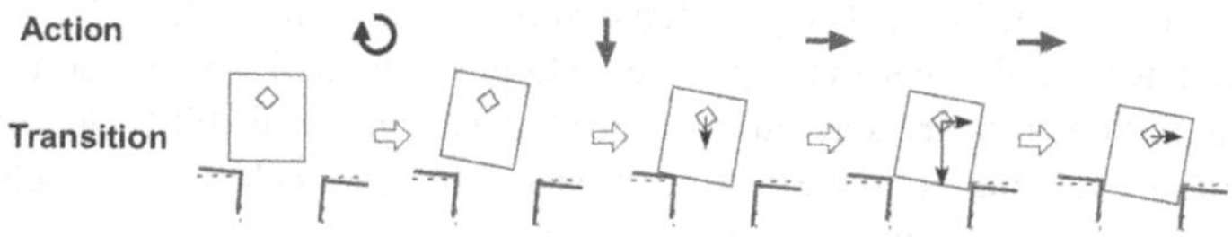

Figure 11. Trajectory of controller K_Q'' in a simulation using plant η_1.

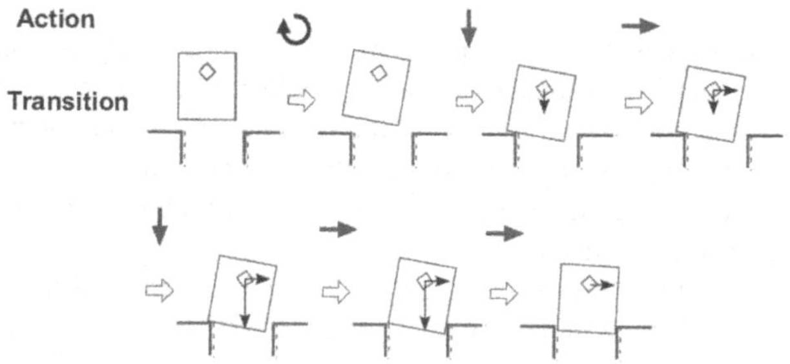

Figure 12. Trajectory of controller K_Q'' in a simulation using plant η_2.

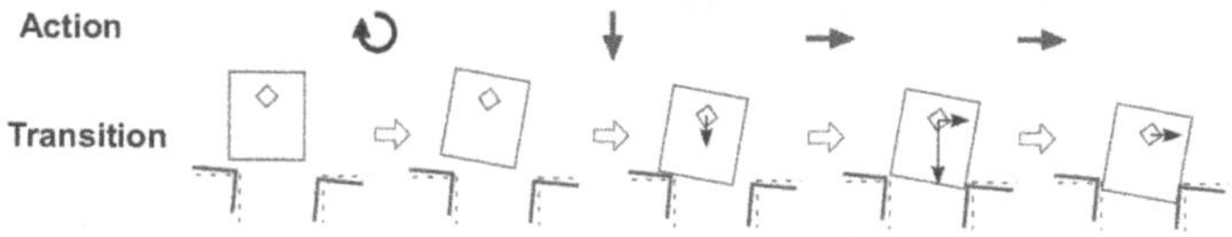

Figure 13. Trajectory of controller K_Q'' in a simulation using plant η_3.

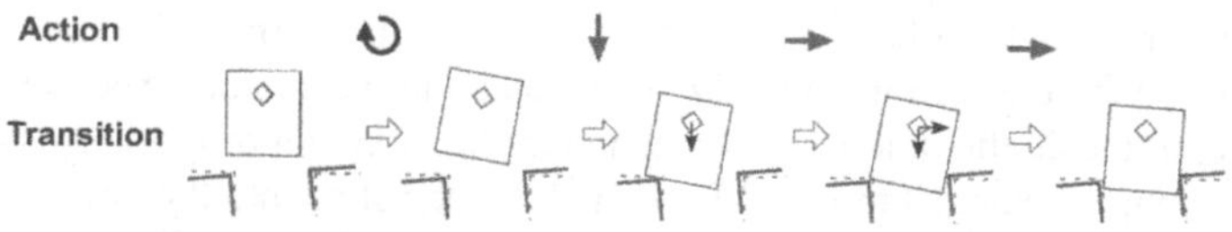

Figure 14. Trajectory of controller K_Q'' in a simulation using plant η_4.

The task is achieved with controller $\boldsymbol{K}_Q''$ in the same environment with error, where $\boldsymbol{K}_Q$ previously did not work at all. This means that the action generated by controller $\boldsymbol{K}_Q''$ is robust against variations as well as human skill. We judge the robot, i.e., controller $\boldsymbol{K}_Q''$, to have obtained a skill, the ability to complete a task when the vision measurement accuracy is low.

EVALUATION OF OBTAINED SKILL

Results of Hardware Experiments

Example results in the hardware experiment using controllers $\boldsymbol{K}_Q$ and $\boldsymbol{K}_Q''$ are shown in Fig. 15. The following variations are used +0.3 [mm] in $\boldsymbol{i}_0$, +1.2 [mm] in $\boldsymbol{j}_0$ and +0. $5\pi/180$ [rad] rotation about the $\boldsymbol{k}_0$-axis. Controller $\boldsymbol{K}_Q$ cannot complete the task due to environmental variations, but controller $\boldsymbol{K}_Q''$ can.

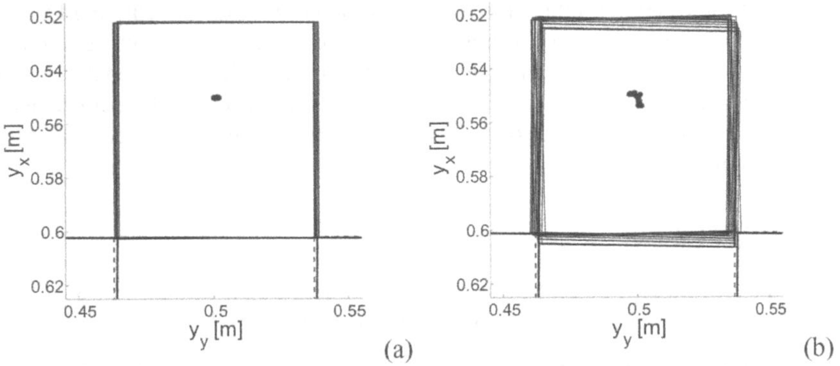

Figure 15. Experimental trajectories using two controllers in an environment with error: (a) controller $\boldsymbol{K}_Q$ and (b) controller $\boldsymbol{K}_Q''$.

Evaluation of Robustness and Control Performance

Robustness and control performance of controllers $\boldsymbol{K}_Q$ and $\boldsymbol{K}_Q''$ are evaluated by simulations and hardware experiments, the peg-in-hole task.

Variation plants, i.e., error in hole position and orientation, are derived from the normal probability distribution in section "Learning of peg-in-hole task with variation". The robustness and the control performance are evaluated respectively by the task achievement ratio and the average step number to the goal. The achievement ratio equals the number of successes divided by the number of trials. Table 2 shows the achievement ratios and the average step number of $\boldsymbol{K}_Q$ and $\boldsymbol{K}_Q''$ as evaluated by simulations and hardware experiments. The simulations and the experiments are executed 10,000 times and 50 times, respectively. The achievement ratios of $\boldsymbol{K}_Q$ are 59 percent and 64 percent in simulation and hardware experiments. Those of $\boldsymbol{K}_Q''$ dramatically increase to 99 percent in numerical simulation and 96 percent in hardware experiments. These results show the robot autonomously generates robust skill using the proposed learning method. The difference in step numbers between hardware experiments and simulations, an increase in hardware steps, may be due to variations, e.g., irregular friction in the environment, joint flexibility, etc. Such variables are not considered in the numerical simulation.

Robust skills are thus autonomously generated by learning in this situation, where variations make a task achievement difficult.

Table 2. Achievement ratios and averaged step numbers of peg-in-hole task with controllers $\boldsymbol{K}_Q$ and $\boldsymbol{K}_Q''$.

	Simulation		Experiment	
Controller	Ratio	Step no.	Ratio	Step no.
$\boldsymbol{K}_Q$	58.7%	19.3	64%	21
$\boldsymbol{K}_Q''$	98.7%	9.83	96%	17

CONCLUSIONS

We have applied reinforcement learning to obtain successful completion of a given task when a robot normally cannot complete the task using controller designed in advance. Peg-in-hole achievement ratios are usually low when we use conventional learning without consideration of plant variations. In the proposed method, using variation consideration, the robot autonomously obtain robust skills which enabled the robot to achieve the task. Simulation and hardware experiments have confirmed the effectiveness of our proposal. Our proposal also ensures robust control by conducting learning stages for a set of plant variations.

REFERENCES

Al-Tamimi, A., F. L. Lewosa and M. Abu-Khalaf. 2007. Model-free Q-learning designs for linear discrete time zero-sum games with application to H-infinity control. Automatica, 43(3): 473–481.

Asada, M. 1999. Issues in applying robot learning and evolutionary methods to real environments. Journal of the Society of Instrument and Control Engineers, 38(10): 650–653 (in Japanese).

Atkeson, C. G. 1998. Nonparametric model-based reinforcement learning. Advances in Neural Information Processing Systems (NIPS).

Bagnell, J. A. and J. C. Schneider. 2001. Autonomous helicopter control using reinforcement learning policy search methods. IEEE International Conference on Robotics and Automation.

Bagnell, J. A., A. Y. Ng, S. Kakade and J. Schneider. 2003. Policy search by dynamic programming. Advances in Neural Information Processing Systems (NIPS).

Başar, T. and P. Bernhard. 1995. H_∞-Optimal Control and Related Minimax Design Problems. Birkhäuser, Boston.

Bertsekas, D. P. and J. N. Tsitsiklis. 1996. Neuro-Dynamic Programming. Athena Scientific, Belmont, MA.

Boyan, J. 2002. Technical update: Least-squares temporal difference learning. Machine Learning, Special Issue on Reinforcement Learning, 49(2–3): 233–246.

Boyan, J. A. and A. W. Moore. 1995. Generalization in reinforcement learning: Safely approximating the value function. Advances in Neural Information Processing Systems (NIPS).

Connolly, C. I., J. B. Burns and R. Weiss. 1990. Path planning using Laplace's equation. IEEE International Conference on Robotics and Automation, pp. 2102–2106.

Fujii, S., K. Senda and S. Mano. 2006. Acceleration of reinforcement learning by estimating state transition probability model. Transactions of the Society of Instrument and Control Engineers, 42(1): 47–53 (in Japanese).

Fukuda, T., W. Srituravanich, T. Ueyama and Y. Hasegawa. 2000. A study on skill acquisition based on environment information (Task Path Planning for Assembly Task Considering Uncertainty). Transactions of the Japan Society of Mechanical Engineers, 66(645): 1597–1604 (in Japanese).

Jakobi, N., P. Husbands and I. Harvey. 1995. Noise and the reality gap: The use of simulation in evolutionary robotics. The 3rd European Conference on Artificial Life.

Kakade, S. and J. Langford. 2002. Approximately optimal approximate reinforcement learning. International Conference on Machine Learning (ICML).

Kober, J. and J. Peters. 2012. Reinforcement learning in robotics: A survey. pp. 579–610. *In*: M. Wiering and M. van Otterlo (eds.). Reinforcement Learning: State-of-the-Art. Springer, Heidelberg.

Lagoudakis, M. G. and R. Parr. 2003. Least-squares policy iteration. Journal of Machine Learning Research, 4: 1107–1149.

Littman, M. L. 1994. Markov games as a framework for multi-agent reinforcement learning. International Conference on Machine Learning (ICML), pp. 157–163.

Mahadevan, S. and M. Maggioni. 2007. Proto-value functions: A Laplacian framework for learning representation and control in Markov decision processes. Journal of Machine Learning Research, 8: 2169–2231.

Miyazaki, F. and S. Arimoto. 1985. Sensory feedback for robot manipulators. Journal of Robotic Systems, 2(1): 53–71.

Morimoto, J. and K. Doya. 2005. Robust reinforcement learning. Neural Computation, 17(2): 335–359.

Powell, W. B. 2007. Approximate Dynamic Programming: Solving the Curses of Dimensionality. Wiley, Hoboken, NJ.

Sato, D. and M. Uchiyama. 2007. Peg-in-hole task by a robot. Journal of the Japan Society of Mechanical Engineers, 110(1066): 678–679 (in Japanese).

Schneider, J. G. 1996. Exploiting model uncertainty estimates for safe dynamic control learning. Advances in Neural Information Processing Systems (NIPS).

Senda, K. 2001. An approach to autonomous space robots. Systems. Control and Information, 45(10): 593–599 (in Japanese).

Senda, K. and Y. Tani. 2011. Optimality principle broken by considering structured plant variation and relevant robust reinforcement learning. IEEE International Conference on Systems, Man, and Cybernetics, pp. 477–483.

Senda, K., T. Kondo, Y. Iwasaki, S. Fujii, N. Fujiwara and N. Suganuma. 2008. Hardware and numerical experiments of autonomous robust skill generation using reinforcement learning. Journal of Robotics and Mechatronics, 20(3): 350–357.

Senda, K., Y. Murotsu, A. Mitsuya, H. Adachi, S. Ito, J. Shitakubo and T. Matsumoto. 2002. Hardware experiments of a truss assembly by an autonomous space learning robot. AIAA Journal of Spacecraft and Rockets, 39(2): 267–273.

Skaar, S. B. and C. F. Ruoff (eds.). 1995. Teleoperation and Robotics in Space. AIAA, Washington, DC.

Sutton, R. S. and A. G. Barto. 1998. Reinforcement Learning: An Introduction. MIT Press, Cambridge, MA.

Szepesvári, Cs. 2010. Algorithms for Reinforcement Learning. Morgan and Claypool, San Rafael, CA.

Whitney, D. E. 1982. Quasi-static assembly of compliantly supported rigid parts. Transactions of ASME, Journal of Dynamic Systems, Measurement, and Control, 104: 65–77.

Xu, X., D. Hu and X. Lu. 2007. Kernel based least squares policy iteration for reinforcement learning. IEEE Transactions on Neural Networks, 18(4): 973–992.

Yamanobe, N., Y. Maeda, T. Arai, A. Watanabe, T. Kato, T. Sato and K. Hatanaka. 2006. Design of force control parameters considering cycle time. Journal of the Robotics Society of Japan, 24(4): 554–562 (in Japanese).

5

Adaptive Control for Multi-Fingered Robot Hands

Satoshi Ueki[1,*] and *Haruhisa Kawasaki*[2]

ABSTRACT

This chapter discusses adaptive control for object manipulation according to the contact condition. When a robot grasps an object, its fingers come into contact with the surface of the object and constrained its hand motion by that contact. Robotic hands are mechanical non-linear systems. In general, treat adaptive control systems as time-varying systems. For dexterous manipulation, the control objective is to control the object position and velocity; that is, to coordinately-control the force at the contact point that occurs for multiple robotic fingers. This is one of the challenging problems in robotics. In the controller presented here, the shapes of the object and positions of the contact points are already known. In fact, position of the contact point could be calculated using the force sensor variable. However, the accuracy is insufficient. Thus, need study of adaptive controls with fewer restrictions.

Keywords: Adaptive control, robot hand, coordinate, manipulation, force control

INTRODUCTION

Multi-fingered robot hands are expected to replace human hands in tasks requiring dexterous manipulation of objects. When a robot grasps an object, its fingers come

[1] Department of Mechanical Engineering, National Institute of Technology, Toyota College, Toyota, Japan.
E-mail: s_ueki@toyota-ct.ac.jp
[2] Department of Mechanical Engineering, Gifu University, Gifu, Japan.
E-mail: h_kawasa@gifu-u.ac.jp
* Corresponding author

into contact with the surface of the object, and its hand motion is constrained by that contact. For dexterous manipulation, a basic understanding of the contact model between a finger and an object, the motion constraint during contact, and the dynamics of these constraints is important. Furthermore, the dynamics of robot hands are non-linear, and require control methods that take constraints into consideration.

When a robot hand grasps and manipulates an object, contact models are important for the analysis of object constraint conditions and robotic hand control. Since contact conditions vary according to point, line, and surface contact, as well as whether or not friction exists at the contact point, the object is affected by diverse constraint conditions. Moreover, kinematic constraints occur according to contact conditions between the robotic finger and the object. Therefore, relative translational velocity and relative angular velocity should be taken into account for robotic hand control.

The cooperative control of multiple fingers on a robotic hand used to perform an orientation by grasping a target object is a basic technique of dexterous manipulation. For many cooperative controls, widely used control laws are based on compliance control (Cutkosky and Kao 1989), impedance control (Kosuge et al. 1995), control of under-actuated fingers (Birglen et al. 2010), dynamic control (Zheng et al. 2000), together with position and force laws grounded in hybrid control (Hayati 1986), the hybrid system approach (Schlegl et al. 2003), and model predictive control (Yingjie and Hosoe 2005). In these control laws, it is assumed that the dynamics of the system control are known. However, it is often difficult to determine the exact dynamics of a system. In addition, the dynamic parameters of the system will vary depending on the object being manipulated. Several approaches have been proposed to accommodate cases where the model parameters of the system are either unknown or fluctuating, including cooperative control using a neural network (Hwang and Toda 1994), learning control (Naniwa and Arimoto 1995), control without object sensing (Ozawa et al. 2005), adaptive sliding control (Su and Stepanenco 1995, Ueki et al. 2006), and adaptive coordinated control (Kawasaki et al. 2006). Adaptive compliant grasp (Zhaopeng et al. 2015), adaptive force control with estimation of environmental stiffness (Wakamatsu et al. 2010), and others have also been proposed.

This chapter describes adaptive control for object manipulation. First, we review the mathematical models and basic theory for stability analysis. In the mathematical models, we explore the equations of motion at the contact point for relative motion between the robotic finger and the object, the dynamic model of a rigid body, and the dynamic model of a robotic hand. We then describe adaptive controllers for object manipulation. We present experimental verification result to demonstrate the effectiveness of the proposed adaptive controller. Finally, conclusions are given.

In general, force sensors are expensive and susceptible to noise. An adaptive control for object manipulation without a force sensor under a condition of rolling-contact can thus prevent a control system from being highly reliable. If a stable controller is configurable without a force sensor, the control law increases the usefulness of the control system. On the other hand, humans use hands to grasp and manipulate objects dexterously using force sensation. To realize dexterous manipulation like that of the human hand, it is necessary to study adaptive control using a force sensor. Moreover, dexterous object manipulation is possible if the finger slides on the object surface. For these reasons, we describe an adaptive control for object manipulation without a force sensor under a condition of rolling-contact, an adaptive control for object manipulation using a force sensor under a condition of rolling-contact, and an adaptive control for object manipulation using a force sensor under a condition of sliding-contact. An introduction to robot hand control and the details of contact models may be found in the references (Mason and Salisbury 1985, Mason 1998, Murray et al. 1993, Kawasaki 2015).

MATHEMATICAL MODELS OF MULTI-FINGERED HANDS

When a robotic hand with k fingers, each with three DOFs (degree of freedom), grasps and manipulates a rigid body (as shown in Fig. 1), in which the i-th robotic finger contacts the object at point C_i, the coordinate systems are defined as follows: Σ_P is the task coordinate system, Σ_O is the object coordinate system fixed on the object, and Σ_{Fi} is the i-th fingertip coordinate system fixed on the i-th fingertip.

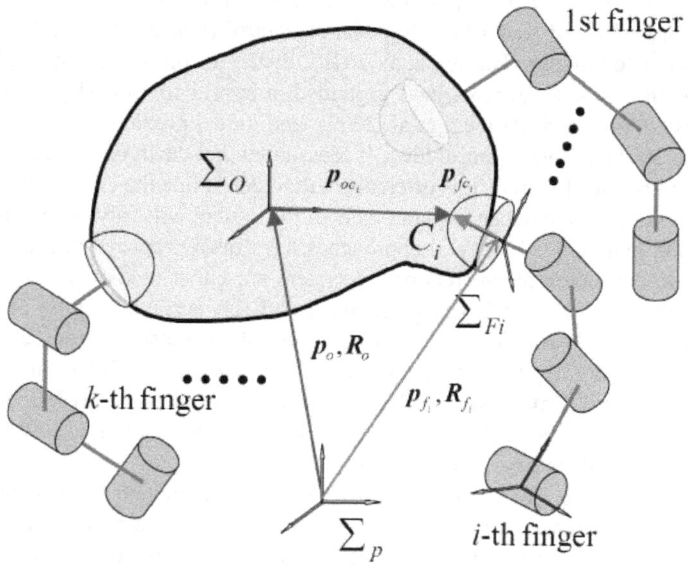

Figure 1. Grasped object and multi-fingered hand coordinate system.

To facilitate the dynamic formulation, we make the following assumptions.

(A1) All the fingertips contact the common object; the grasp is a force closure and manipulable grasp.

(A2) The Jacobian matrix is a non-sigular matrix. That is, the DOF of the finger is not redundant, and there is sufficient DOF to manipulate the object.

(A3) The contact force is all in the friction cone. This indicates that there is no slippage at the contact point, and the contact force has three elements.

(A4) The object and fingertip surfaces are described by twice continuously differentiable hyper surfaces.

The constraint at each contact point is described by the rolling contact. The force generated by the constraint does not work on the system (d'Alembert's principle).

Equation of Motion at the Contact Point

To perform system modeling, it is necessary to describe a contact condition that takes into account the relative motion between the object motion and that of the robot fingertip.

When the *i*-th robot finger manipulates the object, the position vector of the contact point is expressed as

$$\boldsymbol{p}_O + \boldsymbol{R}_O{}^O\boldsymbol{p}_{OCi} = \boldsymbol{p}_{Fi} + \boldsymbol{R}_{Fi}{}^{Fi}\boldsymbol{p}_{FiCi}. \tag{1}$$

Let $\mathbb{R}$ denote the set of all real numbers. Then, $\boldsymbol{p}_O \in \mathbb{R}^3$ is a position vector of the origin of the object coordinate system Σ_O with respect to Σ_p, $\boldsymbol{R}_O \in \mathbb{R}^{3\times3}$ is a rotation matrix from Σ_p to Σ_O, ${}^O\boldsymbol{p}_{OCi} \in \mathbb{R}^3$ is a position vector from Σ_O to the contact point C_i with respect to Σ_O, $\boldsymbol{p}_{Fi} \in \mathbb{R}^3$ is a position vector of the origin of the *i*-th fingertip coordinate system Σ_{Fi} with respect to Σ_p, $\boldsymbol{R}_{Fi} \in \mathbb{R}^{3\times3}$ is a rotation matrix from Σ_p to Σ_{Fi}, and ${}^{Fi}\boldsymbol{p}_{FiCi} \in \mathbb{R}^3$ is a position vector from Σ_{Fi} to the contact point C_i with respect to Σ_{Fi}. The first-order time differentials of Eq. (1) are expressed as

$$\dot{\boldsymbol{p}}_O + \boldsymbol{\omega}_O \times \boldsymbol{R}_O{}^O\boldsymbol{p}_{OCi} + \boldsymbol{R}_O{}^O\dot{\boldsymbol{p}}_{OCi} = \dot{\boldsymbol{p}}_{Fi} + \boldsymbol{\omega}_{Fi} \times \boldsymbol{R}_{Fi}{}^{Fi}\boldsymbol{p}_{FiCi} + \boldsymbol{R}_{Fi}{}^{Fi}\dot{\boldsymbol{p}}_{FiCi}, \tag{2}$$

where $\boldsymbol{\omega}_O \in \mathbb{R}^3$ is an angle velocity of the object, and $\boldsymbol{\omega}_{Fi} \in \mathbb{R}^3$ is an angle velocity of the *i*-th robot fingertip.

Moreover, consider two coordinate systems at the contact point, as shown in Fig. 2. Σ_{COi} is the contact point of the C_i coordinate system on the object with the *z* axis outward and normal to the object surface, and Σ_{CFi} is the contact point of the C_i coordinate system on the robot fingertip with the *z* axis outward and normal to the robot fingertip surface. The rotation matrix at the contact point is expressed as

$$\boldsymbol{R}_O{}^O\boldsymbol{R}_{COi}{}^{COi}\boldsymbol{R}_{CFi} = \boldsymbol{R}_{Fi}{}^{Fi}\boldsymbol{R}_{CFi}, \tag{3}$$

where ${}^O\boldsymbol{R}_{COi} \in \mathbb{R}^{3\times3}$ is a rotation matrix from Σ_O to Σ_{COi}, ${}^{COi}\boldsymbol{R}_{CFi} \in \mathbb{R}^{3\times3}$ is a rotation matrix from Σ_{COi} to Σ_{CFi}, and ${}^{Fi}\boldsymbol{R}_{CFi} \in \mathbb{R}^{3\times3}$ is a rotation matrix from Σ_{Fi} to Σ_{CFi}. The first-order time differentials of Eq. (3) are expressed as

$$\boldsymbol{\omega}_O \times \boldsymbol{R}_O{}^O\boldsymbol{R}_{COi}{}^{COi}\boldsymbol{R}_{CFi} + \boldsymbol{R}_O{}^O\dot{\boldsymbol{R}}_{COi}{}^{COi}\boldsymbol{R}_{CFi} + \boldsymbol{R}_O{}^O\boldsymbol{R}_{COi}{}^{COi}\dot{\boldsymbol{R}}_{CFi} = \boldsymbol{\omega}_{Fi} \times \boldsymbol{R}_{Fi}{}^{Fi}\boldsymbol{R}_{CFi} + \boldsymbol{R}_{Fi}{}^{Fi}\dot{\boldsymbol{R}}_{CFi}. \tag{4}$$

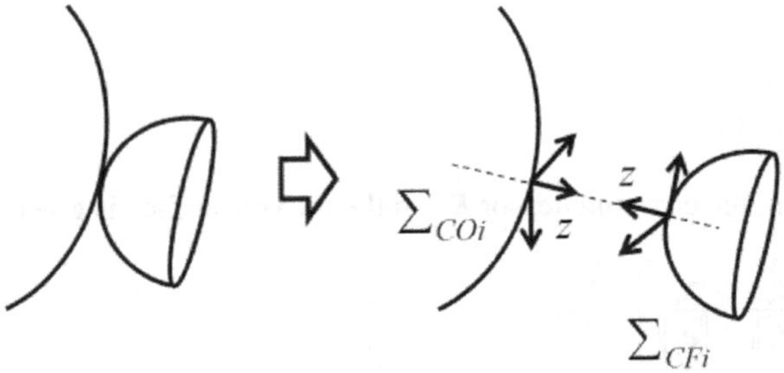

Figure 2. Contact coordinate systems at the contact point.

To consider the details of the relative motion between the object motion and that of the robot fingertip, consider the motion representation by curved surface coordinates. Using the mapping $\boldsymbol{c} : \mathbb{R}^2 \rightarrow \mathbb{R}^3$ from 2D to 3D, as shown in Fig. 3, a point on the 2D surface $\boldsymbol{\alpha} = [u \; v]^T$ is mapped to a point on the object's 3D surface, $\boldsymbol{p}$, as

$$\boldsymbol{p} = \boldsymbol{c}(\boldsymbol{\alpha}), \tag{5}$$

where it is assumed that mapping $\boldsymbol{c}$ is a diffeomorphic mapping, and that the coordinate system on the 2D surface is orthogonal for the sake of simplicity. In this case, the tangent plane of the object is a space spanned by two vectors, $\boldsymbol{c}_u = \frac{\partial \boldsymbol{c}}{\partial u}$ and $\boldsymbol{c}_v = \frac{\partial \boldsymbol{c}}{\partial v}$. The contact coordinate system $\Sigma_{Ci} = [\boldsymbol{x} \; \boldsymbol{y} \; \boldsymbol{z}]$ that consists of the two axes on the tangent plane and the outward normal orthogonal line is represented by

$$[\boldsymbol{x} \quad \boldsymbol{y} \quad \boldsymbol{z}] = \begin{bmatrix} \dfrac{\boldsymbol{c}_u}{\|\boldsymbol{c}_u\|} & \dfrac{\boldsymbol{c}_v}{\|\boldsymbol{c}_v\|} & \dfrac{\boldsymbol{c}_u \times \boldsymbol{c}_v}{\|\boldsymbol{c}_u \times \boldsymbol{c}_v\|} \end{bmatrix}. \tag{6}$$

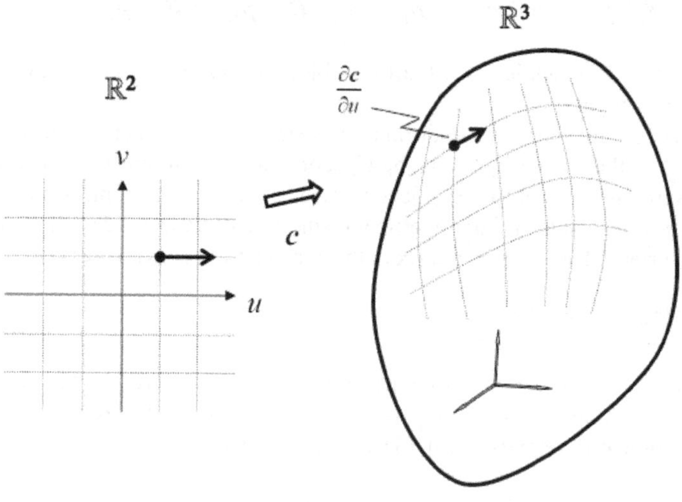

Figure 3. Mapping $\boldsymbol{c}$.

This is called a normalized Gauss frame. Here, the metric tensor $\boldsymbol{M}_C$ on the curved surface is given as

$$\boldsymbol{M}_c = \begin{bmatrix} \|\boldsymbol{c}_u\| & 0 \\ 0 & \|\boldsymbol{c}_v\| \end{bmatrix}. \tag{7}$$

In addition, the curvature tensor $\boldsymbol{K}_C$ on the curved surface is given as

$$\boldsymbol{K}_c = \begin{bmatrix} \boldsymbol{x}^T \\ \boldsymbol{y}^T \end{bmatrix} \begin{bmatrix} \dfrac{\boldsymbol{z}_u}{\|\boldsymbol{c}_u\|} & \dfrac{\boldsymbol{z}_v}{\|\boldsymbol{c}_v\|} \end{bmatrix}, \tag{8}$$

where $z_u = \dfrac{\partial z}{\partial u}$ and $z_v = \dfrac{\partial z}{\partial v}$. This curvature tensor is considered to be the magnitude of the fluctuation of the unit normal to the tangent plane. Twist on the curved surface is defined as the change rate of the curvature along the curved surface, and is represented by

$$\boldsymbol{T}_c = \boldsymbol{y}^T \left[\frac{\boldsymbol{x}_u}{\|\boldsymbol{c}_u\|} \quad \frac{\boldsymbol{x}_v}{\|\boldsymbol{c}_v\|} \right], \tag{9}$$

where $\boldsymbol{x}_u = \dfrac{\partial \boldsymbol{x}}{\partial u}$ and $\boldsymbol{x}_v = \dfrac{\partial \boldsymbol{x}}{\partial v}$. These are collectively referred to as the geometric parameters of the surface. When the object surface is a plane, then $z_u = z_v = \boldsymbol{x}_u = \boldsymbol{x}_v = \boldsymbol{0}$, and the following are obtained: $\boldsymbol{K}_C = \boldsymbol{0}$ and $\boldsymbol{T}_C = \boldsymbol{0}$.

Using the details of relative motion between the motion of the object and that of the robot fingertip, consider the motion of a rigid finger and a rigid object, both with a smooth curved surface, while a point of contact is maintained between them. Using the mapping $\boldsymbol{c}_{Fi} : \mathbb{R}^2 \rightarrow \mathbb{R}^3$ and $\boldsymbol{c}_{Oi} : \mathbb{R}^2 \rightarrow \mathbb{R}^3$ to represent the surface shapes of the object and the finger, and the 2D orthogonal surface coordinates $\boldsymbol{\alpha}_{Fi} = [u_{Fi}\ v_{Fi}]^T$ and $\boldsymbol{\alpha}_{Oi} = [u_{Oi}\ v_{Oi}]^T$, a point of the fingertip surface can be expressed by the finger coordinates and a point of the object surface by the object coordinates as

$${}^{Fi}\boldsymbol{p}_{FiCi} = \boldsymbol{c}_{Fi}(\boldsymbol{\alpha}_{Fi}),\ {}^{O}\boldsymbol{p}_{OCi} = \boldsymbol{c}_{Oi}(\boldsymbol{\alpha}_{Oi}). \tag{10}$$

The deviation of relative attitude between the contact coordinate system on the finger side $\Sigma_{CFi} = [\boldsymbol{x}_{CFi}\ \boldsymbol{y}_{CFi}\ \boldsymbol{z}_{CFi}]$ and that on the object side $\Sigma_{COi} = [\boldsymbol{x}_{COi}\ \boldsymbol{y}_{COi}\ \boldsymbol{z}_{COi}]$ is represented with the angle ϕ_i from $\boldsymbol{x}_{CFi}$ to $\boldsymbol{x}_{COi}$, which is called the angle of contact. Note that the z-axis in each coordinate is in the outward normal direction. Let us denote the relative translational velocity and relative angular velocity of the finger contact point when referring to the object contact point as

$${}^{CFi}\boldsymbol{v}_{Si} = [{}^{CFi}v_{Six}\quad {}^{CFi}v_{Siy}\quad {}^{CFi}v_{Siz}]^T = \boldsymbol{R}_{CFi}^T \left[\left(\dot{\boldsymbol{p}}_{Fi} + \boldsymbol{\omega}_{Fi} \times \boldsymbol{R}_{Fi}{}^{Fi}\boldsymbol{p}_{FiCi} \right) - \left(\dot{\boldsymbol{p}}_O + \boldsymbol{\omega}_O \times \boldsymbol{R}_O{}^{O}\boldsymbol{p}_{OCi} \right) \right] \tag{11}$$

and

$${}^{CFi}\boldsymbol{\omega}_{Si} = [{}^{CFi}\omega_{Six}\ {}^{CFi}\omega_{Siy}\ {}^{CFi}\omega_{Siz}]^T = \boldsymbol{R}_{CFi}^T(\boldsymbol{\omega}_{Fi} - \boldsymbol{\omega}_O), \tag{12}$$

respectively. Then, $\boldsymbol{q}_{Ci} = [\boldsymbol{\alpha}_{Fi}^T\ \boldsymbol{\alpha}_{Oi}^T\ \phi_i]^T \in \mathbb{R}^5$ is the contact coordinates on the two-object surface. Using Eqs. (7)–(9), the geometrical parameters for the contact coordinates of the object side and finger side are represented as $(\boldsymbol{M}_{CFi}, \boldsymbol{K}_{CFi}, \boldsymbol{T}_{CFi})$ and $(\boldsymbol{M}_{COi}, \boldsymbol{K}_{COi}, \boldsymbol{T}_{COi})$, respectively. From Eq. (2) and Eq. (4), the following theorem is verified (Montana 1988).

Theorem 1. (Equation of motion at the contact point)

The motion of contact coordinates are described as

$$\dot{\boldsymbol{q}}_{Ci} = \begin{bmatrix} \boldsymbol{M}_{CFi}^{-1}\left(\boldsymbol{K}_{CFi} + \tilde{\boldsymbol{K}}_{COi}\right)^{-1}\left(\begin{bmatrix} -{}^{CFi}\omega_{Siy} \\ {}^{CFi}\omega_{Six} \end{bmatrix} - \tilde{\boldsymbol{K}}_{COi}\begin{bmatrix} {}^{CFi}v_{Six} \\ {}^{CFi}v_{Siy} \end{bmatrix}\right) \\ \boldsymbol{M}_{COi}^{-1}\boldsymbol{R}_{\phi i}\left(\boldsymbol{K}_{CFi} + \tilde{\boldsymbol{K}}_{COi}\right)^{-1}\left(\begin{bmatrix} -{}^{CFi}\omega_{Siy} \\ {}^{CFi}\omega_{Six} \end{bmatrix} + \boldsymbol{K}_{CFi}\begin{bmatrix} {}^{CFi}v_{Six} \\ {}^{CFi}v_{Siy} \end{bmatrix}\right) \\ {}^{CFi}\omega_{Siz} + \boldsymbol{T}_{CFi}\boldsymbol{M}_{CFi}\dot{\boldsymbol{\alpha}}_{Fi} + \boldsymbol{T}_{COi}\boldsymbol{M}_{COi}\dot{\boldsymbol{\alpha}}_{Oi} \\ \\ \end{bmatrix} \tag{13}$$

$${}^{CFi}v_{Siz} = 0$$

where $\boldsymbol{R}_{\phi} \in \mathbb{R}^{2\times 2}$ *is a rotation matrix that represents the orientation of the finger contact coordinate system* Σ_{CFi} *in reference to the object contact point coordinate system* Σ_{COi}, *and given by*

$$\boldsymbol{R}_{\phi i} = \begin{bmatrix} \cos\phi_i & -\sin\phi_i \\ -\sin\phi_i & -\cos\phi_i \end{bmatrix}, \tag{14}$$

and $\tilde{\boldsymbol{K}}_{COi} = \boldsymbol{R}_{\phi i}\boldsymbol{K}_{COi}\boldsymbol{R}_{\phi i}$.

Equation of Motion of a Rigid Body

According to the Newton-Euler equation and Euler equations, the motion of a rigid body on which gravity acts in 3D space can be written as

$$\boldsymbol{f} = m_O\ddot{\boldsymbol{p}}_{Co} - m_O\tilde{\boldsymbol{g}}, \tag{15}$$

$$\boldsymbol{n} = \frac{d}{dt}\left(\tilde{\boldsymbol{I}}_O\boldsymbol{\omega}_O\right) = \tilde{\boldsymbol{I}}_O\dot{\boldsymbol{\omega}}_O + \boldsymbol{\omega}_O \times \tilde{\boldsymbol{I}}_O\boldsymbol{\omega}_O, \tag{16}$$

where $\boldsymbol{p}_{Co} \in \mathbb{R}^3$ is a position vector at the center of mass, $\boldsymbol{f} \in \mathbb{R}^3$ is a force applied to the object's center of mass, $\boldsymbol{n} \in \mathbb{R}^3$ is an applied moment to the object's center of mass, $\boldsymbol{\omega}_O \in \mathbb{R}^3$ is the angle velocity of the object, $\tilde{\boldsymbol{g}} \in \mathbb{R}^3$ is the gravitational acceleration, m_O is the mass of the object, and $\tilde{\boldsymbol{I}}_O \in \mathbb{R}^{3\times 3}$ is an inertia tensor around the object's center of mass. These are denoted within the reference coordinate system. It is not convenient to deal with the inertia tensor $\tilde{\boldsymbol{I}}_O$, since its value alters with changes in the orientation of the object. Consider the object coordinate system Σ_{Co}, which is attached to the center of mass. The inertia tensor of the rigid body referred to in the object coordinate system ${}^{Co}\tilde{\boldsymbol{I}}_O$ is represented by

$${}^{Co}\tilde{\boldsymbol{I}}_O = \boldsymbol{R}_{Co}^T\tilde{\boldsymbol{I}}_O\boldsymbol{R}_{Co}, \tag{17}$$

where $\boldsymbol{R}_{Co} \in \mathbb{R}^{3\times 3}$ is a rotation matrix that represents the transformational matrix of vector coordinates in the object coordinate system Σ_{Co} into the coordinates of the same vector in the reference coordinate system. The dynamic equation of the object using ${}^{Co}\tilde{\boldsymbol{I}}_O$ is easily handled, because ${}^{Co}\tilde{\boldsymbol{I}}_O$ is the constant. The relation $\tilde{\boldsymbol{I}}_O = \boldsymbol{R}_{Co}{}^{Co}\tilde{\boldsymbol{I}}_O\boldsymbol{R}_{Co}^T$ is derived from Eq. (17). By substituting this into Eq. (16), and then putting it together with Eq. (15), it is expressed as

$$\begin{bmatrix} f \\ n \end{bmatrix} = \begin{bmatrix} m_O \boldsymbol{I}_{3\times3} & \boldsymbol{0} \\ \boldsymbol{0} & \boldsymbol{R}_{Co}{}^{Co}\tilde{\boldsymbol{I}}_O \boldsymbol{R}_{Co}^T \end{bmatrix} \begin{bmatrix} \ddot{p} \\ \dot{\omega}_O \end{bmatrix} + \begin{bmatrix} \boldsymbol{0} \\ \omega_O \times \boldsymbol{R}_{Co}{}^{Co}\tilde{\boldsymbol{I}}_O \boldsymbol{R}_{Co}^T \omega_O \end{bmatrix} + \begin{bmatrix} -m_O \tilde{\boldsymbol{g}} \\ \boldsymbol{0} \end{bmatrix}. \tag{18}$$

Next, we set a new object coordinate system, Σ_O, that is transferred to a position different from the center of mass of the object, as shown in Fig. 4. Using the origin of the new object coordinate system as a position vector, $\boldsymbol{p}_O \in \mathbb{R}^3$, and an inertia tensor of the object around the origin of new object coordinate system, ${}^O\boldsymbol{I}_O$, that is referred to Σ_O, we can derive the equation of object motion using ${}^o\boldsymbol{I}_o$. The position of the origin of the center of mass is expressed as

$$p = p_O + \boldsymbol{R}_O{}^O\boldsymbol{s}_{OCo}, \tag{19}$$

where ${}^O\boldsymbol{s}_{OCo}$ is a position vector from the origin of Σ_O to the origin of Σ_{Co} with respect to Σ_O, and $\boldsymbol{R}_O \in \mathbb{R}^{3\times3}$ is a rotation matrix from Σ_p to Σ_O. Since ${}^O\boldsymbol{s}_{OCo}$ is a constant vector, the first- and second-order time differentials of Eq. (19) are expressed as

$$\dot{p} = \dot{p}_O + \omega_O \times \boldsymbol{R}_O{}^O\boldsymbol{s}_{OCo}, \tag{20}$$

$$\ddot{p} = \ddot{p}_O + \dot{\omega}_O \times \boldsymbol{R}_O{}^O\boldsymbol{s}_{OCo} + \omega_O \times \left(\omega_O \times \boldsymbol{R}_O{}^O\boldsymbol{s}_{OCo}\right). \tag{21}$$

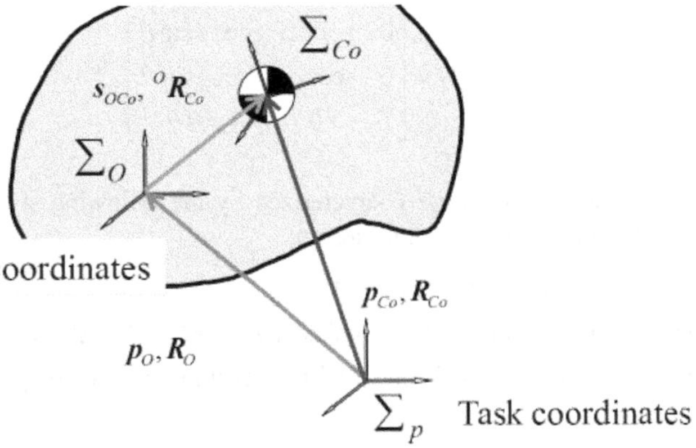

Figure 4. Coordinate system to be set for the target object.

The relationships between force and moment acting on the center of mass, and the equivalent force $\boldsymbol{f}_O$ and moment $\boldsymbol{n}_O$ acting on the origin of the coordinate system Σ_O are given by

$$\boldsymbol{f}_O = \boldsymbol{f}, \tag{22}$$

$$\boldsymbol{n}_O = \boldsymbol{n} + \boldsymbol{s}_{OCo} \times \boldsymbol{f}. \tag{23}$$

Moreover, according to the parallel axis theorem regarding the inertia tensor, there is a relationship as follows

$$ {}^{O}\boldsymbol{I}_O = {}^{Co}\tilde{\boldsymbol{I}}_O + m_O\left({}^{O}\boldsymbol{s}_{OCo}^{T}\,{}^{O}\boldsymbol{s}_{OCo}\boldsymbol{I}_{3\times3} - {}^{O}\boldsymbol{s}_{OCo}\,{}^{O}\boldsymbol{s}_{OCo}^{T}\right). \tag{24} $$

By substituting Eqs. (20)–(24) into Eq. (18), it is expressed as

$$ \boldsymbol{F}_O = \boldsymbol{M}_O\dot{\boldsymbol{v}}_O + \boldsymbol{C}_O\boldsymbol{v}_O + \boldsymbol{g}_O, \tag{25} $$

where $\boldsymbol{v}_O = \begin{bmatrix} \dot{\boldsymbol{p}}_O \\ \boldsymbol{\omega}_O \end{bmatrix}$, $\boldsymbol{M}_O = \begin{bmatrix} m_O\boldsymbol{I}_{3\times3} & m_O[\boldsymbol{R}_O{}^{O}\boldsymbol{s}_{OCo}\times]^T \\ m_O[\boldsymbol{R}_O{}^{O}\boldsymbol{s}_{OCo}\times] & \boldsymbol{R}_O{}^{O}\boldsymbol{I}_O\boldsymbol{R}_O^T \end{bmatrix}$,

$$ \boldsymbol{C}_O = \begin{bmatrix} \boldsymbol{0} & [(\boldsymbol{\omega}_O \times m_O\boldsymbol{R}_O{}^{O}\boldsymbol{s}_{OCo})\times] \\ \boldsymbol{0} & [\boldsymbol{R}_O{}^{O}\boldsymbol{I}_O\boldsymbol{R}_O^T\boldsymbol{\omega}_O\times] \end{bmatrix},\ \boldsymbol{g}_O = \begin{bmatrix} -m_O\tilde{\boldsymbol{g}} \\ -m_O[\boldsymbol{R}_O{}^{O}\boldsymbol{s}_{Oco}\times]\tilde{\boldsymbol{g}} \end{bmatrix},\ \boldsymbol{F}_O = \begin{bmatrix} \boldsymbol{f}_O \\ \boldsymbol{n}_O \end{bmatrix}. $$

Equation (25) is the equation of object motion established by the force and moment acting on the origin of the coordinate system Σ_O. When the orientation of the object is represented by the *zyz*-Euler angle $\boldsymbol{\eta}_O = [\phi\ \theta\ \varphi]^T$, the position and orientation is expressed as $\boldsymbol{r}_O = [\boldsymbol{p}_O^T\ \boldsymbol{\eta}_O^T]^T$, and its time derivative is expressed by $\dot{\boldsymbol{r}}_O = [\dot{\boldsymbol{p}}_O^T\ \dot{\boldsymbol{\eta}}_O^T]^T$. On the other hand, the object velocity can be expressed by $\boldsymbol{v}_O = [\dot{\boldsymbol{p}}_O^T\ \boldsymbol{\omega}_O^T]^T$ when the object angular velocity $\boldsymbol{\omega}_O$ is used. Both have a relation of

$$ \boldsymbol{v}_O = \boldsymbol{T}_O\,\dot{\boldsymbol{r}}_O, \tag{26} $$

where

$$ \boldsymbol{T}_O = \begin{bmatrix} \boldsymbol{I}_{3\times3} & \boldsymbol{0} \\ \boldsymbol{0} & \boldsymbol{\Pi} \end{bmatrix} \in \mathbb{R}^{6\times6},\ \boldsymbol{\Pi} = \begin{bmatrix} 0 & -\sin\phi & \cos\phi\sin\theta \\ 0 & \cos\phi & \sin\phi\sin\theta \\ 1 & 0 & \cos\theta \end{bmatrix}. \tag{27} $$

The dynamic Eq. (25) are characterized by the following structural properties, which are utilized in our controller design.

(P1) $\boldsymbol{M}_O$ is symmetric positive definite.

(P2) The suitable definition of $\boldsymbol{C}_O$ makes matrix $\dot{\boldsymbol{M}}_O - 2\boldsymbol{C}_O$ skew-symmetric.

(P3) The dynamic equation is linear with respect to the dynamic parameter vector, as follows:

$$ \boldsymbol{M}_O\dot{\boldsymbol{v}}_{Or} + \boldsymbol{C}_O\boldsymbol{v}_{Or} + \boldsymbol{g}_O = \boldsymbol{Y}_O(\boldsymbol{r}_O, \dot{\boldsymbol{r}}_O, \boldsymbol{v}_{Or}, \dot{\boldsymbol{v}}_{Or})\boldsymbol{\sigma}_O, \tag{28} $$

where $\boldsymbol{\sigma}_O \in \mathbb{R}^{\rho_O}$ is a dynamic parameter vector of the object, $\boldsymbol{Y}_O(\boldsymbol{r}_O, \dot{\boldsymbol{r}}_O, \boldsymbol{v}_{Or}, \dot{\boldsymbol{v}}_{Or}) \in \mathbb{R}^{6\times\rho_O}$ is a regressor with respect to the dynamic parameters $\boldsymbol{\sigma}_O$, and ρ_O is the number of the object dynamics parameter. The ρ_O is 10 if Eq. (25) is used, and 7 if Eq. (18) is used.

Equation of Motion of both Robot Fingers and Grasped Object

When a robotic hand with k fingers manipulates an object under a condition of rolling contact without slip (“without slip” means ${}^{CFi}v_{Six} = {}^{CFi}v_{Siy} = 0$ in Eq. [11]), the following relation on the rolling velocities of the contact position on both surfaces with respect to Σ_P, is obtained:

$$\boldsymbol{R}_O{}^O\dot{\boldsymbol{p}}_{OCi} = \boldsymbol{R}_{Fi}{}^{Fi}\dot{\boldsymbol{p}}_{FiCi}. \tag{29}$$

Therefore, the relation between the object velocity and angular velocity of the i-th robot finger is expressed as

$$\boldsymbol{W}_{Ci}^T\boldsymbol{v}_O = \boldsymbol{J}_{Ci}\dot{\boldsymbol{q}}_{Fi}, \tag{30}$$

where $\boldsymbol{W}_{Ci} = \begin{bmatrix} I_{3\times3} \\ [R_O{}^O p_{OCi}\times] \end{bmatrix} \in \mathbb{R}^{6\times3}$ is a grasp form matrix, and $\boldsymbol{J}_{Ci} \in \mathbb{R}^{3\times3}$ is a Jacobian matrix at the contact point. For all fingers, Eq. (30) is represented as follows:

$$\boldsymbol{W}_C^T\boldsymbol{v}_O = \boldsymbol{J}_C\dot{\boldsymbol{q}}_F, \tag{31}$$

where $\boldsymbol{W}_C = [\boldsymbol{W}_{C1} \cdots \boldsymbol{W}_{Ck}] \in \mathbb{R}^{6\times3k}$, $\boldsymbol{J}_C = block\ diag[\boldsymbol{J}_{C1} \ldots \boldsymbol{J}_{Ck}] \in \mathbb{R}^{3k\times3k}$, and $\dot{\boldsymbol{q}}_F = [\dot{\boldsymbol{q}}_{F1}^T \ldots \dot{\boldsymbol{q}}_{Fk}^T]^T \in \mathbb{R}^{3k}$. Furthermore, by defining $\dot{\boldsymbol{x}} = [\boldsymbol{v}_O^T\ \dot{\boldsymbol{q}}_F^T]^T$ and $G = [\boldsymbol{W}^T - \boldsymbol{J}_C]$, Eq. (31) can be represented in a more compact form as follows:

$$\boldsymbol{G}\dot{\boldsymbol{x}} = \boldsymbol{0}. \tag{32}$$

The Lagrangian of the target system is given by L. Using the Lagrange multipliers, the Lagrange's equation of motion is derived by applying the variational principle in such a way that

$$\int_{t_1}^{t_2}\left(\delta L + \boldsymbol{u}^T\delta\boldsymbol{x} - \boldsymbol{f}_C\boldsymbol{G}\delta\boldsymbol{x}\right)dt = 0, \tag{33}$$

where $\boldsymbol{f}_C = [\boldsymbol{f}_{C1} \cdots \boldsymbol{f}_{Ck}] \in \mathbb{R}^{3k}$ are the Lagrange multipliers corresponding to the contact force, and $\boldsymbol{u}$ is a non-conservative and external force. Let us define $\boldsymbol{u}$ as $[\boldsymbol{0}^T\ \boldsymbol{\tau}_F^T]^T$, where $\boldsymbol{\tau}_F \in \mathbb{R}^{3k}$ is the input joint torque. The dynamic equation of motion for both the object and the robot hand are given by

$$\boldsymbol{M}_O\dot{\boldsymbol{v}}_O + \boldsymbol{C}_O\boldsymbol{v}_O + \boldsymbol{g}_O = \boldsymbol{W}_C\boldsymbol{f}_C, \tag{34}$$

$$\boldsymbol{M}_F(\boldsymbol{q}_F)\ddot{\boldsymbol{q}}_F + \boldsymbol{C}_F(\boldsymbol{q}_F,\dot{\boldsymbol{q}}_F)\dot{\boldsymbol{q}}_F + \boldsymbol{g}_F(\boldsymbol{q}_F) = \boldsymbol{\tau}_F - \boldsymbol{J}_C^T\boldsymbol{f}_C, \tag{35}$$

where $\boldsymbol{M}_F \in \mathbb{R}^{3k\times3k}$ is the inertia matrix, $\boldsymbol{C}_F\dot{\boldsymbol{q}}_F \in \mathbb{R}^{3k}$ is the velocity square term consisting of centrifugal and Coriolis force, and $\boldsymbol{g}_F$ is the gravitational term. The left side of Eq. (34) is the same structure as right side of Eq. (25). In addition to (P1) – (P3), the dynamic equations (35) are characterized by the following structural properties, which are utilized in our controller design.

(P4) $\boldsymbol{M}_F$ is symmetric positive definite.

(P5) A suitable definition of $\boldsymbol{C}_F$ makes matrix $\dot{\boldsymbol{M}}_F - 2\boldsymbol{C}_F$ skew-symmetric.

(P6) The dynamic equation is linear with respect to the dynamic parameter vector, as follows:

$$\boldsymbol{M}_F\ddot{\boldsymbol{q}}_{Fr} + \boldsymbol{C}_F\dot{\boldsymbol{q}}_{Fr} + \boldsymbol{g}_F = \boldsymbol{Y}_F(\boldsymbol{q}_F,\dot{\boldsymbol{q}}_F,\dot{\boldsymbol{q}}_{Fr},\ddot{\boldsymbol{q}}_{Fr})\,\boldsymbol{\sigma}_F, \tag{36}$$

where $\sigma_F \in \mathbb{R}^{\rho_F}$ is a dynamic parameter vector of the robot hand, $Y_F(q_F, \dot{q}_F, \dot{q}_{Fr}, \ddot{q}_{Fr})$ $\mathbb{R}^{3k\times\rho_F}$ is a regressor with respect to the dynamic parameters σ_F, and ρ_F is the number of robot hand dynamic parameters.

ADAPTIVE CONTROL

Adaptive Control without a Force Sensor

The dynamic equations of the object and robot fingers contain dynamic parameters such as the mass and inertia tensor of the link between object and fingers. When these dynamic parameters are unknown, it is very difficult to precisely control the motion of the object. The purpose of control here is to find a control law such that the object motion follows the desired trajectory asymptotically. The adaptive control is designed in three steps:

(1) Calculate the desired external force, F_O^d, to be applied to the object using an estimated reference model of the object, which is generated by the desired trajectories of the object, r_O^d, $\dot{r}_O^d$, and $\ddot{r}_O^d$. It is assumed that the desired values are bounded and continuous.

(2) Compute the desired contact force f_C^d at the contact point using the desired external force.

(3) Compute an adaptive control law of the *i*-th robot finger using an estimated reference model of the robot finger.

The proposed adaptive coordinated control does not need the measurement of the force at the contact point. It only needs the measurements of position and velocity of the object and all robot fingers.

First, the reference velocity of the object $v_{Or} \in \mathbb{R}^6$ is defined as

$$v_{Or} = T_O(\dot{r}_O^d + \Lambda e_O), \tag{37}$$

where $e_O = r_O^d - r_O$ is the position and orientation error vector of the object, $T_O \in \mathbb{R}^{6\times6}$ is the matrix that converts the velocity of the Euler angles into angular velocity, and $\Lambda_O \in \mathbb{R}^{6\times6}$ is a weighting matrix that is positive and diagonal. Moreover, the residual velocity of the object, $s_O \in \mathbb{R}^6$, is defined as

$$s_O = v_O - v_{Or}. \tag{38}$$

Then, the relation of $s_O = -T_O(\dot{e}_O + \Lambda e_O)$ is obtained. The desired external force to be applied to the object, F_O^d, is generated from the reference model of the object using the following equation:

$$F_O^d = \hat{M}_O(r_O)\dot{v}_{Or} + \hat{C}_O(r_O, \dot{r}_O)v_{Or} + \hat{g}_O(r_O)$$

$$= Y_O(r_O, \dot{r}_O, v_{Or}, \dot{v}_{Or})\hat{\sigma}_O, \tag{39}$$

where $\hat{M}_O$, $\hat{C}_O$ and $\hat{g}_O$ are the estimates of M_O, C_O, and g_O using $\hat{\sigma}_O$. Now, an estimated law of the dynamic parameter vector of the object is given by

$$\dot{\hat{\sigma}}_O = -\Gamma_O Y_O^T(r_O, \dot{r}_O, v_{Or}, \dot{v}_{Or})s_O, \tag{40}$$

where, $\Gamma_O > 0$ is an adaptive gain matrix that is symmetric. The desired external force of the object is sequentially updated based on the estimation of the dynamic parameters of the object. Also, the reference acceleration of the object $\dot{v}_{Or} \in \mathbb{R}^6$ is given by

$$\dot{v}_{Or} = \dot{T}_O(\dot{r}_O^d + \Lambda e_O) + T_O(\ddot{r}_O^d + \Lambda \dot{e}_O). \tag{41}$$

Forces and moments at the contact points are in equilibrium with the external force. Therefore, the desired force at the contact point f_C^d is provided by the following equation:

$$f_C^d = W_C^+ F_O^d + \left(I_{3k\times 3k} - W_C^+ W_C\right) f_{\text{int}}^d . \tag{42}$$

The first term of Eq. (42) is the external force that has an effect on the object motion, and the second term is the internal force that does not have such an effect. The internal force that is bounded and uniformly continuous is a vector in the null space of W_C. Therefore, as long as it meets the contact conditions, the way of setting is unlimited.

The following relationship is established in the joint velocity of the i-th robot finger and the object velocity, as was seen in Eq. (30):

$$\dot{q}_{Fi} = J_{Ci}^{-1} W_{Ci}^T v_O. \tag{43}$$

From this relationship, the reference joint velocity of the i-th robot finger $\dot{q}_{Fri}$ is given by

$$\dot{q}_{Fri} = J_{Ci}^{-1} W_{Ci}^T v_{Or}. \tag{44}$$

Using the estimate $\hat{\sigma}_{Fi}$ of the dynamic parameter vector of the i-th robot finger σ_{Fi}, the reference model of the i-th robot finger is given by

$$\hat{M}_{Fi}(q_{Fi})\ddot{q}_{Fri} + \hat{C}_{Fi}(q_{Fi}, \dot{q}_{Fi})\dot{q}_{Fri} + \hat{g}_{Fi}(q_{Fi}) = Y_{Fi}(q_{Fi}, \dot{q}_{Fi}, \dot{q}_{Fri}, \ddot{q}_{Fri})\hat{\sigma}_{Fi}, \tag{45}$$

where $\hat{M}_{Fi}$, $\hat{C}_{Fi}$, and $\hat{g}_{Fi}$ are the estimates of M_{Fi}, C_{Fi}, and g_{Fi} using $\hat{\sigma}_{Fi}$. The control law of the i-th robot finger is given by

$$\tau_{Fi} = Y_{Fi}(q_{Fi}, \dot{q}_{Fi}, \dot{q}_{Fir}, \ddot{q}_{Fir})\hat{\sigma}_{Fi} + J_{Ci}^T f_{Ci}^d - K_{Fi} s_{Fi}, \tag{46}$$

where $K_{Fi} > 0 \in \mathbb{R}^{3\times 3}$ is a feedback gain matrix, and s_{Fi} is the residual between the actual velocity and the reference velocity defined by

$$s_{Fi} = \dot{q}_{Fi} - \dot{q}_{Fri}; \tag{47}$$

it is represented by

$$s_{Fi} = J_{Ci}^{-1} W_{Ci}^T s_O. \tag{48}$$

The adaptive law of the parameter estimate of the i-th robot finger is given by

$$\dot{\hat{\sigma}}_{Fi} = -\Gamma_{Fi} Y_{Fi}^T(q_{Fi}, \dot{q}_{Fi}, \dot{q}_{Fri}, \ddot{q}_{Fri})s_{Fi}, \tag{49}$$

where $\Gamma_{Fi} > 0$ is an adaptive gain matrix that is symmetric. On the right-hand side of Eq. (46), the first term is a feed-forward input term based on the estimated reference model, the second term is a feed-forward input term corresponding to the desired force at the contact point, and the third term is a feedback input term based on the object trajectory errors. It should be noted that the adaptive control does not require measurements of forces and moments at each contact point. The integrated control law and adaptive law can be written as follows:

$$\boldsymbol{\tau}_F = \boldsymbol{Y}_F(\boldsymbol{q}_F, \dot{\boldsymbol{q}}_F, \dot{\boldsymbol{q}}_{Fr}, \ddot{\boldsymbol{q}}_{Fr})\hat{\boldsymbol{\sigma}}_F + \boldsymbol{J}_C^T \boldsymbol{f}_C^d - \boldsymbol{K}_F \boldsymbol{s}_F \tag{50}$$

$$\dot{\hat{\boldsymbol{\sigma}}}_F = -\boldsymbol{\Gamma}_F \boldsymbol{Y}_F^T(\boldsymbol{q}_F, \dot{\boldsymbol{q}}_F, \dot{\boldsymbol{q}}_{Fr}, \ddot{\boldsymbol{q}}_{Fr})\boldsymbol{s}_F, \tag{51}$$

where $\hat{\boldsymbol{\sigma}}_F = [\hat{\boldsymbol{\sigma}}_{F1}^T \cdots \hat{\boldsymbol{\sigma}}_{Fk}^T]^T$ and $\boldsymbol{\Gamma}_F = block\ diag[\boldsymbol{\Gamma}_{F1} \cdots \boldsymbol{\Gamma}_{Fk}]$. If the grasping conditions (A1)–(A4) are assumed, the following theorem is proved for adaptive control without use of a force sensor.

Theorem 2. *Consider a rigid body that is grasped by k robot fingers, each with three DOFs. For the system in Eqs. (34), (35), and (31), given that Eq. (40) is the parameter estimation law of the object, Eq. (39) is the desired external force of the object, Eq. (42) is the desired force at the contact point, Eq. (48) is the residual, Eq. (51) is the parameter estimate law of the robotic fingers, and Eq. (50) is the control law, the closed-loop system is then asymptotic stable in the following sense:*

(1) $\boldsymbol{r}_O \rightarrow \boldsymbol{r}_O^d$ *and* $\dot{\boldsymbol{r}}_O \rightarrow \dot{\boldsymbol{r}}_O^d$ *as* $t \rightarrow \infty$

(2) *The force error at the contact point* $\boldsymbol{f}_C^d - \boldsymbol{f}_C$ *is bounded.*

The error dynamics of the object is given by

$$\begin{aligned}
\Delta \boldsymbol{F}_O &= \left(\hat{\boldsymbol{M}}_O(\boldsymbol{r}_O)\dot{\boldsymbol{v}}_{Or} + \hat{\boldsymbol{C}}_O(\boldsymbol{r}_O, \dot{\boldsymbol{r}}_O)\boldsymbol{v}_{Or} + \hat{\boldsymbol{g}}_O(\boldsymbol{r}_O)\right) - \left(\boldsymbol{M}_O(\boldsymbol{r}_O)\dot{\boldsymbol{v}}_O + \boldsymbol{C}_O(\boldsymbol{r}_O, \dot{\boldsymbol{r}}_O)\boldsymbol{v}_O + \boldsymbol{g}_O(\boldsymbol{r}_O)\right) \\
&= \left(\hat{\boldsymbol{M}}_O(\boldsymbol{r}_O)\dot{\boldsymbol{v}}_{Or} + \hat{\boldsymbol{C}}_O(\boldsymbol{r}_O, \dot{\boldsymbol{r}}_O)\boldsymbol{v}_{Or} + \hat{\boldsymbol{g}}_O(\boldsymbol{r}_O)\right) - \left(\boldsymbol{M}_O(\boldsymbol{r}_O)(\dot{\boldsymbol{s}}_O + \dot{\boldsymbol{v}}_{Or})\right. \\
&\quad \left. + \boldsymbol{C}_O(\boldsymbol{r}_O, \dot{\boldsymbol{r}}_O)(\boldsymbol{s}_O + \boldsymbol{v}_{Or}) + \boldsymbol{g}_O(\boldsymbol{r}_O)\right) \\
&= \boldsymbol{Y}(\boldsymbol{r}_O, \dot{\boldsymbol{r}}_O, \boldsymbol{v}_{Or}, \dot{\boldsymbol{v}}_{Or})(\hat{\boldsymbol{\sigma}}_O - \boldsymbol{\sigma}_O) - \boldsymbol{M}_O(\boldsymbol{r}_O)\dot{\boldsymbol{s}}_O - \boldsymbol{C}_O(\boldsymbol{r}_O, \dot{\boldsymbol{r}}_O)\boldsymbol{s}_O,
\end{aligned} \tag{52}$$

where $\Delta \boldsymbol{F}_O = \boldsymbol{F}_O^d - \boldsymbol{F}_O = \boldsymbol{W}_C \boldsymbol{f}_C^d - \boldsymbol{W}_C \boldsymbol{f}_C = \boldsymbol{W}_C \Delta \boldsymbol{f}_C$. The error dynamics of the robot fingers is given by

$$\boldsymbol{0} = -\boldsymbol{M}_F \dot{\boldsymbol{s}}_F - \boldsymbol{C}_F \boldsymbol{s}_F + \boldsymbol{Y}_F(\hat{\boldsymbol{\sigma}}_F - \boldsymbol{\sigma}_F) - \boldsymbol{K}_F \boldsymbol{s}_F + \boldsymbol{J}_C^T(\boldsymbol{f}_C^d - \boldsymbol{f}_C). \tag{53}$$

Theorem 2 can be proved using a candidate Lyapunov function given by

$$V = \frac{1}{2}(\boldsymbol{s}_O^T \boldsymbol{M}_O \boldsymbol{s}_O + \Delta \boldsymbol{\sigma}_O^T \boldsymbol{\Gamma}_O^{-1} \Delta \boldsymbol{\sigma}_O + \boldsymbol{s}_F^T \boldsymbol{M}_F \boldsymbol{s}_F + \Delta \boldsymbol{\sigma}_F^T \boldsymbol{\Gamma}_F^{-1} \Delta \boldsymbol{\sigma}_F), \tag{54}$$

where $\Delta \boldsymbol{\sigma}_O = \hat{\boldsymbol{\sigma}}_O - \boldsymbol{\sigma}_O$ and $\Delta \boldsymbol{\sigma}_F = \hat{\boldsymbol{\sigma}}_F - \boldsymbol{\sigma}_F$. A time derivative along the solution of the error equation gives as the following equation:

$$\dot{V} = \boldsymbol{s}_O^T\left(\boldsymbol{M}_O \dot{\boldsymbol{s}}_O + \frac{1}{2}\dot{\boldsymbol{M}}_O \boldsymbol{s}_O\right) + \Delta \dot{\boldsymbol{\sigma}}_O^T \boldsymbol{\Gamma}_O^{-1} \Delta \boldsymbol{\sigma}_O + \boldsymbol{s}_F^T\left(\boldsymbol{M}_F \dot{\boldsymbol{s}}_F + \frac{1}{2}\dot{\boldsymbol{M}}_F \boldsymbol{s}_F\right) + \Delta \dot{\boldsymbol{\sigma}}_F^T \boldsymbol{\Gamma}_F^{-1} \Delta \boldsymbol{\sigma}_F. \tag{55}$$

Substituting Eqs. (52), (53), (40), and (51) into (55), and using (P2) and (P5),

$$\dot{V} = \boldsymbol{s}_O^T(\boldsymbol{Y}_O\Delta\boldsymbol{\sigma}_O - \Delta\boldsymbol{F}_O) + \left(-\boldsymbol{\Gamma}_O\boldsymbol{Y}_O^T\boldsymbol{s}_O\right)^T\boldsymbol{\Gamma}_O^{-1}\Delta\boldsymbol{\sigma}_O + \boldsymbol{s}_F^T\left(\boldsymbol{Y}_F\Delta\boldsymbol{\sigma}_F - \boldsymbol{K}_F\boldsymbol{s}_F + \boldsymbol{J}_C^T\Delta\boldsymbol{f}_C\right)$$
$$+ \left(-\boldsymbol{\Gamma}_F\boldsymbol{Y}_F^T\boldsymbol{s}_F\right)^T\boldsymbol{\Gamma}_F^{-1}\Delta\boldsymbol{\sigma}_F$$
$$= -\boldsymbol{s}_O^T\Delta\boldsymbol{F}_O - \boldsymbol{s}_F^T\boldsymbol{K}_F\boldsymbol{s}_F + \boldsymbol{s}_F^T\boldsymbol{J}_C^T\Delta\boldsymbol{f}_C. \tag{56}$$

From the relationship in Eq. (48),

$$\dot{V} = -\boldsymbol{s}_F^T\boldsymbol{K}_F\boldsymbol{s}_F \le 0. \tag{57}$$

This shows that Eq. (54) is the Lyapunov function; hence, $\boldsymbol{s}_O$, $\boldsymbol{s}_F$, $\Delta\boldsymbol{\sigma}_O$, and $\Delta\boldsymbol{\sigma}_F$ are bounded. To prove asymptotic convergence using a Lyapunov-like lemma, it is necessary to show the uniform continuity of $\dot{V}$. The second-order time differentials of Eq. (54) is given by

$$\ddot{V} = -2\dot{\boldsymbol{s}}_F^T\boldsymbol{K}_F\boldsymbol{s}_F. \tag{58}$$

Therefore, it is necessary to show the boundedness of $\dot{\boldsymbol{s}}_F$.

Because $\boldsymbol{\sigma}_O$ and $\boldsymbol{\sigma}_F$ are constant, $\hat{\boldsymbol{\sigma}}_O$ and $\hat{\boldsymbol{\sigma}}_F$ are bounded. The boundedness of $\boldsymbol{T}_O$ leads to the boundness of $\boldsymbol{e}_O$ and $\dot{\boldsymbol{e}}_O$ by the following equation:

$$\boldsymbol{e}_O(t) = e^{-\Lambda t}\boldsymbol{e}_O(t) - \int_0^t e^{-\Lambda(t-\tau)}\boldsymbol{T}_O^{-1}\boldsymbol{s}_O(\tau)d\tau. \tag{59}$$

These results yield the boundedness of $\boldsymbol{r}_O$, $\dot{\boldsymbol{r}}_O$, and $\boldsymbol{v}_{Or}$ because of the boundedness of the desired trajectory. $\dot{\boldsymbol{T}}_O$ is bounded from the boundedness of $\dot{\boldsymbol{r}}_O$. Hence, $\dot{\boldsymbol{v}}_{Or}$ is bounded. These results yield $\boldsymbol{Y}_O(\boldsymbol{r}_O, \dot{\boldsymbol{r}}_O, \boldsymbol{v}_{Or}, \dot{\boldsymbol{v}}_{Or})$, and $\boldsymbol{F}_O^d$ are bounded. $\boldsymbol{f}_C^d$ is bounded from the boundedness of $\boldsymbol{F}_O^d$ and the second term of Eq. (42); $\dot{\boldsymbol{q}}_{Fr}$ is bounded from Eq. (44); $\dot{\boldsymbol{q}}_F$ is bounded from Eq. (47); and $\dot{\boldsymbol{J}}_C$ and $\dot{\boldsymbol{W}}_C$ are bounded from the boundedness of $\dot{\boldsymbol{q}}_F$ and $\boldsymbol{v}_O$. These results yield a $\ddot{\boldsymbol{q}}_{Fri}$ that is bounded. Then, $\boldsymbol{Y}_F(\boldsymbol{q}_F, \dot{\boldsymbol{q}}_F, \dot{\boldsymbol{q}}_{Fr}, \ddot{\boldsymbol{q}}_{Fr})$ is bounded.

Using the relationship $\Delta\boldsymbol{F}_O = \boldsymbol{W}_C\Delta\boldsymbol{f}_C$, from Eqs. (52) and (53), the following equation is derived.

$$\boldsymbol{W}_C\boldsymbol{J}_C^{-T}\left(\boldsymbol{M}_F\dot{\boldsymbol{s}}_F + \boldsymbol{C}_F\boldsymbol{s}_F - \boldsymbol{Y}_F\Delta\boldsymbol{\sigma}_F + \boldsymbol{K}_F\boldsymbol{s}_F\right) = \boldsymbol{Y}_O\Delta\boldsymbol{\sigma}_O - \boldsymbol{M}_O\dot{\boldsymbol{s}}_O - \boldsymbol{C}_O\boldsymbol{s}_O. \tag{60}$$

Moreover, the time derivative of Eq. (48) for all fingers is given by

$$\dot{\boldsymbol{s}}_F = \boldsymbol{J}_C^{-1}\dot{\boldsymbol{W}}_C^T\boldsymbol{s}_O + \boldsymbol{J}_C^{-1}\boldsymbol{W}_C^T\dot{\boldsymbol{s}}_O - \boldsymbol{J}_C^{-1}\dot{\boldsymbol{J}}_C\boldsymbol{s}_F. \tag{61}$$

Substituting the equation above into Eq. (60),

$$\left(\boldsymbol{M}_O + \boldsymbol{W}_C\boldsymbol{J}_C^{-T}\boldsymbol{M}_F\boldsymbol{J}_C^{-1}\boldsymbol{W}_O^T\right)\dot{\boldsymbol{s}}_O = \boldsymbol{Y}_O\Delta\boldsymbol{\sigma}_O + \boldsymbol{W}_C\boldsymbol{J}_C^{-T}\boldsymbol{Y}_F\Delta\boldsymbol{\sigma}_F$$
$$-\left(\boldsymbol{W}_C\boldsymbol{J}_C^{-T}\boldsymbol{M}_F\boldsymbol{J}_C^{-1}\boldsymbol{W}_O^T + \boldsymbol{C}_O\right)\boldsymbol{s}_O + \boldsymbol{W}_C\boldsymbol{J}_C^{-T}\left(\boldsymbol{M}_F\boldsymbol{J}_C^{-1}\dot{\boldsymbol{J}}_C - \boldsymbol{C}_F - \boldsymbol{K}_F\right)\boldsymbol{s}_F. \tag{62}$$

From $\boldsymbol{M}_O > 0$, $\boldsymbol{M}_F > 0$, and $\boldsymbol{W}_C\boldsymbol{J}_C^{-T}\boldsymbol{M}_F\boldsymbol{J}_C^{-1}\boldsymbol{W}_O^T > 0$,

$$\boldsymbol{M}_O + \boldsymbol{W}_C\boldsymbol{J}_C^{-T}\boldsymbol{M}_F\boldsymbol{J}_C^{-1}\boldsymbol{W}_O^T > 0. \tag{63}$$

Therefore, $\dot{s}_O$ is bounded because of the boundedness of right-hand side of Eq. (62). This result yield an $\dot{s}_F$ that is bounded. The boundedness of $\dot{s}_F$ yields a bounded $f_C^d - f_C$, and $\dot{V}$ is uniformly continuous. It is shown that $\dot{V} \to 0$ as $t \to \infty$ from the Lyapunov-like Lemma. This implies that $s_F \to 0$ as $t \to \infty$. Using Eq. (48), $s_O \to 0$ as $t \to \infty$. This result yields $e_O \to \boldsymbol{0}$ and $\dot{e}_O \to \boldsymbol{0}$ as $t \to \infty$. These results yield $r_O \to r_O^d$ and $\dot{r}_O \to \dot{r}_O^d$ as $t \to \infty$.

The convergence to the desired trajectory does not guarantee that the estimated parameter $\hat{\sigma}(t)$ will converge to the true parameter σ, but the estimated parameter converges to a constant $\sigma(\infty)$ asymptotically. If Y satisfies the *persistently exciting condition* and is uniformly continuous, the estimate parameter will converge to the true value asymptotically (Anderson 1977). By a *persistently exciting condition* on Y, we mean that there exist strictly positive constants α and T such that for any $t > 0$,

$$\int_t^{t+T} Y^T Y d\tau \geq \alpha I. \tag{64}$$

From Eq. (62), it can be shown that $\dot{s}_O$ is uniformly continuous. Hence, $\dot{s}_F$ is uniformly continuous also. Using Barbalat's lemma, $\dot{s}_O \to 0$ and $\dot{s}_F \to 0$ as $t \to \infty$. If $s_O \to 0$, $s_F \to 0$, $\dot{s}_O \to 0$, and $\dot{s}_F \to 0$, then Eq. (60) is represented as

$$\tilde{Y} \Delta\tilde{\sigma} = \boldsymbol{0}, \tag{65}$$

where $\tilde{Y} = [Y_O \;\; W_C J_C^{-T} Y_F]$ and $\Delta\tilde{\sigma} = [\Delta\sigma_O^T \;\; \Delta\sigma_F^T]^T$. Now, $\tilde{Y}$ is also uniformly continuous, and $\Delta\tilde{\sigma}$ is constant because $s_O \to 0$ and $s_F \to 0$. By multiplying $\tilde{Y}^T$ from the left-hand side of Eq. (65) and integrating the equation for a period of time T, we get:

$$\int_t^{t+T} \tilde{Y}^T \tilde{Y} d\tau \Delta\tilde{\sigma} = \boldsymbol{0}. \tag{66}$$

If Eq. (64) is satisfied, then $\Delta\tilde{\sigma} \to \boldsymbol{0}$ as $t \to \infty$, and $\hat{\sigma}_F \to \sigma_F$ and $\hat{\sigma}_O \to \sigma_O$ as $t \to \infty$. Moreover, these results yield $\Delta F_O \to \boldsymbol{0}$ and $\Delta f_C \to \boldsymbol{0}$ as $t \to \infty$.

Adaptive Control using a Force Sensor

Consider an adaptive control law that uses a force sensor in which the contact force converges to the desired contact force asymptotically. The definition of the reference velocity, v_{Or}, and the residual velocity, s_O, of the object are used for Eq. (37) and Eq. (38), respectively. The vector of contact force error,

$$\Delta f_{Ci} = f_{Ci}^d - f_{Ci}, \tag{67}$$

is passed through a low-pass filter given by

$$\dot{v}_{Ci} + \alpha_C v_{Ci} = \alpha_C \Delta f_{Ci}, \tag{68}$$

where $\alpha_C > 0$ is a design parameter of the filter, v_{Ci} is an output signal passed through the filter, and its continuity is guaranteed. The reference model of the object is given as follows on behalf of Eq. (39):

$$\begin{aligned} F_O^d &= \hat{M}_O(r_O)\dot{v}_{Or} + \hat{C}_O(r_O, \dot{r}_O)v_{Or} + \hat{g}_O(r_O) - K_O s_O \\ &= Y_O(r_O, \dot{r}_O, v_{Or}, \dot{v}_{Or})\hat{\sigma}_O - K_O s_O, \end{aligned} \tag{69}$$

where $\boldsymbol{K}_O > 0$ is a symmetric feedback gain matrix. Furthermore, the reference angular velocity of the i-th robot finger, $\dot{\boldsymbol{q}}_{Fri}$, is defined as the following formula, instead of as Eq. (44), and the feedback terms of the force error and integral force errors are added.

$$\dot{\boldsymbol{q}}_{Fri} = \boldsymbol{J}_{Ci}^{-1}\left(\boldsymbol{W}_{Ci}^{T}\boldsymbol{v}_{Or} + \boldsymbol{\Omega}_{Fi}\boldsymbol{v}_{Ci} + \boldsymbol{\Psi}_{Fi}\boldsymbol{\eta}_{Ci}\right), \tag{70}$$

where $\boldsymbol{\eta}_{Ci}$ is an integral of the contact force error defined by

$$\boldsymbol{\eta}_{Ci} = \int_0^t \Delta\boldsymbol{f}_{Ci} dt. \tag{71}$$

In addition, $\boldsymbol{\Omega}_{Fi} > 0$ and $\boldsymbol{\Psi}_{Fi} > 0$ are the symmetric feedback gain matrices of force error and integral force error, respectively. Then, the residuals defined by Eq. (47) are represented as

$$\boldsymbol{s}_{Fi} = \boldsymbol{J}_{Ci}^{-1}\left(\boldsymbol{W}_{Ci}^{T}\boldsymbol{s}_{O} - \boldsymbol{\Omega}_{Fi}\boldsymbol{v}_{Ci} - \boldsymbol{\Psi}_{Fi}\boldsymbol{\eta}_{Ci}\right). \tag{72}$$

The control law of the i-th robot finger is given as follows on behalf of Eq. (46):

$$\boldsymbol{\tau}_{Fi} = \boldsymbol{Y}_{Fi}(\boldsymbol{q}_{Fi}, \dot{\boldsymbol{q}}_{Fi}, \dot{\boldsymbol{q}}_{Fri}, \ddot{\boldsymbol{q}}_{Fri})\hat{\boldsymbol{\sigma}}_{Fi} + \boldsymbol{J}_{Ci}^{T}\boldsymbol{f}_{Ci}^{d} - \boldsymbol{K}_{Fi}\boldsymbol{s}_{Fi} + \beta\boldsymbol{J}_{Ci}^{T}\Delta\boldsymbol{f}_{Ci} \tag{73}$$

It is noted that $\boldsymbol{s}_{Fi}$ contains a feedback term of the contact force error.

The control law and estimation law that are integrated from the i-th finger to the k-th finger are represented as

$$\boldsymbol{\tau}_{F} = \boldsymbol{Y}_{F}(\boldsymbol{q}_{F}, \dot{\boldsymbol{q}}_{F}, \dot{\boldsymbol{q}}_{Fr}, \ddot{\boldsymbol{q}}_{Fr})\hat{\boldsymbol{\sigma}}_{F} + \boldsymbol{J}_{C}^{T}\boldsymbol{f}_{C}^{d} - \boldsymbol{K}_{F}\boldsymbol{s}_{F} + \beta\boldsymbol{J}_{C}^{T}\Delta\boldsymbol{f}_{C}, \tag{74}$$

$$\dot{\hat{\boldsymbol{\sigma}}}_{F} = -\boldsymbol{\Gamma}_{F}\boldsymbol{Y}_{F}^{T}(\boldsymbol{q}_{F}, \dot{\boldsymbol{q}}_{F}, \dot{\boldsymbol{q}}_{Fr}, \ddot{\boldsymbol{q}}_{Fr})\boldsymbol{s}_{F}. \tag{75}$$

The following theorem is established if the grasping conditions (A1)–(A4) are assumed (Ueki et al. 2009).

Theorem 3. *Consider a rigid body that is grasped by k robot fingers, each with three DOFs. For the system in Eqs. (34), (35), and (31), given that Eq. (40) is the parameter estimation law of the object, Eq. (69) is the desired external force of the object, Eq. (42) is the desired force at the contact point, Eq. (61) is the residual, Eq. (75) is the parameter estimate law of the robotic fingers, and Eq. (74) is the control law, β is sufficiently large. The closed-loop system is then asymptotic stable in the following sense:*

(1) $\boldsymbol{r}_O \rightarrow \boldsymbol{r}_O^d$ *and* $\dot{\boldsymbol{r}}_O \rightarrow \dot{\boldsymbol{r}}_O^d$ *as* $t \rightarrow \infty$
(2) $\boldsymbol{f}_C \rightarrow \boldsymbol{f}_C^d$ *as* $t \rightarrow \infty$.

The error dynamics of the object is given by

$$\Delta\boldsymbol{F}_O = \boldsymbol{Y}(\boldsymbol{r}_O, \dot{\boldsymbol{r}}_O, \boldsymbol{v}_{Or}, \dot{\boldsymbol{v}}_{Or})\Delta\boldsymbol{\sigma}_O - \boldsymbol{K}_O\boldsymbol{s}_O - \boldsymbol{M}_O(\boldsymbol{r}_O)\dot{\boldsymbol{s}}_O - \boldsymbol{C}_O(\boldsymbol{r}_O, \dot{\boldsymbol{r}}_O)\boldsymbol{s}_O. \tag{76}$$

The error dynamics of the robot fingers is given by

$$\boldsymbol{0} = -\boldsymbol{M}_F\dot{\boldsymbol{s}}_F - \boldsymbol{C}_F\boldsymbol{s}_F + \boldsymbol{Y}_F\,(\hat{\boldsymbol{\sigma}}_F - \boldsymbol{\sigma}_F) - \boldsymbol{K}_F\boldsymbol{s}_F + \beta_{+1}\boldsymbol{J}_C^T\Delta\boldsymbol{f}_C. \tag{77}$$

where $\beta_{+1} = \beta + 1$. Theorem 3 can be proved using a candidate Lyapunov function given by

$$V = \frac{1}{2}\left\{\beta_{+1}\left(s_O^T M_O s_O + \Delta\sigma_O^T \Gamma_O^{-1}\Delta\sigma_O + \eta_C^T \Psi_F \eta_C + \frac{1}{\alpha_C} v_C^T \Omega_F v_C\right) + s_F^T M_F s_F + \Delta\sigma_F^T \Gamma_F^{-1}\Delta\sigma_F\right\}. \tag{78}$$

A time derivative along the solution of the error equation gives as the following equation:

$$\dot{V} = -\beta_{+1}\left(s_O^T K_O s_O + v_C^T \Omega_F v_C\right) - s_F^T K_F s_F \le 0. \tag{79}$$

This shows that Eq. (78) is the Lyapunov function; hence, s_O, s_F, $\Delta\sigma_O$, $\Delta\sigma_F$, η_C, and v_C are bounded. Therefore, by an analysis similar to that of the case without a force sensor, $\hat{\sigma}_O$, $\hat{\sigma}_F$, e_O, $\dot{e}_O$, r_O, $\dot{r}_O$, v_{Or}, $\dot{T}_O$, $\dot{v}_{Or}$, $Y_O(r_O, \dot{r}_O, v_{Or}, \dot{v}_{Or})$, F_O^d, f_C^d, $\dot{q}_{Fr}$, $\dot{q}_F$, $\dot{J}_C$, and $\dot{W}_C$ are bounded. In using the force sensor, $\ddot{q}_{Fr}$ contains Δf_C. Therefore, the proof from here is different from that of the case of without a force sensor.

The time derivative of Eq. (72) for all fingers is given by

$$\dot{s}_F = J_C^{-1}\left(\dot{W}_C^T s_O + W_C^T \dot{s}_O + \alpha_C \Omega_F v_C - \left(\alpha_C \Omega_F + \Psi_F\right)\Delta f_C - \dot{J}_C s_F\right). \tag{80}$$

Equation (76) is represented as

$$\dot{s}_O = M_O^{-1}\left(Y\Delta\sigma_O - W\Delta f_C - K_O s_O - C_O s_O\right). \tag{81}$$

The time derivative of Eq. (70) is represented as

$$\ddot{q}_{Fr} = a_F + J_C^{-1}\left(\alpha_C \Omega_F + \Psi_F\right)\Delta f_C, \tag{82}$$

where $a_F = J_C^{-1}(W_C^T \dot{v}_{Or} + \dot{W}_C^T v_{Or} - \alpha_C \Omega_F v_C - \dot{J}_C \dot{q}_{Fr})$. Substituting Eqs. (80), (81), and (82) into Eq. (77),

$$\begin{aligned} &Y_F(q_F, \dot{q}_F, \dot{q}_{Fr}, a_F)\,\Delta\sigma_F + \Delta M_F J_C^{-1}\left(\alpha_C \Omega_F + \Psi_F\right)\Delta f_C - K_F s_F + \beta_{+1} J_C^T \Delta f_C \\ &= M_F J_C^{-1} W_C^T M_O^{-1}\left\{Y_O \Delta\sigma_O - W_O \Delta f_C - \left(K_O + C_O\right)s_O\right\} \\ &\quad + M_F J_C^{-1}\left(\dot{W}_O^T s_O + \alpha_C \Omega_F v_C - \left(\alpha_C \Omega_F + \Psi_F\right)\Delta f_C - \dot{J}_C s_F\right) + C_F s_F. \end{aligned} \tag{83}$$

By multiplying $J_C M_F^{-1}$ from the left-hand side of Eq. (83), we get:

$$A\Delta f_C = b, \tag{84}$$

where

$$A = \left(J_C M_F^{-1} \Delta M_F J_C^{-1} + I_{3k\times 3k}\right)\left(\alpha_C \Omega_F + \Psi_F\right) + \beta_{+1} J_C M_F^{-1} J_C^T + W_C^T M_O^{-1} W_O, \tag{85}$$

$$\begin{aligned} b &= J_C M_F^{-1}\left\{-Y_F(q_F, \dot{q}_F, \dot{q}_{Fr}, a_F)\,\Delta\sigma_F + \left(K_F + C_F\right)s_F\right\} \\ &\quad + W_C^T M_O^{-1}\left\{Y_O \Delta\sigma_O - \left(K_O + C_O\right)s_O\right\} + \left(\dot{W}_O^T s_O + \alpha_C \Omega_F v_C - \dot{J}_C s_F\right). \end{aligned} \tag{86}$$

When β_{+1} is sufficiently large and $(\alpha_C \boldsymbol{\Omega}_F + \boldsymbol{\Psi}_F)$ is sufficiently small, the matrix $\boldsymbol{A}$ approaches $\beta_{+1} \boldsymbol{J}_C \boldsymbol{M}_F^{-1} \boldsymbol{J}_C^T + \boldsymbol{W}_C^T \boldsymbol{M}_O^{-1} \boldsymbol{W}_O$, which is nonsingular. In this time, $\Delta \boldsymbol{f}_C$ is bounded because of the boundedness of $\boldsymbol{b}$. This result yields $\boldsymbol{f}_C$, $\Delta \boldsymbol{F}_O$, $\dot{\boldsymbol{s}}_O$, $\dot{\boldsymbol{v}}_O$, $\dot{\boldsymbol{v}}_C$, $\ddot{\boldsymbol{q}}_{Fr}$, and $\dot{\boldsymbol{s}}_F$ that are bounded, sequentially.

Differentiating Eq. (79) with respect to time is given as

$$\ddot{V} = -2\beta_{+1}\left(\dot{\boldsymbol{s}}_O^T \boldsymbol{K}_O \boldsymbol{s}_O + \dot{\boldsymbol{v}}_C^T \boldsymbol{\Omega}_F \boldsymbol{v}_C\right) - 2\dot{\boldsymbol{s}}_F^T \boldsymbol{K}_F \boldsymbol{s}_F. \tag{87}$$

$\ddot{V}$ is bounded because of the boundedness of $\boldsymbol{s}_O$, $\dot{\boldsymbol{s}}_O$, $\boldsymbol{v}_C$, $\dot{\boldsymbol{v}}_C$, $\boldsymbol{s}_F$, and $\dot{\boldsymbol{s}}_F$. This means that $\dot{V}$ is uniformly continuous. It is shown that $\dot{V} \to 0$ as $t \to \infty$ from the Lyapunov-like Lemma. This implies that $\boldsymbol{s}_O \to \boldsymbol{0}$, $\boldsymbol{v}_C \to \boldsymbol{0}$, and $\boldsymbol{s}_F \to \boldsymbol{0}$ as $t \to \infty$. These results yield $\boldsymbol{e}_O \to \boldsymbol{0}$ and $\dot{\boldsymbol{e}}_O \to \boldsymbol{0}$ as $t \to \infty$. These results in turn yield $\boldsymbol{r}_O \to \boldsymbol{r}_O^d$ and $\dot{\boldsymbol{r}}_O \to \dot{\boldsymbol{r}}_O^d$ as $t \to \infty$.

Moreover, these result can lead to the uniform continuity of $\boldsymbol{A}$ and $\boldsymbol{b}$. Therefore, $\Delta \boldsymbol{f}_C$ is uniformly continuous, and $\dot{\boldsymbol{s}}_O$, $\dot{\boldsymbol{s}}_F$, and $\dot{\boldsymbol{v}}_C$ are uniformly continuous because of the uniform continuity of $\Delta \boldsymbol{f}_C$. Using Barbalat's lemma, $\dot{\boldsymbol{s}}_O \to \boldsymbol{0}$, $\dot{\boldsymbol{s}}_F \to \boldsymbol{0}$, and $\dot{\boldsymbol{v}}_C \to \boldsymbol{0}$ as $t \to \infty$. As a result, $\Delta \boldsymbol{f}_C \to \boldsymbol{0}$ as $t \to \infty$.

The asymptotic stability can be shown by the Lyapunov-like lemma. However, the estimated parameter convergence to the true value is not guaranteed. If $\boldsymbol{Y}_i$ $(i = O, F)$ satisfy the persistently exciting condition, that is, there exists strictly positive constants $\alpha > 0$ and T such that for any $t > 0$,

$$\int_t^{t+T} \boldsymbol{Y}_i^T \boldsymbol{Y}_i d\tau \geq \alpha \boldsymbol{I}_i, \quad (i = O, F), \tag{88}$$

where $\boldsymbol{I}_i$ $(i = O, F)$ is the identity matrix of appropriate size, then the parameter error $\Delta \boldsymbol{\sigma}_i$ $(i = O, F)$ will converge to zero exponentially.

Adaptive Control at a Sliding Contact

Consider control when the robot fingers are in point contact with an object; some fingers are in slide-rolling contact on the object surface, while the rest are in twist-rolling contact. In the following, given k robot fingers, the number of fingers in sliding contact is s; it is assumed that the each finger is has three DOF, where the conditions $m = 3k\text{-}2s > 6$ and the grasping conditions (A1)–(A4) are satisfied. Consider that the control of the finger without sliding contact is the same as that with adaptive control using a force sensor in Sec. 3.2, and that control of the finger(s) with sliding contact is such that the desired velocity and position of the finger(s) are expressed with the curved surface coordinates.

First, consider the relationship between the sliding velocity and the velocity with the surface coordinate variable. The motion of contact coordinates Eq. (13) are represented in another form as follows:

$$\boldsymbol{R}_{\phi i} \boldsymbol{M}_{COi} \dot{\boldsymbol{\alpha}}_{Oi} = \begin{bmatrix} {}^{CFi}v_{Six} \\ {}^{CFi}v_{Siy} \end{bmatrix} + \boldsymbol{M}_{CFi} \dot{\boldsymbol{\alpha}}_{Fi}, \tag{89}$$

$$-\boldsymbol{R}_{\phi i} \boldsymbol{K}_{COi} \boldsymbol{M}_{COi} \dot{\boldsymbol{\alpha}}_{Oi} = -\begin{bmatrix} -{}^{CFi}\omega_{Siy} \\ {}^{CFi}\omega_{Six} \end{bmatrix} + \boldsymbol{K}_{CFi} \boldsymbol{M}_{CFi} \dot{\boldsymbol{\alpha}}_{Oi}. \tag{90}$$

Also, Eq. (11) is represented in another form, taking into account ${}^{CFi}v_{Siz} = 0$ as follows:

$$\dot{\boldsymbol{p}}_{Si} = \boldsymbol{D}_i \begin{bmatrix} {}^{CFi}v_{Six} \\ {}^{CFi}v_{Siy} \end{bmatrix} = \boldsymbol{J}_{Ci}\dot{\boldsymbol{q}}_{Fi} - \boldsymbol{W}_{Ci}^T \boldsymbol{v}_O, \tag{91}$$

where $\boldsymbol{D}_i = \boldsymbol{R}_{CFi}\begin{bmatrix} 1 & 0 \\ 0 & 1 \\ 0 & 0 \end{bmatrix}$, $\boldsymbol{R}_{CFi}$ is a rotation matrix from Σ_P to Σ_{CFi}, and $\dot{\boldsymbol{p}}_{Si}$ is a sliding velocity with respect to Σ_P. The rolling angular velocity can be represented by

$$\begin{bmatrix} -{}^{CFi}\omega_{Siy} \\ {}^{CFi}\omega_{Six} \end{bmatrix} = \boldsymbol{C}_i^T\left(\boldsymbol{J}_{\omega i}\dot{\boldsymbol{q}}_{Fi} - \boldsymbol{\omega}_O\right), \tag{92}$$

where $\boldsymbol{J}_{\omega i} \in \mathbb{R}^{3\times3}$ is a Jacobian matrix for angular velocity, $\boldsymbol{C}_i = \boldsymbol{R}_{CFi}\begin{bmatrix} 0 & 1 \\ -1 & 0 \\ 0 & 0 \end{bmatrix}$. Obtaining $\dot{\boldsymbol{q}}_{Fi}$ from Eq. (91), and substituting it with $\boldsymbol{v}_O = [\dot{\boldsymbol{p}}_O^T\ \boldsymbol{\omega}_O^T]^T$ into Eq. (92), it can be expressed by

$$\begin{bmatrix} -{}^{CFi}\omega_{Siy} \\ {}^{CFi}\omega_{Six} \end{bmatrix} = \boldsymbol{C}_i^T\left(\boldsymbol{J}_{\omega i}\boldsymbol{J}_{Ci}^{-1}\boldsymbol{D}_i\begin{bmatrix} {}^{CFi}v_{Six} \\ {}^{CFi}v_{Siy} \end{bmatrix} + \left(\boldsymbol{J}_{\omega i}\boldsymbol{J}_{Ci}^{-1}\boldsymbol{W}_{Ci}^T - [\boldsymbol{0}\quad \boldsymbol{I}_{3\times3}]\right)\boldsymbol{v}_O\right). \tag{93}$$

By substituting Eqs. (89) and (92) into Eq. (90), the following equation is obtained:

$$\dot{\boldsymbol{p}}_{Si} = \boldsymbol{A}_i\dot{\boldsymbol{\alpha}}_{Oi} + \boldsymbol{B}_i\boldsymbol{v}_o, \tag{94}$$

where

$$\boldsymbol{A}_i = \boldsymbol{D}_i\left(\boldsymbol{C}_i^T\boldsymbol{J}_{\omega i}\boldsymbol{J}_{Ci}^{-1}\boldsymbol{D}_i + \boldsymbol{K}_{CFi}\right)^{-1}\left(\boldsymbol{R}_{\phi i}\boldsymbol{K}_{COi}\boldsymbol{R}_{\phi i} + \boldsymbol{K}_{CFi}\right)\boldsymbol{R}_{\phi i}\boldsymbol{M}_{COi}, \tag{95}$$

$$\boldsymbol{B}_i = \boldsymbol{D}_i\left(\boldsymbol{C}_i^T\boldsymbol{J}_{\omega i}\boldsymbol{J}_{Ci}^{-1}\boldsymbol{D}_i + \boldsymbol{K}_{CFi}\right)^{-1}\boldsymbol{C}_i^T\left([\boldsymbol{0}\quad \boldsymbol{I}_{3\times3}] - \boldsymbol{J}_{\omega i}\boldsymbol{J}_{Ci}^{-1}\boldsymbol{W}_{Ci}^T\right). \tag{96}$$

Equation (94) represents the relationship between the sliding velocity, the velocity of the surface coordinate variables, and the object velocity.

Next, consider the friction force on the tangent plane at the contact point. Let us assume that the i-th robot finger is sliding on the object surface at contact point C_i. At this time, the contact force $\boldsymbol{f}_{Ci} \in \mathbb{R}^3$ can be decomposed into a component of the normal direction, $\boldsymbol{f}_{ni}$, and a dynamic friction force on the tangent plane, $\boldsymbol{f}_{ti}$, as follows:

$$\boldsymbol{f}_{Ci} = \boldsymbol{f}_{ni} + \boldsymbol{f}_{ti}. \tag{97}$$

When the unit normal vector at the contact point C_i is expressed by $\boldsymbol{n}_i$, then $\boldsymbol{f}_{ni} = \boldsymbol{n}_i\boldsymbol{n}_i^T\boldsymbol{f}_{Ci}$. It is assumed that the dynamic friction force is represented by

$$\boldsymbol{f}_{ti} = \mu_{Ci}\boldsymbol{n}_i^T\boldsymbol{f}_{Ci}\frac{\dot{\boldsymbol{p}}_{Si}}{\|\dot{\boldsymbol{p}}_{Si}\|} + \mu_{Di}\dot{\boldsymbol{p}}_{Si}, \tag{98}$$

where μ_{Ci} is the Coulomb dynamic friction coefficient, and μ_{Di} is a viscous friction coefficient. Equation (98) is linear regarding the friction parameters μ_{Ci} and μ_{Di}; therefore, it can be represented as

$$\boldsymbol{f}_{ti} = \boldsymbol{Y}_{ti}(\dot{\boldsymbol{p}}_{Si}, \boldsymbol{n}_i, \boldsymbol{f}_{Ci})\boldsymbol{\mu}_i, \tag{99}$$

where $\boldsymbol{\mu}_i = [\mu_{Ci}\ \mu_{Di}]^T$.

In order to compensate for the friction force at the sliding fingers, let us generate the desired friction force on the tangent plane based on an estimate of the friction parameter. First, we define a reference velocity of sliding motion by

$$\boldsymbol{v}_{Sri} = \boldsymbol{A}_i\left(\dot{\alpha}_{Oi}^d + \boldsymbol{\Lambda}_S\left(\alpha_{Oi}^d - \alpha_{Oi}\right)\right) + \boldsymbol{B}_i\boldsymbol{T}_O\dot{\boldsymbol{r}}_o^d, \tag{100}$$

and a residual of the sliding velocity is given by

$$\boldsymbol{s}_{Si} = \boldsymbol{v}_{Si} - \boldsymbol{v}_{Sri}, \tag{101}$$

where α_{Oi}^d is a smooth desired position of the fingertip on the curved surface coordinates, $\boldsymbol{\Lambda}_S > 0$ is a diagonal weight matrix, and $\boldsymbol{v}_{Si} = \dot{\boldsymbol{p}}_{Si} = \boldsymbol{J}_{Ci}\dot{\boldsymbol{q}}_{Fi} - \boldsymbol{W}_{Ci}^T\boldsymbol{v}_O$ is the sliding velocity. Then, a desired friction force and estimation law of the friction parameters can be given by

$$\boldsymbol{f}_{ti}^d = \boldsymbol{Y}_{ti}(\dot{\boldsymbol{p}}_{Si}, \boldsymbol{n}_i, \boldsymbol{f}_{Ci})\hat{\boldsymbol{\mu}}_i, \tag{102}$$

$$\dot{\hat{\boldsymbol{\mu}}}_i = -\boldsymbol{\Gamma}_{ti}\boldsymbol{Y}_{ti}^T(\dot{\boldsymbol{p}}_{Si}, \boldsymbol{n}_i, \boldsymbol{f}_{Ci})\boldsymbol{s}_{Si}, \tag{103}$$

where $\hat{\boldsymbol{\mu}}_i$ is an estimation of $\boldsymbol{\mu}_i$, and $\boldsymbol{\Gamma}_{ti} > 0$ is a symmetrically adaptive gain matrix. It is noted that the desired friction force at the rolling contact is $\boldsymbol{f}_{ti}^d = \boldsymbol{0}$.

Next, consider the desired contact force of the robot finger. The desired external force of the object is provided by Eq. (69) in the same way as adaptive control using a force sensor in Sec. 3.2; that is,

$$\boldsymbol{F}_O^d = \boldsymbol{Y}_O(\boldsymbol{r}_O, \dot{\boldsymbol{r}}_O, \boldsymbol{v}_{Or}, \dot{\boldsymbol{v}}_{Or})\hat{\boldsymbol{\sigma}}_O - \boldsymbol{K}_O\boldsymbol{s}_O. \tag{104}$$

The desired contact force to balance the desired external force should satisfy the following relation:

$$\boldsymbol{F}_O^d = \boldsymbol{W}_C\boldsymbol{f}_C^d. \tag{105}$$

The desired contact force $\boldsymbol{f}_{Ci}^d$ at the contact point C_i is then decomposed into the normal direction and force on the tangent plane, which is represented by

$$\boldsymbol{f}_{Ci}^d = f_{ni}^d\boldsymbol{n}_i + \boldsymbol{f}_{ti}^d, \tag{106}$$

where f_{ni}^d is the size of the desired contact force in the normal direction, and $\boldsymbol{f}_{ti}^d$ is the desired force on the tangent plane. For the sliding contact fingers, the desired external force is in the normal direction, and the desired friction force is in the direction of the tangent. For the rolling-contact fingers, such decomposition is not performed. Then, the desired external force satisfies:

$$\boldsymbol{F}_O^d = \boldsymbol{G}_C\boldsymbol{f}_{Cn}^d + \boldsymbol{W}_C\boldsymbol{f}_t^d, \tag{107}$$

where $f_{Cn}^d \in \mathbb{R}^{3k-2s}$ is a vector that is the summation of the desired contact force of the robot fingers with rolling contact, $G_C \in \mathbb{R}^{6\times(3k-2s)}$ is a matrix with a list of W_C corresponding to f_{Cn}^d, and $f_t^d = [(f_{t1}^d)^T \cdots (f_{tk}^d)^T]^T$. The contact force generates the internal forces that do not affect motion of the object. Therefore, the general solution for f_{Cn}^d is provided by

$$f_{Cn}^d = G_C^+\left(F_O^d - W_C f_t^d\right) + \left(I_{m\times m} - G_C^+ G_C\right) f_{\text{int}}^d, \tag{108}$$

where $G_C^+\left(= G_C^T(G_C G_C^T)^{-1}\right)$ is a generalized inverse matrix of G_C, and f_{int}^d is an arbitrary force vector that generates the internal force, which is continuous and bounded.

Finally, consider the control of the robot fingers. The reference velocity of the i-th robot finger is defined by

$$\dot{q}_{Fir} = J_{Ci}^{-1}\left(W_{Ci}^T v_{Or} + \Omega_{Fi} v_{Ci} + \Psi_{Fi} \eta_{Ci} + v_{Sri}\right). \tag{109}$$

It is given that v_{Ci} is the output after filtering the contact force error $\Delta f_{Ci} = f_{Ci}^d - f_{Ci}$, as in Eq. (68), and η_{Ci} is the integration of a normal component of contact force errors Δf_{Ci}, which is represented by

$$\eta_{Ci} = \eta_{Ci} n_i, \tag{110}$$

$$\eta_{Ci} = \int_0^t n_i^T \Delta f_{Ci} dt. \tag{111}$$

Using the residual of the i-th robot finger variable, $s_{Fi} = \dot{q}_{Fi} - \dot{q}_{Fir}$, the control law and the estimation law of the dynamic parameters of the i-th robot finger are given by

$$\tau_{Fi} = Y_{Fi}(q_{Fi}, \dot{q}_{Fi}, \dot{q}_{Fir}, \ddot{q}_{Fir})\hat{\sigma}_{Fi} + J_{Ci}^T f_{Ci}^d - K_{Fi} s_{Fi} + \beta J_{Ci}^T \Delta f_{Ci}, \tag{112}$$

$$\dot{\hat{\sigma}}_{Fi} = -\Gamma_{Fi} Y_{Fi}^T (q_{Fi}, \dot{q}_{Fi}, \dot{q}_{Fri}, \ddot{q}_{Fri}) s_{Fi}. \tag{113}$$

At this time, the control law and estimation law integrating the 1st to the k-th finger are expressed by

$$\tau_F = Y_F(q_F, \dot{q}_F, \dot{q}_{Fr}, \ddot{q}_{Fr})\hat{\sigma}_F + J_C^T f_C^d - K_F s_F + \beta J_C^T \Delta f_C, \tag{114}$$

$$\dot{\hat{\sigma}}_F = -\Gamma_F Y_F^T (q_F, \dot{q}_F, \dot{q}_{Fr}, \ddot{q}_{Fr}) s_F. \tag{115}$$

The difference of the control at the rolling contact is that the desired contact force is given based on the estimation of the friction parameters. Other configurations are almost the same as the estimation law and control law in Sec. 3.2. The following theorem is then provided by this control (Ueki et al. 2006).

Theorem 4. *Consider a rigid body that is grasped by k robot fingers, each with three DOFs. For the system in Eqs. (34), (35), and (31), given that Eq. (40) is the parameter estimation law of the object, the estimation law of friction parameters with sliding fingers at the contact point is Eq. (103), Eq. (69) is the desired external force of the object, Eq. (108) is the desired force at the contact point, Eq. (115) is the parameter estimate law of the robotic fingers, and Eq. (114) is the control law, then β is sufficiently large, and the closed-loop system is asymptotic stable in the following sense:*

(1) $\boldsymbol{r}_O \rightarrow \boldsymbol{r}_O^d$ *and* $\dot{\boldsymbol{r}}_O \rightarrow \dot{\boldsymbol{r}}_O^d$ *as* $t \rightarrow \infty$.

(2) $\boldsymbol{f}_C \rightarrow \boldsymbol{f}_C^d$ *as* $t \rightarrow \infty$.

(3) $\boldsymbol{\alpha}_{Oi} \rightarrow \boldsymbol{\alpha}_{Oi}^d$ *and* $\dot{\boldsymbol{\alpha}}_{Oi} \rightarrow \boldsymbol{\alpha}_{Oi}^d$ *as* $t \rightarrow \infty$.

Outline of proof. To prove this theorem, a candidate Lyapunov function is given by

$$V = \frac{1}{2}\left\{\beta_{+1}\left(\boldsymbol{s}_O^T \boldsymbol{M}_O \boldsymbol{s}_O + \Delta\boldsymbol{\sigma}_O^T \boldsymbol{\Gamma}_O^{-1} \Delta\boldsymbol{\sigma}_O + \boldsymbol{\eta}_C^T \boldsymbol{\Psi}_F \boldsymbol{\eta}_C + \frac{1}{\alpha_C} \boldsymbol{v}_C^T \boldsymbol{\Omega}_F \boldsymbol{v}_C \right)\right.$$

$$\left. + \boldsymbol{s}_F^T \boldsymbol{M}_F \boldsymbol{s}_F + \Delta\boldsymbol{\sigma}_F^T \boldsymbol{\Gamma}_F^{-1} \Delta\boldsymbol{\sigma}_F + \Delta\boldsymbol{\mu}^T \boldsymbol{\Gamma}_t^{-1} \Delta\boldsymbol{\mu} \right\}, \tag{116}$$

where $\Delta\boldsymbol{\mu} = \hat{\boldsymbol{\mu}} - \boldsymbol{\mu}$. Its time derivative is represented by

$$\dot{V} = -\beta_{+1}\left(\boldsymbol{s}_O^T \boldsymbol{K}_O \boldsymbol{s}_O + \boldsymbol{v}_C^T \boldsymbol{\Omega}_F \boldsymbol{v}_C \right) - \boldsymbol{s}_F^T \boldsymbol{K}_F \boldsymbol{s}_F \le 0. \tag{117}$$

Using a Lyapunov-like lemma, the asymptotic stability can be demonstrated in a procedure similar to the case of Theorem 3.

It should be noted that although this control is complicated, even more dexterous object manipulation is possible if the finger slides on the object surface.

EXPERIMENTAL VERIFICATIONS OF ADAPTIVE CONTROL USING A FORCE SENSOR

Experimental Setup

Experiment of a ball handing by the human-type robot hand named Gifu-Hand III (Mouri et al. 2002) was performed to show the effectiveness of the adaptive control using a force sensor, by comparison of with adaptive law and without adaptive law. In the experiment, as shown in Fig. 5, three fingers grasp the ball of radius of 0.0625 m and mass of 0.05 kg. The joint angle of robot hand is measured by rotary encoder, and the contact force is measured by 6-axis force sensor (NANO sensor, BL AUTOTEC, LTD.). The position and orientation of the object is measured by 3-D position measurement device (OPTRAK, Northern Digital Co.). The resolution of this device is 0.1 mm and sampling frequency is 500 Hz. The orientation vector of the object is represented by the ZYZ-Euler angles. The object coordinate system Σ_O is set in consideration of singular point of To. The contact position ${}^{Fi}\boldsymbol{p}_{FiCi}$ is calculated using contact force by reference (Bicchi et al. 1993). The joint angle velocity, object velocity, object angular velocity, and contact point velocity are calculated using measured value by differential filter. The control sampling is 1000 Hz. The experiment conditions are as follows: the initial values of the unknown dynamic parameters are set to zero, that is $\hat{\boldsymbol{\sigma}}_O = \boldsymbol{0}$ and $\hat{\boldsymbol{\sigma}}_F = \boldsymbol{0}$. The unknown dynamic parameters of the object is mass, center of mass, and inertia tensor. Those of the robot are inertia sensor of link, first moment of link, inertia of motor, and viscosity friction coefficient of motor. The desired trajectory of the ball is given repeatedly by 5 order polynomial in time with initial $\boldsymbol{r}_O^d(0) = \boldsymbol{r}_O(0)$ and $\boldsymbol{r}_O^d(1.5) = \boldsymbol{r}_O(0) + [0.03\ \ 0\ \ 0\ \ 0\ -\pi/9\ \ 0]^T$. $\boldsymbol{f}_{\text{int}}^d$ is given as follows:

$$f_{\text{int}}^{d} = 1.5\left[\frac{p_{F1C1}^{T}}{\|p_{F1C1}\|} \quad \frac{p_{F2C2}^{T}}{\|p_{F2C2}\|} \quad \frac{p_{F3C3}^{T}}{\|p_{F3C3}\|}\right]^{T}.$$

The controller gains were selected to be:
$\boldsymbol{\Gamma}_O = diag[0.02\ \ 0.0001\ ...\ 0.0001]$, $\boldsymbol{K}_O = diag[0.04\ ...\ 0.04\ \ 0.005\ ...\ 0.005]$, $\boldsymbol{\Lambda}_O = diag[40\ ...\ 40\ \ 30\ ...\ 30]$, $\boldsymbol{\Gamma}_F = diag[0.0003\ ...\ 0.0003]$, $\boldsymbol{K}_{Fi} = diag[0.4\ \ 0.4\ \ 0.04]$, $\boldsymbol{\Omega}_F = diag[0.4\ ...\ 0.4]$, $\boldsymbol{\Psi}_F = diag[0.04\ ...\ 0.04]$, $\alpha_C = 0.05/\pi$, and $\beta = 1.2$. In the case of without adaptive law, the experimental conditions are same excluding adaptive gains. The adaptive gains were changed to zero, i.e., $\boldsymbol{\Gamma}_O = \boldsymbol{0}$ and $\boldsymbol{\Gamma}_F = \boldsymbol{0}$.

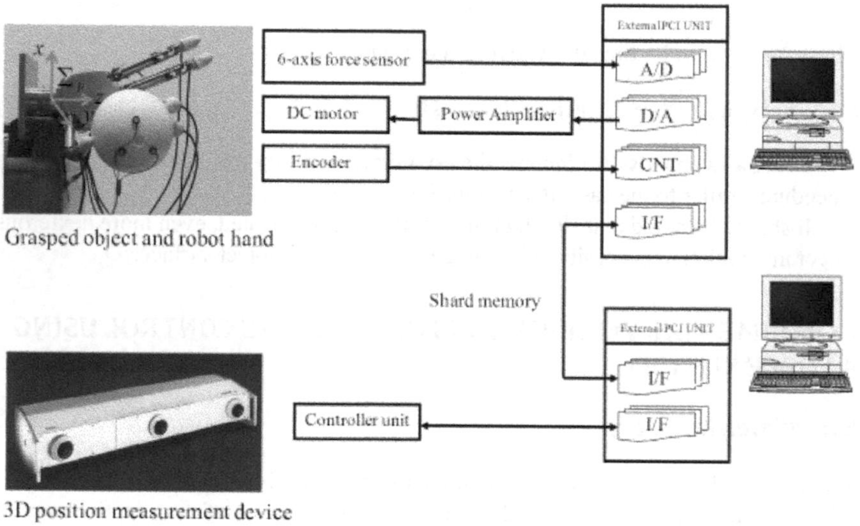

Figure 5. Experiment system consisting of robot hand and 3-D position measurement device.

Experimental Results

Figure 6 shows the desired object position $\boldsymbol{p}_O^d$ and measured object position $\boldsymbol{p}_O$ in case of with adaptive law. Figure 7 shows the norm of object position error $\|\boldsymbol{p}_O^d - \boldsymbol{p}_O\|$. In case of with adaptive law, the object position error decreased by repetition of motion. On the other hand, in case of without adaptive law, the object position error increased by repetition of motion. The contact constraint of rolling contact is nonholonomic constraint, this means nonintegrable. Therefore, the desired motion is not a repetitive motion. Nevertheless, the object position error shows that the adaptation was working successfully.

Figure 8 shows the desired object orientation $\boldsymbol{\eta}_O^d$ and measured object orientation $\boldsymbol{\eta}_O$ in case of with adaptive law. Figure 9 shows the norm of object orientation error $\|\boldsymbol{\eta}_O^d - \boldsymbol{\eta}_O\|$. In case of with adaptive law, the object orientation error decreased by repetition of motion. On the other hand, in case of without adaptive law, the object orientation error increased by repetition of motion. For position and orientation in case of with adaptive law, capabilities of both response and tracking are better compared with case of without adaptive law.

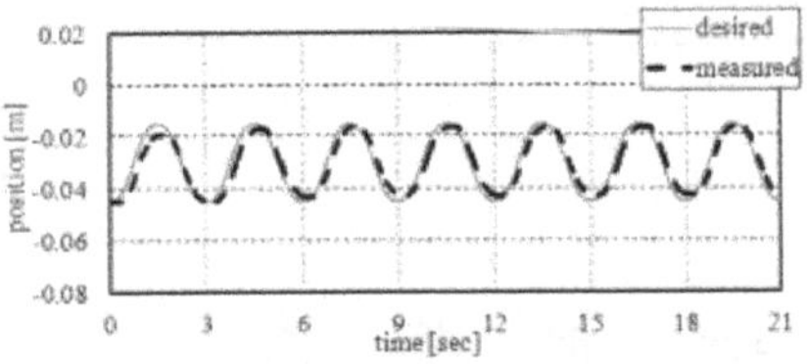

(a) x element

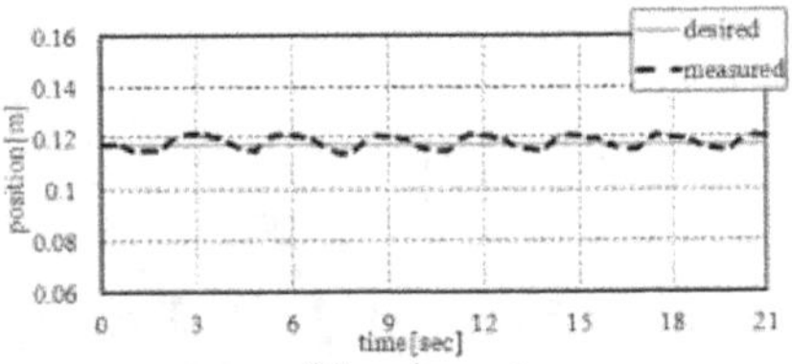

(b) y element

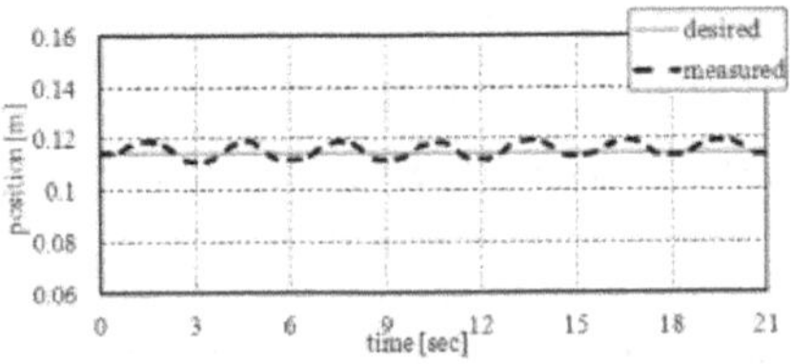

(c) z element

Figure 6. Trajectory of object position with adaptive law.

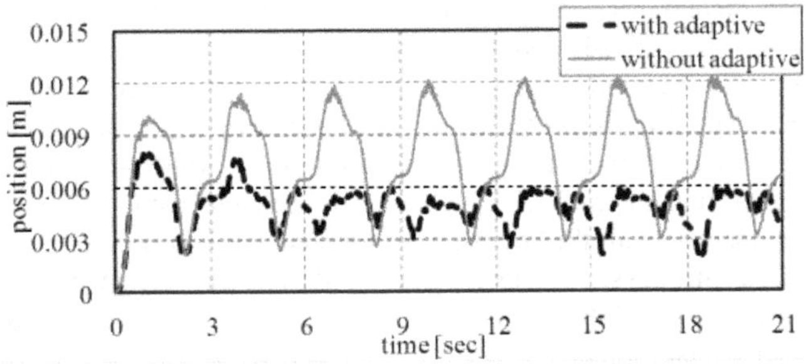

Figure 7. The norm of object position error $\left\| \boldsymbol{p}_O^d - \boldsymbol{p}_O \right\|$.

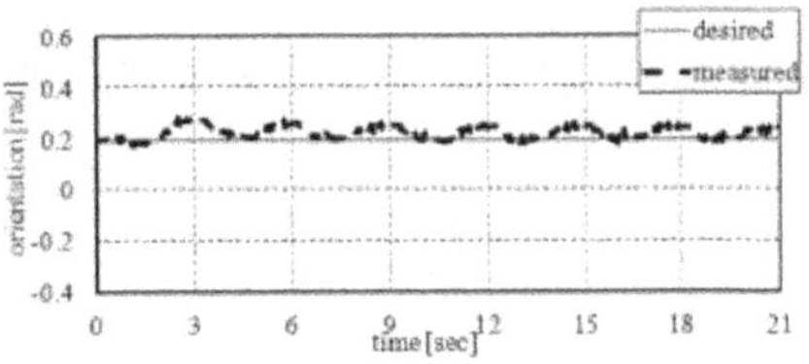

(a) 1st element

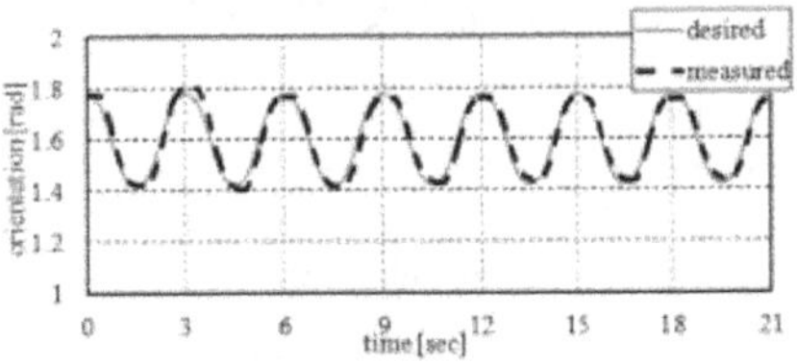

(b) 2nd element

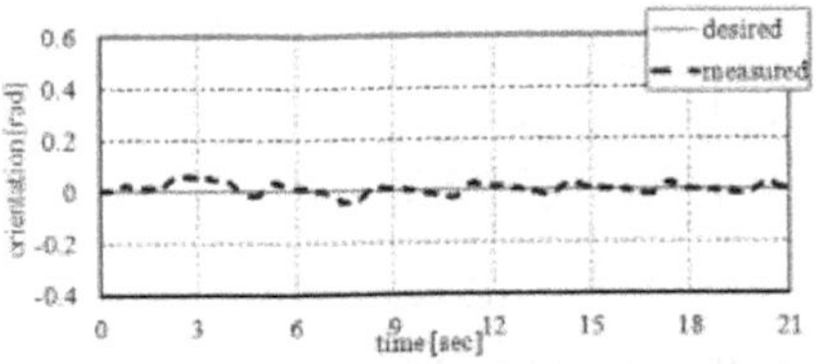

(c) 3rd element

Figure 8. Trajectory of object orientation with adaptive law.

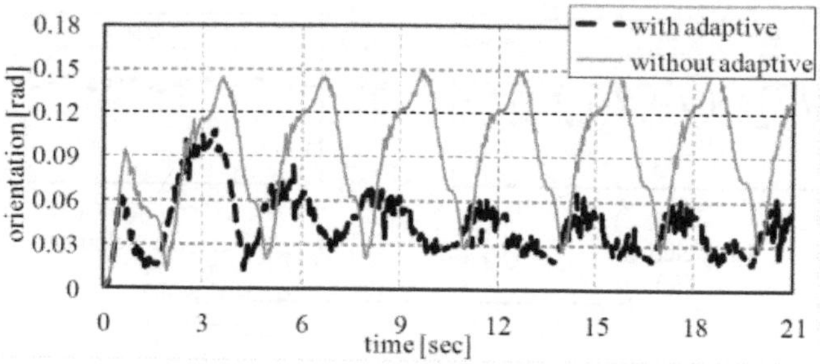

Figure 9. The norm of object orientation error $\|\eta_O^d - \eta_O\|$.

Figures 10–12 show the desired contact force f_C^d and measured contact force f_C of the 1st finger, 2nd finger, and 3rd finger in case of with adaptive law. Figure 13 shows the norm of contact force error $\|f_C^d - f_C\|$. These show that the x and y elements of measured contact force track the desired contact force well, and contact force error decreased by repetition of motion in both cases. The contact force error decrease by the integral of the contact force error η_{Ci} in case of without adaptive law differing from the results of position and orientation. Figure 13 shows that the decrease speed of contact force error in case of with adaptive law is faster than it in case of without adaptive law. However, though measured z element of contact force converges to desired contact force, tracking capability is not sufficient. The z element of contact force correspond to the internal force, and there are several reasons the tracking capability of z element of contact force is not sufficient: the dynamics of the mechanism such as the flexibility of the joint is not modeled; the controller is not a continuous-time control system but a discrete-time control system whose accuracy of trajectory depends on the sampling cycle (Kawasaki and Li 2004); the difference in response occur by the difference in property of individual finger.

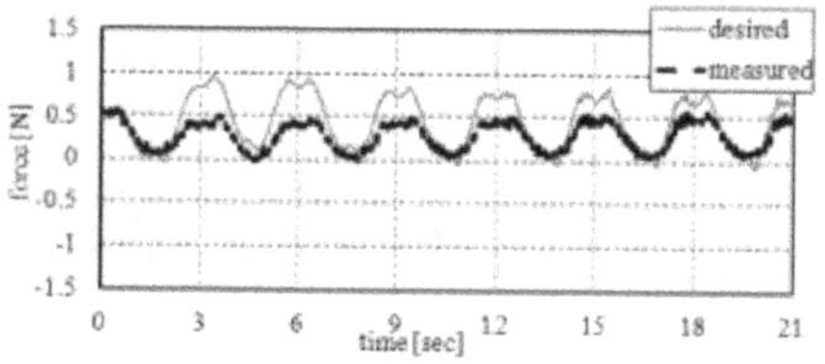

(a) x element

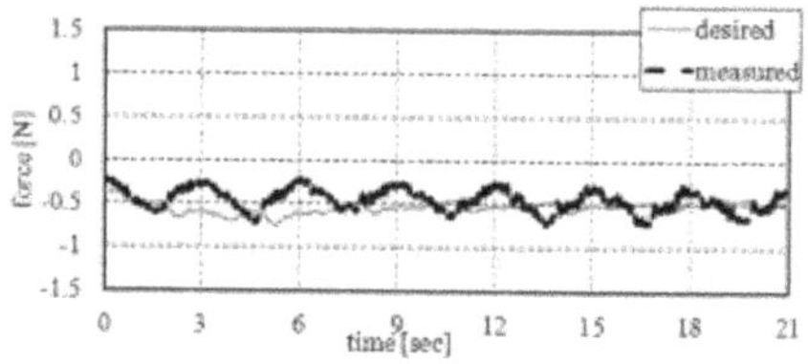

(b) y element

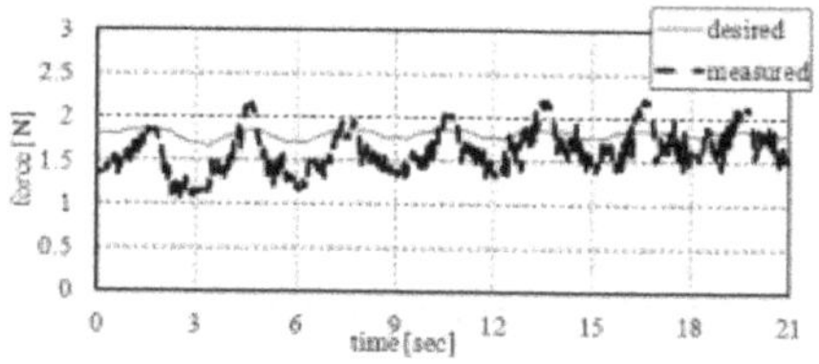

(c) z element

Figure 10. Contact force of 1st finger with adaptive law.

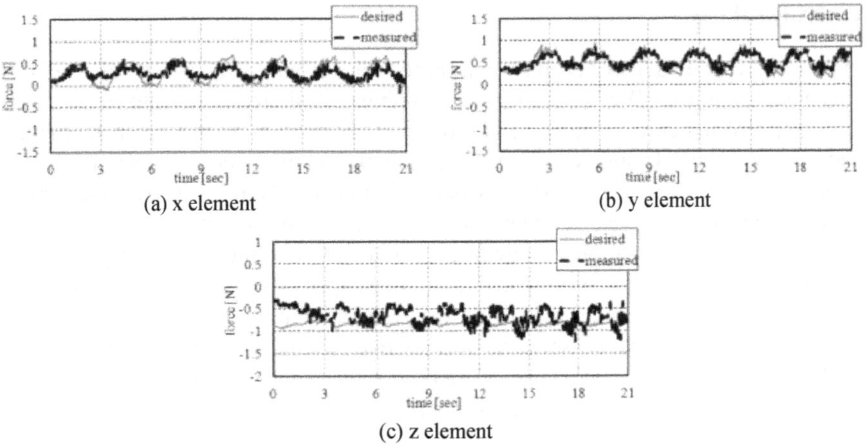

Figure 11. Contact force of 2nd finger with adaptive law.

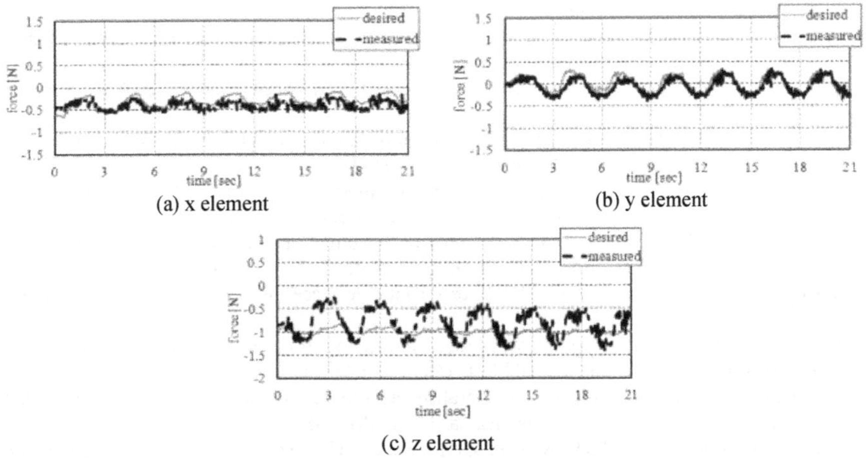

Figure 12. Contact force of 3rd finger with adaptive law.

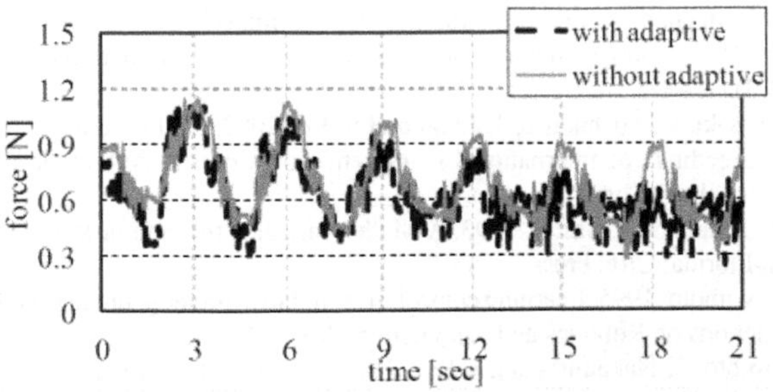

Figure 13. The norm of contact force error $\|f_C^d - f_C\|$.

CONCLUSION

In this chapter, we discussed adaptive control for object manipulation in accordance with the contact condition. When a robot grasps an object, its fingers come into contact with the surface of the object, and its hand motion is constrained by that contact. Robotic hands are mechanical non-linear systems. In general, adaptive control systems are treated as time-varying systems. For dexterous manipulation, the control objective is to control the object position and velocity; that is, to coordinately-control the force at the contact point that occurs for multiple robotic fingers. This is one of the challenging problems in robotics. In the controller presented here, the shapes of the object and positions of the contact points are already known. In fact, the position of the contact point can be calculated using the force sensor variable. However, the accuracy is insufficient. Thus, study of adaptive controls with fewer restrictions is needed. In addition, dynamic control by soft fingers will be an upcoming challenge.

REFERENCES

Anderson, B. O. D. 1977. Exponential stability of linear system arising from adaptive identification. IEEE Transactions on Automatic Control, 22(2): 83–88.

Bicchi, A., J. K. Salisbury and D. L. Brock. 1993. Contact sensing from force measurements. International Journal of Robotics Research, 12(3): 249–262.

Birglen, L., T. Lalibert and C. M. Gosselin. 2010. Under-actuated Robotic Hands. New York: Springer Verlag.

Cutkosky, M. R. and I. Kao. 1989. Computing and controlling the compliance of a robot hand. IEEE Transactions on Robotics and Automation, 5(2): 151–165.

Hayati, S. 1986. Hybrid position/force control of multi-arm cooperating robots. Proceedings of IEEE International Conference on Robotics and Automation, San Francisco, CA, USA, pp. 82–89.

Hwang, Y. Y. and I. Toda. 1994. Coordinated control of two direct-drive robots using neural networks. JSME International Journal Series C, Mechanical Systems, 37: 335–341.

Kawasaki, H. 2015. Robot hands and multi-fingered haptic interfaces. Fundamentals and Applications, World Scientific.

Kawasaki, H. and G. Li. 2004. Expert skill-based gain tuning in discrete-time adaptive control for robots. The Journal of Robotics and Mechatronics, 16(1): 54–60.

Kawasaki, H., S. Ueki and S. Ito. 2006. Decentralized adaptive coordinated control of multiple robot arm without using force sensor. Automatica, 42(3): 481–488.

Kosuge, K., H. Yoshida, T. Fukuda, K. Kanitani, M. Sakai and K. Hariki. 1995. Coordinated motion control of manipulation based on impedance control. Journal of the Robotics Society of Japan, 13(3): 404–410 (in Japanese).

Mason, M. T. and J. K. Salisbury. 1985. Robot Hands and Mechanics of Manipulation. Cambridge, MA, USA: The MIT Press.

Mason, M. T. 1998. Mechanics of Robotic Manipulation. Cambridge, MA, USA: The MIT Press.

Montana, D. J. 1988. The kinematics of contact and grasp. The International Journal of Robotics Research, 7(3): 17–32.

Mouri, T., H. Kawasaki, K. Yoshikawa, J. Takai and S. Ito. 2002. Anthropomorphic robot hand: Gifu Hand III. Proceedings of International Conference on Control, Automation and Systems, pp. 1288–1293, Jeonbuk, Korea.

Murray, R. M., Z. Li and S. S. Sastry. 1993. A Mathematical Introduction to Robotic Manipulation, Boca Raton, Florida: CRC Press.

Naniwa, T. and S. Arimoto. 1995. Learning control for robot task under geometric endpoint constraints. IEEE Transactions on Robotics and Automation, 2(3): 432–441.

Ozawa, R., S. Arimoto, S. Nakamura and J. Bae. 2005. Control of an object with parallel surface by a pair of finger robots without object sensing. IEEE Transactions on Robotics and Automation, 21(5): 965–976.

Schlegl, T., M. Buss and G. Schmidt. 2003. A hybrid system approach toward modeling and dynamical simulation of dextrous manipulation. IEEE/ASME Transactions on Mechatronics, 8(3): 352–361.

Su, Y. and Y. Stepanenco. 1995. Adaptive sliding mode coordinated control of multiple robot arms attached to a constrained object. IEEE Transactions on Systems, Man, and Cybernetics, 25: 871–877.

Ueki, S., H. Kawasaki and T. Mouri. 2006. Adaptive coordinated control of multi-fingered hands with sliding contact. Proceedings of SICE-ICASE International Joint Conference, pp. 5893–5898, Busan, South Korea.

Ueki, S., H. Kawasaki and T. Mouri. 2009. Adaptive coordinated control of multi-fingered robot hand. Journal of Robotics and Mechatronics, 21(1): 36–43.

Wakamatsu, H., M. Yamanoi and K. Tatsuno. 2010. Adaptive force control of robot arm with estimation of environmental stiffness. Proceedings of International Symposium on Micro-NanoMechatronics and Human Science, pp. 226–231, Nagoya, Japan.

Yingjie, Y. and S. Hosoe. 2005. MLD modeling and optimal control of hand manipulation. Proceedings of IEEE/RSJ International Conference on Intelligent Robotics and Systems, pp. 838–843, Alberta, Canada.

Zhaopeng, C., T. Wimbock, M. A. Roa, B. Pleintinger, M. Neves, C. Ott, C. Borst and N. Y. Lii. 2015. An adaptive compliant multi-finger approach-to-grasp strategy for objects with position uncertainties. Proceedings of IEEE International Conference on Robotics and Automation, pp. 4911–4918, Seattle, WA.

Zheng, X., R. Nakashima and T. Yoshikawa. 2000. On dynamic control finger sliding and object motion in manipulation with multi-fingered hands. IEEE Transactions on Robotics and Automation, 16(5): 469–481.

6

Output Feedback Adaptive Control of Uncertain Dynamical Systems with Event-Triggering

Ali Albattat,[1] *Benjamin Gruenwald*[2] and *Tansel Yucelen*[2,*]

ABSTRACT

Networked control for a class of uncertain dynamical systems is studied, where the control signals are computed via processors that are not attached to the dynamical systems and the feedback loops are closed over wireless networks. Since a critical task in the design and implementation of networked control systems is to reduce wireless network utilization while guaranteeing system stability in the presence of system uncertainties, an event-triggered adaptive control architecture is presented in an output feedback setting to schedule the data exchange dependent upon errors exceeding user-defined thresholds. Specifically, using tools and methods from nonlinear systems theory and Lyapunov stability in particular, it is shown that the proposed approach guarantees system stability in the presence of system uncertainties and does not yield to a Zeno behavior. In addition, the effect of user-defined thresholds and output feedback adaptive controller design parameters to the system performance is rigorously characterized and discussed. The efficacy of the proposed event-triggered output feedback adaptive control approach is demonstrated in an illustrative numerical example.

Keywords: Networked control systems, output feedback adaptive control, event-triggering control, system uncertainties, system stability, system performance

[1] Department of Mechanical and Aerospace Engineering, Missouri University of Science and Technology.

[2] Department of Mechanical Engineering, University of South Florida.

* Corresponding author: tyucelen@gmail.com

INTRODUCTION

Networked control of dynamical systems is an appealing methodology in reducing cost for the development and implementation of control systems (Zhang et al. 1993, Walsh and Ye 2001, Tipsuwan and Chow 2003, Zhang and Hristu-Varsakelis 2006, Hespanha et al. 2007, Alur et al. 2007, Wang and Liu 2008, Bemporad et al. 2010). These systems allow the computation of control signals via processors that are not attached to the dynamical systems and the feedback loops are closed over wireless networks. In a networked control setting, since the processors computing control signals are separated from the dynamical systems, not only the feedback control algorithms can be easily modified as necessary but also this setting allows to develop small-size physical systems for low-cost control theory applications.

Motivation and Literature Review

A challenge in the design and implementation of networked control systems is to reduce wireless network utilization. To this end, the last decade has witnessed an increased interest in event-triggering control theory (Tabuada 2007, Mazo, Jr. and Tabuada 2008, Mazo, Jr. et al. 2009, Lunze and Lehmann 2010, Postoyan et al. 2011, Heemels et al. 2012), where it relaxes periodic data exchange demand of the feedback loops closed over wireless networks. Specifically, this theory allows aperiodic data exchange between the processors computing control signals and the dynamical systems, and hence, asynchronous data can be exchanged only when needed.

In networked control systems, another challenge is to guarantee system stability in the presence of system uncertainties. Often when designing feedback controllers for dynamical systems, idealized assumptions, linearization, model-order reduction, exogenous disturbances, and unexpected system changes lead to modeling inaccuracies. If not mitigated, the uncertainties present in the system model can result in poor system performance and system instability (Slotine and Li 1987, Krstic et al. 1995, Zhou et al. 1996, Zhou and Doyle 1998, Narendra and Annaswamy 2012, Ioannou and Sun 2012, Lavretsky and Wise 2012, Astrom and Wittenmark 2013). Therefore, it is essential in the feedback control design process to achieve robust stability and a desired level of system performance when dealing with dynamical systems subject to system uncertainties.

Motivated by these two challenges of networked control systems, this chapter studies control of uncertain dynamical systems over wireless networks with event-triggering. To this end, we consider an adaptive control approach rather than a robust control approach, since the former approach requires less system modeling information than the latter and can address system uncertainties and failures effectively in response to system variations. Notable contributors that utilize event-triggered adaptive control approaches include (Sahoo et al. 2013, Sahoo et al. 2014, Wang and Hovakimyan 2010, Wang et al. 2015, Albattat et al. 2015).

In particular (Sahoo et al. 2013, Sahoo et al. 2014) develop neural networks-based adaptive control approaches to guarantee system stability in the presence of system uncertainties, where these results only consider one-way data transmission from a dynamical system to the controller. Two-way data transmission over a wireless network; that is, from a dynamical system to the controller and from the controller to this dynamical system, is considered in (Wang and Hovakimyan 2010, Wang et al. 2015, Albattat et al. 2015) to guarantee system stability under system uncertainties.

The major difference between the results in (Wang and Hovakimyan 2010, Wang et al. 2015, Albattat et al. 2015) is that the latter does not require the knowledge of a conservative upper bound on the unknown constant gain resulting from the system uncertainty parameterization. Finally, it should be noted that all these approaches documented in (Sahoo et al. 2013, Sahoo et al. 2014, Wang and Hovakimyan 2010, Wang et al. 2015, Albattat et al. 2015) consider an event-triggered state feedback adaptive control approach. Yet, output feedback is required for most applications that involve high-dimensional models such as active noise suppression, active control of flexible structures, fluid flow control systems, and combustion control processes (Khalil 1996, Calise et al. 2001, Hovakimyan et al. 2002, Volyanskyy et al. 2009, Yucelen and Haddad 2012, Lavretsky 2012, Gibson et al. 2014, Lavretsky 2015).

Contribution

In this chapter, networked control for a class of uncertain dynamical systems is studied. Since a critical task in the design and implementation of networked control systems is to reduce wireless network utilization while guaranteeing system stability in the presence of system uncertainties, an event-triggered adaptive control architecture is presented in an output feed-back setting to schedule two-way data exchange dependent upon errors exceeding user-defined thresholds. Specifically, we consider the output feedback adaptive control architecture predicated on the asymptotic properties of LQG/LTR controllers (Lavretsky and Wise 2012, Lavretsky 2012, Gibson et al. 2014, Lavretsky 2015), since this framework has the capability to achieve stringent performance specifications without causing high-frequency oscillations in the controller response, asymptotically satisfies a strictly positive real condition for the closed-loop dynamical system, and is less complex than other approaches to output feedback adaptive control (see, for example (Calise et al. 2001, Hovakimyan et al. 2002, Volyanskyy et al. 2009)).

Building on this output feedback adaptive control architecture as well as our previous event-triggered state feedback adaptive control methodology (Albattat et al. 2015), it is shown using tools and methods from nonlinear systems theory and Lyapunov stability in particular that the proposed feedback control approach guarantees system stability in the presence of system uncertainties. In addition, the effect of user-defined thresholds and output feedback adaptive controller design parameters to the system performance is rigorously characterized and discussed. Moreover, we show that the proposed event-triggered output feedback adaptive control methodology does not yield to a Zeno behavior, which implies that it does not require a continuous two-way data exchange and reduces wireless network utilization. Similar to the state feedback case (Albattat et al. 2015), we also show that the resulting closed-loop dynamical system performance is more sensitive to the changes in the data transmission threshold from the physical system to the adaptive controller (sensing threshold) than the data transmission threshold from the adaptive controller to the physical system (actuation threshold), which implies that the actuation threshold can be chosen large enough to reduce wireless network utilization between the physical system and the adaptive controller without sacrificing closed-loop dynamical system performance. The efficacy of the proposed event-triggered output feedback adaptive control approach is demonstrated in an illustrative numerical example. Although this chapter considers a particular output feedback adaptive control formulation to present its main contributions, the proposed approach can be used in a complimentary way with many other approaches to output feedback adaptive control concerning robotic manipulators (see, for example (Kaneko and Horowitz 1997, Burg et al. 1997, Zergeroglu et al. 1999, Zhang et al. 2000)).

Notation

The notation used in this chapter is fairly standard. Specifically, $\mathbb{R}$ denotes the set of real numbers, $\mathbb{R}^n$ denotes the set of $n \times 1$ real column vectors, $\mathbb{R}^{n\times m}$ denotes the set of $n \times m$ real matrices, $\mathbb{R}_+$ denotes the set of positive real numbers, $\mathbb{R}_+^{n\times n}$ denotes the set of $n \times n$ positive-definite real matrices, $\mathbb{S}^{n\times n}$ denotes the set of $n \times n$ symmetric real matrices, $\mathbb{D}^{n\times n}$ denotes the set of $n \times n$ real matrices with diagonal scalar entries, $(\cdot)^{\mathrm{T}}$ denotes transpose, $(\cdot)^{-1}$ denotes inverse, $\mathrm{tr}(\cdot)$ denotes the trace operator and '$\triangleq$' denotes equality by definition. In addition, we write $\lambda_{\min}(A)$ (respectively, $\lambda_{\max}(A)$) for the minimum and respectively maximum eigenvalue of the Hermitian matrix A, $\|\cdot\|$ for the Euclidean norm, and $\|\cdot\|_{\mathrm{F}}$ for the Frobenius matrix norm. Furthermore, we use "$\vee$" for the "or" logic operator and "$\overline{(\cdot)}$" for the "not" logic operator.

OUTPUT FEEDBACK ADAPTIVE CONTROL OVERVIEW

In this section, we overview the output feedback adaptive control architecture predicated on the asymptotic properties of LQG/LTR controllers (Lavretsky and Wise 2012, Lavretsky 2012, Gibson et al. 2014, Lavretsky 2015), which are needed for the main results of this chapter. For this purpose, consider the uncertain dynamical system given by

$$\dot{x}_{\mathrm{p}}(t) = A_{\mathrm{p}}x_{\mathrm{p}}(t) + B_{\mathrm{p}}\Lambda\,[u(t) + \Delta\,(x_{\mathrm{p}}(t))],\quad x_{\mathrm{p}}(0) = x_{\mathrm{p}0}, \tag{1}$$

$$y_{\mathrm{reg}}(t) = C_{\mathrm{reg}}\,x_{\mathrm{p}}(t), \tag{2}$$

where $A_{\mathrm{p}} \in \mathbb{R}^{n_{\mathrm{p}}\times n_{\mathrm{p}}}$, $B_{\mathrm{p}} \in \mathbb{R}^{n_{\mathrm{p}}\times m}$, and $C_{\mathrm{reg}} \in \mathbb{R}^{m\times n_{\mathrm{p}}}$ are known system matrices, $x_{\mathrm{p}}(t) \in \mathbb{R}^{n_{\mathrm{p}}}$ is the state vector, which is not available for state feedback design, $u(t) \in \mathbb{R}^m$ is the control input, $\Lambda \in \mathbb{R}_+^{m\times m} \cap \mathbb{D}^{m\times m}$ is an unknown control effectiveness matrix, $\Delta : \mathbb{R}^n \to \mathbb{R}^m$ is a system uncertainty, and $y_{\mathrm{reg}}(t) \in \mathbb{R}^m$ is the regulated output vector. In addition, we assume that the uncertain dynamical system given by (1) and (2) has a measured output vector

$$y_{\mathrm{p}}(t) = C_{\mathrm{p}}x_{\mathrm{p}}(t), \tag{3}$$

where $y_{\mathrm{p}}(t) \in \mathbb{R}^{l_{\mathrm{p}}}$, $C_{\mathrm{p}} \in \mathbb{R}^{l_{\mathrm{p}}\times n_{\mathrm{p}}}$, and $l_{\mathrm{p}} \geq m$ such that the elements of $y_{\mathrm{reg}}(t)$ are a subset of the elements of $y_{\mathrm{p}}(t)$. Throughout this chapter, we assume that the triple $(A_{\mathrm{p}}, B_{\mathrm{p}}, C_{\mathrm{p}})$ is minimal, the system uncertainty in (1) can be linearly parameterized as

$$\Delta(x_{\mathrm{p}}(t)) = W_{\mathrm{o}}^{\mathrm{T}}\sigma_{\mathrm{o}}(x_p(t)), \tag{4}$$

where $W_{\mathrm{o}} \in \mathbb{R}^{s\times m}$ is an unknown weight matrix satisfying $\|W_{\mathrm{o}}\|_{\mathrm{F}} \leq \omega^*$, $\omega^* \in \mathbb{R}_+$, and $\sigma_{\mathrm{o}}(x_{\mathrm{p}}(t))$ is a known Lipschitz continuous basis vector satisfying

$$\|\sigma_{\mathrm{o}}(x_{\mathrm{p}}) - \sigma_{\mathrm{o}}(\hat{x}_{\mathrm{p}})\| \leq L_{\sigma}\,\|x_{\mathrm{p}} - \hat{x}_{\mathrm{p}}\|, \tag{5}$$

with $L_{\sigma} \in \mathbb{R}_+$. These assumptions are standard in the output feedback adaptive control literature (see, for example (Lavretsky and Wise 2012, Lavretsky 2012, Gibson et al. 2014, Lavretsky 2015, Kim et al. 2011a, Kim et al. 2011b). For the case when the system uncertainty given by (4) cannot be perfectly parameterized and/or the basis vector does not satisfy (5), note that universal approximation tools such as neural networks can be used in the basis vector on a compact subset of the state space (see, for example (Yesildirek and Lewis 1995, Kim and Lewis 1999)).

Similar to the approaches documented in (Lavretsky 2012, Kim et al. 2011, Kim et al. 2011), we consider a state observer-based nominal control architecture to achieve command following, where control of the regulated outputs that are commanded include integral action and the regulated outputs that are not commanded are subject to proportional control. For this purpose, let

$$y_{\text{reg}}(t) = \begin{bmatrix} y_{\text{reg1}}(t) \\ y_{\text{reg2}}(t) \end{bmatrix} = \begin{bmatrix} C_{\text{reg1}} \\ C_{\text{reg2}} \end{bmatrix} x_{\text{p}}(t), \tag{6}$$

where $y_{\text{reg1}}(t) \in \mathbb{R}^r$, $r \leq m$, is regulated with proportional and integral control to track a given command vector $\text{r}(t) \in \mathbb{R}^r$, $y_{\text{reg2}}(t) \in \mathbb{R}^{m-r}$ is regulated with proportional control, $C_{\text{reg1}} \in \mathbb{R}^{r\times n_\text{p}}$, and $C_{\text{reg2}} \in \mathbb{R}^{(m-r)\times n_\text{p}}$. Now, we define the integrator dynamics as

$$\dot{x}_{\text{int}}(t) = -y_{\text{reg1}}(t) + \text{r}(t) = -C_{\text{reg1}}\, x_{\text{p}}(t) + I_r\, \text{r}(t), \tag{7}$$

where $x_{\text{int}}(t) \in \mathbb{R}^r$ is the integral state vector. Utilizing (1), (2), and (7), the augmented system dynamics are now given by

$$\dot{x}(t) = \underbrace{\begin{bmatrix} A_{\text{p}} & 0 \\ -C_{\text{p1}} & 0 \end{bmatrix}}_{A} x(t) + \underbrace{\begin{bmatrix} B_{\text{p}} \\ 0 \end{bmatrix}}_{B} \Lambda\ [u(t) + \Delta(x_{\text{p}}(t))] + \underbrace{\begin{bmatrix} 0 \\ I_r \end{bmatrix}}_{B_r} \text{r}(t), \tag{8}$$

$$y_{\text{reg}}(t) = \underbrace{\begin{bmatrix} C_{\text{reg}} & 0 \end{bmatrix}}_{C_{\text{Reg}}} x(t), \tag{9}$$

where $x(t) = [x_{\text{p}}(t), x_{\text{int}}(t)]^{\text{T}} \in \mathbb{R}^n$, $A \in \mathbb{R}^{n\times n}$, $B \in \mathbb{R}^{n\times m}$, $B_{\text{r}} \in \mathbb{R}^{n\times r}$, $C_{\text{Reg}} \in \mathbb{R}^{m\times n}$, and $n = n_{\text{p}} + r$. In addition, the augmented measured output vector becomes

$$y(t) = \begin{bmatrix} y_{\text{p}}(t) \\ x_{\text{int}}(t) \end{bmatrix} = \underbrace{\begin{bmatrix} C_{\text{p}} & 0 \\ 0 & I_r \end{bmatrix}}_{C} x(t) \tag{10}$$

where $y(t) \in \mathbb{R}^l$, $C \in \mathbb{R}^{l\times n}$, and $l = l_{\text{p}} + r$.

Next, we define the feedback control law as

$$u(t) = u_n(t) + u_a(t), \tag{11}$$

where $u_n(t)$ is a nominal control law and $u_a(t)$ is an adaptive control law. Using the output feedback adaptive control architecture documented in (Lavretsky and Wise 2012, Lavretsky 2012, Gibson et al. 2014, Lavretsky 2015), we consider the nominal controller given by

$$u_n(t) = -K_x \hat{x}(t), \tag{12}$$

where $K_x \in \mathbb{R}^{m\times n}$ is a feedback matrix and $\hat{x}(t)$ is an estimate of the augmented system state vector $x(t)$ through a state observer to be defined later in this section. In order to determine the structure of the adaptive controller, we rewrite the augmented system dynamics given by (8) and (9) as

$$\dot{x}(t) = Ax(t) + Bu_n(t) + B\Lambda\,(u_a(t) + W^{\mathrm{T}}\sigma\,(x_{\mathrm{p}}(t), u_n(t))) + B_{\mathrm{r}}\mathrm{r}(t), \tag{13}$$

where $W \triangleq [W_{\mathrm{o}}^{\mathrm{T}}, I_{m\times m} - \Lambda^{-1}]^{\mathrm{T}} \in \mathbb{R}^{(n+m)\times m}$ and $\sigma(x_{\mathrm{p}}(t), u_n(t)) \triangleq [\sigma_{\mathrm{o}}(x_{\mathrm{p}}(t))^{\mathrm{T}}, u_n^{\mathrm{T}}(t)]^{\mathrm{T}} \in \mathbb{R}^{n+m}$. Motivating from the structure of the system uncertainties appearing in (13), consider the adaptive controller given by

$$u_a(t) = -\hat{W}^{\mathrm{T}}\sigma\,(\hat{x}_{\mathrm{p}}(t), u_n(t)), \tag{14}$$

where $\sigma(\hat{x}_{\mathrm{p}}(t), u_n(t)) \triangleq [\sigma_{\mathrm{o}}(\hat{x}_{\mathrm{p}}(t))^{\mathrm{T}}, u_n^{\mathrm{T}}(t)]^{\mathrm{T}} \in \mathbb{R}^{n+m}$ and $\hat{W}(t) \in \mathbb{R}^{(n+m)\times m}$ is the estimate of the unknown weight matrix W through the weight update law

$$\dot{\hat{W}}(t) = \Gamma\ \mathrm{Proj}_{\mathrm{m}}[\hat{W}(t), -\sigma(\hat{x}_{\mathrm{p}}(t), u_n(t))\,\tilde{y}^{\mathrm{T}}(t)\,R_0^{-\frac{1}{2}}\,ZS^{\mathrm{T}}], \tag{15}$$

where $\mathrm{Proj}_{\mathrm{m}}$ denotes the projection operator defined for matrices (Lavretsky and Wise 2012, Albattat et al. 2015, Pomet and Praly 1992, Lavretsky et al. 2011), $\Gamma \in \mathbb{R}_+^{(s+m)\times(s+m)} \cap \mathbb{S}^{(s+m)\times(s+m)}$ is a learning rate matrix, $\tilde{y}(t) \in \mathbb{R}^{l}$ given by

$$\tilde{y} \triangleq \hat{y}(t) - y\,(t) = C\,(\hat{x}(t) - x(t)), \tag{16}$$

is the measured output error, and $\hat{x}(t) \in \mathbb{R}^{n}$ is the estimated augmented system state obtained through the state observer given by

$$\dot{\hat{x}}(t) = A\hat{x}(t) + Bu_n(t) + L_v(y(t) - \hat{y}(t)) + B_{\mathrm{r}}\mathrm{r}(t), \tag{17}$$

$$\hat{y}(t) = C\hat{x}(t), \tag{18}$$

with $L_v \in \mathbb{R}^{n\times l}$ being the state observer gain matrix.

Following (Lavretsky and Wise 2012, Lavretsky 2012), the state observer gain matrix is given by

$$L_v = P_v C^{\mathrm{T}} R_v^{-1}, \tag{19}$$

with $P_v \in \mathbb{R}_+^{n\times n}$ being the unique solution to the algebraic Riccati equation

$$0 = \mathrm{P}_v(A + \eta I_{n\times n})^{\mathrm{T}} + (A + \eta I_{n\times n})\,P_v - P_v C^{\mathrm{T}} R_v^{-1} C P_v + Q_v,\ \eta \in \mathbb{R}_+, \tag{20}$$

$$Q_v = Q_0 + \left(\frac{v+1}{v}\right) B_s B_s^{\mathrm{T}},\ Q_0 \in \mathbb{R}_+^{n\times n},\ v \in \mathbb{R}_+, \tag{21}$$

$$R_v = \left(\frac{v}{v+1}\right) R_0,\ R_0 \in \mathbb{R}_+^{l\times l}. \tag{22}$$

In (21), $B_{\mathrm{s}} = [B, B_2]$, where $B_2 \in \mathbb{R}^{n\times(l-m)}$ is a matrix such that $\det(CB_{\mathrm{s}}) \neq 0$ and $C\,(sI_{n\times n} - A)^{-1}B_{\mathrm{s}}$ is minimum phase. Note that $l > m$ is assumed in the above construction, where if $l = m$ then $B_2 = 0$. In addition, the observer closed-loop matrix given by

$$A_v = A - L_v C = A - P_v C^{\mathrm{T}} R_v^{-1} C \tag{23}$$

is Hurwitz for all $v \in \mathbb{R}_+$. Moreover, let $\tilde{P}_v = P_v^{-1}$ and $S = [I_{m\times m}, 0_{m\times(l-m)}]$ to note (Lavretsky and Wise 2012, Lavretsky 2012)

$$\tilde{P}_v B = C^{\mathrm{T}} R_0^{-\frac{1}{2}} Z S^{\mathrm{T}} + O(v), \tag{24}$$

and

$$A_v^{\mathrm{T}} \tilde{P}_v + \tilde{P}_v A_v = -C^{\mathrm{T}} R_v^{-1} C - \tilde{P}_v Q_v \tilde{P}_v - 2\eta \tilde{P}_v < 0. \tag{25}$$

In (15) and (24), $Z = (UV)^{\mathrm{T}}$, where two unitary matrices U and V result from the singular value decomposition $B_s^{\mathrm{T}} C^{\mathrm{T}} R_0^{-1/2} = U\Sigma V$ and Σ is the diagonal matrix of the corresponding singular values. In (24), "$O(\cdot)$" denotes the Bachmann-Lundau asymptotic order notation (Kevorkian and Cole 1996, Murray 1984). For additional details on the output feedback adaptive control architecture overviewed in this section, we refer to (Lavretsky and Wise 2012, Lavretsky 2012) as well as (Gibson et al. 2014, Lavretsky 2015). To summarize, as previously discussed, the considered architecture has the capability to achieve stringent performance specifications without causing high-frequency oscillations in the controller response, asymptotically satisfies a strictly positive real condition for the closed-loop dynamical system, and is less complex than other approaches to output feedback adaptive control.

Finally, for analysis purposes later in this chapter, we define the reference model capturing the ideal closed-loop dynamical system performance given by

$$\dot{x}_{\mathrm{m}} = A_{\mathrm{m}} x_{\mathrm{m}}(t) + B_{\mathrm{m}} \mathrm{r}(t), \quad x_{\mathrm{m}}(0) = x_{\mathrm{m}0}, \tag{26}$$

$$y_{\mathrm{m}} = C_{\mathrm{Reg}} x_{\mathrm{m}}, \tag{27}$$

where $x_{\mathrm{m}}(t) \in \mathbb{R}^n$ is the reference model state, $A_{\mathrm{m}} = A - BK_x$ is Hurwitz, and $B_{\mathrm{m}} = B_{\mathrm{r}}$. In addition, let

$$\tilde{x}(\mathrm{t}) \triangleq \hat{x}(t) - x(t), \tag{28}$$

$$\hat{e}(t) \triangleq \hat{x}(t) - x_{\mathrm{m}}(t), \tag{29}$$

$$\tilde{W}(t) \triangleq \hat{W}(t) - W. \tag{30}$$

be the state estimation error, the state tracking error, and the weight estimation error, respectively. Now, we can write

$$\dot{\hat{e}}(t) = A_{\mathrm{m}} \hat{e}(t) + L_v\,(y(t) - \hat{y}(t)), \tag{31}$$

using (17) and (26), and write

$$\begin{aligned} \dot{\tilde{x}}(t) &= (A - L_v C)\tilde{x}(t) + B\Lambda(\hat{W}^{\mathrm{T}}(t)\,\sigma\,(\hat{x}_{\mathrm{p}}(t), u_n(t)) - W^{\mathrm{T}}\sigma\,(x_{\mathrm{p}}(t), u_n(t))) \\ &= \mathrm{A}_v \tilde{x}(t) + B\Lambda(\tilde{W}^{\mathrm{T}}(t)\,\sigma\,(\hat{x}_{\mathrm{p}}(t), u_n(t)) - \mathrm{g}(\cdot)), \end{aligned} \tag{32}$$

using (13) and (17), where $g(\cdot) \triangleq W^{\mathrm{T}}(\sigma(\hat{x}_{\mathrm{p}}(t), u_n(t)) - \sigma\,(x_{\mathrm{p}}(t), u_n\,(t)))$.

EVENT-TRIGGERED OUTPUT FEEDBACK ADAPTIVE CONTROL

In this section, we present the proposed event-triggered output feedback adaptive control architecture, which allows a desirable command following performance while the proposed controller exchanges data with the uncertain dynamical system through a wireless network. Mathematically speaking, the uncertain dynamical system sends its output signal to the adaptive controller only when a predefined event occurs. The kth time instants of the output transmission is represented by the monotonic sequence $\{s_k\}_{k=1}^{\infty}$, where $s_k \in \mathbb{R}_+$. The controller then uses this triggered system output signal to compute the control signal using the output feedback control architecture. Likewise, the updated feedback control input is transmitted to the uncertain dynamical system only when another predefined event occurs. The jth time instants of the feedback control transmission is then represented by the monotonic sequence $\{r_j\}_{j=1}^{\infty}$, where $r_j \in \mathbb{R}_+$. As depicted in Fig. 1, each system output signal and control input is held by a zero-order-hold operator (ZOH) until the next triggering event for that signal takes place. In this chapter, we do not consider delay in sampling, data transmission, and computation.

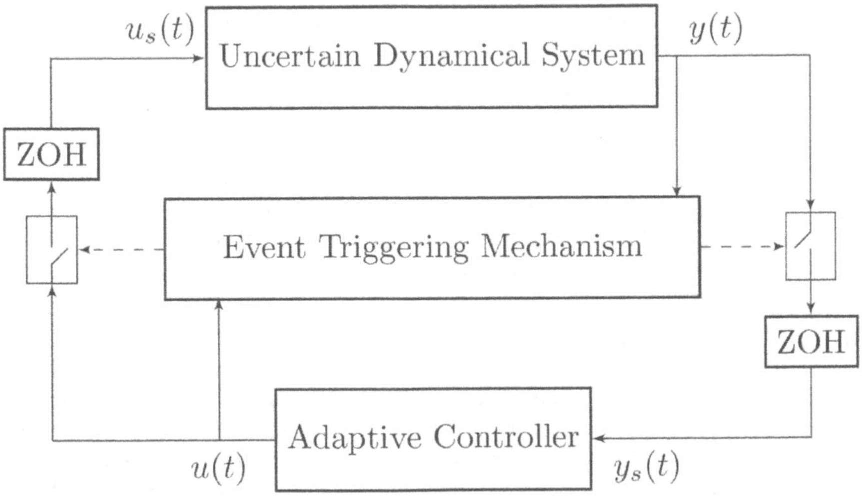

Figure 1. Event-triggered adaptive control system.

Proposed Event-Triggered Adaptive Control Algorithm

Based on the two-way data exchange structure depicted in Fig. 1, consider the augmented uncertain dynamical system given by

$$\dot{x}(t) = Ax(t) + B\Lambda[u_s(t) + \Delta(x_p(t))] + B_r r(t), \tag{33}$$

$$y_{reg}(t) = C_{Reg}x(t), \quad y(t) = Cx(t), \tag{34}$$

where $u_s(t) \in \mathbb{R}^m$ is the sampled control input vector. Under the assumptions stated in Section (**Output Feedback Adaptive Control Overview**) and considering the feedback control law given by (11) subject to the nominal controller given by (12) and the adaptive controller given by (14), the augmented uncertain dynamical system given by (33) and (34) can be equivalently written as

$$\dot{x}(t) = Ax(t) + Bu_n + B\Lambda(u_a(t) + W^{\mathrm{T}}\sigma(x_{\mathrm{p}}(t), u_n(t))) + B\Lambda(u_s(t) - u(t)) + B_{\mathrm{r}}\mathrm{r}(t), \tag{35}$$

$$y_{\mathrm{reg}}(t) = C_{\mathrm{Reg}}\, x(t), \quad y(t) = Cx(t). \tag{36}$$

In addition, we consider

$$\dot{\hat{W}}(t) = \Gamma\, \mathrm{Proj}_{\mathrm{m}}\left[\hat{W}(t), -\sigma(\hat{x}_{\mathrm{p}}(t), u_n(t))\,(\hat{y}(t) - y_s(t))^{\mathrm{T}} R_0^{-\frac{1}{2}} ZS^{\mathrm{T}}\right], \tag{37}$$

for the estimated weight matrix $\hat{W}(t)$ in (14) and

$$\begin{aligned}\dot{\hat{x}}(t) &= A\hat{x}(t) + Bu_n(t) + L_v(y_s(t) - \hat{y}(t)) + B_{\mathrm{r}}\mathrm{r}(t)\\ &= A_{\mathrm{m}}\hat{x}(t) + L_v(y_s(t) - \hat{y}(t)) + B_{\mathrm{r}}\mathrm{r}(t),\end{aligned} \tag{38}$$

$$\hat{y}(t) = C\hat{x}(t), \tag{39}$$

for the state observer, where $y_s(t) \in \mathbb{R}^l$ in (37) and (38) denotes the sampled augmented measured output vector.

The proposed event-triggered output feedback adaptive control algorithm is summarized in Table 1. Specifically, based on the two-way data exchange structure depicted in Fig. 1, the controller generates $u(t)$ and the uncertain dynamical system is driven by the sampled version of this control signal $u_s(t)$ depending on an event-triggering mechanism. Similarly, the controller utilizes $y_s(t)$ that represents the sampled version of the uncertain dynamical system measured output $y(t)$ depending on an event-triggering mechanism. These event-triggering mechanisms are stated next.

Scheduling Two-Way Data Exchange

Let $\epsilon_y \in \mathbb{R}_+$ be a given, user-defined sensing threshold to allow for data transmission from the uncertain dynamical system to the controller. In addition, let $\epsilon_u \in \mathbb{R}_+$ be a given, userdefined actuation threshold to allow for data transmission from the controller to the uncertain dynamical system. Similar in fashion to (Wang and Hovakimyan 2010, Albattat et al. 2015), we now define three logic rules for scheduling the two-way data exchange

Table 1. Event-triggered output feedback adaptive control algorithm.

Augmented unc. dyn. sys.	$\dot{x}(t) = Ax(t) + B\Lambda\,[u_s(t) + \Delta(x_{\mathrm{p}}(t))] + B_{\mathrm{r}}\mathrm{r}(t)$,
	$y_{\mathrm{reg}}(t) = C_{\mathrm{Reg}}x(t)$,
	$y(t) = Cx(t)$
Feedback control law	$u(t) = u_n(t) + u_a(t)$
Nominal control law	$u_n(t) = -K_x\hat{x}(t)$
Adaptive control law	$u_a(t) = -\hat{W}^{\mathrm{T}}\sigma(\hat{x}_{\mathrm{p}}(t), u_n(t))$,
	$\dot{\hat{W}}(t) = \Gamma\,\mathrm{Proj}_{\mathrm{m}}[\hat{W}(t), -\sigma(\hat{x}_{\mathrm{p}}(t), u_n(t))\,(\hat{y}(t) - y_s(t))^{\mathrm{T}} R_0^{-\frac{1}{2}} ZS^{\mathrm{T}}]$
State observer	$\dot{\hat{x}}(t) = A_{\mathrm{m}}\,\hat{x}(t) + L_v(y_s(t) - \hat{y}(t)) + B_{\mathrm{r}}\mathrm{r}(t)$,
	$\hat{y}(t) = C\hat{x}(t)$

$$E_1 : \|y_s(t) - y(t)\| \leq \epsilon_y, \tag{40}$$

$$E_2 : \|u_s(t) - u(t)\| \leq \epsilon_u, \tag{41}$$

$$E_3 : \text{The controller receives } y_s(t). \tag{42}$$

Specifically, when the inequality (40) is violated at the s_k moment of the kth time instant, the uncertain dynamical system triggers the measured output signal information such that $y_s(t)$ is sent to the controller. Likewise, when (41) is violated or the controller receives a new transmitted system output from the uncertain dynamical system (i.e., when $\bar{E}_2 \vee E_3$ is true), then the feedback controller sends a new control input $u_s(t)$ to the uncertain dynamical system at the r_j moment of the j th time instant.

Finally, using the definitions given by (28), (29), and (30), we write

$$\dot{\hat{e}}(t) = A_m \hat{e}(t) + L_v(y_s(t) - \hat{y}(t)),\ \hat{e}(0) = \hat{e}_0, \tag{43}$$

$$\dot{\tilde{x}}(t) = A_v \tilde{x}(t) + L_v(y_s(t) - y(t)) + B\Lambda\,(\tilde{W}^{\mathrm{T}}\sigma\,(\hat{x}_p(t), u_n(t)) + g\,(\cdot)) - B\Lambda\,(u_s(t) - u(t)),$$
$$\tilde{x}(0) = \tilde{x}_0. \tag{44}$$

In the next section, we analyze the stability and performance of the proposed event-triggered output feedback adaptive control algorithm introduced in this section (see Table 1) using the error dynamics given by (43) and (44) well as the data exchange rules E_1, E_2, and E_3 respectively given by (40), (41), and (42).

STABILITY AND PERFORMANCE ANALYSIS

For organizational purposes, this section is divided into three subsections. Specifically, we analyze the uniform ultimate boundedness of the resulting closed-loop dynamical system in Section (Uniform Ultimate Boundedness Analysis), compute the ultimate bound and highlight the effect of user-defined thresholds and the adaptive controller design parameters to this ultimate bound in Section (Ultimate Bound Computation), and show that the proposed architecture does not yield to a Zeno behavior in Section (Zeno Behavior Analysis).

Uniform Ultimate Boundedness Analysis

The following theorem presents the first result of this chapter.

Theorem 1. *Consider the uncertain dynamical system given by (33) and (34), the reference model given by (26) and (27), the state observer given by (38) and (39) with the state observer gain matrix in (19) along with (20), (21), and (22), and the feedback control law given by (11), (12), (14), and (37). In addition, let the data transmission from the uncertain dynamical system to the controller occur when $\bar{E}_1$ is true and the data transmission from the controller to the uncertain dynamical system occur when $\bar{E}_2 \vee E_3$ is true. Then, the closed-loop solution $(\tilde{x}(t), \tilde{W}(t), \hat{e}(t))$ is uniformly ultimately bounded for all initial conditions.*

Proof. Since the data transmission from the uncertain dynamical system to the controller and from the controller to the uncertain dynamical system occur when $\bar{E}_1$ and $\bar{E}_2 \vee E_3$ are true, respectively, note that $\|y_s(t) - y(t)\| \leq \epsilon_y$ and $\|u_s(t) - u(t)\| \leq \epsilon_u$ hold.

Consider the Lyapunov-like function given by

$$\mathcal{V}(\tilde{x}, \tilde{W}, \hat{e}) = \tilde{x}^{\mathrm{T}}\tilde{P}_v\tilde{x} + \mathrm{tr}\left((\tilde{W}\Lambda^{\frac{1}{2}})^{\mathrm{T}}\Gamma^{-1}(\tilde{W}\Lambda^{\frac{1}{2}})\right) + \beta\hat{e}^{\mathrm{T}}P\hat{e}, \tag{45}$$

where $\tilde{P}_v \in \mathbb{R}_+^{n\times n}$ is a solution to (25) with $R_v \in \mathbb{R}_+^{l\times l}$ and $Q_v \in \mathbb{R}_+^{n\times n}$, $v \in \mathbb{R}_+$, $\eta \in \mathbb{R}_+$, $\beta \in \mathbb{R}_+$, and $P \in \mathbb{R}_+^{n\times n} \cap \mathbb{S}^{n\times n}$ is a solution to

$$0 = A_{\mathrm{m}}^{\mathrm{T}} P + PA_{\mathrm{m}} - PBR^{-1}B^{\mathrm{T}}P + Q, \tag{46}$$

with $R \in \mathbb{R}_+^{m\times m}$ and $Q \in \mathbb{R}_+^{n\times n}$. Note that $\mathcal{V}(0, 0, 0) = 0$ and $\mathcal{V}(\tilde{x}, \tilde{W}, \hat{e}) > 0$ for all $(\tilde{x}, \tilde{W}, \hat{e}) \neq (0, 0, 0)$.

The time-derivative of (45) is given by

$$\begin{aligned}
&\dot{\mathcal{V}}(\tilde{x}(t), \tilde{W}(t), \hat{e}(t)) \\
&= 2\tilde{x}^{\mathrm{T}}\tilde{P}_v\dot{\tilde{x}} + 2\mathrm{tr}(\tilde{W}^{\mathrm{T}}(t)\Gamma^{-1}\dot{\tilde{W}}(t)\Lambda) + 2\beta\hat{e}^{\mathrm{T}}P\dot{\hat{e}} \\
&= 2\tilde{x}^{\mathrm{T}}\tilde{P}_v\Big(A_v\tilde{x}(t) + L_v(y_{\mathrm{s}}(t) - y(t)) + B\Lambda(\tilde{W}^{\mathrm{T}}\sigma(\hat{x}_{\mathrm{p}}(t), u_n(t)) + g(\cdot)) \\
&\quad - B\Lambda(u_s(t) - u(t))\Big) + 2\mathrm{tr}(\tilde{W}^{\mathrm{T}}(t)\Gamma^{-1}\dot{\tilde{W}}(t)\Lambda) + 2\beta\hat{e}^{\mathrm{T}}P\dot{\hat{e}} \\
&= -\tilde{x}^{\mathrm{T}}(C^{\mathrm{T}}R_v^{-1}C + \tilde{P}_vQ_v\tilde{P}_v + 2\eta\tilde{P}_v)\tilde{x} + 2\tilde{x}^{\mathrm{T}}\tilde{P}_vL_v(y_{\mathrm{s}}(t) - y(t)) + 2\tilde{x}^{\mathrm{T}}\tilde{P}_vB\Lambda\tilde{W}^{\mathrm{T}}\sigma(\hat{x}_{\mathrm{p}}(t), u_n(t)) \\
&\quad + 2\tilde{x}^{\mathrm{T}}\tilde{P}_vB\Lambda g(\cdot) - 2\tilde{x}^{\mathrm{T}}\tilde{P}_vB\Lambda(u_s(t) - u(t)) + 2\mathrm{tr}\Big(\tilde{W}^{\mathrm{T}}(t)\Gamma^{-1}\dot{\tilde{W}}(t)\Lambda\Big) + 2\beta\hat{e}^{\mathrm{T}}P\dot{\hat{e}} \\
&= -\left(1 + \frac{1}{v}\right)\tilde{x}^{\mathrm{T}}C^{\mathrm{T}}R_0^{-1}C\tilde{x} - \tilde{x}^{\mathrm{T}}\tilde{P}_vQ_0\tilde{P}_v\tilde{x} - \left(1 + \frac{1}{v}\right)\tilde{x}^{\mathrm{T}}\tilde{P}_vB_{\mathrm{s}}B_{\mathrm{s}}^{\mathrm{T}}\tilde{P}_v\tilde{x} - 2\eta\tilde{x}^{\mathrm{T}}\tilde{P}_v\tilde{x} \\
&\quad + 2\tilde{x}^{\mathrm{T}}\Big(C^{\mathrm{T}}R_0^{-\frac{1}{2}}ZS^{\mathrm{T}} + O(v)\Big)\Lambda\tilde{W}^{\mathrm{T}}\sigma(\hat{x}_{\mathrm{p}}(t), u_n(t)) + 2\tilde{x}^{\mathrm{T}}\tilde{P}_vB\Lambda g(\cdot) \\
&\quad + 2\tilde{x}^{\mathrm{T}}\tilde{P}_vL_v(y_{\mathrm{s}}(t) - y(t)) - 2\tilde{x}^{\mathrm{T}}\tilde{P}_vB\Lambda(u_s(t) - \mathrm{u}(t)) + 2\mathrm{tr}\Big(\tilde{W}^{\mathrm{T}}(t)\Gamma^{-1}\dot{\tilde{W}}(t)\Lambda\Big) + 2\beta\hat{e}^{\mathrm{T}}P\dot{\hat{e}} \\
&= -\left(1 + \frac{1}{v}\right)\tilde{x}^{\mathrm{T}}C^{\mathrm{T}}R_0^{-1}C\tilde{x} - \tilde{x}^{\mathrm{T}}\tilde{P}_vQ_0\tilde{P}_v\tilde{x} - \left(1 + \frac{1}{v}\right)\tilde{x}^{\mathrm{T}}\tilde{P}_vB_{\mathrm{s}}B_{\mathrm{s}}^{\mathrm{T}}\tilde{P}_v\tilde{x} - 2\eta\tilde{x}^{\mathrm{T}}\tilde{P}_v\tilde{x} \\
&\quad + 2\tilde{x}^{\mathrm{T}}O(v)\Lambda\tilde{W}^{\mathrm{T}}\sigma(\hat{x}_{\mathrm{p}}(t), u_n(t)) + 2\tilde{x}^{\mathrm{T}}\tilde{P}_vB\Lambda g(\cdot) + 2\tilde{x}^{\mathrm{T}}\tilde{P}_vL_v(y_{\mathrm{s}}(t) - y(t)) \\
&\quad - 2\tilde{x}^{\mathrm{T}}\tilde{P}_vB\Lambda(u_s(t) - u(t)) + 2\mathrm{tr}\Big(\tilde{W}^{\mathrm{T}}(t)\Big(\Gamma^{-1}\dot{\tilde{W}}(t) + \sigma(\hat{x}_{\mathrm{p}}(t), u_n(t))\tilde{y}^{\mathrm{T}}R_0^{-\frac{1}{2}}ZS^{\mathrm{T}}\Big)\Lambda\Big) \\
&\quad + 2\beta\hat{e}^{\mathrm{T}}P(\mathrm{A}_{\mathrm{m}}\hat{e}(t) + L_v(y_{\mathrm{s}}(t) - \hat{y}(t))).
\end{aligned} \tag{47}$$

Now, noting $\|O(v)\| \leq vK$, $K \in \mathbb{R}_+$, and using (37) in (47) yields

$$\begin{aligned}
&\dot{\mathcal{V}}(\tilde{x}(t), \tilde{W}(t), \hat{e}(t)) \\
&\leq -\left(1 + \frac{1}{v}\right)\tilde{x}^{\mathrm{T}}C^{\mathrm{T}}R_0^{-1}C\tilde{x} - \tilde{x}^{\mathrm{T}}\tilde{P}_vQ_0\tilde{P}_v\tilde{x} - \left(1 + \frac{1}{v}\right)\tilde{x}^{\mathrm{T}}\tilde{P}_vB_{\mathrm{s}}B_{\mathrm{s}}^{\mathrm{T}}\tilde{P}_v\tilde{x} - 2\eta\tilde{x}^{\mathrm{T}}\tilde{P}_v\tilde{x} \\
&\quad + 2\tilde{x}^{\mathrm{T}}O(v)\Lambda\tilde{W}^{\mathrm{T}}\sigma(\hat{x}_{\mathrm{p}}(t), u_n(t)) + 2\tilde{x}^{\mathrm{T}}\tilde{P}_vB\Lambda g(\cdot) + 2\tilde{x}^{\mathrm{T}}\tilde{P}_vL_v(y_{\mathrm{s}}(t) - y(t)) \\
&\quad - 2\tilde{x}^{\mathrm{T}}\tilde{P}_vB\Lambda(u_s(t) - u(t)) + 2(y_{\mathrm{s}}(t) - y(t))^{\mathrm{T}}R_0^{-\frac{1}{2}}ZS^{\mathrm{T}}\Lambda\tilde{W}^{\mathrm{T}}(t)\,\sigma(\hat{x}_{\mathrm{p}}(t), u_n(t)) \\
&\quad - \beta\hat{e}^{\mathrm{T}}(t)(-PBR^{-1}\mathrm{B}^{\mathrm{T}}P + Q)\hat{e}(t) - 2\beta\hat{e}^{\mathrm{T}}(t)PL_vC\tilde{x}(t) + 2\beta\hat{e}^{\mathrm{T}}(t)PL_v(y_{\mathrm{s}}(t) - y(t)) \\
&\leq -\left(1 + \frac{1}{v}\right)\lambda_{\min}(R_0^{-1})\,\|C\|_{\mathrm{F}}^2\|\tilde{x}(t)\|^2 - \lambda_{\min}(Q_0)\,\lambda_{\min}^2(\tilde{P}_v)\|\tilde{x}(t)\|^2 - \left(1 + \frac{1}{v}\right)\lambda_{\min}^2(\tilde{P}_v)
\end{aligned}$$

$$\cdot\|B_s\|_F^2\|\tilde{x}(t)\|^2 - 2\eta\,\lambda_{\min}(\tilde{P}_v)\|\tilde{x}(t)\|^2 - 2K\,v\|\Lambda\|_F\|\tilde{W}\|_F\|\sigma(\hat{x}_p(t), u_n(t))\|\,\|\tilde{x}(t)\|$$
$$+ 2\|\tilde{x}\|\,\|\tilde{P}_v B\|_F\|\Lambda\|_F\|g(\cdot)\| + 2\|\tilde{x}\|\,\|\tilde{P}_v\|_F\|L_v\|_F\epsilon_y + 2\|\tilde{x}\|\,\|\tilde{P}_v B\|_F\|\Lambda\|_F\epsilon_u$$
$$+ 2\epsilon_y\lambda_{\min}(R_0^{-\frac{1}{2}})\|ZS^T\|_F\|\Lambda\|_F\|\tilde{W}(t)\|_F\|\,\sigma(\hat{x}_p(t), u_n(t))\| - \beta(\lambda_{\min}(Q) - \lambda_{\max}(R^{-1})$$
$$\cdot\|PB\|_F^2)\|\hat{e}(t)\|^2 + 2\beta\|\hat{e}(t)\|\,\|PL_vC\|_F\|\tilde{x}(t)\| + 2\beta\|\hat{e}(t)\|\,\|P\|_F\|L_v\|_F\epsilon_y. \tag{48}$$

Next, using (5), an upper bound for $\|g(\cdot)\|$ in (48) is given by

$$\begin{aligned}\|g(\cdot)\| &= \|W^T(\sigma(\hat{x}_p(t), u_n(t)) - \sigma(x_p(t), u_n(t))\| \\ &\le \underbrace{W_{\max}L_\sigma}_{K_g}\|\hat{x}_p(t) - x_p(t)\| \\ &\le K_g\|\tilde{x}(t)\|,\end{aligned} \tag{49}$$

where $K_g \in \mathbb{R}_+$ and $\|W\|_F \le W_{\max}$, $W_{\max} \in \mathbb{R}_+$. In addition, noting $\|\hat{x}_p(t)\| \le \|\hat{x}(t)\|$ and using (5), one can compute an upper bound for $\|\sigma(\hat{x}_p(t), u_n(t))\|$ in (48) as

$$\begin{aligned}\|\sigma(\hat{x}_p(t), u_n(t))\| &= \|\sigma(\hat{x}_p(t), u_n(t)) + \sigma(0) - \sigma(0)\| \\ &\le \|\sigma(0)\| + \|\sigma(\hat{x}_p(t), u_n(t)) - \sigma(0)\| \\ &\le b_\sigma + \left\|\begin{matrix}\sigma(\hat{x}_p(t)) - \sigma(0) \\ u_n(t)\end{matrix}\right\| \\ &\le b_\sigma + \sqrt{\|\sigma(\hat{x}_p(t)) - \sigma(0)\|^2 + \|K_x\|^2\|\hat{x}(t)\|^2} \\ &\le b_\sigma + \sqrt{L_\sigma^2\|\hat{x}(t)\|^2 + \|K_x\|^2\|\hat{x}(t)\|^2} \\ &\le b_\sigma + \sqrt{L_\sigma^2 + \|K_x\|^2}\,\|\hat{x}(t)\|.\end{aligned} \tag{50}$$

Furthermore, since A_m is Hurwitz and r(t) is bounded in (38), there exist constants ζ_1 and ζ_2 such that $\|\hat{x}(t)\| \le \zeta_1 + \zeta_2\|y_s(t) - \hat{y}(t)\|$ holds (Lavretsky 2010), where this yields

$$\|\hat{x}(t)\| \le \zeta_1 + \zeta_2\epsilon_y + \zeta_2\|C\|_F\|\tilde{x}(t)\|. \tag{51}$$

Finally, using (51) in (50) gives

$$\begin{aligned}\|\sigma(\hat{x}_p(t), u_n(t))\| &\le b_\sigma + \sqrt{L_\sigma^2 + \|K_x\|^2}(\zeta_1 + \zeta_2\epsilon_y + \zeta_2\|C\|_F\|\tilde{x}(t)\| \\ &= b_1 + b_2\epsilon_y + b_3\|\tilde{x}(t)\|,\end{aligned} \tag{52}$$

where $b_1 \triangleq b_\sigma + \zeta_1\sqrt{L_\sigma^2 + \|K_x\|^2}$, $b_2 \triangleq \zeta_2\sqrt{L_\sigma^2 + \|K_x\|^2}$, and $b_3 \triangleq \zeta_2\|C\|_F\sqrt{L_\sigma^2 + \|K_x\|^2}$.

Noting that $\lambda_{\min}(\tilde{P}_v) \ge \lambda_{\min}(\tilde{P}_0) > 0$ (Lavretsky and Wise 2012) and using the bounds given by (49) and (52) in (48), one can write

$$\dot{\mathcal{V}}(\tilde{x}(t), \tilde{W}(t), \hat{e}(t))$$
$$\le -\left(1 + \frac{1}{v}\right)\lambda_{\min}(R_0^{-1})\,\|C\|_F^2\|\tilde{x}(t)\|^2 - \lambda_{\min}(Q_0)\,\lambda_{\min}^2(\tilde{P}_0)\,\|\tilde{x}(t)\|^2 - \left(1 + \frac{1}{v}\right)\|B_s\|_F^2\,\lambda_{\min}^2(\tilde{P}_0)$$

$$\cdot\|\tilde{x}(t)\|^2 - 2\eta\lambda_{\min}(\tilde{P}_0)\|\tilde{x}(t)\|^2 + 2K\, v\|\Lambda\|_F\|\tilde{W}\|_F(b_1 + b_2\epsilon_y + b_3\|\tilde{x}(t)\|)\,\|\tilde{x}(t)\|$$
$$+ 2\,\|\tilde{P}_v B\|_F\|\Lambda\|_F K_g\|\tilde{x}(t)\|^2 + 2\|\tilde{P}_v\|_F\|L_v\|_F\epsilon_y\|\tilde{x}\| + 2\,\|\tilde{P}_v B\|_F\|\Lambda\|_F\epsilon_u\|\tilde{x}\|$$
$$+ 2\epsilon_y\lambda_{\min}(R_0^{-\frac{1}{2}})\,\|ZS^{\mathrm{T}}\|_F\,\|\Lambda\|_F\,\|\tilde{W}\|_F(b_1 + b_2\epsilon_y + b_3\|\tilde{x}(t)\|) - \beta(\lambda_{\min}(Q) - \lambda_{\max}(R^{-1})$$
$$\cdot\|PB\|_F^2)\,\|\hat{e}^{\mathrm{T}}(t)\|_2^2 + 2\beta\|\hat{e}(t)\|\,\|PL_vC\|_F\|\tilde{x}(t)\| + 2\beta\|P\|_F\|L_v\|_F\epsilon_y\|\hat{e}(t)\|$$
$$= -\Big[\Big(1 + \frac{1}{v}\Big)\lambda_{\min}(R_0^{-1})\,\|C\|_F^2 + \lambda_{\min}(Q_0)\,\lambda_{\min}^2(\tilde{P}_0) + \Big(1 + \frac{1}{v}\Big)\|B_s\|_F^2\,\lambda_{\min}^2(\tilde{P}_0) + 2\eta\lambda_{\min}(\tilde{P}_0)$$
$$- 2K\, v\|\Lambda\|_F\|\tilde{W}\|_F b_3 - 2\,\|\tilde{P}_v B\|_F\|\Lambda\|_F K_g\Big]\|\tilde{x}(t)\|^2 + \Big[2K\, v\|\Lambda\|_F\|\tilde{W}\|_F(b_1 - b_2\epsilon_y)$$
$$+ 2\|\tilde{P}_v\|_F\|L_v\|_F\epsilon_y + 2\,\|\tilde{P}_v B\|_F\,\|\Lambda\|_F\epsilon_u + 2\epsilon_y\lambda_{\min}(R_0^{-\frac{1}{2}})\,\|ZS^{\mathrm{T}}\|_F\,\|\Lambda\|_F\,\|\tilde{W}(t)\|_F\, b_3\Big]\|\tilde{x}(t)\|$$
$$- \beta(\lambda_{\min}(Q) - \lambda_{\max}(R^{-1})\|PB\|_F^2)\|\hat{e}(t)\|_2^2 + 2\beta\|\hat{e}(t)\|\,\|PL_vC\|_F\|\tilde{x}(t)\|$$
$$+ 2\beta\|P\|_F\|L_v\|_F\epsilon_y\|\hat{e}(t)\| + 2\epsilon_y\,\lambda_{\min}(R_0^{-\frac{1}{2}})\,\|ZS^{\mathrm{T}}\|_F\,\|\Lambda\|_F\,\|\tilde{W}(t)\|_F\,(b_1 + b_2\epsilon_y). \tag{53}$$

Moreover, consider $2xy \le \alpha x^2 + \frac{1}{\alpha}y^2$ that follows from Young's inequality (Bernstein 2009) applied to scalars in $x \in \mathbb{R}$ and $y \in \mathbb{R}$, where $\alpha \in \mathbb{R}_+$. Using this inequality for the $2\beta\|\hat{e}(t)\|\,\|PL_vC\|_F\|\tilde{x}(t)\|$ term in (53) yields

$$\dot{\mathcal{V}}(\tilde{x}(t), \tilde{W}(t), \hat{e}(t))$$
$$\le -\Big[\Big(1 + \frac{1}{v}\Big)\lambda_{\min}(R_0^{-1})\,\|C\|_F^2 + \lambda_{\min}(Q_0)\,\lambda_{\min}^2(\tilde{P}_0) + \Big(1 + \frac{1}{v}\Big)\|B_s\|_F^2\,\lambda_{\min}^2(\tilde{P}_0) + 2\eta\lambda_{\min}(\tilde{P}_0)$$
$$- 2K\, v\|\Lambda\|_F\|\tilde{W}\|_F b_3 - 2\,\|\tilde{P}_v B\|_F\|\Lambda\|_F K_g\Big]\|\tilde{x}(t)\|^2 + \Big[2K\, v\|\Lambda\|_F\|\tilde{W}\|_F(b_1 + b_2\epsilon_y)$$
$$+ 2\|\tilde{P}_v\|_F\|L_v\|_F\epsilon_y + 2\,\|\tilde{P}_v B\|_F\,\|\Lambda\|_F\epsilon_u + 2\epsilon_y\lambda_{\min}(R_0^{-\frac{1}{2}})\,\|ZS^{\mathrm{T}}\|_F\,\|\Lambda\|_F\,\|\tilde{W}(t)\|_F\, b_3\Big]\|\tilde{x}(t)\|$$
$$- \beta(\lambda_{\min}(Q) - \lambda_{\max}(R^{-1})\|PB\|_F^2)\|\hat{e}(t)\|_2^2 + \alpha\|PL_vC\|_F^2\|\tilde{x}(t)\|_2^2 + \frac{\beta^2}{\alpha}\,\|\hat{e}(t)\|_2^2$$
$$+ 2\beta\|P\|_F\|L_v\|_F\epsilon_y\|\hat{e}(t)\| + 2\epsilon_y\,\lambda_{\min}(R_0^{-\frac{1}{2}})\,\|ZS^{\mathrm{T}}\|_F\,\|\Lambda\|_F\,\|\tilde{W}(t)\|_F\,(b_1 + b_2\epsilon_y)$$
$$= -\Big[\Big(1 + \frac{1}{v}\Big)\lambda_{\min}(R_0^{-1})\,\|C\|_F^2 + \lambda_{\min}(Q_0)\,\lambda_{\min}^2(\tilde{P}_0) + \Big(1 + \frac{1}{v}\Big)\|B_s\|_F^2\,\lambda_{\min}^2(\tilde{P}_0) + 2\eta\lambda_{\min}(\tilde{P}_0)$$
$$- 2K\, v\|\Lambda\|_F\|\tilde{W}\|_F b_3 - 2\,\|\tilde{P}_v B\|_F\|\Lambda\|_F K_g - \alpha\|PL_vC\|_F^2\Big]\|\tilde{x}(t)\|^2 + \Big[2K\, v\|\Lambda\|_F\|\tilde{W}\|_F(b_1$$
$$+ b_2\epsilon_y) + 2\|\tilde{P}_v\|_F\|L_v\|_F\epsilon_y + 2\,\|\tilde{P}_v B\|_F\,\|\Lambda\|_F\epsilon_u + 2\epsilon_y\lambda_{\min}(R_0^{-\frac{1}{2}})\,\|ZS^{\mathrm{T}}\|_F\,\|\Lambda\|_F\,\|\tilde{W}(t)\|_F\, b_3\Big]$$
$$\cdot\|\tilde{x}(t)\| - \Big[\beta(\lambda_{\min}(Q) - \lambda_{\max}(R^{-1})\|PB\|_F^2) - \frac{\beta^2}{\alpha}\Big]\|\hat{e}(t)\|_2^2 + 2\beta\|P\|_F\|L_v\|_F\epsilon_y\|\hat{e}(t)\|$$
$$+ 2\epsilon_y\lambda_{\min}(R_0^{-\frac{1}{2}})\,\|ZS^{\mathrm{T}}\|_F\,\|\Lambda\|_F\,\|\tilde{W}(t)\|_F\,(b_1 + b_2\epsilon_y)$$
$$\le -d_1\|\tilde{x}(t)\|^2 - d_2\|\hat{e}(t)\|_2^2 + d_3\|\tilde{x}(t)\| + d_4\|\hat{e}(t)\| + d_5 \tag{54}$$

where $d_1 \triangleq \Big(1 + \frac{1}{v}\Big)\lambda_{\min}(R_0^{-1})\|C\|_F^2 + \lambda_{\min}(Q_0)\,\lambda_{\min}^2(\tilde{P}_0) + \Big(1 + \frac{1}{v}\Big)\|B_s\|_F^2\,\lambda_{\min}^2(\tilde{P}_0) + 2\eta\lambda_{\min}(\tilde{P}_0) - 2K\,v\cdot\|\Lambda\|_F\tilde{w}^*\, b_3 - 2\,\|\tilde{P}_v B\|_F\|\Lambda\|_F K_g - \alpha\|PL_vC\|_F^2 \in \mathbb{R}_+$, $d_2 \triangleq \beta(\lambda_{\min}(Q) - \lambda_{\max}(R^{-1})\|PB\|_F^2) - \frac{\beta^2}{\alpha} \in \mathbb{R}_+$, $d_3 \triangleq 2K\,v\|\Lambda\|_F\,\tilde{w}^*\,(b_1 + b_2\epsilon_y) + \|\tilde{P}_v\|_F\|L_v\|_F\epsilon_y + 2\,\|\tilde{P}_v B\|_F\,\|\Lambda\|_F\epsilon_u +$

$2\epsilon_y \lambda_{\min}(R_0^{-\frac{1}{2}})\, ||ZS^{\mathrm{T}}||_{\mathrm{F}}\, ||\Lambda||_{\mathrm{F}}\, \tilde{w}^* b_3$, $d_4 = 2\beta ||P||_{\mathrm{F}} ||L_v||_{\mathrm{F}} \epsilon_y$, and $d_5 \triangleq 2\epsilon_y \lambda_{\min}(R_0^{-\frac{1}{2}})\, ||ZS^{\mathrm{T}}||_{\mathrm{F}}\, ||\Lambda||_{\mathrm{F}}$ $\tilde{w}^* (b_1 + b_2 \epsilon_y)$ with $||\tilde{W}(t)||_{\mathrm{F}} \le \tilde{w}^*$ due to utilizing the projection operator in the weight update law given by (37).

Finally, we rearrange (54) as

$$\dot{V}(\tilde{x}(t), \tilde{W}(t), \hat{e}(t)) \le -\left(\sqrt{d_1}\, ||\tilde{x}(t)|| - \frac{d_3}{2\sqrt{d_1}}\right)^2 - \left(\sqrt{d_2}\, ||\hat{e}(t)|| - \frac{d_4}{2\sqrt{d_2}}\right)^2 + \left(d_5 + \frac{d_3^2}{4d_1} + \frac{d_4^2}{4d_2}\right), \tag{55}$$

which shows that $\dot{V}(\tilde{x}(t), \tilde{W}(t), \hat{e}(t)) \le 0$ when $||\tilde{x}(t)|| \ge \psi_1$ and $||\hat{e}(t)|| \ge \psi_2$, where

$$\psi_1 \triangleq \frac{\frac{d_3}{2\sqrt{d_1}} + \sqrt{d_5 + \frac{d_3^2}{4d_1} + \frac{d_4^2}{4d_2}}}{\sqrt{d_1}} \tag{56}$$

$$\psi_2 \triangleq \frac{\frac{d_4}{2\sqrt{d_2}} + \sqrt{d_5 + \frac{d_3^2}{4d_1} + \frac{d_4^2}{4d_2}}}{\sqrt{d_2}} \tag{57}$$

This argument proves uniform ultimate boundedness of the closed-loop solution $(\tilde{x}(t), \tilde{W}(t), \hat{e}(t))$ for all initial conditions (Lavretsky and Wise 2012, Khalil 1996). ■

In the proof of Theorem 1, it is implicitly assumed that $d_1 \in \mathbb{R}_+$ and $d_2 \in \mathbb{R}_+$, which can be satisfied by suitable selection of the event-triggered output feedback adaptive controller design parameters. Although this theorem shows uniform ultimate boundedness of the closed-loop solution $(\tilde{x}(t), \tilde{W}(t), \hat{e}(t))$ for all initial conditions, it is of practical importance to compute the ultimate bound, which is given next.

Ultimate Bound Computation

For revealing the effect of user-defined thresholds and the event-triggered output feedback adaptive controller design parameters to the system performance, the next corollary presents a computation of the ultimate bound, which presents the second result of this chapter.

Corollary 1. Consider the uncertain dynamical system given by (33) and (34), the reference model given by (26) and (27), the state observer given by (38) and (39) with the state observer gain matrix in (19) along with (20), (21), and (22), and the feedback control law given by (11), (12), (14), and (37). In addition, let the data transmission from the uncertain dynamical system to the controller occur when $\bar{E}_1$ is true and the data transmission from the controller to the uncertain dynamical system occur when $\bar{E}_2 \vee E_3$ is true. Then, the ultimate bound of the system error between the uncertain dynamical system and the reference model is given by

$$\|e(t)\| = \|x(t) - x_{\mathrm{m}}(t)\| \le \tilde{\Phi}[\lambda_{\min}^{-1}(\tilde{P}_v) + \beta^{-1}\lambda_{\min}^{-1}(P)]^{\frac{1}{2}}, \quad t \ge T, \tag{58}$$

where $\tilde{\Phi} \triangleq [\lambda_{\max}(\tilde{P}_v)\psi_1^2 + \beta\lambda_{\max}(P)\psi_2^2 + \Gamma^{-1}\tilde{w}^{*2}\,\|\Lambda\|_{\mathrm{F}}]^{\frac{1}{2}}$.

Proof. It follows from the proof of Theorem 1 that $\dot{\mathcal{V}}(\tilde{x}(t), \tilde{W}(t), \hat{e}(t)) \le 0$ outside the compact set given by $\mathcal{S} \triangleq \{(\tilde{x}(t), \hat{e}(t)) : \|\tilde{x}(t)\| \le \psi_1\} \cap \{(\tilde{x}(t), \hat{e}(t)) : \|\hat{e}(t)\| \le \psi_2\}$. That is, since $\mathcal{V}(\tilde{x}(t), \tilde{W}(t), \hat{e}(t))$ cannot grow outside $\mathcal{S}$, the evolution of $\mathcal{V}(\tilde{x}(t), \tilde{W}(t), \hat{e}(t))$ is upper bounded by

$$\begin{aligned}\mathcal{V}(\tilde{x}(t), \tilde{W}(t), \hat{e}(t)) &\le \max_{(\tilde{x}(t),\hat{e}(t)) \in \mathcal{S}} \mathcal{V}(\tilde{x}(t), \tilde{W}(t), \hat{e}(t)) \\ &= \lambda_{\max}(\tilde{P}_v)\psi_1^2 + \beta\lambda_{\max}(P)\psi_2^2 + \Gamma^{-1}\tilde{w}^{*2}\,\|\Lambda\|_{\mathrm{F}} \\ &= \tilde{\Phi}^2.\end{aligned} \tag{59}$$

It follows from $\tilde{x}^{\mathrm{T}}\tilde{P}_v\tilde{x} \le \mathcal{V}(\tilde{x}, \tilde{W}, \hat{e})$ and $\beta\hat{e}^{\mathrm{T}}P\hat{e} \le \mathcal{V}(\tilde{x}, \tilde{W}, \hat{e})$ that $\|\tilde{x}(t)\|^2 \le \dfrac{\tilde{\Phi}^2}{\lambda_{\min}(\tilde{P}_v)}$ and $\|\hat{e}(t)\|^2 \le \dfrac{\tilde{\Phi}^2}{\beta\lambda_{\min}(P)}$. Finally, since $e(t) = x(t) - \hat{x}(t) + \hat{x}(t) - x_{\mathrm{m}}(t)$, and hence, $\|e(t)\| \le \|x(t) - \hat{x}(t)\| + \|\hat{x}(t) - x_{\mathrm{m}}(t)\| = \|\tilde{x}\| + \|\hat{e}(t)\|$, the bound given by (58) follows. ■

To elucidate the effect of the user-defined thresholds and the event-triggered output feedback adaptive controller design parameters to the ultimate bound given by (58), let $A_{\mathrm{r}} = -5$, $B = 1$, $C = 1$, $W = 1$, $R_o = 1$, $R = 1$, $Q_o = 1$, $Q = 1$, $\Lambda = 1$, $\alpha = 0.5$, and $\beta = 0.25$. In this case, Fig. 2 shows the effect of the variation in v and Γ on the system error bound for $\eta = 5$, $\epsilon_y = 0.1$ and $\epsilon_u = 0.1$. Specifically, one can conclude from this figure that increasing Γ reduces the ultimate bound and the minimum value of this bound is obtained for $v = 0.35$. Figure 3 shows the effect of the variation in v and η on the system error bound for $\Gamma = 100$ and the same previously defined parameters. It is evident from the figure, that increasing η increases the ultimate bound.

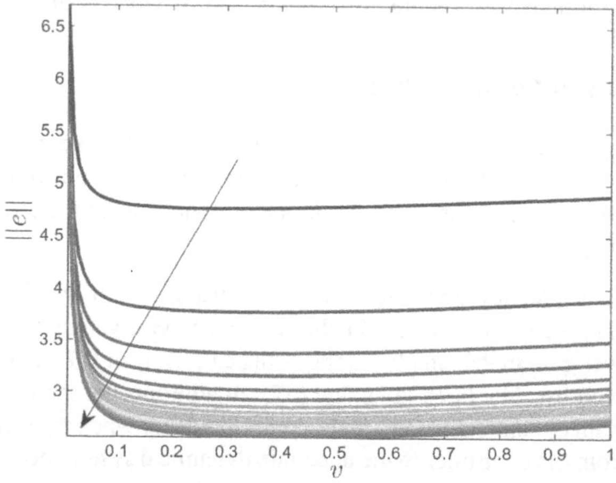

Figure 2. Effect of $\Gamma \in [5,100]$ and $v \in [0.01,1]$ to the ultimate bound (58) for $\eta = 5$, $\epsilon_y = 0.1$ and $\epsilon_u = 0.1$, where the arrow indicates the increase in Γ (dashed line denotes the case with $\Gamma = 100$).

This figure also shows that there exists an optimum value of v for each η value, which allows the selection of the best value of v to avoid increasing the ultimate bound.

Figures 4 and 5 show the effect of the variations in ϵ_y and ϵ_u, respectively. In particular, these figures show that the system error bound is more sensitive to the changes in the data transmission threshold from the physical system to the adaptive controller (sensing threshold, ϵ_y) than the data transmission threshold from the adaptive controller to the physical system (actuation threshold, ϵ_u), which implies that the actuation threshold can be chosen large enough to reduce wireless network utilization between the physical system and the adaptive controller without sacrificing closed-loop dynamical system performance.

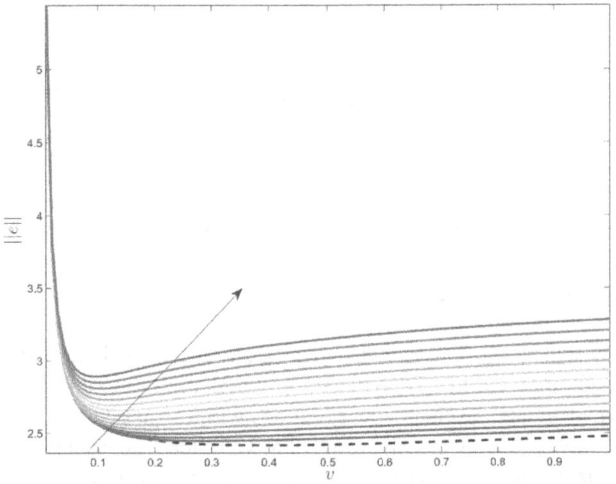

Figure 3. Effect of $\eta \in [5,20]$ to the ultimate bound (58) for $\epsilon_y = 0.1$, $\epsilon_u = 0.1$, $v \in [0.01,1]$ and $\Gamma = 100$, where the arrow indicates the increase in η (dashed line denotes the case with $\eta = 5$).

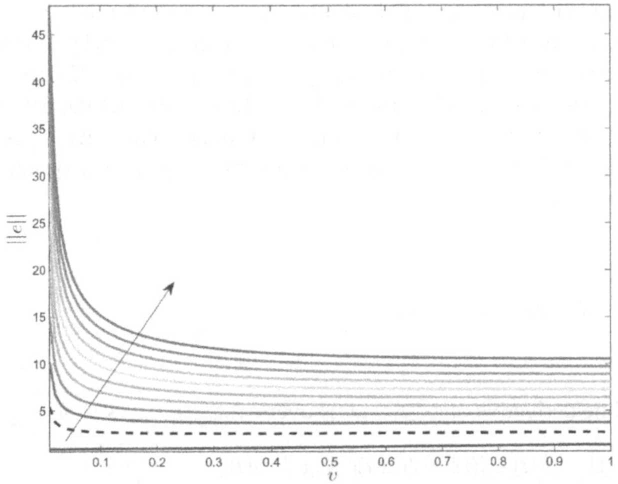

Figure 4. Effect of $\epsilon_y \in [0, 1]$ to the ultimate bound (58) for $\eta = 5$, $\epsilon_u = 0.1$, $v \in [0.01,1]$, and $\Gamma = 100$, where the arrow indicates the increase in ϵ_y (dashed line denotes the case with $\epsilon_y = 0.1$ and blue bottom line denotes the case with $\epsilon_y = 0$).

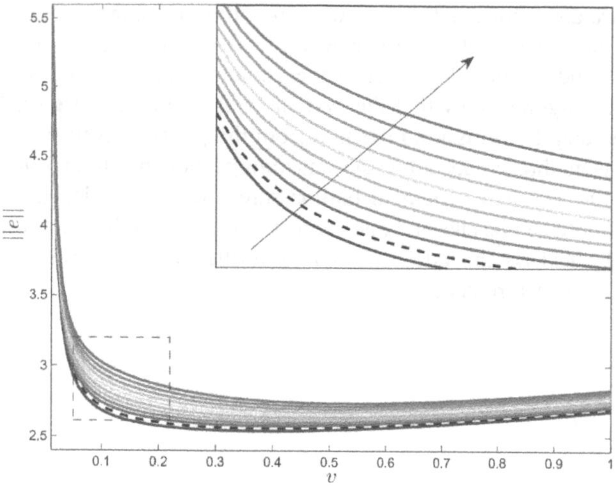

Figure 5. Effect of $\epsilon_u \in [0,1]$ to the ultimate bound (58) for $\eta = 5$, $\epsilon_y = 0.1$, $\nu \in [0.01,1]$, and $\Gamma = 100$, where the arrow indicates the increase in ϵ_u (dashed line denotes the case with $\epsilon_u = 0.1$).

Zeno Behavior Analysis

We now show that the proposed event-triggered output feedback adaptive control architecture does not yield to a Zeno behavior, which implies that it does not require a continuous two-way data exchange and reduces wireless network utilization. For the following corollary presenting the third result of this chapter, we consider $r_1^k \in (s_k, s_{k+1})$ to be the ith time instant when E_2 is violated over (s_k, s_{k+1}), and since $\{s_k\}_{k=1}^{\infty}$ is a subsequence of $\{r_j\}_{j=1}^{\infty}$, it follows that $\{r_j\}_{j=1}^{\infty} = \{s_k\}_{k=1}^{\infty} \cup \{r_i^k\}_{k=1,i=1}^{\infty,m_k}$, where $m_k \in \mathbb{N}$ is the number of violation times of E_2 over (s_k, s_{k+1}).

Corollary 2. Consider the uncertain dynamical system given by (33) and (34), the reference model given by (26) and (27), the state observer given by (38) and (39) with the state observer gain matrix in (19) along with (20), (21), and (22), and the feedback control law given by (11), (12), (14), and (37). In addition, let the data transmission from the uncertain dynamical system to the controller occur when $\bar{E}_1$ is true and the data transmission from the controller to the uncertain dynamical system occur when $\bar{E}_2 \vee E_3$ is true. Then,

$$s_{k+1} - s_k > 0, \quad \forall k \in \mathbb{N}, \tag{60}$$

$$r_{i+1}^k - r_i^k > 0, \quad \forall i \in \{0, ..., m_k\}, \quad \forall k \in \mathbb{N}, \tag{61}$$

holds.

Proof. The time derivative of $\|x_s(t) - x(t)\|$ over $t \in (s_k, s_{k+1})$, $\forall k \in \mathbb{N}$, is given by

$$\begin{aligned}\frac{d}{dt}\|y_s(t) - y(t)\| &\leq \|\dot{y}_s(t) - \dot{y}(t)\| = \|C\dot{x}(t)\| \leq \|C\|_F\|\dot{x}(t)\| \\ &\leq \|C\|_F\big[\|A\|_F\|x(t)\| + \|B\|_F\|\Lambda\|_F\|u_s(t)\| + \|B\|_F\|\Lambda\|_F\|W\|_F\|\sigma(x_p(t))\| \\ &\quad + \|B_r\|_F\|r(t)\|\big].\end{aligned} \tag{62}$$

Now, we determine an upper bound for $\|x(t)\|$ in (62) as

$$\begin{aligned}\|x(t)\| &= \|\tilde{x}(t) + \hat{x}(t)\| \leq \|\tilde{x}(t)\| + \zeta_1 + \zeta_2\epsilon_y + \zeta_2\|C\|_{\mathrm{F}}\|\tilde{x}(t)\| \\ &= \zeta_1 + \zeta_2\epsilon_y + (1 + \zeta_2\|C\|_{\mathrm{F}})\|\tilde{x}(t)\|. \end{aligned} \tag{63}$$

In addition, we determine an upper bound for $\|\sigma(x_{\mathrm{p}}(t))\|$ in (62) as

$$\begin{aligned}\|\sigma(x_{\mathrm{p}}(t))\| &= \|\sigma(x_{\mathrm{p}}(t)) - \sigma(\hat{x}_{\mathrm{p}}(t)) + \sigma(\hat{x}_{\mathrm{p}}(t))\| \\ &\leq L_\sigma\|\tilde{x}_{\mathrm{p}}(t)\| + \|\sigma(\hat{x}_{\mathrm{p}}(t))\| \\ &\leq L_\sigma\|\tilde{x}_{\mathrm{p}}(t)\| + \|\sigma(\hat{x}_{\mathrm{p}}(t)) - \sigma(0)\| + \|\sigma(0)\| \\ &\leq L_\sigma \underbrace{\|\tilde{x}_{\mathrm{p}}(t)\|}_{\leq \|\tilde{x}(t)\|} + L_\sigma \underbrace{\|\hat{x}_{\mathrm{p}}(t)\|}_{\leq \|\hat{x}(t)\|} + b_\sigma \\ &\leq L_\sigma\|\tilde{x}(t)\| + L_\sigma(\zeta_1 + \zeta_2\epsilon_y + \zeta_2\|C\|_{\mathrm{F}}\|\tilde{x}(t)\|) + b_\sigma \\ &= L_\sigma(1 + \zeta_2\|C\|_{\mathrm{F}})\|\tilde{x}(t)\| + L_\sigma(\zeta_1 + \zeta_2\epsilon_y) + b_\sigma. \end{aligned} \tag{64}$$

Substituting (63) and (64) into (62), gives

$$\begin{aligned}\frac{d}{dt}\|y_{\mathrm{s}}(t) - y(t)\| &\leq \|C\|_{\mathrm{F}}\|A\|_{\mathrm{F}}[\zeta_1 + \zeta_2\epsilon_y + (1 + \zeta_2\|C\|_{\mathrm{F}})\|\tilde{x}(t)\|] + \|C\|_{\mathrm{F}}\|B\|_{\mathrm{F}}\|\Lambda\|_{\mathrm{F}}\|u_s(t)\| \\ &\quad + \|C\|_{\mathrm{F}}\|B\|_{\mathrm{F}}\|\Lambda\|_{\mathrm{F}}W_{\max}[L_\sigma(1 + \zeta_2\|C\|_{\mathrm{F}})\|\tilde{x}(t)\| + L_\sigma(\zeta_1 + \zeta_2\epsilon_y) + b_\sigma] \\ &\quad + \|C\|_{\mathrm{F}}\|B_{\mathrm{r}}\|_{\mathrm{F}}\|\mathrm{r}(t)\|. \end{aligned} \tag{65}$$

Since the closed-loop dynamical system is uniformly ultimately bounded by Theorem 1, there exists an upper bound to (65). Letting Φ_1 denote this upper bound and with the initial condition satisfying $\lim_{t\to s_k^+}\|y_s(t) - y(t)\| = 0$, it follows from (65) that

$$\|y_{\mathrm{s}}(t) - y(t)\| \leq \Phi_1(t - s_k),\ \forall t \in (s_k, s_{k+1}). \tag{66}$$

Therefore, when $\overline{E}_1$ is true, then $\lim_{t\to s_{k+1}^-}\|y_s(t) - y(t)\| = \epsilon_y$ and it then follows from (66) that $s_{k+1} - s_k \geq \dfrac{\epsilon_y}{\Phi_1}$.

Similarly, the time derivative of $\|u_{\mathrm{s}}(t) - u(t)\|$ over $t \in (r_i^k, r_{i+1}^k)$, $\forall i \in \mathbb{N}$, is given by

$$\frac{d}{dt}\|u_{\mathrm{s}}(t) - u(t)\| \leq \|\dot{u}_{\mathrm{s}}(t) - \dot{u}(t)\| = \|\dot{u}(t)\| \leq \|\dot{u}_n(t)\| + \|\dot{u}_a(t)\|. \tag{67}$$

Now, we determine an upper bound for $\|\dot{u}_n(t)\|$ in (67) as

$$\begin{aligned}\|\dot{u}_n(t)\| &= \|K_x\dot{\hat{x}}(t)\| \\ &\leq \|K_x\|_{\mathrm{F}}\|\dot{\hat{x}}(t)\| \\ &\leq \|K_x\|_{\mathrm{F}}[\|A\|_{\mathrm{F}}\|\hat{x}(t)\| + \|B\|_{\mathrm{F}}\|u_n(t)\| + \|L_v\|_{\mathrm{F}}\|y_{\mathrm{s}}(t) - \hat{y}(t)\| + \|B_{\mathrm{r}}\|_{\mathrm{F}}\|\mathrm{r}(t)\|] \\ &\leq \|K_x\|_{\mathrm{F}}[\|A\|_{\mathrm{F}}[\zeta_1 + \zeta_2\epsilon_y + \zeta_2\|C\|_{\mathrm{F}}\|\tilde{x}(t)\|] + \|B\|_{\mathrm{F}}\|u_n(t)\| + \|L_v\|_{\mathrm{F}}\|C\|_{\mathrm{F}}\|\tilde{x}(t)\| \\ &\quad + \|L_v\|_{\mathrm{F}}\epsilon_y + \|B_{\mathrm{r}}\|_{\mathrm{F}}\|\mathrm{r}(t)\|]. \end{aligned} \tag{68}$$

Letting β_1 to denote the upper bound of $||\dot{u}_n(t)||$, we determine the upper bound of $||\dot{u}_a(t)||$ in (67) as

$$\begin{aligned}||\dot{u}_a(t)|| &= ||\dot{\hat{W}}^{\mathrm{T}}(t)\,\sigma(\hat{x}_{\mathrm{p}}(t), u_n(t)) + \hat{W}^{\mathrm{T}}(t)\dot{\sigma}\,(\hat{x}_{\mathrm{p}}(t), u_n(t))|| \\ &\le ||SZ^{\mathrm{T}} R_0^{\frac{-1}{2}}(\hat{y}(t) - y_s(t))\sigma^{\mathrm{T}}(\hat{x}_{\mathrm{p}}(t), u_n(t))\,\Gamma\sigma(\hat{x}_{\mathrm{p}}(t), u_n(t))|| \\ &\quad + ||\hat{W}(t)||_{\mathrm{F}}[||\dot{\sigma}_0(\hat{x}_{\mathrm{p}}(t))|| + ||\dot{u}_n(t)||] \\ &\le \lambda_{\max}(\Gamma)||SZ^{\mathrm{T}} R_0^{\frac{-1}{2}}||_{\mathrm{F}}||\sigma(\hat{x}_{\mathrm{p}}(t), u_n(t))||_{\mathrm{F}}^2\,\Phi_1 + ||\hat{W}(t)||_{\mathrm{F}}[\sigma^* + \beta_1],\end{aligned} \tag{69}$$

where $||\dot{\sigma}(\hat{x}_{\mathrm{p}}(t))|| \le \sigma^*$. Substituting (68) and (69) into (67), gives

$$\frac{d}{dt}||u_s(t) - u(t)|| \le \lambda_{\max}(\Gamma)||SZ^{\mathrm{T}} R_0^{\frac{-1}{2}}||_{\mathrm{F}}||\,\sigma\,(\hat{x}_{\mathrm{p}}(t), u_n(t))||^2\Phi_1 + ||\hat{W}(t)||_{\mathrm{F}}[\sigma^* + \beta_1] + \beta_1. \tag{70}$$

Once again, since the closed-loop dynamical system is uniformly ultimately bounded by Theorem 1, there exists an upper bound to (70). Letting Φ_2 denote this upper bound and with the initial condition satisfying $\lim_{t \to r_i^{k+}} ||u_s(t) - u(t)|| = 0$, it follows from (70) that

$$||u_s(t) - u(t)|| \le \Phi_2(t - r_i^k), \quad \forall\, t \in (r_i^k, r_{i+1}^k). \tag{71}$$

Therefore, when $\bar{E}_2 \vee E_3$ is true, then $\lim_{t \to r_{i+1}^{k-}} ||u_s(t) - u(t)|| = \epsilon_u$ and it then follows from (71) that $r_{i+1}^k - r_i^k \ge \dfrac{\epsilon_u}{\Phi_2}$. ■

Corollary 2 shows that the intersample times for the system output vector and feedback control vector are bounded away from zero, and hence, the proposed event-triggered adaptive control approach does not yield to a Zeno behavior.

ILLUSTRATIVE NUMERICAL EXAMPLE

In this section, the efficacy of the proposed event-triggered output feedback adaptive control approach is demonstrated in an illustrative numerical example. For this purpose, we consider the uncertain dynamical system given by

$$\begin{aligned}\begin{bmatrix}\dot{x}_{\mathrm{p}1}(t)\\ \dot{x}_{\mathrm{p}2}(t)\end{bmatrix} &= \begin{bmatrix}0 & 1\\ 0 & 0\end{bmatrix}\begin{bmatrix}x_{\mathrm{p}1}(t)\\ x_{\mathrm{p}2}(t)\end{bmatrix} + \begin{bmatrix}0\\ 1\end{bmatrix}\Lambda\,[u_s(t) + \Delta(x_{\mathrm{p}}(t))], \\ y_{\mathrm{p}}(t) &= \begin{bmatrix}1 & 0\\ 0.5 & 0.5\end{bmatrix}\begin{bmatrix}x_{\mathrm{p}1}(t)\\ x_{\mathrm{p}2}(t)\end{bmatrix}, \quad y_{\mathrm{reg}}(t) = [1\ \ 0]\begin{bmatrix}x_{\mathrm{p}1}(t)\\ x_{\mathrm{p}2}(t)\end{bmatrix}.\end{aligned} \tag{72}$$

For this study, let the uncertain parameters be $\Lambda = 0.5$ and $W = [-2\ ,\ 3]^{\mathrm{T}}$, and we choose $\sigma(x_{\mathrm{p}}(t)) = x_{\mathrm{p}}(t)$ as the basis function.

For the nominal control design, we note

$$A = \begin{bmatrix}0 & 1 & 0\\ 0 & 0 & 0\\ -1 & 0 & 0\end{bmatrix}, \quad B = \begin{bmatrix}0\\ 1\\ 0\end{bmatrix}, \quad B_{\mathrm{r}} = \begin{bmatrix}0\\ 0\\ 1\end{bmatrix},$$

$$C=\begin{bmatrix}1 & 0 & 0\\ 0.5 & 0.5 & 0\\ 0 & 0 & 1\end{bmatrix},\ C_{\text{Reg}}=[1\ \ 0\ \ 0]. \tag{73}$$

for (33) and (34). In particular, a linear quadratic regulator formulation is used to choose Kx of the nominal controller as

$$K_x = R_{\text{lqr}}^{-1}B^{\text{T}}P_{\text{lqr}}, \tag{74}$$

$$0=(A+\eta_{\text{lqr}}I_{n\times n})^{\text{T}}P_{\text{lqr}}+P_{\text{lqr}}(A+\eta_{\text{lqr}}I_{n\times n})-P_{\text{lqr}}BR_{\text{lqr}}^{-1}B^{\text{T}}P_{\text{lqr}}+Q_{\text{lqr}}, \tag{75}$$

where Q_{lqr} = diag([20, 3, 1]), $R_{\text{lqr}} = 0.5$, and $\eta_{\text{lqr}} = 0.2$ is considered, which yields K_x = [9.6, 5.2–3.6]. Next, for the adaptive control design, we choose

$$B_2=\begin{bmatrix}1 & 0\\ 1 & 0\\ 0 & 1\end{bmatrix}, \tag{76}$$

to square up the dynamical system [21], which results in

$$B_s=\begin{bmatrix}0 & 1 & 0\\ 1 & 1 & 0\\ 0 & 0 & 1\end{bmatrix}. \tag{77}$$

In particular, with (77), det(CBs) is nonzero and $G(s) = C(sI_{n\times n} - A)^{-1}B_s$ is minimum phase. To calculate the observer gain L_v given by (19), we set $Q_0 = I$, $R_0 = 30I$, $\eta = 10$, and $v = 0.1$ for (20), (21), and (22), which yields

$$L_v=\begin{bmatrix}20.24 & -18.79 & -0.97\\ 0.72 & 39.84 & -0.48\\ -0.97 & 0.01 & 20.16\end{bmatrix} \tag{78}$$

Finally, note that $d_1 \in \mathbb{R}_+$ and $d_2 \in \mathbb{R}_+$ for $\alpha = 1$ and $\beta = 1$.

Figure 6 presents the results with the proposed event-triggered output feedback adaptive control approach when $\epsilon_y = 0.3$, and $\epsilon_u = 0.3$ are chosen, where the output of the uncertain dynamical system achieves a good command following performance. In Figs. 7 and 8, we fix ϵ_y to 0.3 and change ϵ_u to 0.1 and 0.5, respectively. As expected from the proposed theory, the variation on ϵ_u does not alter the command following performance significantly. In addition, in Figs. 9 and 10, we fix ϵ_u to 0.3 and change ϵ_y to 0.1 and 0.5, respectively, where it can be seen that the variation on ϵ_y alters the command following performance more than the variation in ϵ_u, as discussed earlier in this chapter. Finally, output and control event triggers for the cases in Figs. 6–10 are given in Fig. 11, where it can be seen that increasing ϵ_y (respectively, ϵ_u) yields less output event triggers when ϵ_u (respectively, less control event triggers when ϵ_y) is fixed, which reduces network utilization.

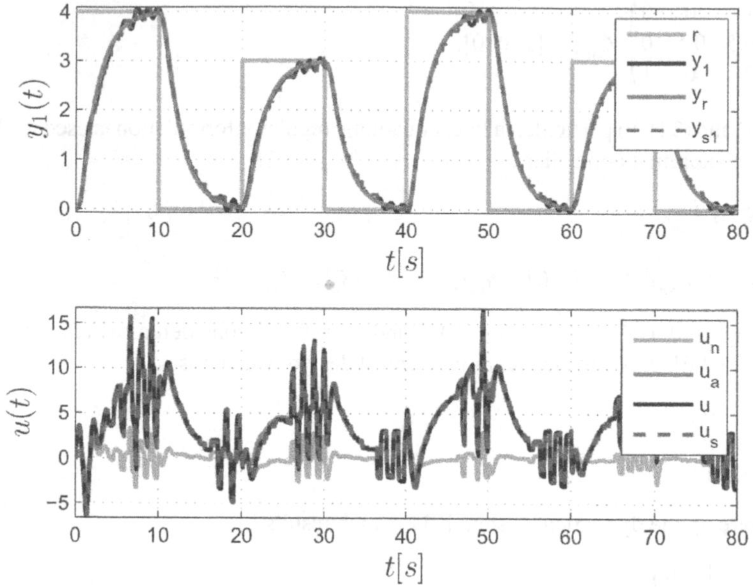

Figure 6. Command following performance for the proposed event-triggered output feedback adaptive control approach with $\Gamma = 50I$, $\epsilon_y = 0.3$, and $\epsilon_u = 0.3$.

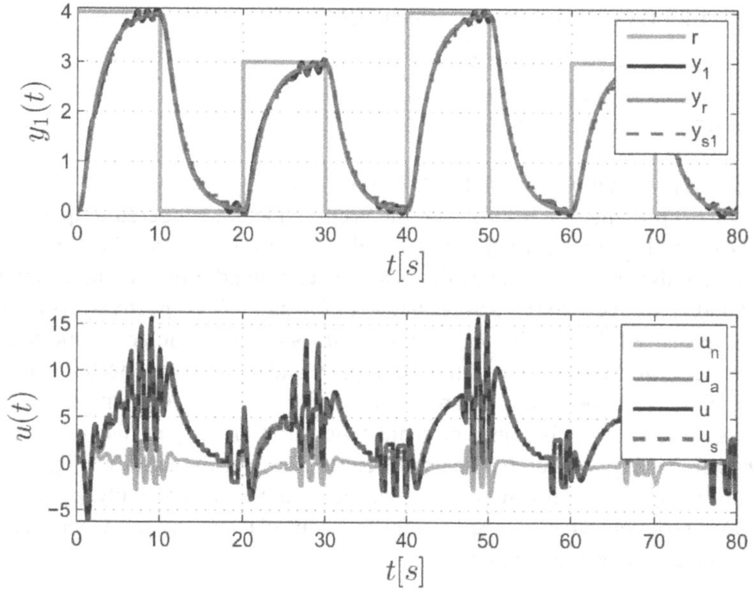

Figure 7. Command following performance for the proposed event-triggered output feedback adaptive control approach with $\Gamma = 50I$, $\epsilon_y = 0.3$, and $\epsilon_u = 0.1$.

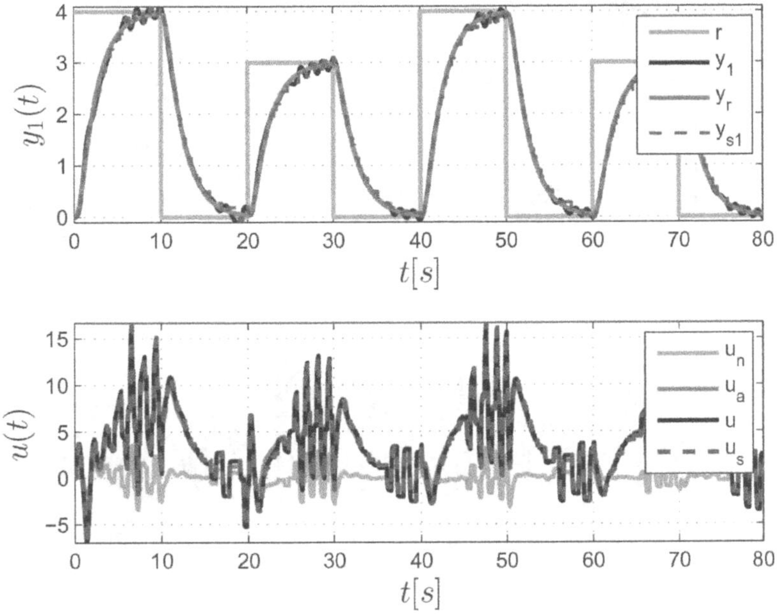

Figure 8. Command following performance for the proposed event-triggered output feedback adaptive control approach with $\Gamma = 50I$, $\epsilon_y = 0.3$, and $\epsilon_u = 0.5$.

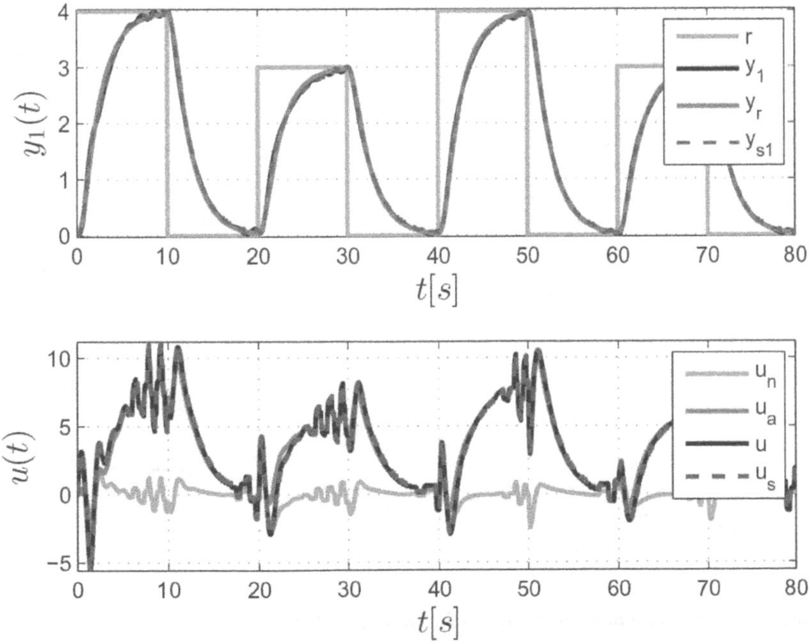

Figure 9. Command following performance for the proposed event-triggered output feedback adaptive control approach with $\Gamma = 50I$, $\epsilon_y = 0.1$, and $\epsilon_u = 0.3$.

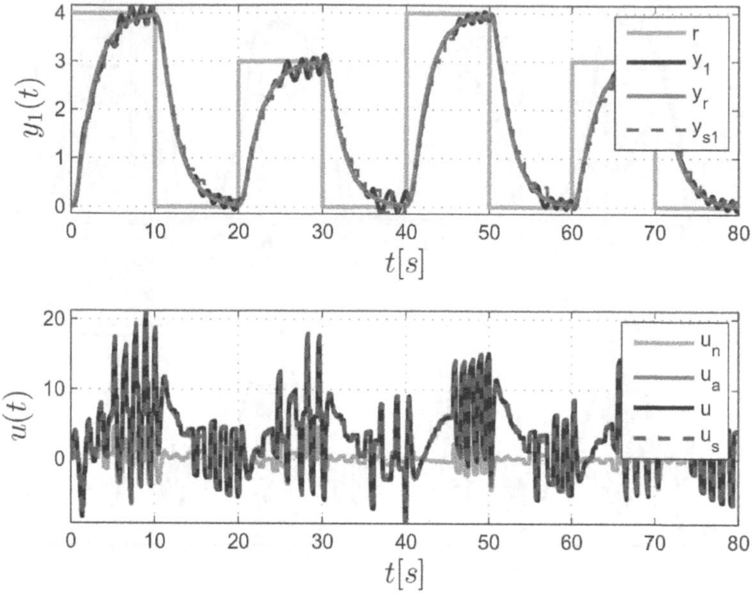

Figure 10. Command following performance for the proposed event-triggered output feedback adaptive control approach with $\Gamma = 50I$, $\epsilon_y = 0.5$, and $\epsilon_u = 0.3$.

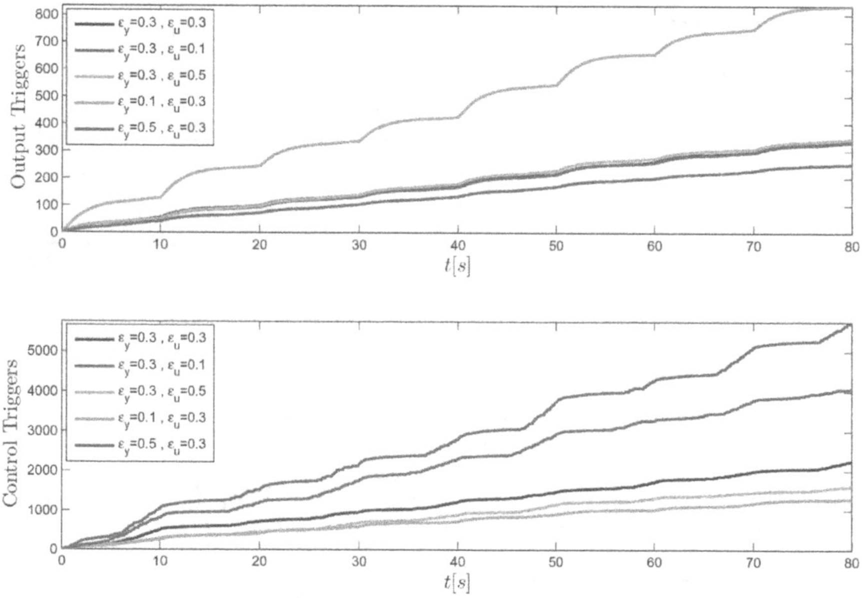

Figure 11. Output and control event triggers for the cases in Figs. 7–10.

CONCLUSION

A critical task in the design and implementation of networked control systems is to guarantee system stability while reducing wireless network utilization and achieving a given system performance in the presence of system uncertainties. Motivating from this standpoint, design and analysis of an event-triggered output feedback adaptive control methodology was presented for a class of uncertain dynamical systems in the presence of two-way data exchange between the physical system and the proposed controller over a wireless network. Specifically, we showed using tools and methods from nonlinear systems theory and Lyapunov stability in particular that the proposed feedback control approach guarantees system stability in the presence of system uncertainties. In addition, we characterized and discussed the effect of user-defined thresholds and output feedback adaptive controller design parameters to the system performance and showed that the proposed methodology does not yield to a Zeno behavior. Finally, we illustrated the efficacy of the proposed adaptive control approach in a numerical example.

REFERENCES

Albattat, A., B. C. Gruenwald and T. Yucelen. 2015. Event-triggered adaptive control, in ASME 2015 Dynamic Systems and Control Conference. American Society of Mechanical Engineers.

Alur, R., K. -E. Arzen, J. Baillieul, T. Henzinger, D. Hristu-Varsakelis and W. S. Levine. 2007. Handbook of networked and embedded control systems. Springer Science & Business.

Astrom, K. J. and B. Wittenmark. 2013. Adaptive Control. Courier Corporation.

Bemporad, A., M. Heemels and M. Johansson. 2010. Networked Control Systems. Springer.

Bernstein, D. S. 2009. Matrix mathematics: Theory, Facts, and Formulas. Princeton University Press.

Burg, T., D. Dawson and P. Vedagarbha. 1997. A redesigned dcal controller without velocity measurements: theory and demonstration. Robotica, 15(03): 337–346.

Calise, A. J., N. Hovakimyan and M. Idan. 2001. Adaptive output feedback control of nonlinear systems using neural networks. Automatica, 37(8): 1201–1211.

Gibson, T. E., Z. Qu, A. M. Annaswamy and E. Lavretsky. 2014. Adaptive output feedback based on closed-loop reference models. arXiv preprint arXiv: 1410.1944.

Heemels, W., K. H. Johansson and P. Tabuada. 2012. An introduction to event-triggered and self-triggered control. pp. 3270–3285. *In*: Decision and Control (CDC), 2012 IEEE 51st Annual Conference on. IEEE.

Hespanha, J. P., P. Naghshtabrizi and Y. Xu. 2007. A survey of recent results in networked control systems. Proceedings of the IEEE, 95(1): 138–162.

Hovakimyan, N., F. Nardi, A. Calise and N. Kim. 2002. Adaptive output feedback control of uncertain nonlinear systems using single-hidden-layer neural networks. Neural Networks, IEEE Transactions on, 13(6): 1420–1431.

Ioannou, P. A. and J. Sun. 2012. Robust adaptive control. Courier Corporation.

Kaneko, K. and R. Horowitz. 1997. Repetitive and adaptive control of robot manipulators with velocity estimation. Robotics and Automation, IEEE Transactions on, 13(2): 204–217.

Kevorkian, J. and J. D. Cole. 1996. Multiple Scale and Singular Perturbation Methods. Springer-Verlag New York.

Khalil, H. K. 1996. Adaptive output feedback control of nonlinear systems represented by input-output models. Automatic Control, IEEE Transactions on, 41(2): 177–188.

Khalil, H. K. 1996. Nonlinear Systems. Upper Saddle River, NJ: Prentice Hall.

Kim, K., A. J. Calise, T. Yucelen and N. Nguyen. 2011a. Adaptive output feedback control for an aeroelastic generic transport model: A parameter dependent Riccati equation approach, in AIAA Guidance. Navigation and Control Conference.

Kim, K., T. Yucelen and A. J. Calise. 2011b. A parameter dependent Riccati equation approach to output feedback adaptive control. *In*: AIAA Guidance, Navigation and Control Conference.

Kim, Y. H. and F. L. Lewis. 1999. Neural network output feedback control of robot manipulators. Robotics and Automation, IEEE Transactions on, 15(2): 301–309.

Krstic, M., P. V. Kokotovic and I. Kanellakopoulos. 1995. Nonlinear and Adaptive Control Design. John Wiley & Sons, Inc.

Lavretsky, E. and K. Wise. 2012. Robust and Adaptive Control: With Aerospace Applications. Springer.

Lavretsky, E. 2010. Adaptive output feedback design using asymptotic properties of lqg/ltr controllers. AIAA Guidance, Navigation, and Control Conference.

Lavretsky, E. 2012. Adaptive output feedback design using asymptotic properties of LQG/LT controllers. Automatic Control, IEEE Transactions on, 57(6): 1587–1591.

Lavretsky, E. 2015. Transients in output feedback adaptive systems with observer-like reference models. International Journal of Adaptive Control and Signal Processing, 29(12): 1515–1525.

Lavretsky, E., T. E. Gibson and A. M. Annaswamy. 2011. Projection operator in adaptive systems.

Lunze, J. and D. Lehmann. 2010. A state-feedback approach to event-based control. Automatica, 46(1): 211–215.

Mazo, Jr. M. and P. Tabuada. 2008. On event-triggered and self-triggered control over sensor/actuator networks. pp. 435–440. *In*: Decision and Control, 2008. CDC 2008. 47th IEEE Conference on. IEEE.

Mazo, Jr. M., A. Anta and P. Tabuada. 2009. On self-triggered control for linear systems: Guarantees and complexity. pp. 3767–3772. *In*: Control Conference (ECC), 2009 European. IEEE.

Murray, J. D. 1984. Asymptotic Analysis. Springer-Verlag New York.

Narendra, K. S. and A. M. Annaswamy. 2012. Stable Adaptive Systems. Courier Corporation.

Pomet, J. -B. and L. Praly. 1992. Adaptive nonlinear regulation: estimation from the Lyapunov equation. IEEE Transactions on Automatic Control, 37(6): 729–740.

Postoyan, R., A. Anta, D. Nesic and P. Tabuada. 2011. A unifying lyapunov-based framework for the event-triggered control of nonlinear systems. pp. 2559–2564. *In*: Decision and Control and European Control Conference (CDC-ECC), 2011 50th IEEE Conference on. IEEE.

Sahoo, A., H. Xu and S. Jagannathan. 2013. Neural network-based adaptive event-triggered control of nonlinear continuous-time systems. IEEE International Symposium on Intelligent Control, pp. 35–40.

Sahoo, A. H. Xu and S. Jagannathan. 2014. Neural network approximation-based event-triggered control of uncertain MIMO nonlinear discrete time systems. IEEE American Control Conference, pp. 2017–2022.

Slotine, J. -J. E. and W. Li. 1987. On the adaptive control of robot manipulators. The International Journal of Robotics Research, 6(3): 49–59.

Tabuada, P. 2007. Event-triggered real-time scheduling of stabilizing control tasks. Automatic Control, IEEE Transactions on, 52(9): 1680–1685.

Tipsuwan, Y. and M. -Y. Chow. 2003. Control methodologies in networked control systems. Control Engineering Practice, 11(10): 1099–1111.

Volyanskyy, K. Y., W. M. Haddad and A. J. Calise. 2009. A new neuroadaptive control architecture for nonlinear uncertain dynamical systems: Beyond-and-modifications. IEEE Transactions on Neural Networks, 20(11): 1707–1723.

Walsh, G. C. and H. Ye. 2001. Scheduling of networked control systems. Control Systems, IEEE, 21(1): 57–65.

Wang, F. -Y. and D. Liu. 2008. Networked control systems. Springer.

Wang, X. and N. Hovakimyan. 2010. $L1$ adaptive control of event-triggered networked systems. IEEE American Control Conference, pp. 2458–2463.

Wang, X., E. Kharisov and N. Hovakimyan. 2015. Real-time $L1$ adaptive control for uncertain networked control systems. IEEE Transactions on Automatic Control, 60(9): 2500–2505.

Yesildirek, A. and F. L. Lewis. 1995. Feedback linearization using neural networks. Automatica, 31(11): 1659–1664.

Yucelen, T. and W. M. Haddad. 2012. Output feedback adaptive stabilization and command following for minimum phase dynamical systems with unmatched uncertainties and disturbances. International Journal of Control, 85(6): 706–721.

Zergeroglu, E., W. Dixon, D. Haste and D. Dawson. 1999. A composite adaptive output feedback tracking controller for robotic manipulators. Robotica, 17(06): 591–600.

Zhang, L. and D. Hristu-Varsakelis. 2006. Communication and control co-design for networked control systems. Automatica, 42(6): 953–958.

Zhang, F., D. M. Dawson, M. S. de Queiroz, W. E. Dixon et al. 2000. Global adaptive output feedback tracking control of robot manipulators. IEEE Transactions on Automatic Control, 45(6): 1203–1208.
Zhang, W., M. S. Branicky and S. M. Phillips. 1993. Stability of networked control systems. Control Systems, IEEE, 21(1): 84–99.
Zhou, K. and J. C. Doyle. 1998. Essentials of robust control. Prentice hall Upper Saddle River, NJ, Vol. 180.
Zhou, K., J. C. Doyle and K. Glover. 1996. Robust and optimal control. Prentice hall New Jersey, Vol. 40.

7

Event Sampled Adaptive Control of Robot Manipulators and Mobile Robot Formations

N. Vignesh, H. M. Guzey* and *S. Jagannathan*

ABSTRACT

In this chapter, the design of adaptive control of both robot manipulator and consensus-based formation of networked mobile robots in the presence of uncertain robot dynamics and with event-based feedback is presented. The linearity in the unknown parameter (LIP) property is utilized to represent the uncertain nonlinear dynamics of the robotic manipulator which is subsequently employed to generate the control torque with event-sampled measurement update. For the case of consensus-based formation control of mobile robots, by utilizing the LIP based representation of the robot dynamics, an adaptive back-stepping based controller with event-sampled feedback is designed. The networked robots communicate their location and velocity information with their neighbors which is ultimately utilized to derive the desired velocities to move the robots to a required consensus-based formation. The control torque is designed to minimize the velocity tracking error by explicitly taking into account the dynamics of the individual robot and by relaxing the perfect velocity tracking assumption. The Lyapunov stability method is utilized in both the applications to develop an event-sampling condition and to demonstrate the tracking performance of the robot manipulator and consensus of the overall formation of the networked

Department of Electrical and Computer Engineering, Missouri University of Science and Technology, Rolla, Missouri, USA.

E-mail: vnxv4; hmgb79; sarangap@mst.edu

* Corresponding author

mobile robots. Finally, simulation results are presented to verify theoretical claims and to demonstrate the reduction in the computations with event-sampled control execution.

Keywords: Event-sampled control, adaptive consensus, mobile robot formations, LIP

INTRODUCTION

Traditional feedback control systems rely on the periodic sampling of the sensors and use this measurement to compute appropriate control input. However, such controllers may be computationally inefficient as the sampling instants are not based on the actual system state and are based on the worst-case analysis. Therefore, in the recent years, event-based sampling has become popular (Tabuada 2007, Sahoo et al. 2015, Zhong et al. 2014, Guinaldo et al. 2012, Wang and Lemmon 2012) wherein the execution time of the control inputs is based on the real-time operation of the system. In the event-sampled framework (Tabuada 2007, Sahoo et al. 2015, Zhong et al. 2014, Guinaldo et al. 2012, Wang and Lemmon 2012), the measured state vector is sampled based on certain state dependent criteria referred to as event-triggering condition and the controller is executed at these aperiodic sampling instants. The event-triggering condition is designed by taking into account the stability and closed-loop performance, and, hence, proven to be advantageous over its periodic counterpart.

Initially, the event-triggered techniques from the literature (Tabuada 2007, Guinaldo et al. 2012, Wang and Lemmon 2012) were designed for ensuring stable operation of the closed-loop system by assuming that a stabilizing controller exists for the system under consideration. Developing an event-triggering condition and establishing the existence of positive inter-event time with the proposed event-sampling condition was the main focus in these works (Tabuada 2007, Guinaldo et al. 2012, Wang and Lemmon 2012). The traditional optimal control problem under the event-sampled framework is studied in Molin and Hirche 2013 while in Sahoo et al. 2015, the event-sampled adaptive controller design was presented for physical systems with uncertain dynamics.

In this chapter, an event-based adaptive controller design is covered for uncertain robotic systems–tracking control of robotic manipulator and adaptive consensus based formation control of mobile robots. The controller design for the robotic manipulators has been studied by several researchers (Lewis et al. 2003) over the past decade by using the traditional periodic feedback approach. The computed torque based controllers were originally popular as they converted the nonlinear dynamics of the robotic manipulator into linear dynamics. The filtered tracking error (Narendra and Annaswamy 2012) was defined to augment the tracking error and the linear controller was designed to stabilize the dynamics of this filtered tracking error. However, this controller required the complete knowledge of the manipulator dynamics. Later, adaptive controller which uses estimated values of the uncertain robot dynamics was presented. The uncertainties were assumed to satisfy linearity-in-the-unknown parameters (LIP) property and the uncertain parameters are adaptively estimated. With the parameter adaptation, the adaptive controllers are computationally intensive when compared to the traditional PID controllers.

Due to the application in the manufacturing and process industry, the control problem for the robotic manipulators is still actively pursued (Cheah et al. 2015, He 2015a, He 2015b, Galicki 2015). The controllers use continuous feedback information of the manipulator to generate the control torque, which are computationally inefficient. Especially, for the application in the manufacturing and process industry, the robot manipulators in large-numbers are used, which do not require periodic update in the control. Event-sampled implementation of the control of these robotic manipulators will significantly reduce the computational costs.

Therefore, in this chapter, an event-sampled adaptive control of robotic manipulator is presented. The uncertain dynamics of the robot manipulator are represented by utilizing the LIP property and the control torque is generated by using the event-sampled feedback information. The tracking performance of the robot manipulator is studied under the event-sampled implementation using Lyapunov stability theory.

Next, the event-based adaptive control of robotic manipulator is extended to the formation control of mobile robots which has been examined in several works by using various approaches including leader-follower control (Fierro and Lewis 1998, Hong et al. 2006), virtual structure (Low 2011) or behavior-based approaches (Balch and Arkin 1998), to name a few. Of all these approaches, consensus-based formation control (Ren et al. 2005b, Min et al. 2011, Semsar-Kazerooni and Khorasani 2009, Bauso et al. 2009, Tian and Liu 2008, Olfati-Saber and Murray 2004) is considered to be more robust and reliable due to scalability and its inherent properties that enable the robots to maintain their formation even if one of the robots experiences a failure.

In earlier works (Ren and Beard 2005a, Olfati-Saber and Murray 2004, Qu 2009, Bauso et al. 2009, Tian and Liu 2008, Sheng et al. 2012), consensus-based schemes have been studied for generalized linear systems with known system dynamics and applied to systems with time varying communication graphs (Ren and Beard 2005a), bounded disturbances (Bauso et al. 2009), and communication delays during state information sharing (Olfati-Saber and Murray 2004, Tian and Liu 2008). Further, the consensus-based optimal formation control scheme was also presented in Semsar-Kazerooni and Khorasani 2008.

Traditionally, adaptive controllers require more computations when compared to the PID controllers. Moreover, since the mobile robots use their neighbor information to reach consensus, they share information among each other through a communication network. Therefore, utilizing the communication network continuously will lead to higher communication cost and inefficient use of the communication resource. It may also lead to network congestion and undesired performance of the controller. Therefore, event-sampled execution of the controller in this case not only reduces computations, but also reduces the overhead in communication cost.

Motivated by the aforementioned challenges, an event-sampled adaptive consensus based formation control is proposed for mobile robots. The robots communicate their location and the velocity information with the neighborhood robots, which are used to obtain the consensus based formation errors. The formation errors are used to obtain the desired velocities for each robot which would drive the robots to a desired formation. These velocities are required to be tracked by each mobile robot. However, due to the robot dynamics, a velocity tracking error exists. Using the LIP-based representation of the mobile robot dynamics, the control inputs are obtained to minimize this velocity tracking error with event-sampled feedback. It is worth mentioning that the velocity tracking errors at each robot acts as a virtual controller for the formation error system. Thus using the back-stepping controller design, if the velocity tracking error is reduced, the formation error reduces and the robots reach a desired formation. It should to noted that, in contrast to the existing consensus based formation control approaches (Ren and

Beard 2005a, Olfati-Saber and Murray 2004, Qu 2009, Bauso et al. 2009, Tian and Liu 2008, Sheng et al. 2012), the dynamics of the mobile robot is explicitly taken into account, relaxing the perfect velocity tracking assumption.

For both the robotic systems, an adaptive event-sampling condition is required to determine the sampling instants to generate the feedback information in order to update the controllers. Since the unknown parameters are tuned at the event sampled instants, the computations are reduced when compared to traditional adaptive control schemes, but it introduces aperiodic parameter updates. Therefore, an event-sampling condition is derived using the stability conditions directly to ensure that the performance of the adaptive controller is not deteriorated due to the intermittent feedback. Finally, the extension of Lyapunov direct method is used to prove the local uniform ultimate boundedness (UUB) of the tracking errors and the parameter estimation error in each application with event-sampled feedback.

The contributions of this chapter include—(a) the design of novel event-based adaptive control of robotic manipulator and adaptive consensus based formation control of mobile robots; (b) development of novel event-sampling condition to determine the sampling instants and generate feedback signals for the controller; and (c) demonstration of stability of the robotic systems with the proposed event-sampled controller using Lyapunov stability theory.

In this chapter, $\mathfrak{R}^n$ is used to denote n dimensional Euclidean space. Euclidean-norm is used for vectors and for matrices, Frobenius norm is used.

BACKGROUND AND PROBLEM FORMULATION

In this section, a brief background on the event-sampled control implementation is provided first. Later, the dynamics of the robotic systems are introduced and the event-based tracking control problem for the robotic manipulator is formulated. Finally, the adaptive consensus based formation control problem, with event-sampled feedback, for mobile robots is presented. Here the state vector is considered measurable for both the robot manipulator and mobile robot.

Event-sampled Control

The traditional periodic/continuous feedback based control problem is well-established in the literature and several techniques are available to analyze the performance of these controllers. However, the event sampled controller implementation is relatively new and involves many challenges. Here, the event-based control problem is introduced by highlighting the challenges involved in the design with respect to controller adaptation and system stability.

In an event sampled framework, the system state vector is sensed continuously and released to the controller only at the event sampled instants. To denote the sampling instants we define an increasing sequence of time instants $\{t_k\}_{k=1}^{\infty}$, referred to as event sampled instants satisfying $t_{k+1} > t_k$, $\forall k = 1,2,\cdots$, and $t_0 = 0$, the initial sampling instant. The sampled state $x(t_k)$, is released to the controller and the last sampled state at the controller denoted by $\breve{x}(t)$ is updated. Then it is held by using a zero order hold (ZOH) at the controller until the next update and it is denoted as

$$\breve{x}(t) = x(t_k), t_k < t \leq t_{k+1}, \forall k = 1,2,\cdots. \tag{1}$$

The error, $e_{ET}(t)$, introduced due to the event sampled state can be written as

$$e_{ET}(t) = x(t) - \breve{x}(t),\ t_k < t \leq t_{k+1},\ \forall k = 1,2,\cdots, \tag{2}$$

where $e_{ET}(t)$ is referred to as event sampling error. Thus, the event sampling error is reset to zero with sampling and update in the state, that is, $e_{ET} = 0$, $t = t_k$, $\forall k = 1,2,\cdots$.

For the event-triggered controllers, an event-sampling mechanism/condition is required to determine the sampling instants, without jeopardizing the system stability. Also, if the controller parameters are adaptive and learn from the feedback information, the parameter adaptation process is also dependent on the event-based sampling instants. Therefore, the event-sampling mechanism should be carefully designed so that event-based feedback does not impede the adaptation process of the controller as well.

Remark 1. In the inter-event period, the control input is held by a ZOH and hence the control policy is piecewise continuous. It can also be viewed as continuously updated signal driven by the event-sampled measurement errors given by (2).

Next, the dynamics of the robotic system will be presented first followed by the control problem formulation with event-sampled feedback.

Robot Manipulator Dynamics

The dynamics of an n-link robot manipulator can be expressed as (Lewis et al. 2003)

$$M(q)\ddot{q} + V_m(q,\dot{q})\dot{q} + G(q) + F(\dot{q}) + \tau_d(t) = \tau(t), \tag{3}$$

with $q(t) \in \Re^n$, the joint variable vector, $M(q)$ the inertia matrix, $V_m(q,\dot{q})$ the Coriolis/centripetal matrix, $G(q)$ the gravity vector, and $F(\dot{q})$ the friction, $\tau_d(t)$ represents the unknown bounded disturbance, and $\tau(t)$ is the control torque, with appropriate dimensions. It should be noted that the robot manipulators of the form given by (3), satisfy certain properties. With the following assumptions, the tracking control problem is formulated next.

Assumption 1 (Lewis et al. 2003). The mass matrix $M(q)$ is a positive definite matrix and satisfies $B_{m_1} I \leq M(q) \leq B_{m_2} I$, with B_{m_1}, B_{m_2} being known positive constants, and Coriolis matrix $V_m(q,\dot{q})$ is bounded by $V_b\|\dot{q}\|$ such that V_b is a constant. The gravity vector is bounded by $g_B > 0$ and the bound on the friction term is assumed to be of the form $B_F\|\dot{q}\| + B_f$, such that B_F, B_f are positive constants. The matrix $\dot{M} - 2V_m$ is skewsymmetric. The disturbance vector τ_d is bounded such that $\|\tau_d\| \leq \tau_{dM}$, with τ_{dM} a known positive constant.

A nonlinear state space representation of the robot manipulator, which can be obtained with the joint angle and its velocity as state variables, is given by

$$\begin{bmatrix} \dot{q} \\ \ddot{q} \end{bmatrix} = \begin{bmatrix} \dot{q} \\ -M^{-1}(q)(V_m(q,\dot{q})\dot{q} + F(\dot{q}) + G(q)) \end{bmatrix} + \begin{bmatrix} 0 \\ M^{-1}(q) \end{bmatrix} \tau(t) + \begin{bmatrix} 0 \\ -M^{-1}(q) \end{bmatrix} \tau_d(t) \tag{4}$$

Assumption 2. The desired motion trajectory for the robotic manipulator $q_d(t) \in \Re^n$ is prescribed such that $\|Q_d\| \le q_B$, where $Q_d(t) = [q_d(t)\, \dot{q}_d(t)\, \ddot{q}_d(t)]^T$ and q_B is a known scalar bound.

Now, we can define the tracking error as

$$e(t) = q_d(t) - q(t) \tag{5}$$

and the filtered tracking error as (Lewis et al. 2003)

$$r(t) = \dot{e}(t) - \lambda e(t) \tag{6}$$

where λ is a symmetric, positive definite matrix of appropriate dimension.

Remark 2. It can be seen that choosing λ as a diagonal matrix with positive entries ensures that the $\dot{e}(t)$ will be stable and $e(t)$ will remain bounded as long as the designed control torque $\tau(t)$ ensures that the filtered tracking error remain bounded.

On differentiating (6), the dynamics of the filtered tracking error can be obtained as $\dot{r}(t) = \ddot{e}(t) + \lambda\dot{e}(t)$. Using the robot manipulator dynamics (4) and the definition of the tracking error (5), we get the following filtered error dynamics (Lewis et al. 2003)

$$M\dot{r} = -V_m r - \tau + f(x) + \tau_d \tag{7}$$

where $f(x) = M(q)(\ddot{q}_d + \lambda\dot{e}) + V_m(q,\dot{q})(\dot{q}_d + \lambda e) + F(\dot{q}) + G(q)$ and x is a vector that includes all the signals required to compute $f(.)$ and it is defined as $x = [e^T\, \dot{e}^T\, q_d^T\, \dot{q}_d^T\, \ddot{q}_d^T]^T$. If the function $f(x)$ is known, it is straight-forward to generate a control torque such that the filtered tracking error dynamics in (7) are stable. However, it is important to note that the parameters in $f(x)$ are uncertain and tend to vary with time. This calls for the design of adaptive controllers, which are based on representing the uncertain function $f(x)$ using LIP property.

Assumption 3. The mass of each link, denoted by m_i, is considered unknown and the friction are considered negligible. The nonlinearity $f(x)$, is Lipschitz continuous function on a compact set, such that $\|f(x) - f(y)\| \le L_\phi \|x - y\|$ and satisfies LIP (Khalil and Grizzle 1996).

Let us consider the dynamics of the two-link robot manipulator (Lewis et al. 2003) with the following matrices, $M(q) = \begin{bmatrix} (m_1+m_2)l_1^2 + m_2 l_2^2 + 2m_2 l_1 l_2 \cos q_2 & m_2 l_2^2 + m_2 l_1 l_2 \cos q_2 \\ m_2 l_2^2 + m_2 l_1 l_2 \cos q_2 & m_2 l_2^2 \end{bmatrix}$, with m_i being the mass of each robot link, l_i being the length of each link. $V(q,\dot{q}) = [-m_2 l_1 l_2 (2\dot{q}_1\dot{q}_2 + \dot{q}_2^2)\sin q_2 \quad m_2 l_1 l_2 \dot{q}_1^2 \sin q_2]^T$ and $G(q) = [(m_1+m_2)g l_1 \cos q_1 + m_2 g l_2 \cos(q_1+q_2) \quad m_2 g l_2 \cos(q_1+q_2)]^T$, with g representing the acceleration due to gravity. With the dynamics in (3), the nonlinear function $f(x)$, with the masses m_1, m_2 considered uncertain, can be represented in the LIP form, such that (Lewis et al. 2003), $f(x) = \theta_R^T \phi(x)$, with $\theta_R = [m_1\ m_2]^T$, being the uncertain parameters and the corresponding regression function $\phi(x) = \begin{bmatrix} \phi_1 & \phi_2 \\ \phi_3 & \phi_4 \end{bmatrix}$, where $\phi_1 = l_1^2\ddot{q}_1 + l_1 g\, c_1$, $\phi_3 = 0$, $\phi_4 = l_1 l_2\, c_2\, \ddot{q}_1 + l_1 l_2\, s_2\, \dot{q}_1^2 + l_2 g\, c_{12} + l_2^2(\ddot{q}_1 + \ddot{q}_2)$, and $\phi_2 = l_2^2(\ddot{q}_1 + \ddot{q}_2) + l_1 l_2\, c_2 (2\ddot{q}_1 + \ddot{q}_2) + l_1\ddot{q}_1 - l_1 l_2\, s_2\, \dot{q}_2^2 - 2 l_1 l_2\, s_2\, \dot{q}_1\dot{q}_2 + l_2 g\, c_{12} + l_1 g c_1$, and c_i, s_i represent cosine and sine functions of the parameter q_i, $c_{12} = \cos(q_1 + q_2)$.

Remark 3. For the robotic manipulator, the mass matrix changes with the object carried by the manipulator and it depends on the application.

For the robot manipulator, using the event-sampled state estimate $\hat{f}(\breve{x})$, of the function $f(x)$, an adaptive control torque input can be generated as

$$\tau = \hat{f}(\breve{x}) + k_v \breve{r}, \qquad t_k \le t < t_{k+1} \tag{8}$$

where k_v is a positive definite design matrix and $\breve{r}(t) = r(t_k) = r(t) - e_{ET}(t), t \in [t_k, t_{k+1})$. With the adaptive control (8), the filtered tracking-error dynamics become

$$M\dot{r} = -V_m r - \hat{f}(\breve{x}) - k_v \breve{r} + f(x) + \tau_d, \quad t_k \le t < t_{k+1} \tag{9}$$

Defining the function estimation error as $f(\breve{x}) = f(\breve{x}) - \hat{f}(\breve{x})$, we get the filtered tracking error dynamics

$$M\dot{r} = -V_m r - K_v r + \tilde{f}(\breve{x}) + \tau_d - k_v e_{ET} + f(x) - f(\breve{x}), \qquad t_k \le t < t_{k+1} \tag{10}$$

Here the filtered tracking error dynamics are driven by the function estimation and the event sampled measurement errors. The LIP property of the robotic manipulator dynamics can be exploited to estimate the function $\hat{f}(\breve{x})$, which can be utilized to obtain the adaptive control torque input (8). If the parameter adaptation is accurate, the estimation error will become zero and the filtered tracking error will remain bounded due to the disturbance torque.

Therefore, the objective is to design an adaptive control torque input such that the filtered tracking error dynamics are stable. In order to minimize the event sampling error, which is a part of estimation error, a suitable event trigger condition needs to be designed while ensuring stability. Next, the background on the formation control for the mobile robot will be presented and adaptive control problem with event-sampled feedback will be formulated.

Mobile Robot Dynamics

Consider the non-holonomic robot shown in Fig. 1, where x_r, y_r denote Cartesian positions with respect to the robot frame, d is the distance between the rear-axis and the center of mass of the robot, r, R are the radius of the rear wheels, and half of the robot width, respectively.

The equations of motion about the center of mass, C, for the ith robot in a networked robot formation are written as

$$\dot{q}_i = \begin{bmatrix} \dot{x}_{ci} \\ \dot{y}_{ci} \\ \dot{\theta}_i \end{bmatrix} = \begin{bmatrix} \cos\theta_i & -d_i \sin\theta_i \\ \sin\theta_i & d_i \cos\theta_i \\ 0 & 1 \end{bmatrix} \begin{bmatrix} v_i \\ \omega_i \end{bmatrix} = S_i(q_i)\overline{v}_i, \tag{11}$$

where d_i is the distance from the rear axle to the robot's center of mass; $q_i = [x_{ci}\, y_{ci}\, \theta_i]^T$ denotes the Cartesian position of the center of mass and orientation of the ith robot; v_i, and ω_i represent linear and angular velocities, respectively, and $\overline{v}_i = [v_i\, \omega_i]^T$ for the ith robot.

Mobile robotic systems, in general, can be characterized as having an n-dimensional configuration space $\mathcal{C}$ with generalized coordinates $(q_1, \ldots q_n)$ subject to ℓ constraints

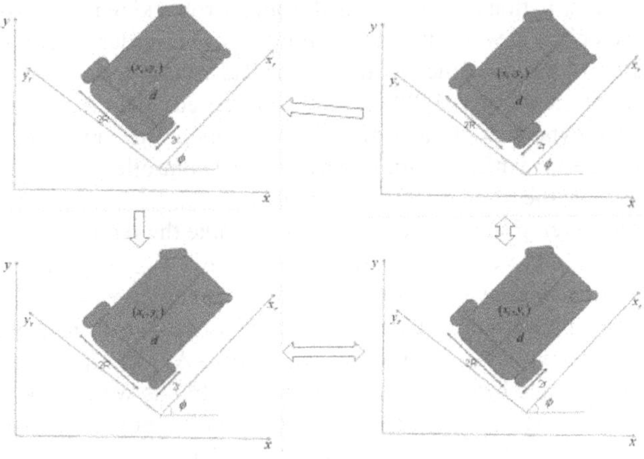

Figure 1. Differentially driven mobile robot.

(Fierro and Lewis 1998). Applying the transformation, the dynamics of the mobile robot are given by

$$\bar{M}_i\dot{\bar{v}}_i + \bar{V}_{mi}(q_i,\dot{q}_i)\bar{v}_i + \bar{F}_i(\bar{v}_i) + \bar{\tau}_{di} = \bar{\tau}_i, \tag{12}$$

where $\bar{M}_i \in \Re^{pxp}$ is a constant positive definite inertia matrix, $\bar{V}_{mi} \in \Re^{pxp}$ is the bounded centripetal and Coriolis matrix, $\bar{F}_i \in \Re^p$ is the friction vector, $\bar{\tau}_{di} \in \Re^p$ represents unknown bounded disturbances such that $\|\bar{\tau}_{di}\| \le d_M$ for a known constant, d_M, $\bar{B}_i \in \Re^{pxp}$ is a constant, nonsingular input transformation matrix, $\bar{\tau}_i = \bar{B}_i\tau_i \in \Re^p$ is the input vector, and $\tau_i \in \Re^p$ is the control torque vector. For complete details on (12) and the parameters, refer to (Fierro and Lewis 1998).

Assumption 4. The robotic system (12) satisfies the following properties: $\bar{M}_i$ is a known positive definite matrix and it is bounded by B_{iM} and $0 < \bar{M}_i^{-1} < B_{im}$, the norm of $\bar{V}_{mi}$, and $\|\bar{\tau}_{di}\| \le d_M$ are all bounded. The matrix $\dot{\bar{M}}_i - 2\bar{V}_{mi}$ is the skew-symmetric (Fierro and Lewis 1998).

Remark 4. In (Ren et al. 2005b), a controller was designed to ensure that all regulation errors for the linear systems achieved a common value. Due to the nonholonomic constraints considered in this paper, the formation consensus error is defined as the difference between the robot's own regulation error and the regulation error of one of its neighbor, referred to as robot *j*. The benefit of such consensus based formation controller is that the ith robot will be able to reach consensus with another neighbor when the communication is not available with the jth robot anymore. As shown in (Ren et al. 2005b), average consensus is achieved if the information exchange topology is both strongly connected and balanced. In the case that the information exchange topology has a spanning tree, the final consensus value is equal to the weighted average of initial conditions of those agents that have a directed path to all the other agents (Ren et al. 2005b). In this work, we will assume that the information exchange topology forms a spanning tree.

The main focus of this section is to formulate the back-stepping control problem for the mobile robots to minimize the consensus based formation error. For this, the consensus based formation error is obtained from the robot kinematics and it is utilized to derive the velocities at which the robots should move in-order to achieve the desired consensus. Due to the dynamics of the individual robot, perfect velocity tracking assumption (Fierro and Lewis 1998) becomes undesired. Therefore, explicitly taking into account the dynamics of each robot, controllers are designed to minimize the velocity tracking error which in turn acts as a virtual controller to the formation error dynamics and helps the robots reach consensus.

In order to develop such controller, we first define the regulation errors for each robot on positions and velocities as $\Delta x_i = x_{ci} - x_{ci}^r$, $\Delta y_i = y_{ci} - y_i^r$, $\Delta\theta_i = \theta_i - \theta_i^r$ with x_i^r, y_i^r, θ_i^r being the reference positions and bearing angles which forms a desired formation shape for the group of robots. Now, the consensus errors between the ith robot and j^{th} robot are defined as $\delta_{xi} = \Delta x_i - \Delta x_j$, $\delta_{yi} = \Delta y_i - \Delta y_j$ and $\delta_{\theta i} = \Delta\theta_i - \Delta\theta_j$, for x and y directions and the bearing angle, respectively. In this work, it will be assumed that the desired heading angles, θ_i^r, are common for each robot in the formation so that each robot is oriented in the same direction, which yields $\delta_{\theta i} = \theta_i - \theta_j$.

Next, the consensus formation error is transformed into the reference frame attached to the mobile robot using the transformation

$$e_{iF} = \begin{bmatrix} e_{i1} \\ e_{i2} \\ e_{i3} \end{bmatrix} = \begin{bmatrix} \cos\theta_i & \sin\theta_i & 0 \\ -\sin\theta_i & \cos\theta_i & 0 \\ 0 & 0 & 1 \end{bmatrix} \begin{bmatrix} \delta_{xi} \\ \delta_{yi} \\ \delta_{\theta i} \end{bmatrix}. \tag{13}$$

Taking the derivative of (13) reveals

$$\dot{e}_{iF} = \begin{bmatrix} -\sin\theta_i\omega_i\delta_{xi} + \cos\theta_i\dot{\delta}_{xi} + \cos\theta_i\omega_i\delta_{yi} + \sin\theta_i\dot{\delta}_{yi} \\ -\cos\theta_i\omega_i\delta_{xi} - \sin\theta_i\dot{\delta}_{xi} - \sin\theta_i\omega_i\delta_{yi} + \cos\theta_i\dot{\delta}_{yi} \\ \dot{\delta}_{\theta i} \end{bmatrix} \tag{14}$$

Using the expression (11) in (14) gives

$$\dot{e}_{iF} = \begin{bmatrix} -\sin\theta_i\omega_i\delta_{xi} + \cos\theta_i\left(\cos\theta_i v_i - d_i\sin\theta_i\omega_i - \cos\theta_j v_j + d_j\sin\theta_j\omega_j\right) \\ +\cos\theta_i\omega_i\delta_{yi} + \sin\theta_i\left(\sin\theta_i v_i + d_i\cos\theta_i\omega_i - \sin\theta_j v_j - d_j\cos\theta_j\omega_j\right) \\ \\ -\cos\theta_i\omega_i\delta_{xi} - \sin\theta_i\left(\cos\theta_i v_i - d_i\sin\theta_i\omega_i - \cos\theta_j v_j + d_j\sin\theta_j\omega_j\right) \\ -\sin\theta_i\omega_i\delta_{yi} + \cos\theta_i\left(\sin\theta_i v_i + d_i\cos\theta_i\omega_i - \sin\theta_j v_j - d_j\cos\theta_j\omega_j\right) \\ \\ \dot{\delta}_{\theta i} \end{bmatrix} \tag{15}$$

On simplification using the trigonometric identities and using (13), yields the formation error dynamics as

$$\begin{bmatrix} \dot{e}_{i1} \\ \dot{e}_{i2} \\ \dot{e}_{i3} \end{bmatrix} = \begin{bmatrix} e_{i2}\omega_i + v_i - v_j \cos(\theta_i - \theta_j) \\ -e_{i1}\omega_i + v_j \sin(\theta_i - \theta_j) \\ \omega_i - \omega_j \end{bmatrix}, \tag{16}$$

Remark 5. It can be observed that (16) resembles the trajectory tracking error system from a single robot control architecture tracking a virtual reference cart. In this work, instead of tracking a virtual cart, the mobile robots attempt to reach consensus with their neighbors to a desired formation, and each $e_{i(-)}$ represents the consensus error instead of the trajectory tracking error (Fierro and Lewis 1998).

Under the perfect velocity tracking assumption, the consensus-based formation control velocity is given by

$$\overline{v}^F_{id} = \begin{bmatrix} v^F_{id} \\ \omega^F_{id} \end{bmatrix} = \begin{bmatrix} -k_1 e_{i1} + v_j \cos(\theta_i - \theta_j) \\ \omega_j - k_2 v_j e_{i2} - k_3 \sin(\theta_i - \theta_j) \end{bmatrix}. \tag{17}$$

with k_1, k_2, $k_3 > 0$, being the design constants. Next, the back-stepping technique is utilized, wherein the desired velocity (17) is utilized in (16) to drive the consensus error dynamics. We denote the velocity tracking error as e^F_{iv}, for formation control and defining $e^F_{iv} = \overline{v}_i - \overline{v}^F_{id}$ reveals $\overline{v}_i = \overline{v}^F_{id} + e_{iv}$, and the consensus error system (16) becomes

$$\begin{bmatrix} \dot{e}_{i1} \\ \dot{e}_{i2} \\ \dot{e}_{i3} \end{bmatrix} = \begin{bmatrix} e_{i2}\omega_i + \left(e^F_{iv1} - k_1 e_{i1} + v_j \cancel{\cos}\left(\theta_i - \theta_j\right)\right) - v_j \cancel{\cos}\left(\theta_i - \theta_j\right) \\ -e_{i1}\omega_i + v_j \sin\left(\theta_i - \theta_j\right) \\ \left(e^F_{iv2} + \cancel{\omega_j} - k_2 v_j e_{i2} - k_3 \sin\left(\theta_i - \theta_j\right)\right) - \cancel{\omega_j} \end{bmatrix}, \tag{18}$$

Simplifying the expression in (18), leads to the consensus error dynamics as

$$\begin{bmatrix} \dot{e}_{i1} \\ \dot{e}_{i2} \\ \dot{e}_{i3} \end{bmatrix} = \begin{bmatrix} e_{i2}\omega_i - k_1 e_{i1} - e^F_{iv1} \\ -e_{i1}\omega_i + v_j \sin\left(\theta_i - \theta_j\right) \\ -k_2 v_j e_{i2} - k_3 \sin(\theta_i - \theta_j) - e^F_{iv2} \end{bmatrix}. \tag{19}$$

Using (12) and the definition of the velocity tracking error, its dynamics are obtained as

$$\overline{M}_i \dot{e}^F_{iv} = -\overline{M}_i \dot{\overline{v}}^F_{id} - \overline{V}_{mi}(q_i, \dot{q}_i)\left(\overline{v}^F_{id} + e^F_{iv}\right) - \overline{F}_i(\overline{v}_i) - \overline{\tau}_{di} + \overline{\tau}_i. \tag{20}$$

Defining $f(z_i) = \overline{V}_{mi}(q_i, \dot{q}_i)(\overline{v}^F_{id} + e^F_{iv}) + \overline{F}_i(\overline{v}_i)$ and using (20)

$$\overline{M}_i \dot{e}^F_{iv} = -\overline{M}_i \dot{\overline{v}}^F_{id} - f(z_i) - \overline{\tau}_{di} + \overline{\tau}_i \tag{21}$$

where $z_i = [e^F_{iv1}\, e^F_{iv2}\, v_i\, \omega_i]$ is the set of signals required to construct the uncertain function $f(z_i)$ (Fierro and Lewis 1998) as

$$f(z_i) = \begin{bmatrix} 0 & -d_i\omega_i\left(m_i - 2mW_i\right) \\ d_i\omega_i\left(m_i - 2mW_i\right) & 0 \end{bmatrix} \begin{bmatrix} v_{id} + e^F_{iv1} \\ \omega_{id} + e^F_{iv2} \end{bmatrix} + \begin{bmatrix} Fvv_i + Fdsign\left(v_i\right) \\ Fv\omega_i + Fdsign\left(\omega_i\right) \end{bmatrix}, \tag{22}$$

with $e_{iv}^F = [e_{iv1}^F \; e_{iv2}^F]$, Fv, Fd being the unknown friction coefficients, m_i, mW_i being the total mass of the robot and the mass of a wheel respectively. Defining $\ell_i = d_i(m_i - 2mW_i)$, we get the uncertain nonlinear function as

$$f(z_i) = \begin{bmatrix} -\ell_i \omega_i \left(\omega_{id} + e_{iv2}^F\right) + Fvv_i + Fdsign\left(v_i\right) \\ \ell_i \omega_i \left(v_{id} + e_{iv1}^F\right) + Fv\omega_i + Fdsign\left(\omega_i\right) \end{bmatrix}. \tag{23}$$

Finally, recalling the velocity tracking error dynamics (21), define the control torque such that the robot tracks its desired velocity as

$$\overline{\tau}_i = \overline{M}_i \dot{\overline{v}}_{id}^F - K_v e_{iv}^F + \hat{f}\left(z_i\right) - \gamma\left(e_{i1}, e_{i2}, e_{i3}\right), \tag{24}$$

where $\gamma(e_{i1}, e_{i2}, e_{i3})$ is a function of the consensus error states (19) required for the purpose of stability, and $\hat{f}(z_i)$ is the estimated unknown dynamics which will be parametrized next.

Assumption 5. The nonlinearity $f(z_i)$, is Lipschitz continuous function on a compact set, such that $\|f(x_i) - f(y_i)\| \le L_\phi \|x_i - y_i\|$ and satisfies LIP. Also, the derivative of the desired velocities $\dot{\overline{v}}_{id}^F$ satisfy Lipschitz continuity such that $\left\|\dot{\overline{v}}_{id}^F(x_i) - \dot{\overline{v}}_{id}^F(y_i)\right\| \le L_v \left\|x_i - y_i\right\|$.

Substituting (24) into (21) reveals the closed loop velocity tracking error dynamics

$$\overline{M}_i \dot{e}_{iv}^F = -K_v e_{iv}^F + \overline{\tau}_{di} - \gamma\left(e_{i1}, e_{i2}, e_{i3}\right) + \tilde{f}\left(z_i\right). \tag{25}$$

where the function estimation error is given by $\tilde{f}(.) = \hat{f}(.) - f(.)$.

Remark 6. For the case of mobile robots, the friction coefficients and the Coriolis matrices are assumed to be uncertain. Using the event-sampled feedback of the sensor measurements, and the LIP based representation of the uncertain nonlinearity these quantities are adaptively tuned to generate the desired control torque.

In the next section, the event-sampling mechanism will be designed first to determine the event-based sampling instants and then the controller design for the tracking control of the robotic manipulator followed by the formation control of the mobile robots will be presented.

CONTROLLER DESIGN

In this section the adaptive controller design, with event-sampled feedback, for the robotic systems will be presented and the parameter adaptation laws will be derived from the Lyapunov stability analysis.

Robot Manipulator

For the controller design, the event-sampling mechanism which decides the feedback instants is crucial and is usually designed based on the closed-loop stability analysis. The event-sampling mechanism is designed such that the measurement error (2) satisfy a state-dependent threshold for every inter-event period, which is of the form

$$\left\|e_{ET}(t)\right\|^2 \le \sigma \mu_k \left\|r(t)\right\|^2, \quad t_k \le t < t_{k+1}, \; k = 1,2,3.. \tag{26}$$

with $0 < \sigma < 1$, and μ_k is a positive design parameter. Due to the parameter adaptation at periodic sampling instants, the adaptive controllers involve more computations when compared to traditional controllers. However by using the event-sampled feedback, the objective is to reduce these computations without compromising the stability while ensuring acceptable tracking performance.

Using the results of Assumption 2, the unknown dynamics of the robot manipulator can be represented by using standard LIP formulation. The resulting dynamics can be used to design the feedback-linearizing controller to achieve satisfactory tracking performance of the robot manipulator.

Consider the uncertain robot manipulator dynamics in (4), using Assumption 2, the unknown nonlinear dynamics can be represented as $f(x) = \theta_R^T \phi(x)$, with θ_R being the unknown masses in the robot manipulator dynamics and $\phi(x)$ is the regression vector obtained from the system dynamics (7). The estimated dynamics with event-sampled states is given as

$$\hat{f}(\breve{x}) = \hat{\theta}_R^T \phi(\breve{x}), \quad t_k \le t < t_{k+1} \tag{27}$$

where $\hat{\theta}_R$ is the estimate of the unknown parameter matrix in the manipulator dynamics. Defining the parameter estimation error $\tilde{\theta}_R^T = \theta_R^T - \hat{\theta}_R^T$ and using filtered tracking error dynamics in (10) and the estimation of the unknown parameters, we have

$$M\dot{r} = (-V_m - k_v)r + \hat{\theta}_R^T \phi(\breve{x}) - \theta_R^T \phi(x) - k_v e_{ET} + \tau_d, \quad t_k \le t < t_{k+1} \tag{28}$$

The filtered error dynamics can be rewritten as

$$M\dot{r} = (-V_m - k_v)r + \tilde{\theta}_R^T \phi(\breve{x}) + \theta_R^T [\phi(x) - \phi(\breve{x})] - k_v e_{ET} + \tau_d, \quad t_k \le t < t_{k+1}. \tag{29}$$

For the stability analysis of the event-sampled controller, the closed-loop system with event-sampled measurement error is analyzed in the following theorem, and the adaptation rule for tuning the unknown parameters (27) with event-sampled states is presented.

With the following definition on input-to-state stability (ISS), it will be shown in Theorem 1 that the closed-loop system admits an ISS Lyapunov function in the presence of bounded measurement error. Later, the assumption on the boundedness of the measurement error introduced by the event-sampled feedback will be relaxed in the subsequent theorem. The ISS result will be used only to establish the existence of positive inter event time, which is required to demonstrate that the event-sampling mechanism does not exhibit zeno-behavior (Tabuada 2007).

Definition 1. (Khalil and Grizzle 1996) Let $L(t, Z)$ be a continuously differentiable function such that

$$\underline{L}(\|Z\|) \le L(t,Z) \le \overline{L}(\|Z\|) \tag{30}$$

$$\dot{L}(t,Z,e) \le -\Phi(Z), \quad \forall \|Z\| \ge \gamma(\|e\|) > 0 \tag{31}$$

where $\underline{L}$, $\overline{L}$ are class κ_∞ functions, e is a bounded measurement error and γ is a class κ function and $\Phi(Z)$ is a continuous positive definite function on $\Re^n$. Then the system $\dot{Z} = f(t, Z, e)$, $\forall Z(0) \in A$ is input-to-state stable.

Theorem 1 (Input-to-state stability). *Consider the filtered tracking error dynamics of the robotic manipulator given in (29) with τ_d being the disturbance torque. Let the Assumptions 1–3 hold. Using the LIP representation of the uncertain robot manipulator dynamics given in (27), generate the control torque from (8). Let the parameters be tuned using the following adaptation rule*

$$\dot{\hat{\theta}}_R = \Gamma\phi(\breve{x})\breve{r}^T - \kappa\Gamma\|\breve{r}\|\hat{\theta}_R \tag{32}$$

where $\Gamma > 0, \kappa > 0, k_v > \frac{1}{2}$. *Let $\bar{B}_{ETM}$ be the bound on the measurement error e_{ET} which is allowed to satisfy the condition $||e_{ET}|| \leq \bar{\mu}_k||r||$, such that $0 < \bar{\mu}_k < 1$, $\forall k$, and let θ_{RM} be the bound on the unknown mass matrix. The tracking and the parameter estimation errors are bounded and the closed-loop system is input to state stable, with the input being a function of the measurement error e_{ET}.*

Proof. Consider the Lyapunov candidate function of the form, $L(r, \tilde{\theta}_R^T) = \frac{1}{2}r^T Mr + \frac{1}{2} tr(\tilde{\theta}_R^T \Gamma^{-1}\tilde{\theta}_R)$, the first derivative using the filtered tracking error dynamics (29) and the parameter update rule (32) is given by

$$\begin{aligned}\dot{L}(r,\tilde{\theta}_R) &= r^T(-V_m - k_v)r + \frac{1}{2}r^T\dot{M}r + r^T\tilde{\theta}_R^T\phi(\breve{x}) + r^T\{\theta_R[\phi(x)-\phi(\breve{x})] - k_v e_{ET} + \tau_d\} \\ &+ tr(\tilde{\theta}_R^T\Gamma^{-1}(-\Gamma\phi(\breve{x})\breve{r}^T + \kappa\Gamma\|\breve{r}\|\hat{\theta}_R))\end{aligned} \tag{33}$$

With the skew-symmetry property in Assumption 1 and the Lipschitz continuity of the regression function, we obtain

$$\dot{L}(r,\tilde{\theta}_R) = -r^T k_v r + r^T\{\theta_R L_\phi e_{ET} - k_v e_{ET} + \tau_d\} + tr((-\tilde{\theta}_R^T\phi(\breve{x})\breve{r}^T + r^T\tilde{\theta}_R^T\phi(\breve{x}) + \tilde{\theta}_R^T\kappa\|\breve{r}\|\hat{\theta}_R)) \tag{34}$$

Applying norm operator and using the bounds on the disturbance torque from Assumption 1, we obtain

$$\leq -k_v\|r\|^2 + \|\tilde{\theta}_R\|\|\phi(\breve{x})\|\|e_{ET}\| + \kappa\theta_{RM}\|\breve{r}\|\|\tilde{\theta}_R\| - \kappa\|\breve{r}\|\|\tilde{\theta}_R\|^2 + \|r\|\zeta_1(\bar{B}_{ETM}) \tag{35}$$

where, $\zeta_1(\bar{B}_{ETM}) = \theta_{RM}L_\phi\bar{B}_{ETM} + k_v\bar{B}_{ETM} + \tau_{dM}$.

Using the Young's inequality and the relation $||e_{ET}|| \leq \bar{\mu}_k||r||$, we obtain

$$\leq -\|r\|^2(k_v - \frac{1}{2}) - \|\tilde{\theta}_R\|^2\kappa\|r\| + \|\tilde{\theta}_R\|^2\kappa\|e_{ET}\| + \delta_1\|\tilde{\theta}_R\|\|r\| + \|\phi(\breve{x})\|\|\tilde{\theta}_R\|\|e_{ET}\| + \eta_B \tag{36}$$

where $\delta_1 = (2\kappa\theta_{RM})$, $\eta_B = \frac{\|\zeta_1(\bar{B}_{ETM})\|^2}{2}$. From the Lipschitz continuity of the regression vector, we have

$$\begin{aligned}&\leq -\|r\|^2(k_v - \frac{1}{2}) - \|\tilde{\theta}_R\|^2\kappa\|r\| + \|\tilde{\theta}_R\|^2\kappa\bar{\mu}_k\|r\| + \delta_1\|\tilde{\theta}_R\|\|r\| + L_\phi\|r\|\|\tilde{\theta}_R\|\bar{B}_{ETM} \\ &+ \|r\|\|\tilde{\theta}_R\|L_\phi\bar{\mu}_k\bar{B}_{ETM} + \eta_B\end{aligned} \tag{37}$$

Grouping similar terms, with $\delta_1 = (\kappa\theta_{RM}) + L_\phi \bar{B}_{ETM} + L_\phi \bar{\mu}\bar{B}_{ETM}$, we get

$$\leq -\|r\|^2 (k_v - \frac{1}{2}) - \|\tilde{\theta}_R\|^2 \kappa\|r\| + \|\tilde{\theta}_R\|^2 \kappa\bar{\mu}_k \|r\| + \delta_1 \|\tilde{\theta}_R\| \|r\| + \eta_B \tag{38}$$

On simplifying (38), using the notations $\bar{\kappa} = \kappa\,(1 - \bar{\mu}_k)$, $\bar{k}_v = k_v - \frac{1}{2}$, we arrive at

$$\leq -\|r\|(\|r\|\bar{k}_v + \bar{\kappa}\|\tilde{\theta}_R\|^2 - \delta_1\|\tilde{\theta}_R\|) + \eta_B \tag{39}$$

By completing the squares, we get the Lyapunov first derivative as

$$\leq -\|r\|(\|r\|\bar{k}_v + \bar{\kappa}[\|\tilde{\theta}_R\| - \frac{\delta_1}{2\bar{\kappa}}]^2 - \frac{\delta_1^2}{4\bar{\kappa}}) + \eta_B \tag{40}$$

The Lyapunov function derivative in (40) is less than zero as long as r, $\tilde{W}$ are outside their respective ultimate bounds. The ultimate bounds are a function of η_B, δ_1, which in turn are obtained as a function of the measurement error and disturbance torque input. Hence, (40) is in a form similar to the input-to-state stability characterization given by (30) and (31).

Remark 7. The result in Theorem 1 shows that the closed-loop system in the presence of a bounded measurement error is locally ISS. That is, the continuous closed-loop system admits an ISS Lyapunov function. However, for the event-sampled implementation of the controller, the boundedness of the event-sampled measurement error is required to be proven. In the following theorem, the closed-loop signals are indeed shown to be bounded using which the measurement error and the existence of $0 < \bar{\mu}_k < 1$, satisfying the event trigger condition $\|e_{ET}\| \leq \bar{\mu}_k\|r\|$ will be demonstrated.

In the following theorem, the event-sampling mechanism is designed and stability of the robotic manipulator is analyzed by using the Lyapunov stability theory, in the ideal case, where the sensor signal is noise free, the epsilon modification term in (41) is not used and there is no disturbance torque input.

Theorem 2. *(Ideal case) Consider the filtered tracking error dynamics of the robotic manipulator given in (29) with $\tau_d = 0$. Let the Assumptions 1 and 3 hold. Apply the LIP representation of the uncertain robot manipulator dynamics given in (27), and use the estimated values of the unknown parameter to generate the control torque (8). Let the condition (26) be utilized to determine the event-sampling instants and (41) be used to tune the unknown parameters at the event-sampling instants with $\kappa = 0$, $\Gamma > 0$. Let θ_{RM} be the bound on the unknown mass matrix, L_ϕ, is the Lipschitz constant for the regression vector. The tracking and the parameter estimation errors are locally UUB, both at the event-sampling instants and during the inter-event period.*

Proof. Case 1. At the event-sampling instants $t = t_k$, $\forall k$, the event-sampling error $e_{ET} = 0$, and the tracking error and regression function in (29) becomes $\breve{r} = r, \phi(\breve{x}) = \phi(x)$. The adaptation law becomes

$$\dot{\hat{\theta}}_R = \Gamma\phi(x)r^T \tag{41}$$

Consider the following Lyapunov candidate function $L(r,\tilde{\theta}_R^T) = L(r) + L(\tilde{\theta}_R)$, such that

$$L(r,\tilde{\theta}_R^T) = \frac{1}{2}r^T Mr + \frac{1}{2}tr(\tilde{\theta}_R^T\Gamma^{-1}\tilde{\theta}_R) \tag{42}$$

Taking the first derivative of (42), we get

$$\dot{L}(r,\tilde{\theta}_R^T) = r^T M\dot{r} + tr(\tilde{\theta}_R^T \Gamma^{-1} \dot{\tilde{\theta}}_R) + \frac{1}{2} r^T \dot{M} r \tag{43}$$

Using the filtered tracking error dynamics in (29), we have

$$\dot{L}(r,\tilde{\theta}_R^T) = r^T(-V_m - k_v)r + r^T \tilde{\theta}_R^T \phi(x) + tr(\tilde{\theta}_R^T \Gamma^{-1} \dot{\tilde{\theta}}_R) + \frac{1}{2} r^T \dot{M} r \tag{44}$$

Utilizing the skew-symmetry property mentioned in Assumption 1, we obtain

$$\dot{L}(r,\tilde{\theta}_R^T) = -r^T k_v r + tr(\tilde{\theta}_R^T \Gamma^{-1} \dot{\tilde{\theta}}_R + r^T \tilde{\theta}_R^T \phi(x)) \tag{45}$$

Using the property of trace of a matrix and with the definition $\tilde{\theta}_R = \theta_R - \hat{\theta}_R$, we obtain

$$\dot{L}(r,\tilde{\theta}_R^T) = -r^T k_v r + tr(\tilde{\theta}_R^T (\phi(x) r^T - \Gamma^{-1} \dot{\hat{\theta}}_R)) \tag{46}$$

Utilizing the parameter update rule in (41), we get the first derivative of the Lyapunov function as

$$\dot{L}(r,\tilde{\theta}_R) = -r^T k_v r \tag{47}$$

It can be concluded from (47), that the filtered tracking and the parameter estimation errors are bounded when $e_{ET} = 0$. It can be observed from (47), $\dot{L} \le 0$, negative semi-definite when $e_{ET} = 0$, during the event-sampling instants. Taking the second derivative of the Lyapunov function using (47), we get the expression $\ddot{L} \le -2r^T k_v \dot{r}$, r bounded, $\dot{r}$ bounded, therefore $\dot{L}$ is uniformly continuous and from Barbalat's Lemma, $r \to 0$ as the event-sampling instants $t_k \to \infty$.

Case 2: In the inter-event period $t_k \le t < t_{k+1}$, the measurement error due to event-sampled feedback is non-zero, $e_{ET} \neq 0$. The parameters are not updated during this period and therefore $\dot{\hat{\theta}}_R = 0$. Using the Lyapunov candidate function (42), the first derivative is obtained as

$$\dot{L}(r,\tilde{\theta}_R^T) = r^T M\dot{r} + tr(\tilde{\theta}_R^T \Gamma^{-1} \dot{\tilde{\theta}}_R) + \frac{1}{2} r^T \dot{M} r \tag{48}$$

Substituting the filtered tracking error dynamics (29), with $\tau_d = 0$, $e_{ET} \neq 0$, and since $\dot{\hat{\theta}}_R = 0$, we have

$$\dot{L}(r,\tilde{\theta}_R^T) = r^T(-V_m - k_v)r + r^T \tilde{\theta}_R{}^T \phi(\breve{x}) + r^T \theta_R{}^T [\phi(x) - \phi(\breve{x})] - r^T k_v e_{ET} + \frac{1}{2} r^T \dot{M} r \tag{49}$$

Utilizing the skew-symmetry property mentioned in Assumption 1, we obtain

$$\dot{L}(r,\tilde{\theta}_R^T) = -r^T k_v r - r^T \hat{\theta}_R{}^T \phi(\breve{x}) + r^T \theta_R{}^T \phi(x) - r^T k_v e_{ET} \tag{50}$$

Utilizing the definition of the parameter estimation error, we have

$$\dot{L}(r,\tilde{\theta}_R^T) = -r^T k_v r - r^T \hat{\theta}_R{}^T \phi(\breve{x}) + r^T (\hat{\theta}_R{}^T + \tilde{\theta}_R{}^T) \phi(x) - r^T k_v e_{ET} \tag{51}$$

On grouping the estimated parameters together, we obtain

$$\dot{L}(r,\tilde{\theta}_R^T) = -r^T k_v r + r^T \hat{\theta}_R^T [\phi(x) - \phi(\breve{x})] + r^T \tilde{\theta}_R^T \phi(x) - r^T k_v e_{ET} \tag{52}$$

Using the Assumption on the regression vector and applying the norm operator, we obtain

$$\dot{L}(r,\tilde{\theta}_R^T) \le -k_v \|r\|^2 + \|r^T\| (\|\hat{\theta}_R^T\| L_\phi + k_v) \|e_{ET}\| + \|r^T\| \|\tilde{\theta}_R^T\| \|\phi(x)\| \tag{53}$$

Adding and subtracting $\|r^T\|\ \|\tilde{\theta}_R^T\|\ \|\phi(\overline{x})\|$ to get

$$\le -k_v \|r\|^2 + \|r^T\| (\|\hat{\theta}_R^T\| L_\phi + k_v) \|e_{ET}\| + \|r^T\| \|\tilde{\theta}_R^T\| (\|\phi(x) - \phi(\breve{x})\|) + \|r^T\| \|\tilde{\theta}_R^T\| \|\phi(\breve{x})\| \tag{54}$$

Using the Lipschitz assumption on the regression function, we get

$$\le -k_v \|r\|^2 + \|r^T\| (\|\hat{\theta}_R^T\| L_\phi + k_v + \|\tilde{\theta}_R^T\| L_\phi) \|e_{ET}\| + \|r^T\| \|\tilde{\theta}_R^T\| \|\phi(\breve{x})\| \tag{55}$$

Applying the definition of the parameter estimation error and triangle inequality to get

$$\dot{L}(r,\tilde{\theta}_R^T) \le -k_v \|r\|^2 + \|r^T\| (2\|\hat{\theta}_R^T\| L_\phi + k_v + \theta_{RM} L_\phi) \|e_{ET}\| + \|r^T\| \|\tilde{\theta}_R^T\| \|\phi(\breve{x})\| \tag{56}$$

Using Young's inequality, $2ab \le a^2 + b^2$, $\forall a,b > 0$, we have the relation

$$\dot{L}(r,\tilde{\theta}_R^T) \le -k_v \|r\|^2 + \frac{\|r\|^2}{2} + \frac{(\theta_{RM}{}^2 L_\phi{}^2 + k_v{}^2 + 4\|\hat{\theta}_R^T\|^2 L_\phi{}^2) \|e_{ET}\|^2}{2} + \|r\|^2 + \frac{\|r\|^2}{2} + \frac{\|\tilde{\theta}_R^T\|^2 \|\phi(\breve{x})\|^2}{2} \tag{57}$$

By grouping the similar terms in (57), we arrive at

$$\dot{L}(r,\tilde{\theta}_R^T) \le -(k_v - 2)\|r\|^2 + (\theta_{RM}{}^2 L_\phi{}^2 + k_v{}^2 + 4\|\hat{\theta}_R^T\|^2 L_\phi{}^2) \frac{\|e_{ET}\|^2}{2} + \frac{\|\tilde{\theta}_R^T\|^2 \|\phi(\breve{x})\|^2}{2} \tag{58}$$

Using the event-sampling condition (26), we have

$$\dot{L}(r,\tilde{\theta}_R^T) \le -(k_v - 2)\|r\|^2 + (\theta_{RM}{}^2 L_\phi{}^2 + k_v{}^2 + 4\|\hat{\theta}_R^T\|^2 L_\phi{}^2) \frac{\mu_k \|r\|^2}{2} + \frac{\|\tilde{\theta}_R^T\|^2 \|\phi(\breve{x})\|^2}{2} \tag{59}$$

Choosing $\mu_k = \dfrac{2}{(\theta_{RM}{}^2 L_\phi{}^2 + k_v{}^2 + 4\|\hat{\theta}_R^T\|^2 L_\phi{}^2)}$, for each inter event period, the Lyapunov first derivative can be obtained as

$$\dot{L}(r,\tilde{\theta}_R^T) \le -(k_v - 3)\|r\|^2 + \frac{\|\tilde{\theta}_R^T\|^2 \|\phi(\breve{x})\|^2}{2}. \tag{60}$$

With $k_v > 3$, and using the fact that $\tilde{\theta}_R(t) = \tilde{\theta}_R(t_k)$, $\phi(\overline{x}(t)) = \phi(t_k)$, $\forall t \in [t_k, t_{k+1})$, it can be concluded that the filtered tracking error and the parameter estimation error are bounded during the inter-event period since the regression function with previously sampled state vector is proven to be bounded.

Remark 8. It should be noted that the filtered tracking and the parameter estimation errors are bounded both at the event-sampling instants and during the inter-event period. However, in order to avoid the parameter drift and to relax PE condition, additional terms are included in the parameter update rule (Narendra and Annaswamy 2012).

Remark 9. It should be noted that the bounds obtained for the filtered tracking error from Theorem 2 for both the cases are independent of the measurement error demonstrating that the closed-loop system is bounded.

Remark 10. Since the target parameters are constant and the estimated parameters are not tuned during the inter-event period, the parameter estimation error dynamics are zero during this period. From Theorem 2, it can be concluded that if the parameters are tuned using the proposed adaptation rule at the event-sampling instants, the parameter estimation error converges to its ultimate bounds as the number of event sampled instants increase. However, for the analysis of the tracking error of the robotic manipulator with event-sampled implementation, it is required to show that the tracking error is bounded during the inter-event period and the event-sampling instants.

In the following theorem, the evolution of the tracking error with event-sampled implementation of the controller is presented with the adaptation rule (32) with the additional term to relax the PE condition and to ensure that the parameter estimation errors are bounded.

Theorem 3. *Consider the filtered tracking error dynamics of the robotic manipulator given in (29) with* τ_d *being the disturbance torque. Let the Assumptions 1-3 hold. Using the LIP based representation of the uncertain robot manipulator dynamics (27), generate the control torque (8). Let the unknown parameters be tuned at the event-sampling instants using (32), where* $\Gamma > 0$, $\kappa > 0$, $k_v \geq 3$. *Let the measurement error satisfy the condition in (26), with* $0 < \mu_k < 1$, *for each inter-event period and* $0 < \sigma < 1$. *The tracking and the parameter estimation errors are locally UUB both at the event-sampling instants and during the inter-event period. In addition the measurement error introduced (2) due to the event-sampled feedback is bounded during the time interval* $[t_k, t_{k+1})$ *for all k. Moreover, the bounds for the filtered tracking error and the parameter estimation error are obtained independent of the measurement error.*

Proof. Case 1. At the event-sampling instants $t = t_k$, $\forall k$, the event-sampling error $e_{ET} = 0$, and the tracking error and regression function in (29) becomes $\breve{r} = r, \phi(\breve{x}) = \phi(x)$. Consider the Lyapunov candidate function (42). Its first derivative using the filtered tracking error dynamics and the parameter update law, is obtained as

$$\dot{L}(r,\tilde{\theta}_R) = -r^T k_v r + r^T \tilde{\theta}_R^T \phi(x) + r^T \{\zeta_1\} + tr(\tilde{\theta}_R^T(-\phi(x)r^T + \kappa\|r\|\hat{\theta}_R)) \tag{61}$$

with $\zeta_1 = \tau_d$. Using the property of trace operator, we get

$$\dot{L}(r,\tilde{\theta}_R) \leq -r^T k_v r + r^T \tilde{\theta}_R^T \phi(x) + r^T \zeta_1 + tr(-\tilde{\theta}_R^T \phi(x) r^T + \tilde{\theta}_R^T \kappa\|r\|\theta_R - \tilde{\theta}_R^T \kappa\|r\|\tilde{\theta}_R) \tag{62}$$

Applying the norm operator and using the bounds on the target parameter, we obtain

$$\dot{L}(r,\tilde{\theta}_R) \leq -k_v\|r\|^2 + \|r\|\|\zeta_1\| + \|\tilde{\theta}_R\|\kappa\|r\|\theta_{RM} - \kappa\|r\|\|\tilde{\theta}_R\|^2 \tag{63}$$

On completing the squares, we get

$$\leq -\|r\|[k_v\|r\| + \kappa(\|\tilde{\theta}_R\| - \frac{\theta_{RM}^2}{2})^2 - \frac{(\kappa\theta_{RM}^2 + 4\|\zeta_1\|)}{4}] \tag{64}$$

Using the Lyapunov first derivative in (64), the bounds on the tracking and the parameter estimation errors are obtained as

$$\|r\| > \frac{\kappa\theta_{RM}^2 + 4\|\zeta_1\|}{4k_v}, \text{ or, } \|\tilde{\theta}_R\| > \sqrt{\frac{\kappa\theta_{RM}^2 + 4\|\zeta_1\|}{4\kappa}} + \frac{\theta_{RM}}{2}$$

Case 2: In the inter-event period $t_k \leq t < t_{k+1}$, the measurement error due to event-sampled feedback is non-zero, $e_{ET} \neq 0$. The adaptive parameters are not updated during this period and therefore $\dot{\hat{\theta}}_R = 0$. Using the Lyapunov candidate function (42), the first derivative is obtained as

$$\dot{L}(r, \tilde{\theta}_R{}^T) = r^T M\dot{r} + \frac{1}{2} r^T \dot{M} r + tr(\tilde{\theta}_R{}^T \Gamma^{-1} \dot{\tilde{\theta}}_R) \tag{65}$$

Using the filtered tracking error dynamics and the skew-symmetry property in Assumption 1, we have

$$\dot{L}(r, \tilde{\theta}_R{}^T) = -r^T k_v r - r^T \hat{\theta}_R{}^T \phi(\breve{x}) + r^T [\hat{\theta}_R{}^T + \tilde{\theta}_R{}^T]\phi(x) - r^T k_v e_{ET} + r^T \tau_d + tr(\tilde{\theta}_R{}^T \Gamma^{-1} \dot{\tilde{\theta}}_R) \tag{66}$$

Since the parameter adaptation is not carried out during the inter-event period, the parameter error dynamics are zero. Now, applying the norm operator, we get

$$\dot{L}(r, \tilde{\theta}_R{}^T) \leq -k_v\|r\|^2 + \|r^T\|\|\hat{\theta}_R{}^T\|\|\phi(x) - \phi(\breve{x})\| + \|r^T\|\|\tilde{\theta}_R{}^T\|\|\phi(x)\| + \|r^T\|k_v\|e_{ET}\| + \|r^T\|\tau_{dM} \tag{67}$$

Adding and subtracting $\|r^T\|\ \|\tilde{\theta}_R^T\|\ \|\phi(\breve{x})\|$ and using the Lipschitz inequality to get

$$\leq -k_v\|r\|^2 + \|r^T\|(\|\hat{\theta}_R{}^T\|L_\phi + k_v + \|\tilde{\theta}_R{}^T\|L_\phi)\|e_{ET}\| + \|r^T\|\|\tilde{\theta}_R{}^T\|\|\phi(\breve{x})\| + \|r^T\|\tau_{dM} \tag{68}$$

Applying the definition of parameter estimation error and using triangle inequality, we get

$$\leq -k_v\|r\|^2 + \|r^T\|(2\|\hat{\theta}_R{}^T\|L_\phi + k_v + \theta_{RM}L_\phi)\|e_{ET}\| + \|r^T\|\|\tilde{\theta}_R{}^T\|\|\phi(\breve{x})\| + \|r^T\|\tau_{dM} \tag{69}$$

Using the Lipschitz continuity of the regression vector and applying Young's inequality, we get

$$\leq -k_v\|r\|^2 + ((4\|\hat{\theta}_R{}^T\|^2 + \theta_{RM}^2)L_\phi{}^2 + k_v{}^2)\frac{\|e_{ET}\|^2}{2} + \frac{\|r^T\|^2}{2} + \frac{\|\tilde{\theta}_R{}^T\|^2\|\phi(\breve{x})\|^2}{2} + 2\|r^T\|^2 + \frac{\tau_{dM}{}^2}{2} \tag{70}$$

On simplifying the expression, we arrive at

$$\leq -(k_v - 2)\|r\|^2 + \frac{((4\|\hat{\theta}_R{}^T\|^2 + \theta_{RM}^2)L_\phi{}^2 + k_v{}^2)\|e_{ET}\|^2}{2} + \frac{\|\tilde{\theta}_R{}^T\|^2\|\phi(\breve{x})\|^2}{2} + \frac{\tau_{dM}{}^2}{2} \tag{71}$$

Using the event-sampling condition, $\|e_{ET}\|^2 \le \sigma\mu_k\|r\|^2$, we get

$$\le -(k_v-2)\|r\|^2 + (2\|\hat{\theta}_R^{\,T}\|^2 L_\phi^{\,2} + \theta_{RM}^2 L_\phi^{\,2} + k_v^{\,2})\frac{\sigma\mu_k}{2}\|r\|^2 + \frac{\|\tilde{\theta}_R^{\,T}\|^2\|\phi(\breve{x})\|^2}{2} + \frac{\tau_{dM}^{\,2}}{2} \tag{72}$$

By choosing μ_k, as in (59), for each inter-event period, we have the Lyapunov first derivative as

$$\dot{L}(r,\tilde{\theta}_R^{\,T}) \le -(k_v-3)\|r\|^2 + \frac{\|\tilde{\theta}_R^{\,T}\|^2\|\phi(\breve{x})\|^2}{2} + \frac{\tau_{dM}^{\,2}}{2} \tag{73}$$

From (73), it can be seen that the filtered tracking error is bounded during the inter-event period as long as $k_v > 3$, and since the unknown parameters are not updated, they remain bounded during the inter-event period such that

$$\tilde{\theta}_R(t) = \tilde{\theta}_R(t_k), \phi(\breve{x}(t)) = \phi(t_k), t \in [t_k, t_{k+1}). \tag{74}$$

Therefore, from Case (i) and Case (ii), it can be concluded that the filtered-tracking error and the parameter estimation error remain bounded during the inter-event period. Due to the event-sampling condition, the measurement error introduced by the event-sampled feedback is also bounded during the inter-event period.

Remark 11. Notice that the value of μ_k obtained from (71) requires the information regarding the estimated parameters to determine the event-based sampling instants and also the bounds on the unknown mass matrix. This requires an additional adaptive estimator at the event-sampling mechanism that runs in parallel with the estimator at the controller. Also, it can be seen from the value of μ_k, that as the bounds on the mass matrix is increased, the event-sampling threshold is decreased causing frequent events.

Remark 12. From the results in Theorem 3, it can be seen that the measurement error is bounded for all the inter-event period due to the event-sampling condition (26) with the value of μ_k obtained from (72). By choosing the k_v as required in the Lyapunov conditions in (73), we obtain $0 < \mu_k < 1$, $\forall k$. With the states of the closed-loop system defined as $Z = [r\ \tilde{\theta}_R]^T$, and using the Theorems 1 and 3, we have the ISS characterization in (30) and (31). Thus combining the results from Theorem 1, 3, the existence of a positive minimum inter event time can be established (Tabuada 2007).

Remark 13. The ultimate bound obtained in (64) can be made arbitrarily small by adjusting the controller gains K_v, κ. An increase in the disturbance input will increase the bounds on the tracking and parameter estimation errors.

Remark 14. Once the tracking error converges to a small value, a dead-zone operator can be used to minimize the events. This way the feedback information is not utilized frequently and computations can be reduced further. This is the trade-off between the tracking accuracy and computational cost, due to the event-sampled feedback.

Remark 15. With the event-sampling condition of the form (26), one can conclude that the tracking error is bounded during the inter-event period. By choosing a suitable Lyapunov function for the tracking error and by selecting the event-sampling condition such that the inequality $\|r(t)\|^2 \le (1-\tau_s)\alpha\|r(t_k)\|^2$, $t_k \le t < t_{k+1}$, is satisfied during the inter-event period, the tracking error can be made to decrease both during the event-sampling period and the inter event period (Wang and Lemmon 2011).

Next, the adaptive control design of the mobile robot is presented.

Mobile Robots

For the controller design of mobile robots, the event-sampling mechanism which decides the event-based feedback instants is crucial and is usually designed based on the closed-loop stability analysis. The event-sampling mechanism is designed such that the event-sampling error (2) satisfy a state-dependent threshold for every inter-event period for each robot, which is of the form

$$\left\|\mathrm{E}_{iET}\right\| \le \sigma_i \mu_{ik} \left\|\mathrm{E}_i\right\|, \qquad t_k \le t < t_{k+1},\ k = 1,2,3.. \tag{75}$$

with $0 < \sigma_i < 1$, and μ_{ik} is a positive design parameter and E_{iET}, E_i are functions of event-sampling error and the formation, and velocity tracking errors respectively.

Using the results of Assumption 4, the unknown dynamics of the mobile robot can be represented by using standard LIP formulation. The unknown dynamics (23) is parametrized as

$$f(z_i) = \Theta_i^T \psi_i(z_i) \tag{76}$$

where $\Theta_i = \begin{bmatrix} \Theta_{1i} & 0_{1\times 3} \\ 0_{1\times 3} & \Theta_{2i} \end{bmatrix} \in \Re^{2\times 6}$ is the unknown coefficients matrix with $\Theta_{1i} = [-\ell_i\ Fv\ Fd]$, $\Theta_{2i} = [\ell_i\ Fv\ Fd]$ with $\|\Theta_i\| \le \Theta_{iM}$ and $\psi_i = \left[\left(\omega_{id} + e_{iv2}^F\right)\ \ v_i\ \ sign(v_i)\ \ \left(v_{id} + e_{iv1}^F\right)\ \ \omega_i\ \ sign(\omega_i)\right]^T$ is the regression vector. The unknown parameters can be estimated as $\hat{\Theta}_i$ and estimation of the unknown dynamics (76) with event sampled feedback can be given as

$$\hat{f}(z_i) = \hat{\Theta}_i^T \psi_i(\breve{z}_i), \quad t_k \le t < t_{k+1}. \tag{77}$$

with $\breve{z}_i = z_i + e_{iET}$, being the event-sampled signals at the ith mobile robot. The unknown parameter estimation error is defined as $\tilde{\Theta}_i = \Theta_i - \hat{\Theta}_i$ and the estimation error dynamics can be given as $\dot{\tilde{\Theta}}_i = -\dot{\hat{\Theta}}_i$.

Remark 16. Calculation of the control torque requires computation of $\dot{\bar{v}}_{id}^F$, which is a function of the dynamics of robot j, $\dot{\bar{v}}_j$. Therefore, the control law (24) not only compensates for the dynamics of the ith robot, but also for the dynamics of the formation. Formation dynamics can become significant when robots are traveling at high speeds or in robots with high moments of inertia. To calculate $\dot{\bar{v}}_{id}^F$, it is assumed that robot j communicates its state vector, which includes x_{ci}, y_{ci}, θ_i and linear, angular velocities, to the ith robot using wireless communication at the event-sampled instants decided by its event-sampling mechanism.

The event-sampled control torque using (77) is obtained as

$$\bar{\tau}_i = \bar{M}_i \dot{\breve{\bar{v}}}_{id}^F - K_v \breve{e}_{iv}^F + \hat{f}(\breve{z}_i) - \breve{\gamma}(e_{i1}, e_{i2}, e_{i3}), \qquad t_k \le t < t_{k+1}. \tag{78}$$

with $\breve{e}_{iv}^{F} = e_{iv}^{F} + e_{iET}$ and $\breve{\gamma}$ is the stabilizing term with measurement error due to event-sampled mechanism. Substituting (78) into (21) reveals the closed-loop velocity tracking error dynamics

$$\bar{M}_i \dot{e}_{iv}^{F} = -K_v e_{iv}^{F} + \bar{\tau}_{di} - \breve{\gamma}(e_{i1}, e_{i2}, e_{i3}) + \tilde{f}(\breve{z}_i) + K_v e_{iET} + [f(z_i) - f(\breve{z}_i)] - \bar{M}_i[\dot{\breve{v}}_{id}^{F} - \dot{\breve{v}}_{id}^{F}],$$
$$t_k \le t < t_{k+1}. \tag{79}$$

where $\tilde{f}(\breve{z}_i) = \tilde{\Theta}_i \psi_i(\breve{z}_i)$.

Now, to define the formation error dynamics with event-sampled measurement error, consider (17), during the kth inter-event period, the desired virtual control equations are obtained as

$$\breve{v}_{id}^{F} = \begin{bmatrix} \breve{v}_{id}^{F} \\ \breve{\omega}_{id}^{F} \end{bmatrix} = \begin{bmatrix} -k_1 \breve{e}_{i1} + \breve{v}_j \cos(\breve{\theta}_i - \breve{\theta}_j) \\ \breve{\omega}_j - k_2 \breve{v}_j \breve{e}_{i2} - k_3 \sin(\breve{\theta}_i - \breve{\theta}_j) \end{bmatrix}, \quad t_k \le t < t_{k+1}. \tag{80}$$

where $\breve{e}_{i1}, \breve{e}_{i2}, \breve{v}_j, \breve{\theta}_i, \breve{\theta}_j, \breve{\omega}_j$ are event-sampled formation error along x_{ci}, y_{ci}, event-sampled linear velocity of the jth robot, event-sampled bearing angle of the ith, j^{th} robot and the angular velocity of the jth robot, respectively. Defining the event-sampled signals $(\breve{\bullet}) = (\bullet) + \varepsilon_{\bullet}$, $\forall t \in [t_k, t_{k+1})$, (80) can be rewritten with the measurement error as

$$\breve{v}_{id}^{F} = \begin{bmatrix} \breve{v}_{id}^{F} \\ \breve{\omega}_{id}^{F} \end{bmatrix} = \begin{bmatrix} -k_1 e_{i1} + v_j \cos(\theta_i - \theta_j) + v_{i\varepsilon} \\ \omega_j - k_2 v_j e_{i2} - k_3 \sin(\theta_i - \theta_j) + \omega_{i\varepsilon} \end{bmatrix}, \quad t_k \le t < t_{k+1}. \tag{81}$$

where $v_{i\varepsilon}, \omega_{i\varepsilon}$ are given by $v_{i\varepsilon} = k_3[\sin(\theta_i - \theta_j) - \sin(\breve{\theta}_i - \breve{\theta}_j)] + \varepsilon_{\omega j} - k_2 v_j \varepsilon_{i2} - k_2 \varepsilon_{vj} e_{i2} - k_2 \varepsilon_{vj} \varepsilon_{i2}$, and $\omega_{i\varepsilon} = -v_j[\cos(\theta_i - \theta_j) - \cos(\breve{\theta}_i - \breve{\theta}_j)] + \varepsilon_{vj} \cos(\breve{\theta}_i - \breve{\theta}_j) - k_1 \varepsilon_{i1}$. To get the closed-loop formation error dynamics in the presence of measurement error, use (81) in (16) to get

$$\begin{bmatrix} \dot{e}_{i1} \\ \dot{e}_{i2} \\ \dot{e}_{i3} \end{bmatrix} = \begin{bmatrix} e_{i2}\omega_i - k_1 e_{i1} + v_{i\varepsilon} - e_{iv1}^{F} \\ -e_{i1}\omega_i + v_j \sin(\theta_i - \theta_j) \\ -k_2 v_j e_{i2} - k_3 \sin(\theta_i - \theta_j) + \omega_{i\varepsilon} - e_{iv2}^{F} \end{bmatrix} \tag{82}$$

The following theorem provides the stability analysis for the control law that does not require the perfect velocity tracking assumption. As noted before, the theorem does not assume that complete knowledge of the robot dynamics is available to compute the control law (24), and the controller is updated only during the event-based sampling instants. Similar to the case of the robot manipulator, it can also be shown that the mobile robot closed-loop dynamics is input-to-state stable, the proof of which is not included due to space consideration.

Theorem 4. *(Ideal case) Given the consensus error dynamics (79) for the ith robot in the network with the disturbance torque $\bar{\tau}_d = 0$, let the consensus-based formation controller (78) be applied to the ith robot. Let the control torque be defined by (78) with*

$$\gamma(e_{i1}, e_{i2}, e_{i3}) = \bar{M}_i \begin{bmatrix} -e_{i1} \\ -\dfrac{1}{k_2} \sin e_{i3} \end{bmatrix} \tag{83}$$

and the event-sampling condition (75) is used to determine the sampling instants. Further, when the unknown parameters are tuned at the sampling instants using the parameter adaptation rule

$$\dot{\hat{\Theta}}_i = \Lambda_1 \psi_i(\breve{z}_i)\breve{e}_{iv}^{F^T} + \Lambda_1 \kappa_i \hat{\Theta}_i \tag{84}$$

where $\Lambda_1 = B_{im}\Lambda^{-1}, B_{im}, \Lambda > 0$, and $\kappa_i = 0$. *The velocity tracking error system (79) and consensus error system (19) remain bounded and the* i^{th} *robot tracks its desired velocity and reaches consensus with its neighbor robot j when the gains are chosen such that* $k_1 > \frac{k_2^2}{2} - \frac{5}{2}, \frac{k_3}{k_2} > \frac{3}{2}, B_{im}K_v > 3, k_2 > 0.$

Proof. Case 1. With the event-sampled measurement error, $e_{IET} = 0, \forall i, \kappa_i = 0$. The adaptation rule (84) is rewritten as

$$\dot{\hat{\Theta}}_i = \Lambda_1 \psi_i(z_i)e_{iv}^{F^T} \tag{85}$$

Consider the following Lyapunov candidate

$$L_i = \frac{1}{2}\left(e_{i1}^2 + e_{i2}^2\right) + \frac{1}{k_2}\left(1 - \cos e_{i3}\right) + \frac{1}{2}e_{iv}^{F\,T}e_{iv}^F + \frac{1}{2}tr\left\{\tilde{\Theta}_i^T \Lambda \tilde{\Theta}_i\right\}. \tag{86}$$

Taking the derivative of (86), we get

$$\dot{L}_i = e_{i1}\dot{e}_{i1} + e_{i2}\dot{e}_{i2} + \frac{\dot{e}_{i3}\sin e_{i3}}{k_2} + e_{iv}^{F\,T}\dot{e}_{iv}^F + tr\left\{\tilde{\Theta}_i^T \Lambda \dot{\hat{\Theta}}_i\right\} \tag{87}$$

Substituting the consensus error system (19) and velocity tracking error dynamics (79) keeping the measurement error and the disturbance torque zero, reveals

$$\dot{L}_i = e_{i1}\left(e_{i2}\omega_i - k_1 e_{i1} - e_{iv1}^F\right) + e_{i2}\left(-e_{i1}\omega_i + v_j \sin\left(\theta_i - \theta_j\right)\right) +$$

$$\frac{1}{k_2}\left(-k_2 v_j e_{i2} - k_3 \sin\left(\theta_i - \theta_j\right) - e_{iv2}^F\right)\sin e_{i3}$$

$$+ e_{iv}^{F\,T}\left(-\bar{M}_i^{-1}K_v e_{iv}^F - \bar{M}_i^{-1}\gamma\left(e_{i1}, e_{i2}, e_{i3}\right) + \bar{M}_i^{-1}\tilde{\Theta}_i^T \psi\left(z_i\right)\right) + tr\left\{\tilde{\Theta}_i^T \Lambda \dot{\hat{\Theta}}_i\right\} \tag{88}$$

Using the definition of e_{i3} in (88), we obtain the first derivative as

$$\begin{aligned}\dot{L}_i &= -k_1 e_{i1}^2 - e_{i1}e_{iv1}^F + e_{i2}v_j \sin e_{i3} - v_j e_{i2}\sin e_{i3} - \frac{k_3}{k_2}\sin^2 e_{i3} - \frac{\sin e_{i3}e_{iv2}^F}{k_2} \\ &+ e_{iv}^{F\,T}\left(-\bar{M}_i^{-1}K_v e_{iv}^F - \bar{M}_i^{-1}\gamma\left(e_{i1}, e_{i2}, e_{i3}\right) + \bar{M}_i^{-1}\tilde{\Theta}_i^T \psi\left(z_i\right)\right) + tr\left\{\tilde{\Theta}_i^T \Lambda \dot{\hat{\Theta}}_i\right\}\end{aligned} \tag{89}$$

Utilizing the stabilizing term (83) in the control torque, we get

$$\dot{L}_i = -k_1 e_{i1}^2 - e_{i1}e_{iv1}^F - \frac{k_3}{k_2}\sin^2 e_{i3} - \frac{\sin e_{i3}e_{iv2}^F}{k_2} - \bar{M}_i^{-1}K_v\left\|e_{iv}^F\right\|^2 - e_{iv}^{F\,T}\begin{bmatrix} -e_{i1} \\ -\frac{1}{k_2}\sin e_{i3}\end{bmatrix} +$$

$$tr\left\{e_{iv}^{F\,T}\bar{M}_i^{-1}\tilde{\Theta}_i^T\psi(z_i)-\tilde{\Theta}_i^T\Lambda\dot{\hat{\Theta}}_i\right\} \tag{90}$$

Using the bound on the mass matrix from Assumption 4 and trace property, we get

$$\dot{L}_i \le -k_1e_{i1}^2 - e_{i1}e_{iv1}^F - \frac{k_3}{k_2}\sin^2 e_{i3} - \frac{\sin e_{i3}e_{iv2}^F}{k_2} - \bar{M}_i^{-1}K_v\left\|e_{iv}^F\right\|^2 + e_{iv1}^F e_{i1} + \frac{e_{iv2}^F}{k_2}\sin e_{i3} +$$

$$tr\left\{e_{iv}^{F\,T}B_{im}\tilde{\Theta}_i^T\psi(z_i)-\tilde{\Theta}_i^T\Lambda\dot{\hat{\Theta}}_i\right\} \tag{91}$$

Using the parameter adaptation rule defined in (85) to get

$$\dot{L}_i \le -k_1e_{i1}^2 - \frac{k_3}{k_2}\sin^2 e_{i3} - B_{im}K_v\left\|e_{iv}^F\right\|^2 \le 0 \tag{92}$$

Since (92) is not a function of e_{i2} and the parameter estimation error, $\dot{L}_i$ is negative semi-definite, and the consensus errors and velocity tracking error are bounded. Therefore, Barbalat's Lemma can be used to show that the formation errors and the velocity tracking errors approach zero.

First, take the derivative of (92) while using (16) and (79), and observing the boundedness of all signals reveals that $\ddot{L}_i$ is also bounded. Therefore, $\dot{L}_i$ converges to zero and thus e_{i1}, $\dot{e}_{i1}$, e_{i3}, $\dot{e}_{i3}$, $\|e_{iv}^F\|$, and $\|\dot{e}_{iv}^F\|$ all converge to zero as well. Finally, examining the definition in (16) while noting that $\dot{e}_{i3} \to 0$ reveals that e_{i2} must also converge to zero. Therefore, the velocity tracking error system (79) and consensus error system (19) converge to the origin asymptotically, and the ith robot tracks its desired velocity and reaches consensus with its neighbor robot j.

Case 2: When the measurement error, $e_{iET} \neq 0$, $\forall t \in [t_k, t_{k+1})$, $\dot{\hat{\Theta}} = 0$, therefore, using the Lyapunov candidate function (86) and the first derivative is obtained as

$$\dot{L}_i = e_{i1}\dot{e}_{i1} + e_{i2}\dot{e}_{i2} + \frac{\dot{e}_{i3}\sin e_{i3}}{k_2} + e_{iv}^{F\,T}\dot{e}_{iv}^F + tr\left\{\tilde{\Theta}_i^T\Lambda\dot{\hat{\Theta}}_i\right\} \tag{93}$$

Utilizing the consensus error dynamics and the velocity tracking error dynamics with event-sampled measurement error from (82) and (79), we get

$$\dot{L}_i = e_{i1}\left(e_{i2}\omega_i - k_1e_{i1} + v_{i\varepsilon} - e_{iv1}^F\right) + e_{i2}\left(-e_{i1}\omega_i + v_j\sin(e_{i3})\right) +$$

$$\frac{\sin e_{i3}}{k_2}\left(-k_2v_je_{i2} - k_3\sin(e_{i3}) + \omega_{i\varepsilon} - e_{iv2}^F\right)$$

$$-e_{iv}^{F^T}\bar{M}_i^{-1}K_ve_{iv}^F - e_{iv}^{F^T}\bar{M}_i^{-1}\breve{\gamma}(e_i) + e_{iv}^{F^T}\bar{M}_i^{-1}\breve{f}(\breve{z}_i) - e_{iv}^{F^T}\bar{M}_i^{-1}K_ve_{iET} +$$

$$e_{iv}^{F^T}\bar{M}_i^{-1}[f(z_i) - f(\breve{z}_i)] - [\dot{\bar{v}}_{id}^F - \dot{\breve{\bar{v}}}_{id}^F]. \tag{94}$$

Using the definition of e_{i3} and expanding the expression, we obtain

$$= e_{i1}e_{i2}\omega_i - k_1 e_{i1}^2 + e_{i1}v_{i\varepsilon} - e_{i1}e_{i2}\omega_i + v_j e_{i2}\sin e_{i3} - v_j e_{i2}\sin e_{i3} - \frac{k_3}{k_2}\sin^2(e_{i3}) +$$

$$(\frac{\omega_{i\varepsilon}}{k_2} - \frac{e_{iv2}^F}{k_2})\sin e_{i3} - e_{i1}e_{iv1}^F$$

$$-e_{iv}^{F^T}\bar{M}_i^{-1}K_v e_{iv}^F - e_{iv}^{F^T}\bar{M}_i^{-1}\bar{\gamma}(e_i) + e_{iv}^{F^T}\bar{M}_i^{-1}\tilde{f}(\breve{z}_i) - e_{iv}^{F^T}\bar{M}_i^{-1}K_v e_{iET} +$$

$$e_{iv}^{F^T}\bar{M}_i^{-1}[f(z_i) - f(\breve{z}_i)] - [\dot{\bar{v}}_{id}^F - \dot{\breve{\bar{v}}}_{id}^F].$$

$$\dot{L}_i = -k_1 e_{i1}^2 + e_{i1}v_{i\varepsilon} - \frac{k_3}{k_2}\sin^2(e_{i3}) + \frac{1}{k_2}\omega_{i\varepsilon}\sin e_{i3} - \bar{M}_i^{-1}K_v\left\|e_{iv}^F\right\|^2 - \frac{1}{k_2}\sin e_{i3}e_{iv2}^F - e_{i1}e_{iv1}^F - [\dot{\bar{v}}_{id}^F - \dot{\breve{\bar{v}}}_{id}^F]$$
$$-e_{iv}^{F^T}\bar{M}_i^{-1}\bar{\gamma}(e_{i1}, e_{i2}, e_{i3}) + e_{iv}^{F^T}\bar{M}_i^{-1}\tilde{\Theta}_i^T\psi(\breve{z}_i) - e_{iv}^{F^T}\bar{M}_i^{-1}K_v e_{iET} + e_{iv}^{F^T}\bar{M}_i^{-1}[f(z_i) - f(\breve{z}_i)]. \quad (95)$$

Using (83) in (95) and from Assumption 5, we obtain

$$\dot{L}_i \le -k_1 e_{i1}^2 - \frac{k_3}{k_2}\sin^2(e_{i3}) - \bar{M}_i^{-1}K_v\left\|e_{iv}^F\right\|^2 + e_{i1}v_{i\varepsilon} + \frac{\omega_{i\varepsilon}}{k_2}\sin e_{i3} + \frac{e_{iv2}^F}{k_2}\sin e_{i3} - \frac{e_{iv2}^F}{k_2}[\sin e_{i3} - \sin\breve{e}_{i3}]$$
$$+e_{iv1}^F e_{i1} + e_{iv}^{F^T}\bar{M}_i^{-1}\tilde{\Theta}_i^T\psi(\breve{z}_i) + e_{iv}^{F^T}\bar{M}_i^{-1}(L_{i\phi} - K_v)e_{iET} + e_{iv1}^F\varepsilon_{i1} - \frac{e_{iv2}^F}{k_2}\sin e_{i3} - e_{i1}e_{iv1}^F + L_v\left\|\varepsilon_i\right\|.$$

with $\varepsilon_i = [\varepsilon_{i1}\ \varepsilon_{i2}]$. Utilizing the definition of the event-sampled formation errors, we get

$$\dot{L}_i \le -k_1 e_{i1}^2 - \frac{k_3}{k_2}\sin^2(e_{i3}) - \bar{M}_i^{-1}K_v\left\|e_{iv}^F\right\|^2 + \frac{1}{k_2}\sin e_{i3} - \frac{1}{k_2}e_{iv2}^F[\sin e_{i3} - \sin\breve{e}_{i3}] - e_{i1}k_2 v_j\varepsilon_{i2} - e_{i1}k_2\varepsilon_{vj}e_{i2}$$
$$+e_{iv}^{F^T}\bar{M}_i^{-1}\tilde{\Theta}_i^T\psi(\breve{z}_i) + e_{iv}^{F^T}\bar{M}_i^{-1}(L_{i\phi} - K_v)e_{iET} + e_{iv1}^F\varepsilon_{i1} + e_{i1}k_3[\sin e_{i3} - \sin\breve{e}_{i3}] + e_{i1}\varepsilon_{\omega j} - e_{i1}k_2\varepsilon_{vj}\varepsilon_{i2}$$
$$-\frac{1}{k_2}\sin e_{i3}v_j[\cos e_{i3} - \cos\breve{e}_{i3}] + \frac{1}{k_2}\sin e_{i3}\varepsilon_{vj}\cos\breve{e}_{i3} - \frac{1}{k_2}\sin e_{i3}k_1\varepsilon_{i1} + L_v\varepsilon_i. \quad (96)$$

Using the definition of $\breve{v}_j$ in (96) to obtain

$$\dot{L}_i \le -k_1 e_{i1}^2 - \frac{k_3}{k_2}\sin^2(e_{i3}) - \bar{M}_i^{-1}K_v\left\|e_{iv}^F\right\|^2 - \frac{e_{iv2}^F}{k_2}[\sin e_{i3} - \sin\breve{e}_{i3}] - e_{i1}k_2(\breve{v}_j + \varepsilon_{vj})\varepsilon_{i2} - e_{i1}k_2\varepsilon_{vj}e_{i2}$$
$$+e_{iv}^{F^T}\bar{M}_i^{-1}\tilde{\Theta}_i^T\psi(\breve{z}_i) + e_{iv}^{F^T}\bar{M}_i^{-1}(L_{i\phi} - K_v)e_{iET} + e_{iv1}^F\varepsilon_{i1} + e_{i1}k_3[\sin e_{i3} - \sin\breve{e}_{i3}] + e_{i1}\varepsilon_{\omega j} - e_{i1}k_2\varepsilon_{vj}\varepsilon_{i2}$$
$$-\frac{1}{k_2}\sin e_{i3}(\breve{v}_j + \varepsilon_{vj})[\cos e_{i3} - \cos\breve{e}_{i3}] + \frac{1}{k_2}\sin e_{i3}\varepsilon_{vj}\cos\breve{e}_{i3} - \frac{1}{k_2}\sin e_{i3}k_1\varepsilon_{i1} + L_v\varepsilon_i. \quad (97)$$

Applying the norm operator, we obtain the inequality as

$$\dot{L}_i \le -k_1 e_{i1}^2 - \frac{k_3}{k_2}\sin^2(e_{i3}) - \bar{M}_i^{-1}K_v\left\|e_{iv}^F\right\|^2 + \frac{1}{k_2}e_{iv2}^F\left\|[\sin e_{i3} - \sin\breve{e}_{i3}]\right\| + e_{i1}k_2\varepsilon_{i2}\breve{v}_j - e_{i1}k_2e_{i2}\varepsilon_{vj}$$
$$+\left\|e_{iv}^{F^T}\right\|\left\|\bar{M}_i^{-1}\tilde{\Theta}_i^T\psi(\breve{z}_i)\right\| + \left\|e_{iv}^{F^T}\right\|\left\|\bar{M}_i^{-1}(L_{i\phi} - K_v)e_{iET}\right\| + e_{iv1}^F\varepsilon_{i1} + e_{i1}\left\|k_3[\sin e_{i3} - \sin\breve{e}_{i3}]\right\| + \left\|e_{i1}\right\|\left\|\varepsilon_{\omega j}\right\|$$
$$+\frac{1}{k_2}\left\|\sin e_{i3}\right\|\left\|(\breve{v}_j + \varepsilon_{vj})\right\|\left\|[\cos e_{i3} - \cos\breve{e}_{i3}]\right\| + \frac{1}{k_2}\left\|\sin e_{i3}\right\|\left\|\varepsilon_{vj}\cos\breve{e}_{i3}\right\| + \frac{k_1\left\|\varepsilon_{i1}\right\|}{k_2}\left\|\sin e_{i3}\right\| + \left\|L_v\right\|\left\|\varepsilon_i\right\|. \tag{98}$$

By utilizing the boundedness of the trigonometric terms, (98) is simplified as

$$\le -k_1 e_{i1}^2 - \frac{k_3}{k_2}\sin^2(e_{i3}) - \bar{M}_i^{-1}K_v\left\|e_{iv}^F\right\|^2 + \left\|e_{iv1}^F\right\|\left(\frac{1}{k_2} + \left\|\bar{M}_i^{-1}\tilde{\Theta}_i^T\psi(\breve{z}_i)\right\|\right) +$$

$$e_{i1}k_2(\varepsilon_{i2}\breve{v}_j + \varepsilon_{i2}\varepsilon_{vj} - \breve{e}_{i2}\varepsilon_{vj}) + \left\|L_v\right\|\left\|\varepsilon_i\right\|$$

$$+\left\|e_{iv}^{F^T}\right\|\left(\left\|\bar{M}_i^{-1}(L_{i\phi} - K_v)e_{iET}\right\| + \left\|\varepsilon_{i1}\right\|\right) + \left\|e_{i1}\right\|\left(k_3 + \left\|\varepsilon_{\omega j}\right\|\right) + \left(\frac{\left\|(\breve{v}_j + \varepsilon_{vj})\right\|}{k_2} + \frac{\left\|\varepsilon_{vj}\right\|}{k_2} + \frac{k_1\left\|\varepsilon_{i1}\right\|}{k_2}\right)\left\|\sin e_{i3}\right\|.$$

Using Young's inequality and combining similar terms reveals

$$\le -\left(k_1 - \frac{k_2^2}{2} - \frac{3}{2}\right)e_{i1}^2 - \left(\frac{k_3}{k_2} - \frac{1}{2}\right)\sin^2(e_{i3}) - (\bar{M}_i^{-1}K_v - 2)\left\|e_{iv}^F\right\|^2 +$$

$$\frac{1}{2}\left\|\bar{M}_i^{-1}(L_{i\phi} - K_v)\right\|^2\left\|e_{iET}\right\|^2 + \left(\frac{1}{2} + \frac{k_1}{k_2}\right)\left\|\varepsilon_{i1}\right\|^2$$

$$+\frac{k_2^2\breve{v}_j^2}{2}\left\|\varepsilon_{i2}\right\|^2 + \frac{1}{2}\left\|e_{i1}\varepsilon_{i2}\right\|^2 + \frac{1}{2}\left\|\varepsilon_{\omega j}\right\|^2 + \left(\frac{k_2^2}{2} + \frac{\left\|\breve{e}_{i2}\right\|^2}{2} + \frac{5}{k_2}\right)\left\|\varepsilon_{vj}\right\|^2 + \frac{1}{2}\left\|\varepsilon_i\right\|^2 + \bar{B}_i.$$

with the bounding term $\bar{B}_i = \frac{1}{2}k_3^2 + \frac{1}{2}\left\|\bar{M}_i^{-1}\tilde{\Theta}_i^T\psi(\breve{z}_i)\right\|^2 + \frac{1}{2k_2^2} + \frac{\left\|2\breve{v}_j\right\|^2}{k_2} + \frac{1}{2}\left\|L_v\right\|^2$. It should be noted that $\breve{v}_j$ represents the last updated velocity information from the jth robot and it is held until the next event at the jth robot.

For simplicity, defining $\bar{\mathrm{k}}_1 = \left(k_1 - \frac{k_2^2}{2} - \frac{3}{2}\right)$, $\bar{\mathrm{k}}_2 = \left(\frac{k_3}{k_2} - \frac{1}{2}\right)$, $\bar{\mathrm{k}}_3 = (\bar{M}_i^{-1}K_v - 2)$, we obtain

$$\dot{L}_i \le -\bar{\mathrm{k}}_1 e_{i1}^2 - \bar{\mathrm{k}}_2\sin^2(e_{i3}) - \bar{\mathrm{k}}_3\left\|e_{iv}^F\right\|^2 + \frac{1}{2}\left\|\bar{M}_i^{-1}(L_{i\phi} - K_v)\right\|^2\left\|e_{iET}\right\|^2 + \left(1 + \frac{k_1}{k_2}\right)\left\|\varepsilon_{i1}\right\|^2$$
$$+\left(\frac{k_2^2\breve{v}_j^2}{2} + \frac{1}{2}\left\|\breve{e}_{i1}\right\|^2 + \frac{1}{2}\right)\left\|\varepsilon_{i2}\right\|^2 + \frac{1}{2}\left\|\varepsilon_{i1}\varepsilon_{i2}\right\|^2 + \frac{1}{2}\left\|\varepsilon_{\omega j}\right\|^2 + \left(\frac{k_2^2}{2} + \frac{\left\|\breve{e}_{i2}\right\|^2}{2} + \frac{5}{k_2}\right)\left\|\varepsilon_{vj}\right\|^2 + \bar{B}_i. \tag{99}$$

Defining $\mathrm{E} = [e_{i1}^2 \quad \sin^2(e_{i3}) \quad \left\|e_{iv}^F\right\|^2]^T$, $\mathrm{K}_i = [\bar{\mathrm{k}}_1 \quad \bar{\mathrm{k}}_2 \quad \bar{\mathrm{k}}_3]$, $\mathrm{E}_{ET} = [\left\|e_{iET}\right\|^2 \quad \left\|\varepsilon_{i1}\right\|^2 \quad \left\|\varepsilon_{i2}\right\|^2 \quad \left\|\varepsilon_{i1}\varepsilon_{i2}\right\|^2]^T$, we get

$$\dot{L}_i \leq -\mathrm{K}_i\mathrm{E} + [\frac{1}{2}\left\|\bar{M}_i^{-1}(L_{i\phi} - K_v)\right\|^2 \quad (1+\frac{k_1}{k_2}) \quad (\frac{k_2^2\breve{v}_j^2}{2} + \frac{1}{2} + \frac{1}{2}\|\breve{e}_{i1}\|^2) \quad \frac{1}{2}]\mathrm{E}_{ET} + \frac{1}{2}\left\|\varepsilon_{\omega j}\right\|^2$$

$$+(\frac{1}{2} + \frac{k_2^2}{2} + \frac{\|\breve{e}_{i2}\|^2}{2} + \frac{5}{k_2})\left\|\varepsilon_{vj}\right\|^2 + \bar{B}_i. \tag{100}$$

For the system of N_i robots in the neighborhood of the ith robot, we have the Lyapunov function derivative as the sum of all the Lyapunov derivatives of the form (100), which yields

$$\dot{L} \leq \sum_{i=1}^{N} -\mathrm{K}_i\mathrm{E}_i + \mathrm{K}_{iET}\mathrm{E}_{iET} + \bar{B}_i. \tag{101}$$

where $\mathrm{E}_i = [e_{i1}^2 \ \sin^2(e_{i3}) \ \|e_{iv}^F\|^2]^T$, $E_{iET} = [\|e_{iET}\|^2 \ \|\varepsilon_{i1}\|^2 \ \|\varepsilon_{i2}\|^2 \ \|\varepsilon_{i1}\,\varepsilon_{i2}\|^2]^T$, $\mathrm{K}_i = [\bar{\mathrm{k}}_1 \ \bar{\mathrm{k}}_2 \ \bar{\mathrm{k}}_3]$, and $\mathrm{K}_{iET} = [\frac{1}{2}\|B_{im}(L_{i\phi} - K_v)\|^2 + N_i(\frac{1}{2} + (\frac{1}{2} + \frac{k_2^2}{2} + \frac{\|\breve{e}_{j2}\|^2}{2} + \frac{5}{k_2}))\ (1 + \frac{k_1}{k_2})\ (\frac{k_2^2\breve{v}_j^2}{2} + \frac{1}{2} + \frac{1}{2}\|\breve{e}_{i1}\|^2)\ \frac{1}{2}]$, with N_i being the total number of robots in the neighborhood of the ith robot. Now, using the event-sampling condition (75) in (101) results in

$$\dot{L}_i \leq -\left\|\mathrm{K}_i\right\|\left\|\mathrm{E}_i\right\| + \left\|\mathrm{K}_{iET}\right\|\mu_i\left\|\mathrm{E}_i\right\| + \bar{B}_i. \tag{102}$$

Choosing $\mu_{ik} = \dfrac{1}{\left\|\mathrm{K}_{iET}\right\|}$, the Lyapunov derivative is further simplified such that

$$\dot{L}_i \leq -(\left\|\mathrm{K}_i\right\| - 1)\left\|\mathrm{E}_i\right\| + \bar{B}_i. \tag{103}$$

From (103), it can be seen that the formation and velocity tracking errors are bounded during the inter-event period provided $k_1 > \frac{k_2^2}{2} - \frac{5}{2}, \frac{k_3}{k_2} > \frac{3}{2}, B_{im}K_v > 3, k_2 > 0$, and since the unknown parameters are not updated, they remain bounded during the inter-event period such that

$$\tilde{\Theta}_i(t) = \tilde{\Theta}_i(t_k), \phi(\breve{z}(t)) = \phi(t_k), \quad t \in [t_k, t_{k+1}). \tag{104}$$

Therefore, from Case (i) and Case (ii), it can be concluded that the velocity-tracking errors, formation and the parameter estimation errors remain bounded during the inter-event period. Due to the event-sampling condition, the measurement error introduced by the event-sampled feedback is also bounded during the inter-event period.

Remark 17. The evolution of the velocity tracking error and the formation error with event-sampled implementation of the controller is presented with the adaptation law (84) with an additional term to relax the PE condition and to ensure that the parameter estimation errors are bounded. Similar to the ideal case, analytical results can be obtained for $\bar{\tau}_d, \kappa_i \neq 0$. The bounds on the closed loop signals for this case will be obtained as a function of the $\bar{\tau}_d$ and the epsilon modification (Narendra and Annaswamy 2012) term in the adaptation rule.

Remark 18. From the results in Theorem 4, it can be seen that the measurement error is bounded for all the inter-event period due to the event-sampling condition (75) with the value of μ_k obtained from (102). Similar to the case of robot manipulator, with the states of the closed-loop system defined as $Z = [e_i\ e_{vi}^F\ \tilde{\Theta}]^T$, the ISS characterization can be obtained. With the ISS characterization and the Assumption 5 on Lipschitz continuity of the error dynamics, the existence of a positive minimum inter event time can be established (Tabuada 2007).

Remark 19. Notice that the value of μ_k obtained from (102) requires the information regarding the last updated information of the jth robot to determine the event-based sampling instants. Also, for the event-sampling condition, the formation errors and the measurement errors for the ith robot can be calculated with the previously obtained information from the jth robot, the details of which are not included in this chapter. This makes the event-sampling mechanism decentralized.

Remark 20. From the results in Theorem 4, it can be seen that the measurement error is bounded for all the inter-event period due to the event-sampling condition (75) with the value of μ_k obtained from (102). By choosing the gains as required in the Lyapunov conditions in (103), we obtain $0 < \mu_k < 1, \forall k$.

Remark 21. Once the velocity tracking and the formation errors converge to a small value, a dead-zone operator can be used to minimize the events. This way the feedback information is not utilized frequently and computations, communication can be reduced further.

Next, the simulation results are presented to verify the proposed event-sampled controller design for the robotic systems.

RESULTS AND DISCUSSIONS

In this section, the simulation results for the event-sampled control of a 2-link robotic manipulator are presented first, followed by the results of event-sampled formation control to illustrate the effectiveness of the proposed analytical design.

Example 1. Consider the dynamical equation of the robotic manipulator with two-joint angles q_1, q_2. The following parameters are used in the simulation results presented here. $m_1 = 1, m = 2.3, L_1 = 1, L_2 = 1, g = 9.8$. The following parameters are used in the controller $k_v = 30I; \lambda = 0.5I, \kappa = 0.1$ and the event-sampling mechanism with $\sigma = 0.05$, $L_\phi = 5$, $\theta_{RM} = 5$. The initial joint angles and the unknown parameters were chosen randomly in the interval [0,1]. For the desired trajectory, $\sin(\omega t)$, $\cos(\omega t)$ were chosen with $\omega = 0.4, \tau_d = [0.8\sin(2\omega t)\ 0.7\cos(4\omega t)]^T$. It should be noted from the event-sampling condition (59) that as the bounds on the mass matrix is increased, the event-sampling threshold is decreased causing frequent events.

The performance of the adaptive controllers with continuous feedback is presented first Figs. 2 and 3. This will be used to analyze the performance of the event-sampled controllers. Due to the continuous feedback, the tracking error and the parameter estimation error converges to their respective ultimate bounds very quickly. In order to compare the performance of the event-sampled controllers, simulations are carried out without changing any controller parameters and initial conditions.

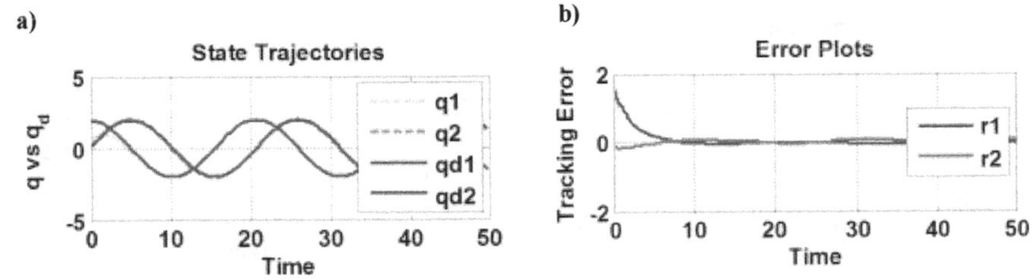

Figure 2(a). Joint angle trajectories. **(b)** Tracking errors (Continuous feedback).

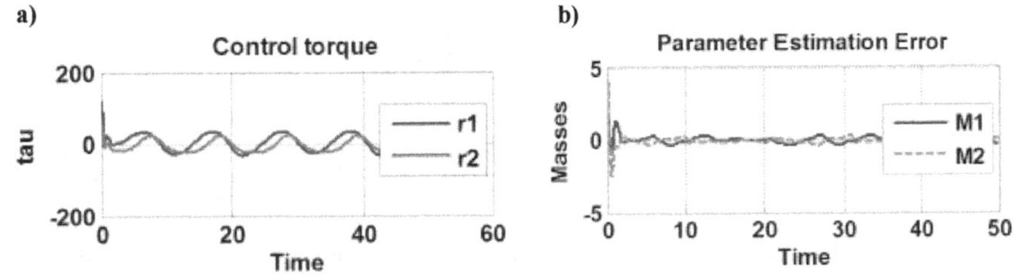

Figure 3(a). Control torque. **(b)** Estimation error (Continuous feedback).

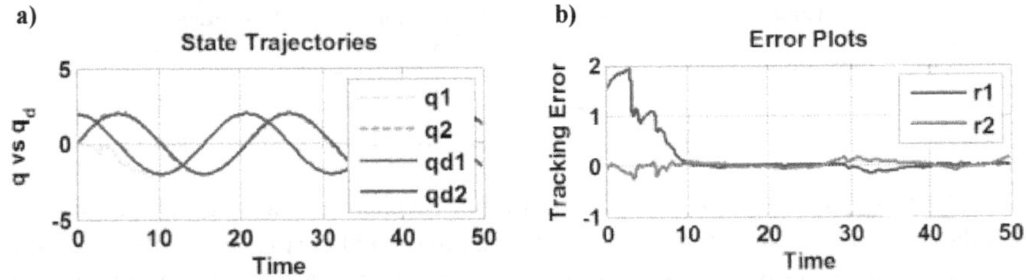

Figure 4(a). Joint angle trajectories. **(b)** Tracking errors (Event-sampled feedback).

The designed event-sampled controllers were able to stabilize the robotic manipulator and the parameter tuning rule was able to learn the uncertain dynamics of the robotic manipulator. The desired and actual joint angles are plotted together in Fig. 4a while the tracking error trajectory is plotted in Fig. 4b. The ultimate bounds of the tracking error were slightly larger when compared to the continuous feedback case (Figs. 2a and 2b).

The control torque and parameter estimation error with event-sampled feedback is shown in Figs. 5a and 5b. It can be observed that with the event-sampled feedback, the bound on parameter estimation error is slightly higher compared to continuous feedback (Figs. 3a and 3b). This is due to intermittent feedback based on the event-sampling instants.

Figure 6 depicts the performance of the event-sampling mechanism. It can be seen that the computations, for a single robotic manipulator operated for 50 seconds, are reduced by approximately 33%. During this adaptation period, out of the 5000 sensor samples 3365 samples are utilized to generate the control torque, without compromising the tracking performance. If the bound on the unknown mass matrix is increased, the event-sampling threshold is reduced creating more events.

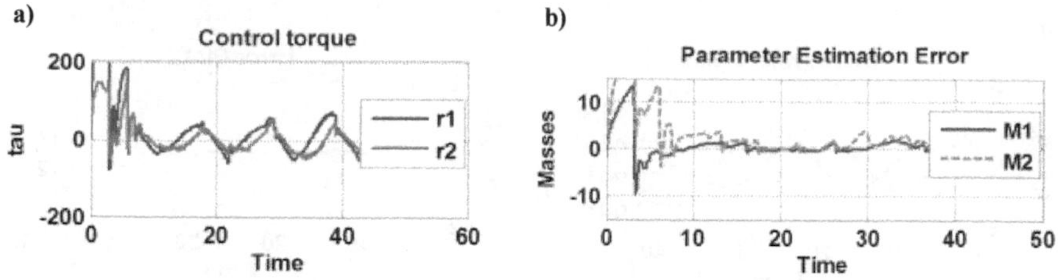

Figure 5(a). Control torque. **(b)** Estimation error (Event-sampled feedback).

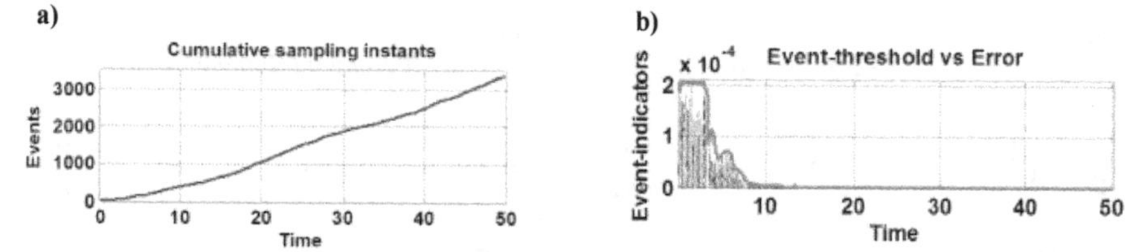

Figure 6. Event-sampling mechanism **(a).** Total events, and **(b)** Measurement error vs. Threshold.

Table 1. Number of sampling instants with different threshold parameter σ.

σ	Total Sensor Samples	Event-based Samples	Percentage Reduction
0.03	5000	3739	25.22
0.04	5000	3647	27.06
0.05	5000	3365	32.7

The design parameter σ in the event-sampling mechanism can be used to change the threshold on the measurement error during the inter-event period. When the threshold is large, the measurement error is allowed to increase to a larger value and therefore, the events are spaced out. This reduces the computations, while the tracking error and the parameter estimation errors remain bounded. The Table 1, lists the variation in the event-based sampling instants as the design parameter σ is increased. It can be seen that as the threshold is increased the event-based sampling instants are reduced considerably. The simulation figures presented before are obtained with $\sigma = 0.05$, where all the signals are very close to those obtained with continuous feedback. This indicates the efficiency of using the event-based feedback. It should also be noted that increasing σ further, reduced the event-based sampling instants. However, the tracking performance is still achieved at the expense of more control effort.

Example 2. To illustrate the effectiveness of the proposed event-based adaptive controller, a group of four non-holonomic mobile robots is considered. The robots are initiated from an arbitrary position and move to a desired shape of a square. The desired and initial positions, initial bearing angles and the initial velocities of the non-holonomic mobile robots are given by

$x_1(t_0) = 15.6$, $x_2(t_0) = 13.1$, $x_3(t_0) = 9.6$, $x_4(t_0) = 12$, $x_1^d = .5$, $x_2^d = .5$, $x_3^d = -.5$, $x_4^d = -.5$, $y_1^d = .5$, $y_2^d = -.5$, $y_1(t_0) = -10.8$, $y_2(t_0) = -15.6$, $y_3(t_0) = -16.8$, $y_4(t_0) = -12$, $y_3^d = .5$, $y_4^d = -.5$, $\theta_1(t_0) = 0$, $\theta_2(t_0) = 0$, $\theta_3(t_0) = 0$, $\theta_4(t_0) = 0$, $\theta_1^d = \pi$, $\theta_2^d = \pi$, $\theta_3^d = \pi$, $\theta_4^d = \pi$.

The controller gains are selected as $K_v = 80, k_1 = 2, k_2 = 1, k_3 = 0.5$. The parameters for the robot dynamics are selected as $m = 5$ kg, $I = 3$ kg², $R = 0.15$ m, r = 0.08 m, d = 0.4 m for each robot and the mass matrix is calculated as $\bar{M} = \begin{bmatrix} 810 & 0 \\ 0 & 133.02 \end{bmatrix}$, and $\|\bar{M}^{-1}\| = 0.0075$. Figure 7 depicts the motion of four independent non-holonomic mobile robots with continuous feedback. They are initiated in a non-square shape which can be seen on the plot.

Given their desired locations, they form a square shape by minimizing the consensus error along x, y, θ in the first case and in the second case the robots align along a straight line. The friction coefficients and the Coriolis matrix are considered uncertain as described in the problem formulation. The initial movements of the robots are oscillatory because of the unknown dynamics. With the parameter adaptation using $\Lambda = 40$, $\kappa_i = 50$, the controllers of each robot learns the unknown parameters. Once the uncertain parameters are tuned, the robots reach consensus.

The robots reach consensus around 30th second as it can be viewed in Figs. 8 and 9. The difference between the virtual controller (which is the desired linear and angular velocities to reach consensus) and the actual linear and angular velocities which is the velocity tracking error is plotted in Fig. 9.

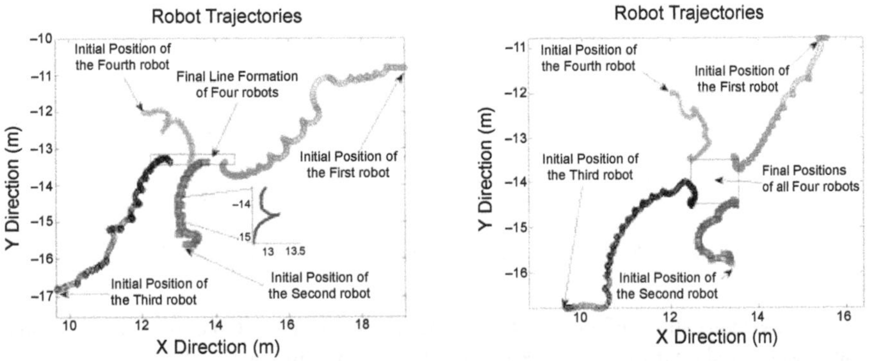

Figure 7. Mobile robots moving to their desired formation.

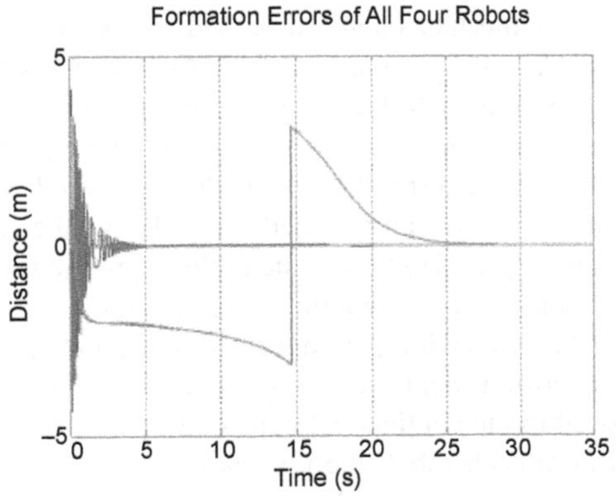

Figure 8. Formation errors on bot x and y directions.

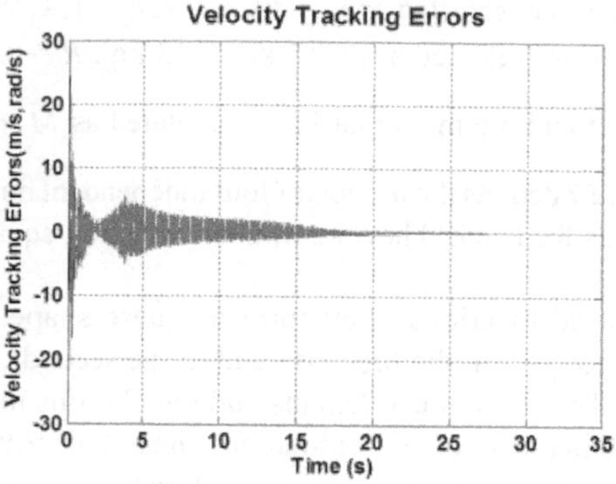

Figure 9. Velocity tracking errors.

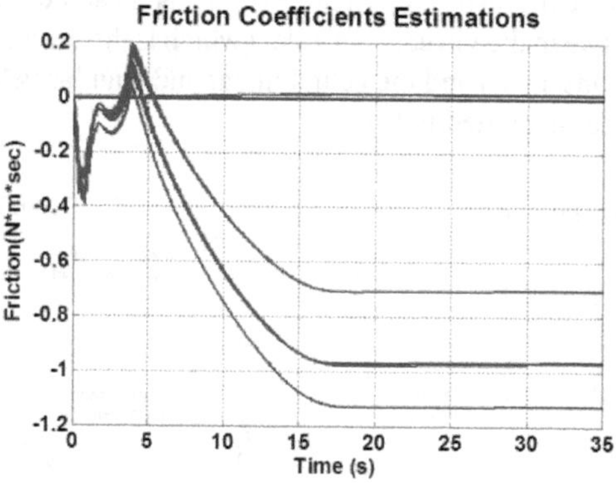

Figure 10. Frictions parameters estimates their desired formation.

The viscous and Coulomb frictions coefficients of each robot converge to a steady state value as shown in Fig. 10. Since the robots may move in different terrain, the friction terms may change over time. Therefore, adaptation by the controller is valuable and is achieved with the designed controller.

The event-sampling mechanism was designed with k_1, k_2, k_3, k_v as selected in the controller and $\sigma = 0.5$, $N_i = 1$, $L_{\phi i} = 5$, $B_{im} = 0.075$. To compare with the controller with continuous feedback, all the controller gains and initial values of the parameters and the initial conditions of the robots were unchanged. With the proposed event-sampled feedback, the mobile robots were able to reach the desired consensus as seen in Fig. 11. The formation errors remain bounded in Fig. 12 and the velocity tracking error remains bounded as in Fig. 13. However, due to the aperiodic feedback, these bounds are slightly larger compared to their continuous counterpart.

The parameter adaptation with aperiodic event-based updates resulted in a bounded parameter estimation error as depicted in Fig. 14. Clearly, the bounds on these errors were higher compared to the continuously updated case. Due to the designed event-sampling mechanism at each robot, the total number of parameter updates and the communication instants are considerably reduced as seen in Fig. 15.

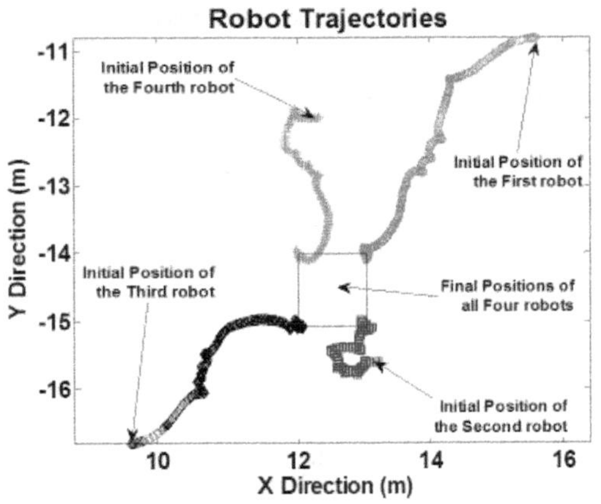

Figure 11. Mobile robots moving to their desired formation.

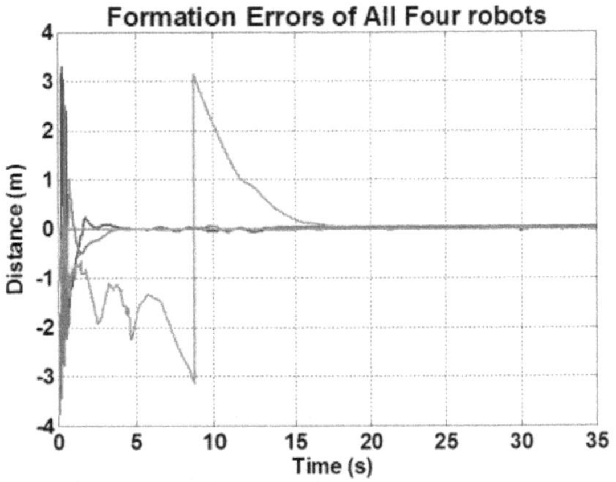

Figure 12. Formation errors on bot x and y directions.

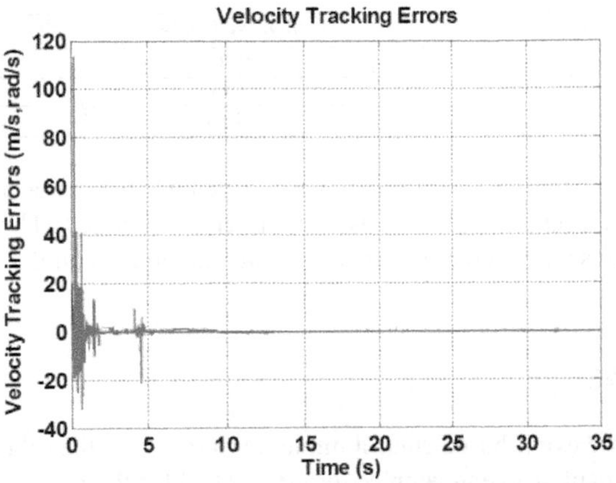

Figure 13. Velocity tracking errors.

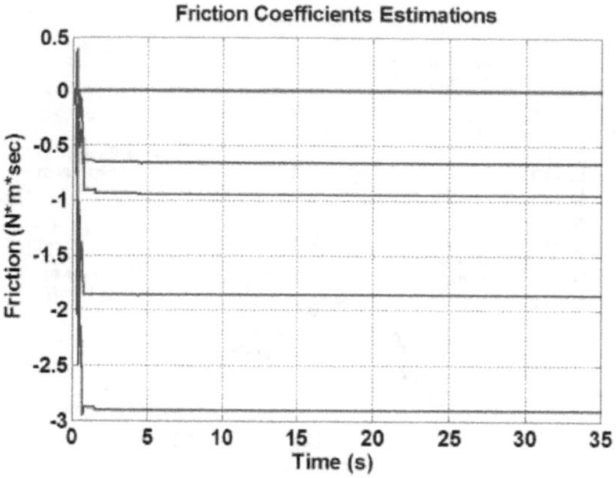

Figure 14. Friction coefficient estimates.

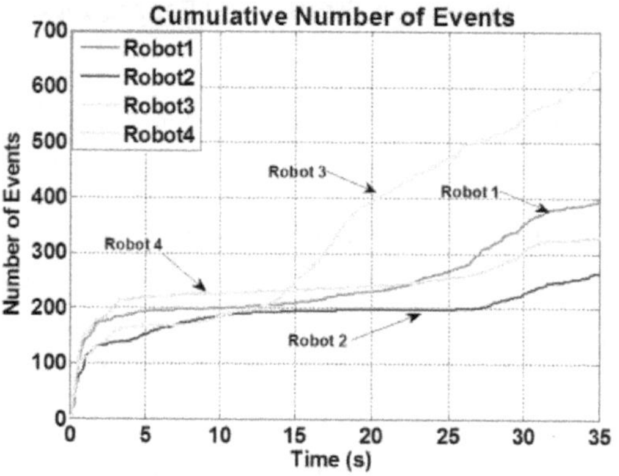

Figure 15. Cumulative number of events of each robot.

Table 2. Number of sampling instants (average of four robots) with different threshold parameter σ.

σ	Total Sensor Samples	Event-based Samples	Percentage Reduction
0.4	5237	690	86.82
0.2	5357	806	84.96
0.1	5517	1138	79.37

In the event-sampling condition, when the design parameter σ_i was varied, the threshold on the event-sampled measurement error changed. This resulted in the reduction of events with an increase in σ_i whereas the bounds on the closed-loop error signals go up.

CONCLUSIONS

In this chapter, the event-based control implementations for the robotic systems were presented. The adaptive event-sampled torque control for the robotic manipulator was able to drive the joint angles to track the desired trajectories with bounded error due to event sampled measurement errors. The bounds are also found to be a function of the disturbance bounds. The event-sampling mechanism was able to generate additional events so that the tracking error remains bounded in the presence of disturbance.

Next the event-sampled adaptive control of mobile robot formation was derived. The uncertain friction and Coriolis terms are represented in the LIP form and are adaptively tuned by the controller at event-sampled instants. Using the dynamics as well as the kinematics of the mobile robot, the consensus error and the velocity tracking errors were controlled. The event-sampling condition at each robot and the parameter adaptation rules were derived using the Lyapunov stability analysis. The analytical results were verified using the simulation examples and the efficiency of the event-sampled controller execution was demonstrated.

REFERENCES

Balch, T. and R. C. Arkin. 1998. Behaviour-based formation control for multi-robot teams. Robotics and Automation, IEEE Transactions on, 14(6): 926–939.

Bauso, D., L. Giarré and R. Pesenti. 2009. Consensus for networks with unknown but bounded disturbances. SIAM Journal on Control and Optimization, 48(3): 1756–1770.

Cheah, C. C., X. Li, X. Yan and D. Sun. 2015. Simple PD control scheme for robotic manipulation of biological cell. Automatic Control, IEEE Transactions on, 60(5): 1427–1432.

Fierro, R. and F. L. Lewis. 1998. Control of a non-holonomic mobile robot using neural networks. Neural Networks, IEEE Transactions on, 9(4): 589–600.

Galicki, M. 2015. Finite-time control of robotic manipulators. Automatica, 51: 49–54.

Guinaldo, M., D. Lehmann, J. Sanchez, S. Dormido and K. H. Johansson. 2012. December. Distributed event-triggered control with network delays and packet losses. pp. 1–6. *In*: Decision and Control (CDC), 2012 IEEE 51st Annual Conference on, IEEE.

He, W., A. O. David, Z. Yin and C. Sun. 2015a. Neural network control of a robotic manipulator with input deadzone and output constraint. Systems, Man and Cybernetics: Systems, IEEE Transactions on (In Press).

He, W., Y. Dong and C. Sun. 2015b. Adaptive neural impedance control of a robotic manipulator with input saturation. Systems, Man and Cybernetics: Systems, IEEE Transactions on (In Press).

Hong, Y., J. Hu and L. Gao. 2006. Tracking control for multi-agent consensus with an active leader and variable topology. Automatica, 42(7): 1177–1182.

Khalil, H. K. and J. W. Grizzle. 1996. Nonlinear systems (Vol. 3). New Jersey: Prentice hall.

Lewis, F. L., D. M. Dawson and C. T. Abdallah. 2003. Robot manipulator control: theory and practice. CRC Press.

Low, C. B. 2011. December. A dynamic virtual structure formation control for fixed-wing UAVs. pp. 627–632. *In*: Control and Automation (ICCA), 2011 9th International Conference on, IEEE.

Min, H. K., F. Sun, S. Wang and H. Li. 2011. Distributed adaptive consensus algorithm for networked Euler-Lagrange systems. Control Theory & Applications, IET, 5(1): 145–154.

Molin, A. and S. Hirche. 2013. On the optimality of certainty equivalence for event-triggered control systems. Automatic Control, IEEE Transactions on, 58(2): 470–474.

Narendra, K. S. and A. M. Annaswamy. 2012. Stable adaptive systems. Courier Corporation.

Olfati-Saber, R. and R. M. Murray. 2004. Consensus problems in networks of agents with switching topology and time-delays. Automatic Control, IEEE Transactions on, 49(9): 1520–1533.

Qu, Z. 2009. Cooperative control of dynamical systems: applications to autonomous vehicles. Springer Science & Business Media.

Ren, W. and R. W. Beard. 2005a. Consensus seeking in multi-agent systems under dynamically changing interaction topologies. Automatic Control, IEEE Transactions on, 50(5): 655–661.

Ren, W., R. W. Beard and E. M. Atkins. 2005b. June. A survey of consensus problems in multi-agent coordination. pp. 1859–1864. *In*: American Control Conference, 2005. Proceedings of the 2005, IEEE.

Sahoo, A., H. Xu and S. Jagannathan. 2015. Near optimal event-triggered control of nonlinear discrete-time systems using neuro-dynamic programming. Neural Networks and Learning systems, IEEE Transactions on (In press).

Semsar-Kazerooni, E. and K. Khorasani. 2008. Optimal consensus algorithms for cooperative team of agents subject to partial information. Automatica, 44(11): 2766–2777.

Semsar-Kazerooni, E. and K. Khorasani. 2009. June. Analysis of actuator faults in a cooperative team consensus of unmanned systems. pp. 2618–2623. *In*: American Cont. Conf., 2009. ACC'09. IEEE.

Sheng, L., Y. J. Pan and X. Gong. 2012. Consensus formation control for a class of networked multiple mobile robot systems. Journal of Control Science and Engineering, pp. 1–12.

Tabuada, P. 2007. Event-triggered real-time scheduling of stabilizing control tasks. Automatic Control, IEEE Transactions on, 52(9): 1680–1685.

Tian, Y. P. and C. L. Liu. 2008. Consensus of multi-agent systems with diverse input and communication delays. Automatic Control, IEEE Transactions on, 53(9): 2122–2128.

Wang, X. and M. D. Lemmon. 2011. Event-triggering in distributed networked control systems. Automatic Control, IEEE Transactions on, 56(3): 586–601.

Zhong, X., Z. Ni, H. He, X. Xu and D. Zhao. 2014. July. Event-triggered reinforcement learning approach for unknown nonlinear continuous-time system. pp. 3677–3684. *In*: Neural Networks (IJCNN), 2014 International Joint Conference on, IEEE.

8

Design, Integration and Analysis of a Hybrid Controller for Multi Degrees of Freedom Serial Mechanisms

Dan Zhang[1] and *Bin Wei*[2,*]

ABSTRACT

This paper proposes a hybrid controller by combining a PID controller and a model reference adaptive controller (MRAC), and also compares the convergence performance of the PID controller, model reference adaptive controller, and PID+MRAC hybrid controller. The results show that the convergence speed and its performance for the MRAC controller and PID+MRAC controller is better than that of the PID controller. The convergence performance for the hybrid control is better than that of the MRAC control.

Keywords: PID control, model reference adaptive control, hybrid control, convergence speed, robotic arm

INTRODUCTION

PID control is the widely used control method in many industries. For example, in many robotic arm used industries, PID control applies to each joint to control the whole robotic arm. By adjusting the PID gains of the PID controller, we can have the desired

[1] Department of Mechanical Engineering, York University, Toronto, Ontario, M3J 1P3, Canada. E-mail: dzhang99@yorku.ca

[2] Faculty of Engineering and Applied Science, University of Ontario Institute of Technology, 2000 Simcoe Street North, Oshawa, Ontario, Canada. E-mail: Bin.Wei@uoit.ca

* Corresponding author

output performance. Model reference adaptive control is another control method that was proposed early by Landau (Landau 1979) and it has been developed afterwards (Dubowsky and Desforges 1979, Cao and Hovakimyan 2008, Priyank Jain and Nigam 2013, Nhan T. Nguyen et al. 2008, Idan et al. 2002, Li and Cheah 2012, Rossomando et al. 2011, Sharifi et al. 2014, Ortega and Panteley 2014). The reason that one needs to apply adaptive control, especially the model reference adaptive control approach is that traditional controllers cannot compensate the payload variations, i.e., when the end-effector grasps different payload masses, the joint output will vary, which affects the positioning accuracy of the end-effector. However for the model reference adaptive control, the above problem can be effectively resolved and payload variation effect can be compensated. For example, the MRAC controller proposed by R. Horowitz and later associated developments by other authors (Horowitz and Tomizuka 1986, Horowitz 1983) contains an adaptive algorithm block and a position feedback loop which provides the difference between desired and actual position of the joints. This difference is acted upon by the integral portion of a seemingly PID control structure before feedback values of position and velocity are subtracted from it. In this chapter, by combining the PID controller and MRAC controller, a hybrid controller is proposed and designed. In Horowitz's MRAC controller (Horowitz 1983), it assumes that M and N are constant during adaptation. For the 1-DOF link, because its M and N matrices of the dynamic equation are constant (M is constant, N is 0), it can directly combine the PID and MRAC controllers to design the PID+MRAC controller. The convergence performance is compared among these three controllers. The results show that the convergence speed and its performance for MRAC controller and PID+MRAC controller is better than that of the PID controller. The convergence performance for the hybrid control is better than that of the MRAC control. For the MRAC, the joint output gradually goes towards to the desired position while for the PID+MRAC, it overshoots the desired position and then gradually comes back to the desired position. But for more than 1-DOF link, it is not applicable anymore because the M and N matrices of the dynamic equation are not constant. For the PID control, we need to use the Lagrange dynamic model, but for the MRAC, we need to use the Gibbs-Appell dynamic formulation. Since they are not compatible, we cannot combine the PID and MRAC in this case. On the positive side, however, Sadegh proposed an improved MRAC that can remove the condition that the M and N matrices are constant (Sadegh and Horowitz 1987, Sadegh and Horowitz 1990), so that the Lagrange dynamic equation can be used. By using Sadegh's improved adaptive algorithm and structure, and by combining the PID and MRAC controllers, a hybrid controller is designed for the more than one-DOF link (e.g., 2-DOF and 3-DOF links) case. The results show that the convergence speed for the hybrid controller is faster than that of the MRAC controller. The hybrid controller and MRAC controller are both better than that of the PID controller.

PID CONTROLLER, MRAC CONTROLLER, AND PID+MRAC HYBRID CONTROLLER

PID Controller

The PID controller is illustrated in Fig. 1. The output of the plant will compare with the desired model rp and then will have an error. This error will go through the PID control

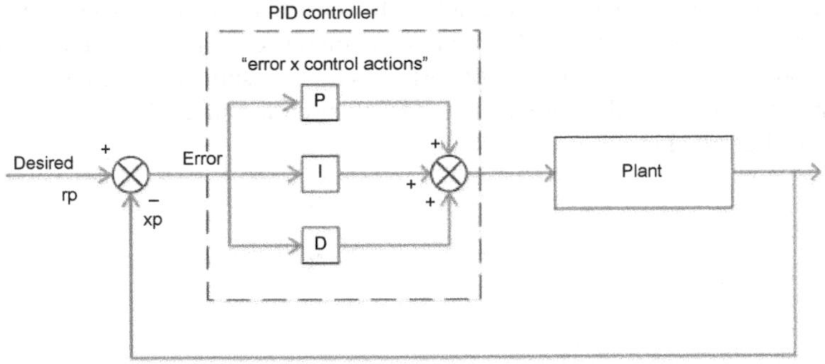

Figure 1. PID controller.

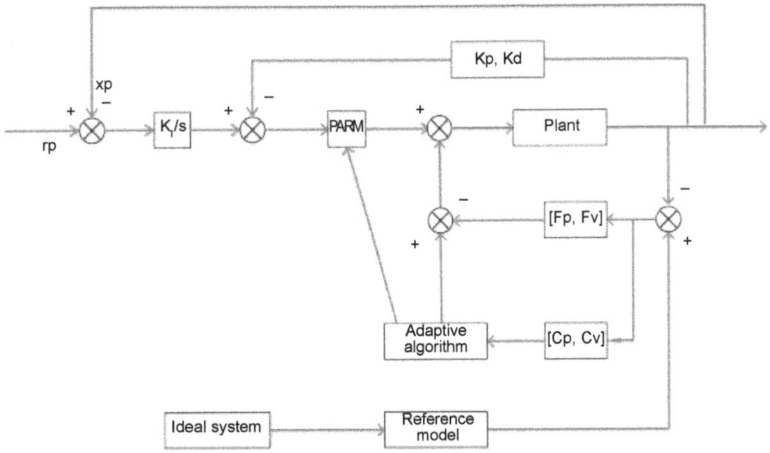

Figure 2. MRAC controller.

and through "error times control actions". The output of the PID controller will be the input to the plant model, and this circle will continue until the error between the actual output from the plant and the desired model converges to 0. This is the basic working principle of the PID control.

MRAC Controller

For the MRAC controller, the following Fig. 2 shows such a system.

One can see that this system does not contain any PID control. The output from the plant will compare with the reference model, which will produce an error. This error will be used by the adaptive algorithm block and then produce the input elements to the plant. In the meantime, the output of the plant will compare with the desired model rp and will produce another error. This error will go through the integration action and then subtract the feedback processed position and velocity by the Kp and Kd elements. This process is very similar to the PID control, but it is not a PID control. The output from this process, times the elements from the adaptive algorithm, plus the elements from the adaptive algorithm, will be the input to the plant. This process will continue until the error between the output of the plant and the reference model converges to 0.

The ideal system is isolated from the plant, in the sense that feedback values of the plant variables are not used to process the input to the reference model. The reference model input is processed from its own output variables by the "similar PID controller". The ideal system is completely unaffected by the plant performance. Sadegh's improved MRAC is illustrated in Fig. 3:

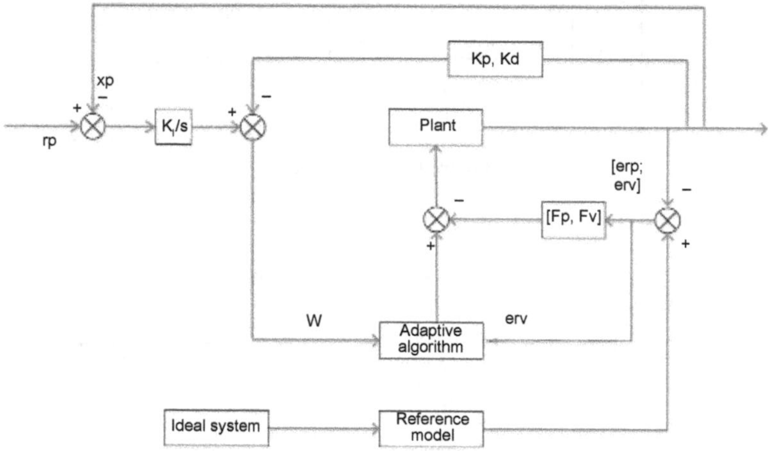

Figure 3. Improved MRAC controller.

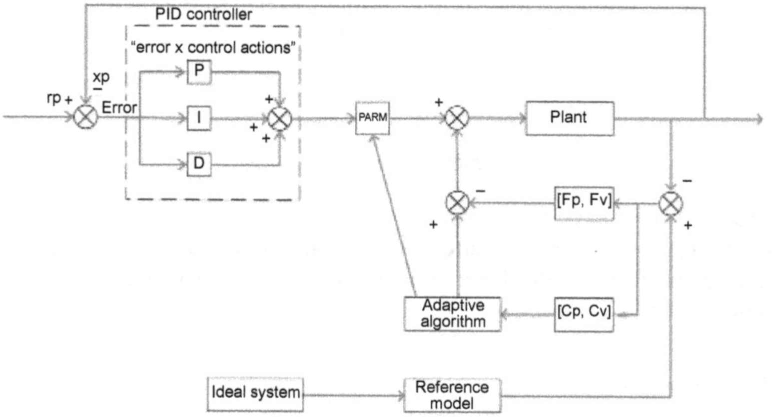

Figure 4. PID+MRAC hybrid controller for 1-DOF link.

PID+MRAC Hybrid Controller

By combining the PID controller and MRAC controller, the PID+MRAC hybrid controller is obtained as shown in Fig. 4. Same with the MRAC, the only difference between this hybrid PID+MRAC and MRAC is that the output of the plant will compare with the desired model rp and will produce an error. This error will go through the PID controller. The output of the PID controller, times the elements from the adaptive algorithm, plus the elements from the adaptive algorithm, will be the input to the plant. For 2-DOF and 3-DOF link cases, the hybrid controller is shown in Fig. 5.

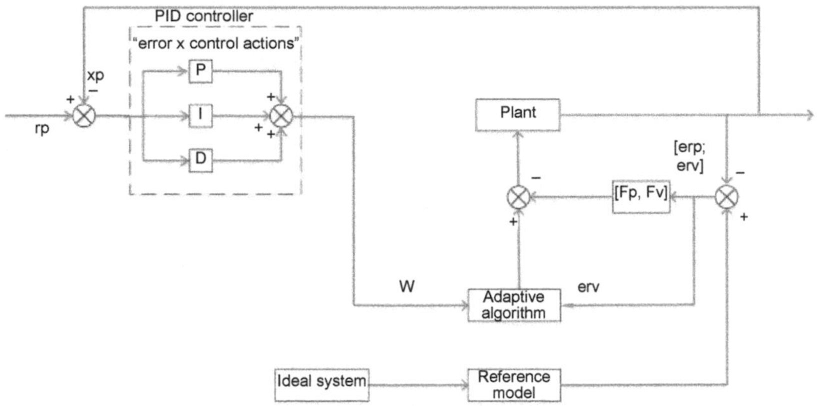

Figure 5. PID+MRAC hybrid controller for more than 1-DOF link.

CONVERGENCE PERFORMANCE COMPARISON BETWEEN PID, MRAC AND HYBRID CONTROLLER

One-DOF Link Case

Here the one link manipulator as shown in Fig. 6 will be used as an example. In order to implement PID control of the one link manipulator case, the dynamic equation has to be derived. By using the Lagrange method (John J. Craig 2005), the dynamic equation is presented as follows:

The kinetic and potential energy of this link are as follows:

$$K_1 = \frac{1}{2} m_1 (l_1 \dot{\theta}_1)^2, P_1 = m_1 g (l_1 \sin \theta_1) \tag{1}$$

The total kinetic and potential energy are:

$$K = K_1 = \frac{1}{2} m_1 (l_1 \dot{\theta}_1)^2, P = P_1 = m_1 g (l_1 \sin \theta_1) \tag{2}$$

According to the Lagrange method,

$$L = K - P = \frac{1}{2} m_1 (l_1 \dot{\theta}_1)^2 - m_1 g (l_1 \sin \theta_1) \tag{3}$$

Thus the torque applied to the joint can be determined by:

$$\tau_1 = \frac{d}{dt} \frac{\partial L}{\partial \dot{\theta}_1} - \frac{\partial L}{\partial \theta_1}. \tag{4}$$

where $\frac{\partial L}{\partial \dot{\theta}_1} = m_1 l_1^2 \dot{\theta}_1, \frac{d}{dt} \frac{\partial L}{\partial \dot{\theta}_1} = m_1 l_1^2 \ddot{\theta}_1, \frac{\partial L}{\partial \theta_1} = -m_1 g l_1 \cos \theta_1$

Rewriting equation (4) results in:

$$\tau_1 = (m_1 l_1^2) \ddot{\theta}_1 + (m_1 l_1 \cos \theta_1) g = M \ddot{\theta}_1 + 0 + (m_1 l_1 \cos \theta_1) g = M \ddot{\theta}_1 + 0 + G g \tag{5}$$

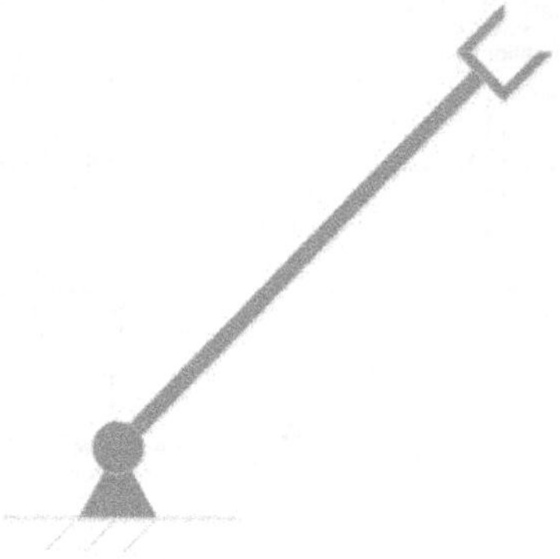

Figure 6. One link manipulator.

Through applying PID control, the controller output is the torque, i.e.,

$$K_p e + K_i \int e dt + K_d \dot{e} = \tau_1 \tag{6}$$

where error $e = r_p - x_p$

We know from the one-link manipulator M and N matrices, $M = m_1 l_1^2$, $N = 0$, the output from the manipulator (i.e., acceleration of the joint) can be determined as follows:

$$K_p e + K_i \int e dt + K_d \dot{e} = \tau_1 = (m_1 l_1^2)\ddot{\theta}_1 + (m_1 l_1 \cos\theta_1) g \Rightarrow \ddot{\theta}_1 = M^{-1}(K_p e + K_i \int e dt + K_d \dot{e}) \tag{7}$$

After deriving the acceleration of joint 1, take the integral with respect to time to obtain the velocity of joint 1 and take another integral to obtain the position of joint 1.

$$\dot{\theta}_1 = \int \ddot{\theta}_1 dt, \theta_1 = \int \dot{\theta}_1 dt \tag{8}$$

After the simulation, as shown in Fig. 5, the blue line represents the joint output under the PID control.

For the MRAC, as with the PID control, the output from the controller will be:

$$ControllerOut = \tau_1 = \hat{M} u + \hat{V} - F_p e - F_v \dot{e} \tag{9}$$

where $u = K_I \int (r_p - x_p) - K_p x_p - K_d x_v$

The manipulator dynamic equation is:

$$\tau_1 = (m_1 l_1^2)\ddot{\theta}_1 + (m_1 l_1 \cos\theta_1) g = Ma + 0 + Gg \tag{10}$$

So the output from the manipulator (i.e., acceleration of the joint) is written as:

$$\hat{M} u + \hat{V} - F_p e - F_v \dot{e} = \tau_1 = Ma + V \Rightarrow a = \ddot{\theta}_1 = M^{-1}(\hat{M} u + \hat{V} - F_p e - F_v \dot{e} - V) \tag{11}$$

After the simulation, as shown in Fig. 7, the red line represents the joint output under the MRAC control, and yellow line represents the hybrid control. From this

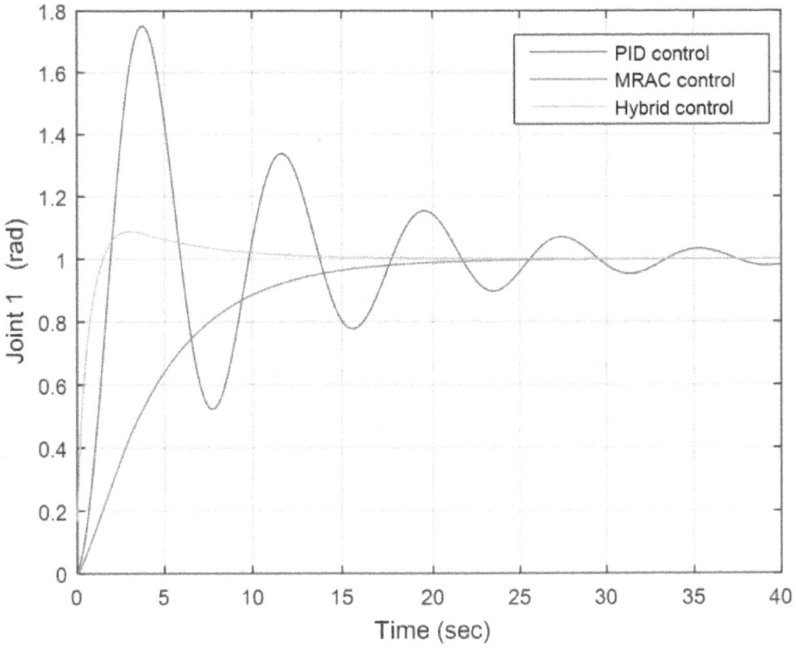

Figure 7. Joint output under PID, MRAC and hybrid control.

figure, one can see that for the PID control, it will take roughly 40 seconds to converge to 0, for the MRAC control, it will take about 20 seconds to converge to the desired position, which is half the time of the PID control. Finally for the hybrid control, it takes about 10 seconds to converge to the desired position, which halves the time of MRAC control. Furthermore, another difference between the MRAC and the hybrid control is that the MRAC control gradually converges to the desired position whereas for the hybrid control, it first overshoots the desired position very quickly and then gradually converges to the desired position. From above analysis, one can see that the convergence performance for the hybrid control is better than that of the MRAC control, and the MRAC control is better than the PID control.

Two-DOF Link Case

The two-link manipulator is shown in Fig. 8, based on the Lagrange method, the dynamic equation is derived as follows:

$$\begin{bmatrix} \tau_1 \\ \tau_2 \end{bmatrix} = M\ddot{\theta} + N = \begin{bmatrix} m_{11} & m_{12} \\ m_{12} & m_{22} \end{bmatrix} \begin{bmatrix} \ddot{\theta}_1 \\ \ddot{\theta}_2 \end{bmatrix} + \begin{bmatrix} n_{11} \\ n_{21} \end{bmatrix} \tag{12}$$

where

$$m_{11} = (m_1 + m_2)l_1^2 + m_2 l_2^2 + 2m_2 l_1 l_2 \cos\theta_2, m_{12} = m_2 l_2^2 + m_2 l_1 l_2 \cos\theta_2, m_{22} = m_2 l_2^2$$

$$n_{11} = 2(-m_2 l_1 l_2 \sin\theta_2)\dot{\theta}_1\dot{\theta}_2 + (-m_2 l_1 l_2 \sin\theta_2)\dot{\theta}_2^{\,2}, n_{21} = m_2 l_1 l_2 \sin\theta_2 \dot{\theta}_1^{\,2}$$

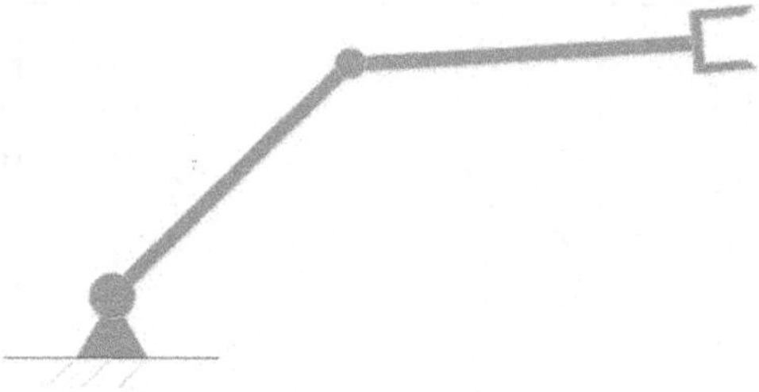

Figure 8. Two-link manipulator.

By re-parametrization of the above dynamic equation,

$$\begin{bmatrix} (m_1+m_2)l_1^2+m_2l_2^2+2m_2l_1l_2\cos\theta_2 & m_2l_2^2+m_2l_1l_2\cos\theta_2 \\ m_2l_2^2+m_2l_1l_2\cos\theta_2 & m_2l_2^2 \end{bmatrix}\begin{bmatrix} u_1 \\ u_2 \end{bmatrix}+ \tag{13}$$

$$\begin{bmatrix} 2(-m_2l_1l_2\sin\theta_2)\dot{\theta}_1\dot{\theta}_2+(-m_2l_1l_2\sin\theta_2)\dot{\theta}_2^2 \\ m_2l_1l_2\sin\theta_2\dot{\theta}_1^2 \end{bmatrix}$$

$$= W\cdot\begin{bmatrix} \Theta_1 \\ \Theta_2 \\ \Theta_3 \end{bmatrix}$$

By choosing $\Theta_1=(m_1+m_2)l_1^2+m_2l_2^2, \Theta_2=m_2l_2^2, \Theta_3=m_2l_1l_2$

$$\Rightarrow W=\begin{bmatrix} u_1 & u_2 & 2u_1\cos\theta_2+u_2\cos\theta_2-2\dot{\theta}_1\dot{\theta}_2\sin\theta_2-\dot{\theta}_2\dot{\theta}_2\sin\theta_2 \\ 0 & u_1+u_2 & u_1\cos\theta_2+\dot{\theta}_1\dot{\theta}_1\sin\theta_2 \end{bmatrix}$$

$$\tau=W\begin{bmatrix} \Theta_1 \\ \Theta_2 \\ \Theta_3 \end{bmatrix}-F_v\cdot erv-F_p\cdot erp \tag{14}$$

$$M\ddot{\theta}+N=\tau=W\begin{bmatrix} \Theta_1 \\ \Theta_2 \\ \Theta_3 \end{bmatrix}-F_v\cdot erv-F_p\cdot erp \tag{15}$$

$$\Rightarrow \ddot{\theta}=(W\begin{bmatrix} \Theta_1 \\ \Theta_2 \\ \Theta_3 \end{bmatrix}-F_v\cdot erv-F_p\cdot erp-N)/M \tag{16}$$

After deriving the acceleration of joint, take the time integral to obtain the velocity of joint and take another integral to obtain the position of joint.

$$\begin{bmatrix} \dot{\theta}_1 \\ \dot{\theta}_2 \end{bmatrix}=\int\begin{bmatrix} \ddot{\theta}_1 \\ \ddot{\theta}_2 \end{bmatrix}dt, \begin{bmatrix} \theta_1 \\ \theta_2 \end{bmatrix}=\int\begin{bmatrix} \dot{\theta}_1 \\ \dot{\theta}_2 \end{bmatrix}dt \tag{17}$$

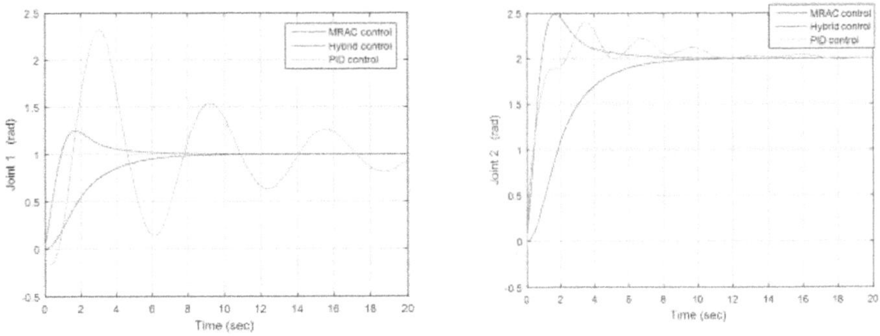

Figure 9. Joints 1 and 2 output under PID, MRAC and hybrid control.

As shown in Fig. 9, the results show that the convergence speed for the hybrid controller is faster than that of the MRAC controller. The hybrid controller and MRAC controller are both better than that of the PID controller. By using the same method, it can be extended to multi-DOF serial manipulators.

Three-DOF Link Case

The three-link manipulator is shown in Fig. 10, based on the Lagrange method, the dynamic equation is derived as follows:

$$\begin{bmatrix} \tau_1 \\ \tau_2 \\ \tau_3 \end{bmatrix} = M\ddot{\theta} + N = \begin{bmatrix} m_{11} & m_{12} & m_{13} \\ m_{12} & m_{22} & m_{23} \\ m_{13} & m_{23} & m_{33} \end{bmatrix} \begin{bmatrix} \ddot{\theta}_1 \\ \ddot{\theta}_2 \\ \ddot{\theta}_3 \end{bmatrix} + \begin{bmatrix} n_{11} \\ n_{21} \\ n_{31} \end{bmatrix} \tag{18}$$

where

$$m_{11} = m_1a^2 + m_2r_1^2 + m_3r_1^2, m_{12} = r_1(m_2b + m_3r_2)\cos(\theta_2 - \theta_1), m_{13} = m_3r_1c\cos(\theta_1 + \theta_3)$$

$$m_{22} = m_2b^2 + m_3r_2^2, m_{23} = m_3r_2c\cos(\theta_2 + \theta_3), m_{33} = m_3c^2$$

$$n_{11} = -r_1(m_2b + m_3r_2)\sin(\theta_2 - \theta_1)\dot{\theta}_2^2 - m_3r_1c\sin(\theta_1 + \theta_3)\dot{\theta}_3^2$$

$$n_{21} = r_1(m_2b + m_3r_2)\sin(\theta_2 - \theta_1)\dot{\theta}_1^2 - m_3r_2c\sin(\theta_2 + \theta_3)\dot{\theta}_3^2$$

$$n_{31} = -m_3r_1c\sin(\theta_1 + \theta_3)\dot{\theta}_1^2 - m_3r_2c\sin(\theta_2 + \theta_3)\dot{\theta}_2^2$$

$$a = \frac{l_1}{2}, b = \frac{l_2}{2}, c = \frac{l_3}{2}, r_1 = l_1, r_2 = l_2, r_3 = l_3$$

By re-parametrization of the above dynamic equation,

$$\begin{bmatrix} m_1a^2 + m_2r_1^2 + m_3r_1^2 & r_1(m_2b + m_3r_2)\cos(\theta_2 - \theta_1) & m_3r_1c\cos(\theta_1 + \theta_3) \\ r_1(m_2b + m_3r_2)\cos(\theta_2 - \theta_1) & m_2b^2 + m_3r_2^2 & m_3r_2c\cos(\theta_2 + \theta_3) \\ m_3r_1c\cos(\theta_1 + \theta_3) & m_3r_2c\cos(\theta_2 + \theta_3) & m_3c^2 \end{bmatrix} \begin{bmatrix} u_1 \\ u_2 \\ u_3 \end{bmatrix} +$$

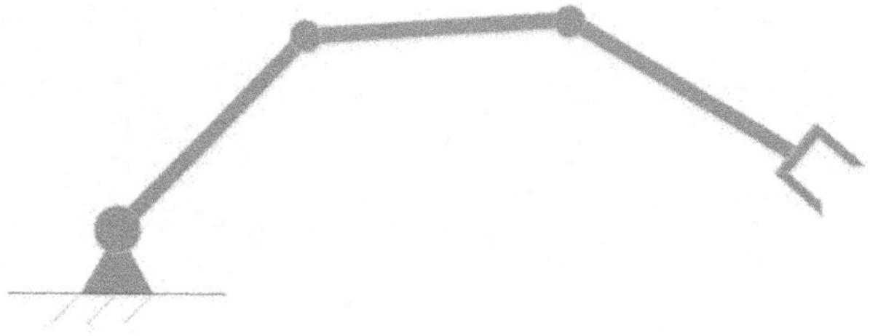

Figure 10. Three-link manipulator.

$$\begin{bmatrix} -r_1(m_2b+m_3r_2)\sin(\theta_2-\theta_1)\dot{\theta}_2^{2} - m_3r_1c\sin(\theta_1+\theta_3)\dot{\theta}_3^{2} \\ r_1(m_2b+m_3r_2)\sin(\theta_2-\theta_1)\dot{\theta}_1^{2} - m_3r_2c\sin(\theta_2+\theta_3)\dot{\theta}_3^{2} \\ -m_3r_1c\sin(\theta_1+\theta_3)\dot{\theta}_1^{2} - m_3r_2c\sin(\theta_2+\theta_3)\dot{\theta}_2^{2} \end{bmatrix} = W \cdot \begin{bmatrix} \Theta_1 \\ \Theta_2 \\ \Theta_3 \\ \Theta_4 \\ \Theta_5 \\ \Theta_6 \end{bmatrix} \tag{19}$$

By choosing $\Theta_1 = m_1a^2 + m_2r_1^2 + m_3r_1^2, \Theta_2 = r_1(m_2b+m_3r_2), \Theta_3 = m_3r_1c, \Theta_4 = m_2b^2 + m_3r_2^2, \Theta_5 = m_3r_2c, \Theta_6 = m_3c^2$

$$\Rightarrow W = \left[\begin{matrix} u_1 & u_2\cos(\theta_2-\theta_1)-\sin(\theta_2-\theta_1)\dot{\theta}_2^{2} & u_3\cos(\theta_1+\theta_3)-\sin(\theta_1+\theta_3)\dot{\theta}_3^{2} \\ 0 & u_1\cos(\theta_2-\theta_1)+\sin(\theta_2-\theta_1)\dot{\theta}_1^{2} & 0 \\ 0 & 0 & u_1\cos(\theta_1+\theta_3)-\sin(\theta_1+\theta_3)\dot{\theta}_1^{2} \end{matrix}\right.$$

$$\left.\begin{matrix} 0 & 0 & 0 \\ u_2 & u_3\cos(\theta_2+\theta_3)-\sin(\theta_2+\theta_3)\dot{\theta}_3^{2} & 0 \\ 0 & u_2\cos(\theta_2+\theta_3)-\sin(\theta_2+\theta_3)\dot{\theta}_2^{2} & u_3 \end{matrix}\right]$$

$$\tau = W \begin{bmatrix} \Theta_1 \\ \Theta_2 \\ \Theta_3 \\ \Theta_4 \\ \Theta_5 \\ \Theta_6 \end{bmatrix} - F_v \cdot erv - F_p \cdot erp \tag{20}$$

$$M\ddot{\theta}+N=\tau=W\begin{bmatrix}\Theta_1\\ \Theta_2\\ \Theta_3\\ \Theta_4\\ \Theta_5\\ \Theta_6\end{bmatrix}-F_v\cdot erv-F_p\cdot erp \tag{21}$$

$$\Rightarrow\ddot{\theta}=(W\begin{bmatrix}\Theta_1\\ \Theta_2\\ \Theta_3\\ \Theta_4\\ \Theta_5\\ \Theta_6\end{bmatrix}-F_v\cdot erv-F_p\cdot erp-N)/M \tag{22}$$

After deriving the acceleration of joint, take the time integral to obtain the velocity of joint and take another integral to obtain the position of joint.

$$\begin{bmatrix}\dot{\theta}_1\\ \dot{\theta}_2\\ \dot{\theta}_3\end{bmatrix}=\int\begin{bmatrix}\ddot{\theta}_1\\ \ddot{\theta}_2\\ \ddot{\theta}_3\end{bmatrix}dt,\quad \begin{bmatrix}\theta_1\\ \theta_2\\ \theta_3\end{bmatrix}=\int\begin{bmatrix}\dot{\theta}_1\\ \dot{\theta}_2\\ \dot{\theta}_3\end{bmatrix}dt \tag{23}$$

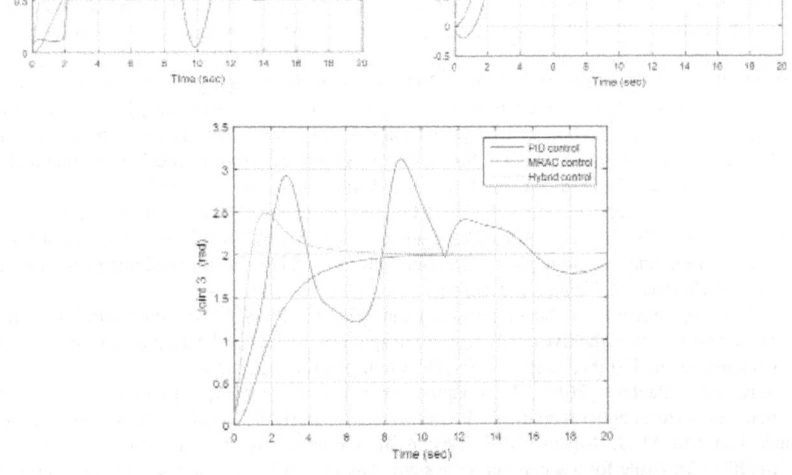

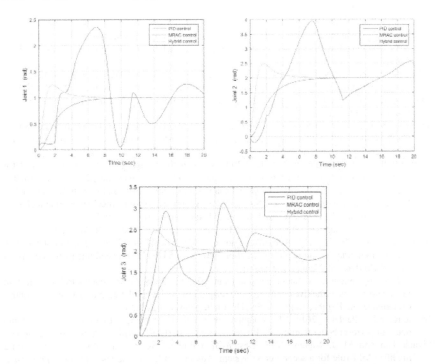

Figure 11. Joints 1, 2 and 3 output under PID, MRAC and hybrid control.

CONCLUSION

This paper proposes a hybrid controller by combining a PID controller and a MRAC controller, and also compares the convergence performance of PID, MRAC, and PID+MRAC hybrid controllers for one-DOF, two-DOF and three-DOF manipulators. For the one-DOF case, the results show that the convergence speed and its performance for MRAC controller and PID+MRAC controller is better than that of the PID controller. Whereas for the MRAC controller and PID+MRAC controller, the convergence performance for the hybrid control is better than that of the MRAC control. For the MRAC, the joint output gradually goes towards to the desired position while for the PID+MRAC, it overshoots the desired position and then gradually comes back to the desired position. For more than one-DOF case, the results show that the convergence speed for the hybrid controller is faster than that of the MRAC controller. The hybrid controller and MRAC controller are both better than that of the PID controller. This study will provide a guideline for future research in the direction of new controller designs for manipulators in terms of convergence speed and other performances.

ACKNOWLEDGMENT

The authors would like to thank the Natural Sciences and Engineering Research Council of Canada (NSERC) for the financial support. The authors gratefully acknowledge the financial support from Canada Research Chairs program.

REFERENCES

Cao, C. and N. Hovakimyan. 2008. Design and analysis of a Novel L1 adaptive control architecture with guaranteed transient performance. IEEE Transactions on Automatic Control, 53(2): 586–591.

Dubowsky, S. and D. Desforges. 1979. The application of model-referenced adaptive control to robotic manipulators. Journal of Dynamic Systems Measurement and Control, 101: 193–200.

Horowitz, R. 1983. Model reference adaptive control of mechanical manipulators, PhD Thesis, University of California.

Horowitz, R. and M. Tomizuka. 1986. An adaptive control scheme for mechanical manipulators—Compensation of nonlinearity and decoupling control. Journal of Dynamic Systems, Measurement and Control, 108(2): 1–9.

Idan, M., M. D. Johnson and A. J. Calise. 2002. A hierarchical approach to adaptive control for improved flight safety. AIAA Journal of Guidance, Control and Dynamics, 25(6): 1012–1020.

John J. Craig. 2005. Introduction to robotics: mechanics and control. 3rd ed., Pearson/Prentice Hall.

John J. Craig, Ping Hsu and S. Shankar Sastry. 1986. Adaptive control of mechanical manipulators. IEEE International Conference on Robotics and Automation, pp. 190–195.

Landau, Y. D. 1979. Adaptive control: the model reference approach. Marcel Dekker, New York.

Li, X. and C. C. Cheah. 2012. Adaptive regional feedback control of robotic manipulator with uncertain kinematics and depth information. pp. 5472–5477. *In*: Proceedings of the American Control Conference, Montr´eal, Canada.

Nhan T. Nguyen, Kalmanje Krishnakumar and Jovan Boskovic. 2008. An optimal control modification to model-reference adaptive control for fast adaptation. AIAA Guidance, Navigation and Control Conference and Exhibit, August, 2008, Honolulu, Hawaii, pp. 1–19.

Ortega, R. and E. Panteley. 2014. L1—Adaptive control always converges to a linear PI control and does not perform better than the PI. Proceedings of 19th IFAC World Congress, pp. 6926–6928.

Priyank Jain and M. J. Nigam. 2013. Design of a model reference adaptive controller using modified MIT rule for a second order system. Advance in Electronic and Electric Engineering, 3(4): 477–484.

Rossomando, F. G., C. Soria, D. Patiño and R. Carelli. 2011. Model reference adaptive control for mobile robots in trajectory tracking using radial basis function neural networks. Latin American Applied Research, 41: 177–182.

Sadegh, N. and R. Horowitz. 1987. Stability analysis of an adaptive controller for robotic manipulators. Proceedings of 1987 IEEE International Conference on Robotics and Automation, pp. 1223–1229.

Sadegh, N. and R. Horowitz. 1990. Stability and robustness analysis of a class of adaptive controllers for robotic manipulators. International Journal of Robotics Research, 9(3): 74–92.

Sharifi, M., S. Behzadipour and G. R. Vossoughi. 2014. Model reference adaptive impedance control in Cartesian coordinates for physical human–robot interaction. Advanced Robotics, 28: 19, 1277–1290.

9

Adaptive Control of Modular Ankle Exoskeletons in Neurologically Disabled Populations#

Anindo Roy,[1,*] *Larry W. Forrester*[2] and *Richard F. Macko*[3]

ABSTRACT

Advances in our understanding of systems control engineering are now being integrated into modular robotics exoskeletons in order to customize cooperative motor learning and functional neuroplasticity for individuals with sensorimotor deficits after stroke. Major advances have been made with upper extremity robotics, which have been tested for efficacy in multi-site trials across the subacute and chronic phases of stroke. In contrast, use of lower extremity robotics to promote locomotor re-learning has been more recent and presents unique challenges by virtue of the complex multi-segmental mechanics of gait.

In this chapter, we present an approach to using adaptive control algorithms to precisely customize an impedance-controlled ankle exoskeleton ("anklebot")

[1] Department of Neurology, University of Maryland School of Medicine; VA Maryland Exercise and Robotics Center of Excellence, VA Maryland Health Care System, Baltimore VA Medical Center; University of Maryland Rehabilitation and Orthopaedics Institute, Baltimore, MD.

[2] VA Maryland Exercise and Robotics Center of Excellence; Geriatric Research Education and Clinical Center, VA Maryland Health Care System, Baltimore VA Medical Center, Baltimore, MD.

[3] Geriatric Research Education and Clinical Center, VA Maryland Health Care System, Baltimore VA Medical Center; Department of Neurology, University of Maryland School of Medicine; University of Maryland Rehabilitation and Orthopaedics Institute, Baltimore, MD.

* Corresponding author: aroy@som.umaryland.edu

This material is based upon work supported by the Department of Veterans Affairs, Veterans Health Administration, Office of Research and Development, VA Maryland Health Care System, Baltimore VA Medical Center.

for deficit severity adjusted locomotor training after stroke. Our objective is to implement adaptive control anklebot as a therapeutic tool to individualize gait training using cooperative motor learning precisely timed to address the inherent heterogeneity of hemiparetic (half-paralyzed) gait and diversity of stroke deficits. Individuals with stroke have reduced strength and motor control that often reduces ground foot clearance and/or push-off propulsion during walking, impairing gait function and balance. Underlying our approach is a novel gait event-triggered, sub-task control algorithm that enables precise timing of robotic assistance to key functional deficits of hemiparetic gait, as well as sagittal-plane biomechanical models capable of predicting necessary levels of robotic support contingent on the severity of ankle motor deficits, and their dynamic change across training. We outline gait phase-specific models and show that these models can predict the parameters of an impedance controller to generate a desired level of swing and anterior-posterior push-off during assisted walking with anklebot, personalized to the directionality and magnitude of gait deficit disorders. Critical to the conceptual model of activity dependent neuroplasticity underlying recovery, we employ an adaptive control approach in which training parameters are incrementally progressed towards more normal gait, contingent on subject performance and tolerance. We validate the sub-event detection and sub-task control method and the biomechanical models for swing, loading response, and stance phases of gait, as well as provide proof-of-concept of efficacy of adaptive control robotics in stroke.

Keywords: Rehabilitation robotics, gait biomechanics, motor learning, hemiparetic stroke, event-based control

INTRODUCTION

Stroke is the leading cause of long-term disability in the United States and its prevalence will increase with an aging population (Heart disease and stroke statistics: American Heart Association 2009). The impact on walking is significant, negatively affecting an individual's mobility (Forster and Young 1995) and ability to perform everyday activities, including optimal participation in community life. Gait and balance deficits limit functional mobility and contribute to more than 70% of stroke survivors sustaining a fall within six months (Forster and Young 1995), leading to higher risks for hip and wrist fractures (Dennis et al. 2002, Kanis et al. 2001, Ramnemark et al. 1998). These disabling consequences of stroke not only limit participation in community life but also set the stage for a sedentary lifestyle that reinforces learned nonuse and further declines in mobility and balance functions.

Physical therapy remains the standard of care to retrain patients and restore walking ability, but many individuals still have residual deficits, even after completion of all conventional rehabilitation therapy. Moreover, increasing clinical and experimental data suggest that conventional rehabilitation does not provide adequate task-repetitive practice to optimize motor learning and recovery across the continuum of care. This, and advances in our understanding of neuroplasticity and motor learning post-stroke, has prompted the development and utilization of robotic tools for delivering movement therapy. Robots are now being leveraged to change the natural history of stroke recovery. The prevailing rationale is that robots lend themselves to the process of motor learning by facilitating massed repetitive practice, providing precise measurement capability and the potential to customize training progression and motor learning paradigms.

In terms of clinical translation, the greatest advances have been made with the use of upper extremity upper extremity robotics, which have been tested for efficacy in multi-site trials (Burgar et al. 2011, Lo et al. 2010). In contrast, the implementation of lower extremity robotics to promote locomotor re-learning through massed practice has been more recent and presents unique challenges by virtue of the complex dynamics of gait, including the coordination of both legs and the multi-segmental balance control inherent to upright locomotion.

The US Department of Veterans Affairs Baltimore Medical Center, in collaboration with MIT, has developed an impedance-controlled, highly backdrivable, 2 degree-of-freedom (DOF) actuated ankle robot ("anklebot") to improve walking and balance functions after stroke (Roy et al. 2009), by means of increasing the paretic ankle contribution into task-oriented functional activities (Fig. 1). The rationale to focus on the ankle was due to the critical role it plays in the biomechanics of gait and balance. For example, the ankle musculature provides propulsion during mid-to-late stance, ground clearance during swing, and "shock absorption" during the loading response phase of gait. Following a stroke, some (or all) of these normal aspects of gait are disrupted. For example, "drop foot" is a common impairment caused by a weakness in the dorsiflexor muscles that lift the foot. A major complication of drop foot is the slapping of the foot after heel strike (foot slap). In addition to inadequate dorsiflexion (foot "up"), the paretic ankle may also suffer from excessive inversion (foot "inward"). This begins in the swing phase and results in toe contact (as opposed to heel contact) and lateral instability in stance. The anklebot possesses the potential to control both since it can be actuated in 2-DOFs (Roy et al. 2009), and is able to independently

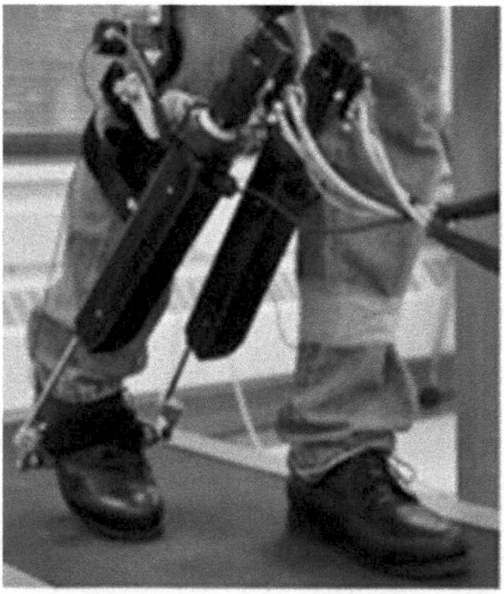

Figure 1. Stroke participant undergoing anklebot-assisted gait training on a treadmill. The anklebot is providing plantar flexion push-off at the paretic ankle during mid-to-terminal stance, beginning with heel-off and peaking just prior to toe-off.

modulate stance, swing, and specific sub-tasks within the gait cycle to better address the heterogeneity of stroke recovery.

To date, our studies with the anklebot have demonstrated in different phases of stroke (chronic and early sub-acute) that seated visually-guided and evoked isolated ankle training with the anklebot improves paretic ankle motor control (smoother, faster and more accurate ankle targeting) that translates to faster steady-state floor walking speeds (Forrester et al. 2011) and improved gait symmetry (Forrester et al. 2013). Notably, the average gain (20%) in floor-walking speed after 6-week seated anklebot training was comparable or greater than that reported in other task-specific, robotic gait training studies in chronic stroke (e.g., Hornby et al. 2008). In this chapter we expand beyond seated posture to integrate the anklebot directly into task-specific locomotor training to examine whether perhaps we can further enhance gait function. Specifically, we describe our invention and pre-clinical testing of a new gait event-triggered, deficit-adjusted control approach that incorporates offline sagittal-plane biomechanical models (specifically, for swing, loading response, stance phases), for using the anklebot to train upright locomotion on a treadmill. This novel approach seeks to optimize walking recovery based on profiles of biomechanical impediments in hemiparetic gait (e.g., foot drop vs. impaired push-off vs. improper loading response) by precisely timing robotic support to gait phases (e.g., swing) that correspond to key hemiparetic deficits (e.g., foot drop). The biomechanical models developed here are used for initial parameterization of the anklebot and for systematic progression of training parameters based on prior and ongoing performance.

The work described in this chapter is part of a larger ongoing study that investigates the hypothesis that a modular, deficit-adjusted approach to using the impedance-controlled anklebot for task-oriented locomotor training will lead to sustainable gains in mobility function in those with residual hemiparetic gait deficits. The underlying rationale is the need to promote volitional control of the paretic leg through versatile interaction during robot-assisted gait—one that does not impose prescribed movement as a substitute for volitional control, but rather responds to disparate forms and degrees of impairments with the ability to dynamically adjust the level of support concomitant with the patient's capacity and recovery. Our *overall* objective is to determine whether sensorimotor task-oriented practice in walking with precisely-timed and appropriately-scaled robotic support delivered to the paretic ankle at critical instances of the gait cycle, can improve key functional deficits of hemiparetic gait. As a first step, here we present a novel control approach for safe, effective first use of anklebot-assisted treadmill gait training in those with hemiparetic gait deficits: (1) a control algorithm that enables precise sub event-triggered robotic support during the gait cycle; (2) sagittal-plane biomechanical models for swing, loading response, and stance phases of gait to scale robotic support to deficit profiles; and (3) proof-of-concept by validating the control algorithm and biomechanical models during anklebot-assisted treadmill walking; first, with a healthy participant and then, with a chronic stroke survivor.

OVERVIEW OF CURRENT GAIT TRAINING APPROACHES

In conventional gait physiotherapy, the therapist pushes or slides the patient's swing leg forward, either on the ground or on a treadmill. There is clear evidence from recent studies (e.g., Macko et al. 2005) that higher intensities of walking practice (resulting in more trained repetitions) result in better outcomes for patients after stroke; however, one disadvantage of treadmill-based therapy might be the necessary effort required by therapists to set the paretic limbs and to control weight shift, thereby possibly limiting the intensity of therapy especially in more severely disabled patients. To counter this problem, in the past decade and a half, several groups have designed and deployed automated electromechanical devices that provide non-ambulatory patients intensive practice of complex gait cycles, thereby reducing therapist effort compared to treadmill training with partial body weight support alone. Most of these devices fall in one of two categories based on their principle of operation: (1) The leg is propelled by the robotic exoskeleton orthosis acting on the patient's leg (e.g., Jezernik et al. 2003), or (2) Actuated foot plates are attached to the patients' foot simulating phases of gait (e.g., Hesse and Uhlenbrock 2000). More recent devices include "Haptic Walker" (Schmidt 2004), lower extremity powered exoskeleton (LOPES) (Ekkelenkamp et al. 2007), and MIT-Skywalker (Bosecker and Krebs 2009).

SEATED VISUALLY-GUIDED ANKLEBOT ISOLATED JOINT TRAINING

To date, we have studied the effects of anklebot mass on hemiparetic gait kinematics (Khanna et al. 2010), demonstrated its safety and feasibility for use in chronic stroke, and its use as a clinical assessment tool in the measurement of passive ankle stiffness (Roy et al. 2011)—a potential marker related to motor deficit severity. Clinical investigations using seated computer-video interfaced anklebot training in chronic hemiparetic stroke provided evidence of paretic ankle motor learning, both in the short-term (48-hours) (Roy et al. 2011) and at 6 weeks (Forrester et al. 2011). Specifically, chronic stroke survivors who underwent 6 weeks of seated anklebot training with their paretic ankle engaged in a visuomotor task customized to individual ankle deficits, showed reduced ankle impairments (increased strength, voluntary range of motion), improved paretic ankle motor control (smoother and faster ankle movements), and higher unassisted overground gait speed (Forrester et al. 2011). Notably, the mean increase (20%) in independent floor-walking speed after 6 weeks of seated anklebot training is comparable to that reported for other task-specific, locomotor training approaches in the chronic stroke population (e.g., Hornby et al. 2008). These findings resulting from seated anklebot intervention support a rationale for a modular, impairment focused approach towards upright anklebot training.

THE ANKLEBOT AS A GAIT TRAINING DEVICE

Rationale

The ability to walk and maintain balance is essential for successful and safe execution of activities of daily life. Following a stroke, the leading cause of long-term disability in the United States (Heart disease and stroke statistics: American Heart Association 2009), the impact on walking is often significant, negatively affecting an individual's mobility (Forster and Young 1995) at home and in the community. Ambulation speed is widely used as a measure of functional recovery in post-stroke rehabilitation (Andriacchi et al. 1977), and may be thought of as a "global" top-level outcome that is influenced by a multitude of spatial-temporal kinematic (e.g., cadence, cycle time, stance and swing duration, stride length) and kinetic (ground reaction forces [GRF], joint torques) variables. One such key contributor to walking speed is the anterior-posterior (A-P) push-off propulsion during the mid-to-terminal stance phase of gait (Bowden et al. 2006). Subjects with hemiparetic (hemiparetic) gait often generate less propulsion in the paretic leg compared to the non-paretic leg (Turns et al. 2007) that compromises economical gait and leads to slower walking speeds (Perry et al. 1995). Hence, asymmetric paretic leg A-P propulsion has been a subject of numerous investigations (e.g., Bowden et al. 2006, Turns et al. 2007, Neptune et al. 2001) and has been correlated to gait function (Neptune et al. 2001). Another factor that impedes safe walking post-stroke is the inability to clear the toes during the swing phase of gait (foot drop). This further result in two complications: dragging the toes during walking ("toe drag") and loading response with the foot inverted that result in lateral instability during stance, a major cause of ankle injuries. The promise of using the driving stimulus of the treadmill to train locomotor patterning motivated major innovations in the area of powered gait orthoses and other robotic gait trainers including the modular anklebot. Consistent with its design, there is also ongoing work aimed at integrating the anklebot module into task-specific gait training, both on a treadmill and eventually, over ground in more real-world contexts.

Clinical Considerations

One clinical and biomechanical concern is the impact of the asymmetric mass of the robot. We have initiated studies to develop training applications for subjects with stroke. A pilot study investigated the effects of ankle robot mass on gait patterns of chronic stroke survivors to facilitate the design of treadmill-based anklebot training (Khanna et al. 2010). Results indicated that the added inertia and friction of the unpowered anklebot did not significantly alter the paretic and nonparetic step time and stance as percentage of the gait cycle, either in over ground or treadmill conditions. Regardless of loading conditions, the inter-limb symmetry as characterized by relative stance durations was greater on the treadmill than over ground. Unpowered robot loading significantly reduced the nonparetic knee peak flexion on the treadmill and paretic peak dorsiflexion over ground. Overall, the results showed that the added mass of the anklebot does not significantly alter the hemiparetic gait pattern.

In addition to the above, any robot must be capable of delivering tailored therapies in order to address the wide variety of gait deficits and compensatory gait strategies encountered in the clinic. From a control standpoint, its closed loop must also be Hurwitz stable to ensure stability of the man-machine interface. Both these issues can be addressed, at least in part, by taking advantage of the anklebot's 2-DOF actuation (Roy et al. 2009), its impedance control, and the sub-event triggered timing algorithm that we have developed. The 2-DOF actuation enables customization of robotic support to different profiles of hemiparetic gait deficits in the sagittal and/or frontal planes, while impedance control makes the robot minimally intrusive by providing assistance "as-needed", i.e., allowing and encouraging volitional movements much like a therapist, rather than constraining the natural leg dynamics. Stability is addressed along with the question of when and how much robotic support to provide during assisted walking across a wide range of hemiparetic gait deficit profiles.

The Control Approach: A Deficit-Adjusted Solution

We prioritize precise timing of robotic support at critical gait sub-events of interest in order to both prevent destabilization and afford a deficit-adjusted approach. Accordingly, we developed a sub event-triggered method (Fig. 2) in which the anklebot delivers torques at the paretic ankle during one or more of two key time epochs, each with unique functional needs: (1) Concentric plantar flexion (foot "down") torque to enable push-off propulsion during terminal stance, starting with heel-off and peaking just prior to toe-off; (2) Concentric dorsiflexion torque to facilitate swing clearance, starting at toe-off and continuing until mid-swing; and (3) Velocity-dependent viscous (resistive) torque to attenuate or absorb the impact force during loading response. anklebot torque, both assistive and resistive, are generated by kinematic control of the anklebot, i.e., by having the total net torque (human plus robot) track appropriate task-specific programmable positional reference trajectory. The impedance controller generates ankle torques proportional to the magnitude of the positional and velocity error between the commanded and actual ankle trajectory via the torsional stiffness (K) and damping (b) settings (Fig. 3). Notably, we are not explicitly controlling the torque output (i.e., force control) or the position (i.e., position control) to actuate the anklebot. The precise timing of robotic actuation to gait events is achieved via footswitch insoles that detect the occurrence of the specified events (Fig. 3). This enables "on-the fly" adjustment of robotic outputs across strides, thereby accounting for step-to-step variability by ensuring stability of the human-robot interface during assisted locomotor training that is critical for safety.

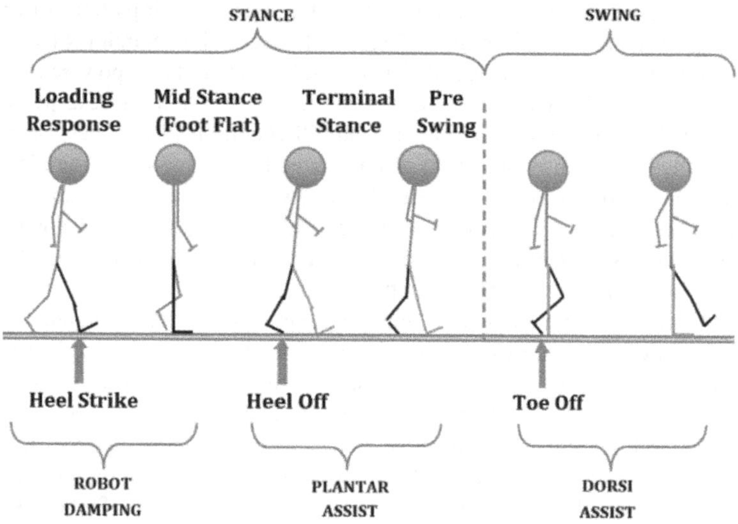

Figure 2. Conceptual diagram showing three key sub-events during a gait cycle. Depending on the type of gait deficit(s), the anklebot is programmed to generate assistive (or resistive) torques at the paretic ankle during one or more gait sub-events.

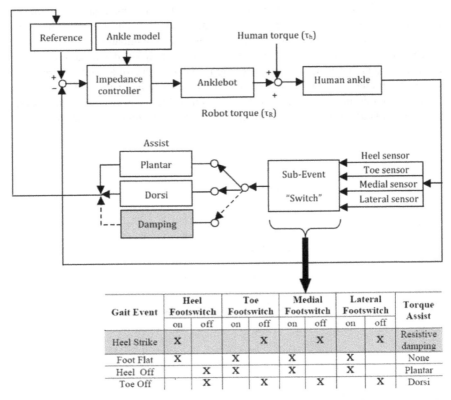

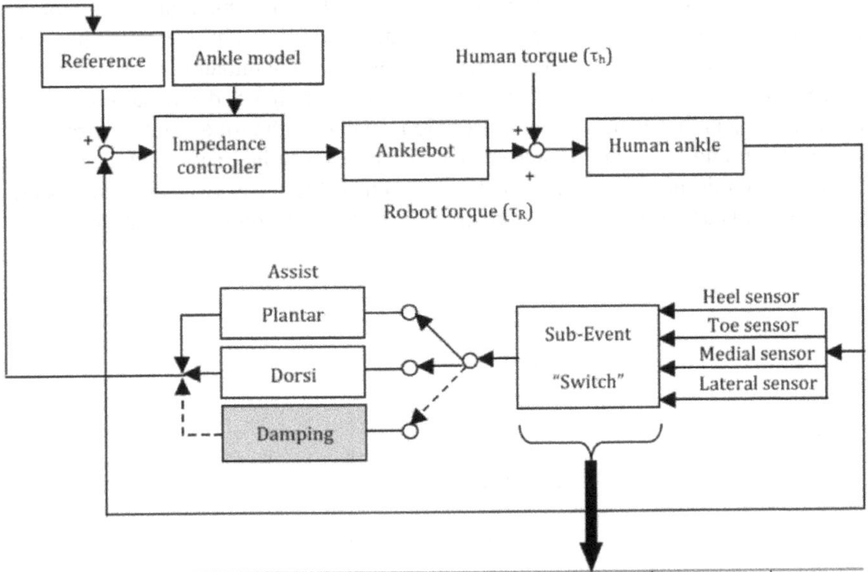

Gait Event	Heel Footswitch		Toe Footswitch		Medial Footswitch		Lateral Footswitch		Torque Assist
	on	off	on	off	on	off	on	off	
Heel Strike	X			X		X		X	Resistive damping
Foot Flat	X		X		X		X		None
Heel Off		X	X		X		X		Plantar
Toe Off		X		X		X		X	Dorsi

Figure 3. Conceptual block diagram showing the sub event-triggered approach to detect gait sub-events of interest and precisely time robot actuation. In the figure, the sub-event of interest (shown for illustration) is loading response, specifically heel strike event (dashed line). Biomechanical models specific to each phase of interest are used to scale robotic torques to deficit severity via the controller settings, *K* and *b*.

Precise Timing of Robotic Support

An event-triggered solution: In order to precisely time robotic support, we employed footswitches embedded inside the individual's shoes to detect the occurrence of key gait sub-events. Although not novel in itself (e.g., Li et al. 2011), this approach is effective, at least in principle, because it allows the human to "inform" (via 5-pin LEMO footswitch insoles worn bilaterally) the anklebot of the occurrence of gait sub-events, thereby enabling accurate and automatic adjustment of timing of robotic outputs across strides. Each footswitch (B&L Engineering, Santa Ana, CA) contains 4 individual switches: at the heel, forefoot, medial and lateral zones at the level of metatarsals. An event is detected when the cumulative footswitch voltage equals a pre-defined value unique to each sub-event (Fig. 4). For example, heel-strike is detected when all switches except the heel switch are open. Following this, the loading continues and is characterized by the closing of medial-lateral switches followed by the forefoot switch at which point the foot is in mid-stance. Late stance is detected by the opening of the heel switch followed by the medial-lateral switches marking the beginning of terminal stance. The forefoot switch then opens marking the beginning of swing phase. For optimal sensitivity, tolerance bands were created around each sub-event's threshold voltage, then calibrated to ensure consistent, accurate capture of each sub-event. It is possible, especially during the early phases of therapy or for severely impaired patients, that the footswitch on the affected (robot donning) footswitch may not detect sub-events and actuate the anklebot with sufficient precision owing to abnormal foot patterns with respect to ground contact (e.g., severe foot drop may result in dragging of the toes that could lead to missing toe-off sub-events). To counter this, we have developed capacity to achieve this using the footswitch on the unaffected (non-robot donning) leg.

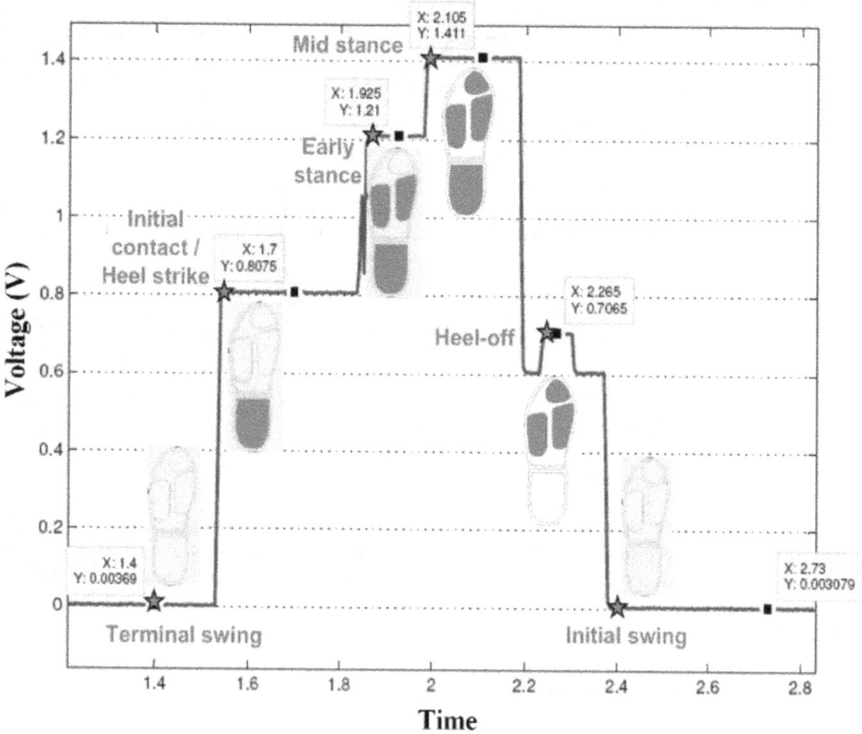

Figure 4. Footswitch voltages during a typical gait cycle. Shaded regions within the footprints represent instants when one or more switches close due to the occurrence of a gait sub-event (filled asterisks).

Performance Metrics: Timing of Robotic Assist

The following metrics were devised to quantify robustness and accuracy of the timing of robotic assist:

1) Uniqueness (U) is the difference between the total number gait cycles and the total number of gait cycles with one and only one robot torque pulses (simply defined, a pulse is a torque-time profile bounded by two consecutive zero values, the first zero characterizing initiation of robotic actuation and the second zero characterizing completion of actuation), i.e.,

$$U = |N_{\text{cycle}} - N_{\text{single,torque}}|, \tag{1}$$

where N_{cycle} is the number of gait cycles and $N_{single,torque}$ is the total number of gait cycles that consist of one and only one torque pulse. Since torque is not being commanded explicitly (i.e., we are not using a force controller), its magnitude is determined by the impedance controller, while its duration is determined by the occurrence of the sub-event/s. The rationale to count the number of gait cycles that consist of only a single torque pulse (as defined above) is to detect and avoid via fine tuning, missed or redundant triggers (see "look-no-further" criterion below) that could cause disruption to the human-robot synchronicity during walking. Because sub-event detection is based on pre-defined voltage thresholds and/or tolerance windows, there is a possibility to erroneously actuate the anklebot, i.e., either not actuate it at all (excessively narrow band) or actuate it multiple times within the same gait cycle (excessively wide band). This measure quantifies any missed or extra sub-events within a gait cycle. For example, if a desired sub-event was not detected in one or more gait cycles, $N_{single,torque} < N_{cycle}$ so that $U > 0$ and similarly, if a sub-event occurred multiple times during a gait cycle, $N_{single,torque} > N_{cycle}$ so that $U < 0$. Ideally, we desire $U = 0$. In other words, if some gait cycles had missed trigger/s and others had multiple trigger/s, $N_{single,torque} \neq 0$, so that $U \neq 0$. Hence, this metric quantifies the robustness of correct timing of robotic actuation (completely robust if $U = 0$, otherwise not).

Since this approach targets specific deficits of hemiparetic gait (e.g., foot drop, weak propulsion), we focus only on detecting sub-events that correspond to those deficits (e.g., toe-off, heel-off). In order to circumvent potential problems from redundant or "false" triggers for a sub-event within the same gait cycle, a "look-no-further" criterion is embedded in the detection algorithm—that is, the first time a sub-event of interest is detected, the software program is commanded to stop the detection process for a certain period of time ($\Delta T_{\text{sub-event}}$ ms). Note that in order to ensure that this criterion works across gait cycles only when the first event is detected accurately. Hence, the criterion is programmed to initiate itself in real-time only when the first gait sub-event of interest is accurately detected. $\Delta T_{\text{sub-event}}$ is determined *a priori* by averaging the sub-event duration across multiple (typically, 60) steady-state gait cycles with the anklebot operating in a "record-only" mode. Further, fine tuning of the tolerance windows and the "look-no-further" delay value are available to accommodate changes in gait performance step-to-step or over the course of a training program. Finally, the value of $\Delta T_{\text{sub-event}}$ depends on each patient's spatial-temporal gait parameters (e.g., cycle time, swing/stance duration) captured with the robot donned in an unassisted trial, as well as the commanded angle that is set based on deficit severity and ongoing recovery.

2) Initiation delay (ΔT) is the latency between the occurrence of a sub-event and subsequent initiation of the robot-generated torque pulse. Mathematically,

$$\Delta T_{avg} = \sum_{i=1}^{N_{cycle}} (t_{i,torque} - t_{i,event}) / N_{cycle} \qquad (2)$$

where ΔT_{avg} (ms) is the mean delay averaged across the total number of gait cycles, and $t_{i,\,event}$ (ms) and $t_{i,\,torque}$ (ms) are the instants of occurrence of the event and initiation of torque in the ith gait cycle, respectively. Here, initiation of robot torque is defined to be the time instant when the torque value first attains 2% of the measured (offline) peak torque in a given cycle. In order to prevent destabilization of the man-machine interface, ΔT_{avg} should be "small" in some sense (e.g., $\Delta T_{avg} << T_{sub\text{-}event,avg}$); if not, delay compensation using one or more traditional control methods may become necessary.

Summary

A novel gait event-triggered control approach (US patent pending, #61/906,453) to time robotic support to gait events that correspond to key functional hemiparetic gait deficits (Fig. 2) is presented. The adaptive, deficit-adjusted control approach is capable of being integrated into any modular ankle exoskeleton to deliver torques at a single joint or a plurality of joints during one or more key time epochs, each with unique functional needs: (1) Plantar-flexion torque to enhance push-off propulsion during mid-to-terminal stance, starting with heel-off and peaking just prior to toe-off; (2) Dorsiflexion torque to facilitate swing clearance during initial swing, starting at toe-off and continuing until mid-swing; and (3) Restorative viscous torque to lessen impact force during loading response.

PHASE-SPECIFIC BIOMECHANICAL MODELS

While we may have a reliable algorithm for accurate sub-event detection that in turn precisely dictates the anklebot actuation, one would still need to determine the magnitude of robotic torques. From a clinical standpoint, this is necessary to customize robotic support to deficit severity; in control terms, this is equivalent to tuning the impedance controller gains (K, b) in real-time based on each subject's actual performance across training. As a first attempt, we developed simple sagittal-plane models of ankle dynamics, specific to each phase of interest (and thus key deficits). The purpose of these models is to simply approximate normative ankle dynamics for initial parameterization of the controller and hence the robot's output during assistive walking. Moreover, the model predictions will be used to inform robot torques over the course of training in order to modify key spatial-temporal parameters of hemiparetic gait towards more normal values as tolerated by the subject and concomitant with any gains in volitional ankle motor control and/or gait performance.

Robot-Incorporated Model for Initial Contact-Loading Response of Gait

A two-segment[1] inverted pendulum model (e.g., Iqbal and Roy 2004) with the rigid link acting as pivot over a moving base of support is used to model ankle dynamics during initial stance, in particular, from initial contact through loading response just prior to foot-flat (mid-stance) (Fig. 5). The problem to be solved is: given a human with body height H (m) and mass M (kg), and assuming zero volitional torque[2] ($\tau_h = 0$ N-m), determine the minimum damping b_{min} (N-m-sec/rad) needed to constrain

[1] For simplicity, the two segments considered to be a rigid inverted pendulum are the head-arm-trunk plus leg, and the base of support that includes foot plus ankle joint (Iqbal and Roy 2004), as two separate rigid bodies.

[2] Clinically, a subject with lack of trace dorsiflexion, i.e., no palpable or observable muscle contractile activity in the gravity eliminated position.

the peak ankle angular speed v_m (rad/s) to be less than some desired (e.g., normative) value V_m (rad/s) to lessen impact forces during loading response. Mathematically,

$$b_{min} = \min(b) \text{ where } b = \{b: v_m(b) \le V_m, t_{HS} \le t \le t_{FF}\}, V_m > 0 \quad (3)$$

where b (N-m-s/rad) is the robot-induced damping (i.e., the derivative gain of the controller), and t_{HS} (s) and t_{FF} (s) represent instants of heel-strike and foot flat, respectively. In other words, we are considering the period of initial contact through loading response (mid-stance). Ignoring the foot's mass and length relative to body mass (M) and height (H), the equation of motion is:

$$I\, d^2\theta/dt^2 + b(d\theta/dt - v_{HS})\cos\theta - Mgk\sin\varphi = 0 \quad (4)$$

where I is the moment of inertia (kg-m^2), v_{HS} is the initial speed at heel-strike (rad/s), k is the distance between body center of mass (COM) and the ankle (m), g is the acceleration due to gravity (9.81 m-s^{-2}), and φ (rad) and θ (rad) represent the inverted pendulum angular position measured from vertical and the ankle angle with respect to the ground, respectively. Linearizing Eq. (4) about $\theta = 0$ and assuming that $|\varphi| \ll 1$ about the vertical, we obtain:

$$I\, d^2\theta/dt^2 + b(d\theta/dt - v_{HS}) - Mgk\varphi = 0. \quad (5)$$

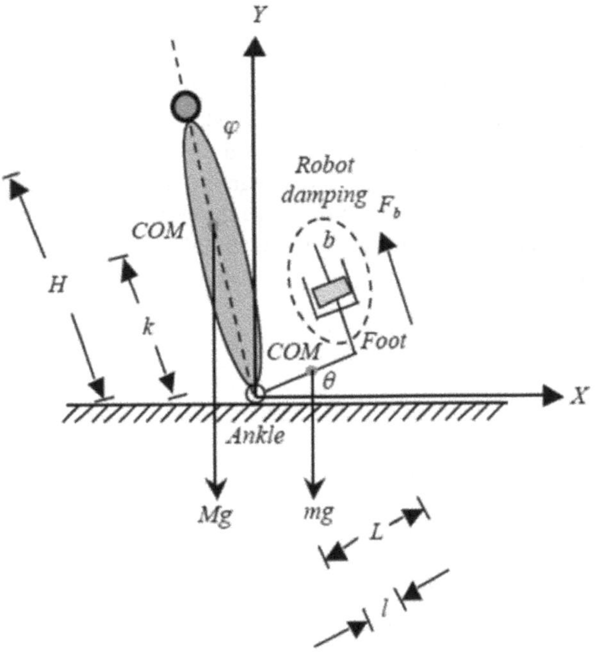

Figure 5. Two-segment inverted pendulum showing heel-strike with robot attached anterior to shank (displacement of inverted pendulum from vertical is exaggerated for illustration purposes only). Robot damping lessens impact forces via resistive viscous torque at the ankle, whose magnitude is proportional to the ankle angular speed and damping. H: body height, M: body mass, L: length of foot, l: distance between ankle and foot COM, k: distance between ankle and body COM, b: robot damping, F_b: robot-generated viscous force, φ: angle of inverted pendulum with respect to vertical, θ: ankle angle, g: gravitational acceleration (9.81 ms^{-2}).

Note that $|\varphi|<<1$ is a realistic assumption based on healthy bipedal gait biomechanics. Equation (5) is solved for $\dot{\theta}(t, b)$ and a desired upper-bound V_m placed on the peak speed $v_m(b)$, i.e., $||\theta(t, b)||_\infty$, i.e.,

$$0 \leq v_m(b) = v_{HS} + \frac{\alpha g M H \varphi}{b} \leq V_m, \quad t_{HS} < t \leq t_{FF} \tag{6}$$

where $\alpha = k/H$. The minimum damping needed is therefore,

$$b_{\min} \geq \frac{\alpha g M H \varphi}{V_m - v_{HS}}, \quad t \in [t_{HS}, t_{FF}] \quad \text{(N-m-s/rad)}. \tag{7}$$

As one would expect, the minimum damping is inversely proportional to the desired upper-bound on the peak angular speed, i.e., the higher the damping, the less is the peak angular speed (and hence the impact force), and vice versa.

Robot-Incorporated Model for the Swing Phase of Gait

Our objective during this phase is to provide supplemental robotic support to enable sufficient ground clearance. In other words, assuming zero voluntary ankle torque (τ_h = 0 N-m) we wish to determine the minimum stiffness K_{min} needed for the peak ankle angle during swing to attain a desired (e.g., a more normal) value. Mathematically,

$$K_{min} = \min(\mathrm{K}) \text{ where } \mathrm{K} = \{K: \theta_{max}(K) = \gamma\theta_d, t_{TO} \leq t \leq t_{HS}\}, \tag{8}$$

where K (N-m/rad) is the robot stiffness (i.e., the proportional gain of the controller), θ_d (rad) ≥ 0 and θ_{max} (rad) are the desired and actual peak angles during swing, respectively, t_{TO} (s) is instant of toe-off so that the time period of interest $t_{TO} \leq t < t_{HS}$ is the swing phase, and $0 < \gamma = \theta_{max}/\theta_d \leq 1$. Since θ_d is nonnegative and γ strictly positive, we desire a nonnegative peak swing angle. We will derive the system dynamics in the s-domain using the closed-loop transfer function (Fig. 6a) stimulated with a "ramp up - hold ramp down" reference trajectory (Fig. 6b,c):

$$\theta^*(t) = \frac{\theta_d}{T} t\{u(t) + u(t - t_h - 2T)\}$$
$$+ \theta_d(1 - \frac{1}{T})u(t - t_h) + \theta_d(1 - \frac{t}{T})u(t - t_h - T)$$

where T (sec) is the ramp duration to the desired peak angle θ_d (rad). Ignoring the foot length for simplicity, and assuming $t_h << T << 1$ and a [0/1] Pade′ approximant for e^{-Ts}, the ankle trajectory in the s-domain is obtained, i.e.,

$$\theta(s) = \frac{\theta_d(bs + K)}{s(Ts + 1)(I_h s^2 + (b + b_h)s + (K + K_h))}, \tag{9}$$

where I_h is the ankle moment of inertia (kg-m^2), and K_h (N-m/rad) and b_h (N-m-s/rad) are the intrinsic stiffness and damping of the ankle in dorsiflexion, respectively. It can be verified from Eq. (9) that the closed-loop is Hurwitz stable since the poles satisfy Re[s] ≤ 0 for all $K \geq 0$. Taking the inverse Laplace transform of Eq. (9) yields the swing trajectory $\theta(t)$:

$$\theta(t) \approx \frac{\theta_d K}{(K+K_h)}(1-e^{-\omega_n \zeta t}\cos(\omega_n\sqrt{1-\zeta^2}\cdot t)\ \text{(rad)}, \qquad (10)$$

where ω_n (rad/s) and $0<\zeta\leq 1$ are the system natural frequency and damping coefficient that depend solely on the intrinsic ankle properties (I_h, K_h, and b_h), i.e., $\omega_n=\sqrt{(K+K_h)/I}$ rad/s and $\zeta=(b+b_h)/2\sqrt{I(K+K_h)}$. To evaluate the peak swing angle, we compute $||\theta(t, K)||_\infty$ from Eq. (10) and set it equal to $\gamma\theta_d$, yielding:

$$K_{\min}=\gamma K_h/(1-\gamma),\ t\in[t_{TO}, t_{HS}) \qquad \text{(N-m/rad)}. \qquad (11)$$

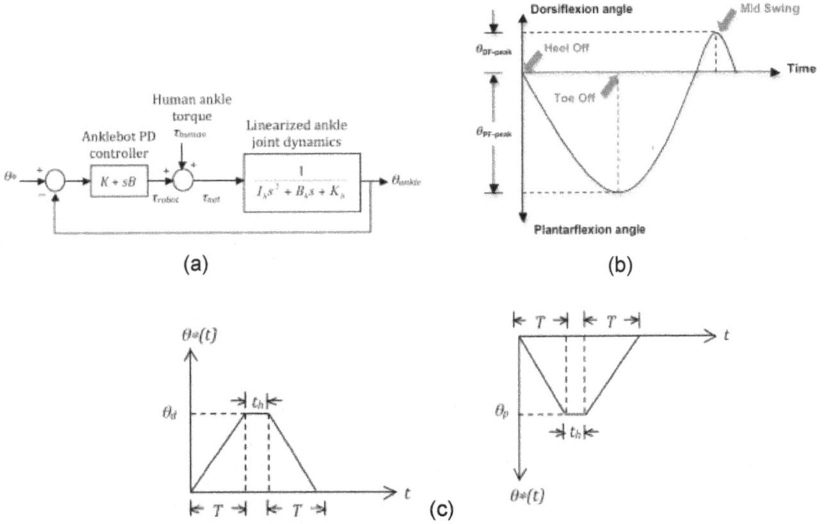

Figure 6. (a) Block diagram of anklebot control system (Dennis et al. 2002) showing major components: the proportional-derivative controller parameters (K, B), linearized second-order ankle dynamics, and the kinematic reference. The sensor "event switch" is not shown and unity feedback is assumed; (b) Positional reference trajectory θ^* that can be shaped using a few spatial-temporal parameters. Although the figure shows a continuous sinusoid-like waveform, in practice they are used as two distinct reference trajectories: one for plantar-flexion assist and the other for dorsiflexion assist; (c) The "ramp up-hold-ramp down" positional trajectory approximating the dorsiflexion and plantar-flexion waveforms (b). It is synthesized using three parameters (T, t_h, θ_d). For "small" ramp period T, the waveform reduces to a unit step signal ($T \leq t \leq T+t_h$).

As one would expect, the K_{min} is inversely proportional to γ, i.e., greater swing clearance necessitates higher robot stiffness, and vice versa. Moreover, the model correctly predicts less robot torque (i.e., lower K_{min}) for a more compliant ankle (i.e., lower K_h), and vice versa. Unlike the model for the loading response phase (Eq. (7)), the output of the swing phase model (i.e., minimum robot stiffness (K_{min}) needed for desired swing clearance) does not depend on whole-body parameters (body mass and height), but rather only on the intrinsic stiffness of the paretic ankle, K_h (Fig. 7) that can be measured using experimental procedures described in (Roy et al. 2011). We have incorporated the capability to manually adjust the ramp up hold ramp down parameters θ_d, T, and t_h, and thus K_{min}, in real-time, i.e., during assisted gait, so that the therapist can set them based on actual observations of gait deficits, recovery, and/or compensatory strategies.

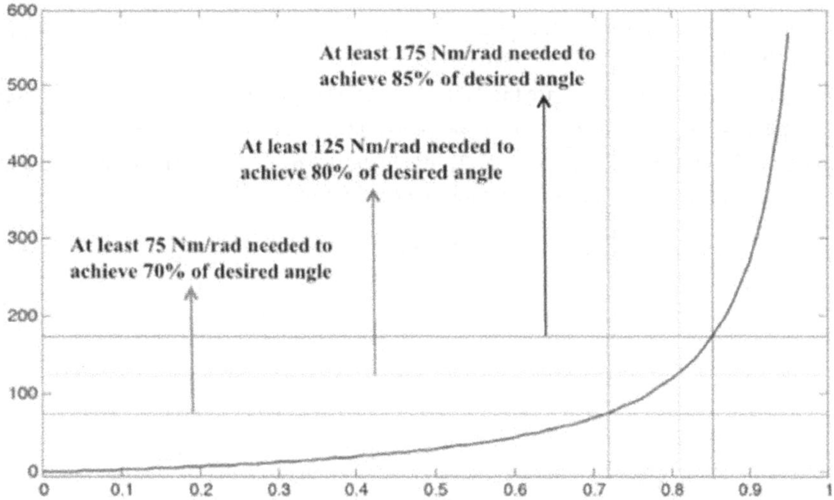

Figure 7. Model predictions on the minimum robot stiffness K_{min} needed for peak swing angles to achieve different values of desired angles plotted as a function of γ with $K_h = 30$ N-m/rad (see **Results** for details).

Robot-Incorporated Model for Push-off Stance Propulsion

The objective during this phase is to provide supplemental robotic support to generate sufficient push-off propulsion during the late stance period of gait. To develop the model, a sagittal-plane, two-segment inverted pendulum dynamics of the human "body" (head-arm-trunk plus leg minus foot) with the massless linkage acting as pivot over a moving base of support, i.e., foot (Fig. 8) system is used to derive the model (see Appendix), similar to the one used for swing and loading response phases of gait, but "placed" differently in the gait cycle (i.e., a posture during late stance just following heel-off event as shown in Fig. 8). The problem to be solved is as follows: given a subject's body height (H), mass (M), and steady-state self-selected walking speed (s), determine: (a) P_x^+ (output, N-s) resulting from a given value of K (input, Nm/rad); and (b) minimum robot stiffness (input K_{min}, Nm/rad) needed to generate desired (e.g., normative) P_x^+ (output, N-s); mathematically,

$$K_{\min} = \min(K) \text{ where } K = \{K : P_x^+ \geq P_d\}, \tag{12}$$

where P_x^+ and P_d are the actual and desired peak propulsive impulses during mid-to-terminal stance, respectively, i.e.,

$$P_x^+ := \int_{t^+} |\vec{F}_x|\, dt \tag{13}$$

and t^+ are time epochs of positive impulse when A-P GRF > 0 (in contrast to braking impulse when A-P GRF < 0), i.e.,

$$t^+ = \{t \in [t_{HO}, t_{TO}) : \vec{F}_x > 0\} \tag{14}$$

where t_{HO} and t_{TO} denote instants of heel-off (characterized by opening of heel switch with the medial, lateral and forefoot switches closed) and toe-off (characterized by opening of all individual foot switches) respectively, so that $[t_{HO}, t_{TO})$ represents mid-to-terminal stance duration. The push-off model equations links a key anklebot parameter (K) to a key gait function output (P_x^+) using the following inequalities:

$$P_x^+ \geq \{K(\theta_{PF}^*\Delta T_{LS} - I_\theta) - I_a(\Delta\dot{\theta} + \Delta\dot{\varphi}) + b\Delta\theta_{LS} + m_f g c_a \Delta T_{LS} - l_f \int_{t_{HO}}^{t_{TO^-}} |F_y|\, dt\}/l_f. \quad (15)$$

The minimum stiffness, K_{min} needed to achieve $P_x^+ \geq P_d$ can be obtained by imposing the condition of Eq. (12) in inequality (15):

$$K(\theta_{PF}^*\Delta T_{LS} - I_\theta) - I_a(\Delta\dot{\theta} + \Delta\dot{\varphi}) + b\Delta\theta_{LS} + m_f g c_a \Delta T_{LS} \geq l_f(P_d + \int_{t_{HO}}^{t_{TO^-}} |F_y|\, dt). \quad (16)$$

where ΔT_{LS} is late stance duration (a proxy for gait speed), I_θ is ankle angle vs. time area-under-curve, $\Delta\theta_{LS}$ is the ankle angular displacement during late stance, and $\Delta\dot{\theta}(s, K)$, $\Delta\dot{\varphi}$ are differences in ankle and body angular velocities from heel-off to toe-off, respectively. Note that inequality (16) is nonlinear in K (since ΔT_{LS}, I_θ, $\Delta\theta_{LS}$, etc. vary with K) and can be computed numerically for given P_d (see model simulation in

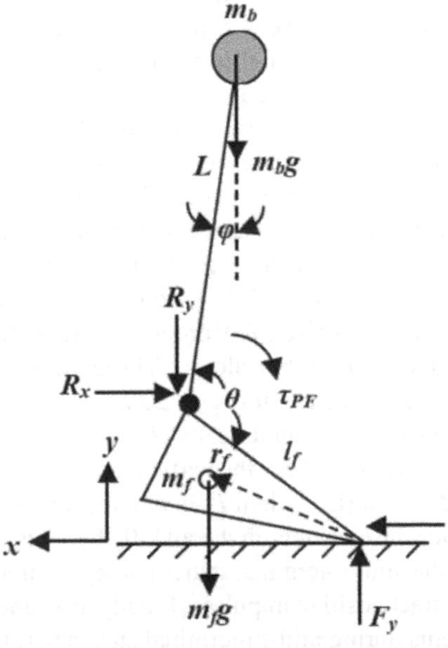

Figure 8. Two-segment inverted pendulum stick figure showing heel-strike (displacement of inverted pendulum from vertical is exaggerated for illustration purposes only). Robot damping lessens impact forces via resistive viscous torque at the ankle, whose magnitude is proportional to the ankle angular speed and damping. H: body height, M: body mass, L: length of foot, l: distance between ankle and foot COM, k: distance between ankle and body COM, b: robot damping, F_b: robot-generated viscous force, φ: angle of inverted pendulum with respect to vertical, θ: ankle angle, g: gravitational acceleration (9.81 ms^{-2}).

Fig. 15a). Inequalities (15) and (16) are *two variants of the desired model*: Inequality (15) predicts mid-to-terminal stance A-P positive propulsive impulse, P_x^+ for inputs robot stiffness, K. Inequality (16) may be used to solve for the minimum stiffness, K_{min} needed for mid-to-terminal stance A-P positive propulsive impulse to be equal to or greater than a desired value (e.g., normative), P_d. The first variant of the model is validated, by comparing its predictions to experimental data (see **Results**).

Calculating Model Constants

The push-off propulsion model given by inequality (16) consists of a number of constants that need to be calculated. Each constant (except robot damping, b) can be classified as either: (1) individual whole-body (e.g., H, M) or ankle joint (e.g., x_a, y_a, c_a, m_f, I_a) dimensions that are independent of K; or (2) gait characteristics (e.g., $\Delta\theta_{LS}$, I_θ, ΔT_{LS}) that are dependent on K; or (3) reference parameters (e.g., θ^*_{PF}) that are independent of K. To validate the model, these constants are calculated using body mass and height, joint dimensions relative to body mass and height (Dempster and Gaughran 1967, Iqbal and Roy 2009), and force plate data acquired during unassisted gait (see Experiments below).

Experiments to validate the push-off model

We computed the A-P propulsive impulses from dual force plate (Bertec, Columbus OH) GRF data collected during floor walking (averaged across 3 trials) at self-selected comfortable speed (see **Results**). The force plate data consisted of raw GRF time series in the Cartesian coordinate system (X: anterior-posterior, Y: vertical, Z: medial-lateral). Two sets of trials were conducted: the first set was with the anklebot in a "record-only" mode, to acquire unassisted GRFs and calculate volitional (human only) propulsive impulse. The second set of trials consisted of walking with anklebot plantar-flexion assistance at 5 stiffness settings (K = 50 N-m/rad to 250 N-m/rad in increments of 50 N-m/rad). During these trials, the anklebot provided plantar-flexion assistance, commencing at heel-off and peaking just prior to toe-off. Heel-off was detected by the footswitch (Fig. 9a) when the voltage dropped from ~1.5 V (full-load) to ~0.7 V (forefoot + toe switches closed). The difference between unassisted and assisted A-P propulsion was calculated to approximate the propulsive impulse due to the robot alone.[3] Throughout, the commanded peak plantar-flexion angle was held constant at $\theta^*_{PF} = 20°$.

The A-P propulsion impulse was numerically computed from unfiltered GRF traces using a custom MATLAB® program: first, the vertical ground reaction force (F_y) vector was used to identify mid stance (i.e., when F_y is at a local maximum, the leg is fully loaded) and terminal stance (i.e., when F_y first equals 0, the leg is fully unloaded) events (Fig. 9b). Second, within the mid-to-terminal stance phase, the time epochs during which $F_x > 0$ are identified to extract positive impulses. Finally, the time integral (trapezoidal method) $F_x > 0$ components during mid-to-terminal stance period yield P_x^+ (Fig. 9c).

[3] Since torque from human muscle was not measured, we assume same muscle activation levels at different stiffnesses, i.e., $\tau_h \approx \tau_h$, $K = 0$. We acknowledge the simplistic nature of this assumption; however, it enables a basic experiment to validate the model as a first pass by attributing the torque difference solely to robot stiffness, i.e., $\tau_a \approx \tau_{PF}$.

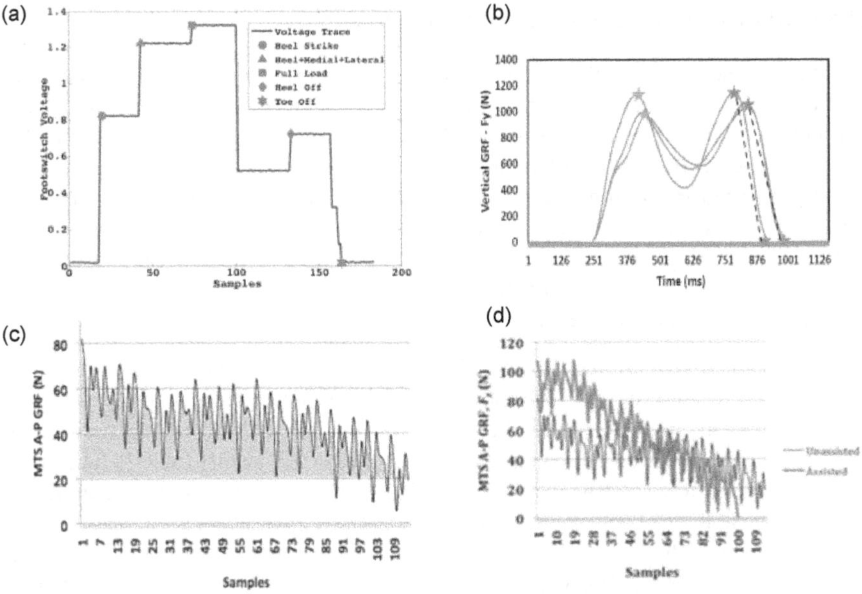

Figure 9. (a) Footswitch voltage trace of a single gait cycle during unassisted walking over force plates at self-selected comfortable speed. The plot shows heel-strike, mid-stance (full load), and toe-off sub-events. Late stance is defined as the duration between heel-off and toe-off. During assisted trials, the anklebot provides plantar-flexion torque during this period to enhance push-off propulsion; (b) Unassisted vertical GRF time profiles collected from force plates during 3 trials of walking. The biphasic traces show heel strike (green asterisk), mid stance (red asterisk) and start of swing (blue asterisk). These time points are used to identify mid-to-terminal stance; (c) Exemplar unassisted mid-to-terminal stance A-P GRF ($t = 0$ is mid-stance). The shaded region is the area under the curve, i.e., the integral of GRF or, positive propulsive impulse; (d) Exemplar unfiltered traces of mid-to-terminal stance A-P GRF collected from force plate during assisted and unassisted floor walking ($t = 0$ is mid-stance). A key model constant, $\int F_y dt$ is calculated from these data.

RESULTS

In the early phase of this study, we are testing the feasibility of our approach. Here as proof of concept, we present data from 2 participants. First, a 40-year old healthy individual who imitated hemiparetic gait and, in particular, foot drop during unassisted vs. assisted treadmill gait. The data from this person were used to: (a) quantify the precision of footswitches in detection of gait sub-events and determine whether that information correctly actuated the anklebot, and (b) validate the loading response and swing phase models. Our second participant was a chronic stroke survivor with pronounced foot drop. The data from this person were used to provide initial evidence of causality of our deficit-adjusted approach in that, whether (a) a few sessions of anklebot-assisted treadmill gait training that targets foot drop elicits gains in the peak swing angle during unassisted treadmill walking, and (b) if there are short-term and across training retention in those gains. The protocol was approved by the University of Maryland Baltimore Institutional Review Board, the Baltimore VA R&D Committee. All participants gave informed consent.

Accuracy of the Event-Triggered Actuation

The healthy participant walked at a self-selected comfortable speed on the treadmill, first without anklebot assistance, then followed by the anklebot assisting in swing phase. For both conditions the subject was asked to keep her foot relaxed and let it "dangle" following toe-off (but with sufficient hip and knee flexion to avoid scuffing) so as to imitate foot drop and minimize any volitional torque, thereby making the experimental condition consistent with the assumption made in the models, i.e., $\tau_h = 0$ (N-m) in Eqs. (7) and (11). During unassisted gait, the anklebot was in a record-only mode and data from this trial was used to compute footswitch thresholds for each sub-event, that were then used to actuate the anklebot during assisted walking.

1. Thresholds: Validation of footswitch voltages was performed for four sub-events: toe-off, heel-strike, heel-off, and foot-flat during unassisted treadmill walking. Specifically, we computed the error between the actual versus theoretical (manufacturer-specified) voltage threshold for each sub-event (Table 1). The errors were of the order of 10–102 mV depending on the sub-event. In particular, it was very small for the heel-off and foot-flat (0.1% and 0.7% relative error, respectively) but higher for heel-strike (12%). Errors may have resulted from the movement of foot (and hence the footswitch) within the shoe itself and can be compensated for by setting tolerance bands around the voltage threshold.
2. Timing: After the correct thresholds were obtained, we needed to verify if: (1) footswitches were accurately detecting sub-events and doing so consistently across gait cycles; and (2) in response to sub-event detection, the anklebot was being actuated accurately. Our findings were that: (a) there was one and only one torque pulse for each toe-off (i.e., $U = 0$); (b) the torques were initiated almost simultaneously ($\Delta T = 9.1 \pm 0.6$ ms) with the detection of toe-off (Fig. 10).

Table 1. Validation of foot switch outputs.

Event	Measured, mV	Theoretical, mV	Δ, mV (%)
Toe-off	107–110	0.00	107–110
Heel strike	897–899	800	97–99 (12%)
Heel-off	699–700	700	0–1 (0.1%)
Foot flat	1489–1491	1500	9–11 (0.7%)

Simulations and Validation of Robot-Incorporated Biomechanical Models

After verifying that the sub-event detection and anklebot actuation were robustly precise, the next step was to validate the swing and loading response models.

1. Swing phase: During assisted treadmill walking, a peak dorsiflexion angle of 5° was commanded with actuation starting at toe-off. Specifically, the objective was to provide robotic assistance needed for the ankle to not only cross the neutral angle during swing but, in fact, attain 70% (chosen arbitrarily) of the desired value (i.e., 3.5°). The passive ankle stiffness was estimated using prior experimental procedures (K_h = 30 N-m/rad) (Roy et al. 2011). Using $\gamma = 0.7$ in Eq. (11), the model predicted K_{min} = 70 N-m/rad. Figure 11 shows 6 gait cycles with (70 N-m/rad) and without (0 N-m/rad) robot assistance. During assisted walking, the ankle

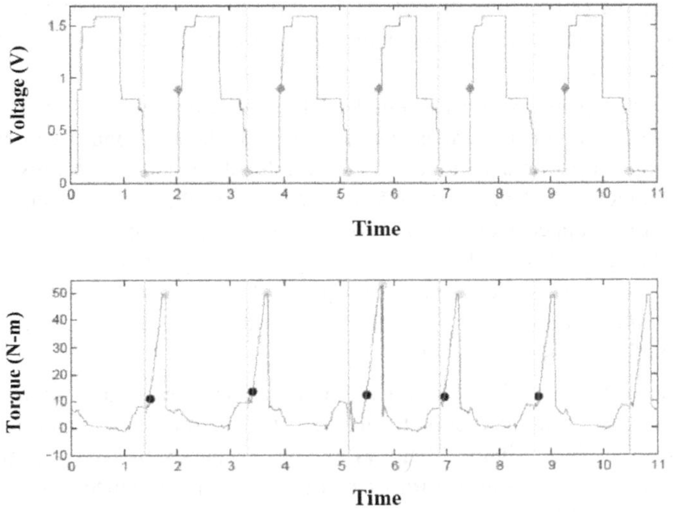

Figure 10. Five cycles of assisted treadmill gait: (a) Footswitch trace during (circles: toe-off, diamonds: heel-strike); (b) Robot torques (dark circles: instant of torque initiation, light circles: peak torque). The anklebot delivered dorsiflexion torques during swing, starting at toe-off.

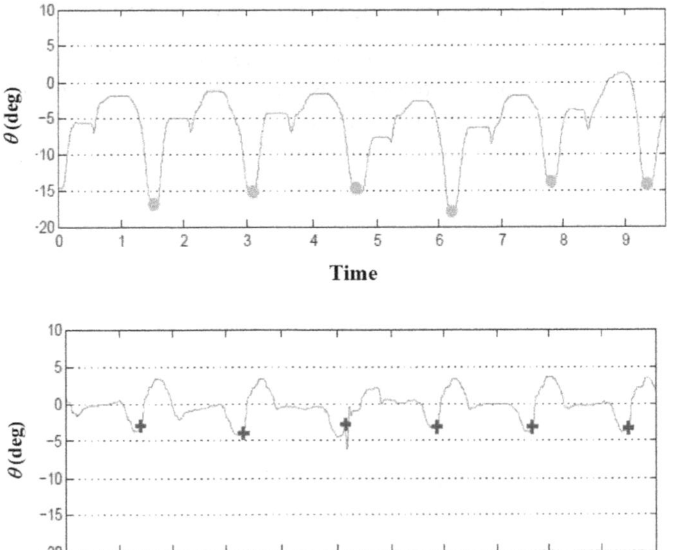

Figure 11. Validation of swing phase model using data from a healthy subject imitating drop foot. (*Top*) Trace of ankle angle without robotic assistance and (*Bottom*) With robotic assistance, during the swing phase. The anklebot delivered dorsiflexion assist following toe-off (unassisted: circles, assisted: squares). Using $K \geq K_{min}$, the ankle crosses the neutral into dorsiflexion during swing attaining a peak value close to the desired peak angle.

crossed the neutral in each cycle (3.2 ± 0.6°) but not without assistance (–2.4 ± 1.6°). The average peak swing angle (3.2°) was remarkably close to the desired value (3.5°).

2. Loading response phase: For this test, the participant was asked to "slap" her robot-donned foot "as hard as possible" with the heel in contact with the base of support so as to simulate foot slap. The anklebot generated restorative torques at different levels of damping (0, 3, 5 N-m-s/rad). To validate the model, we imposed an arbitrary upper-bound (200°/s) on the peak angular speed during the period of initial contact through flat foot. Using the appropriate constants ($M = 80$ kg, $H = 1.74$ m, $\varphi = 0.01$ rad, and $\alpha = 0.7$ (Iqbal and Roy 2004)) in Eq. (7), the predicted minimum damping was $b_{min} = 3.01$ N-m-s/rad as compared to the experimental value of 3 N-m-s/rad (Fig. 12).
3. Stance phase: A healthy 18-year old adult participated in the experiments. The subject gave informed consent prior to participation. Subject anthropometry is listed in Table 2. Figure 9d shows exemplar unassisted ($K = 0$ N-m/rad) and assisted ($K = 50$ N-m/rad) traces of mid-to-terminal stance A-P GRF. Using methods described previously, we computed the propulsion due to robot alone (assisted minus unassisted), at each K. For model validation, the constants (Table 3) were calculated from footswitch traces and robot-measured ankle kinematics during unassisted walking over force plates (Fig. 13). Table 3 shows key model constants across the selected stiffness range with constant damping, $b = 1$ N-m-s/rad. These are used in inequality (15) to yield P_x^+ for each K.

Figure 14 shows 3 key model constants in Table 3 as a function of robot stiffness, K. Our findings are that higher stiffness results in: (a) lower late stance durations (ΔT_{LS}), with a nonlinear inverse-Sigmoid type ΔT_{LS} vs. K trend; (b) smaller ankle angle area-

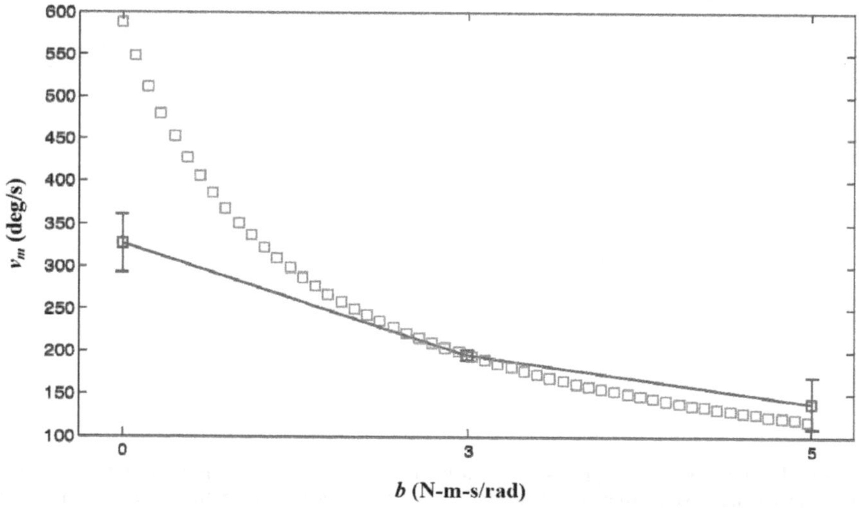

Figure 12. Validation of loading response phase model. The figure shows the predicted (open squares) and actual (mean±SD) peak angular speed during simulated "foot slap" at different levels of robot damping. The anklebot provides restorative torque to lessen the angular speed (and hence impact force) at initial contact. Note the excellent match between the experimental and model for b_{min} > ~2 N-m-s/rad.

Table 2. Subject whole-body, ankle joint and "upper body" anthropometric data.

Parameter, Symbol	Value (SI Units)
Body height, H	1.78 (m)
Body mass, M	75 (kg)
Body linkage length*, L	1.73 (m)
Body (minus foot) mass, m_b	74 (kg)
Foot mass, m_f	1.0 (kg)
Ankle distance from forefoot, l_f	0.215 (m)
Ankle (X,Y) coordinates from forefoot (x_a, y_a)	(0.2, 0.075) (m)
Foot COM horizontal distance from ankle, c_a	0.035 (m)
Foot COM distance from forefoot, r_f	0.167 (m)
Foot moment of inertia, I_f	8×10^{-3} (kg-m^2)
Foot moment of inertia about ankle, I_a	10^{-2} (kg-m^2)
Body COM to ankle distance, r_b	1.0 (m)
Body moment of inertia about COM, I_b	12.66 (kg-m^2)
Body moment of inertia about ankle, $I_b+m_b r_b^2$	87.66 (rad-sec)

*Computed using [Schmidt 2004, Ekkelenkamp et al. 2007] relative to body mass (M) and height (H).

Table 3. Model constants for each stiffness value in the push-off model.

Stiffness, K (Nm/rad)	ΔT_{LS} (sec)*	$\Delta\theta_{LS}$ (rad)**	$\Delta\dot{\theta}$ (rad/s)**	I_θ (rad-s)**	$\int F_y dt$ (N-s)*
50	0.1552	–0.5763	0.2867	–0.1391	68.186
100	0.1581	–0.0730	0.0771	–0.0959	68.919
150	0.1496	–0.2314	0.1092	–0.0601	68.107
200	0.1306	–0.4009	0.2578	–0.0624	56.624
250	0.1282	–0.8981	0.6885	–0.0397	49.572

*Calculated using foot switch time series, and anklebot-recorded angle and angular velocity time series. **Calculated from vertical GRF time series. Integral term numerically computed using MATLAB® "trapz" function.

under-curve (I_θ), with a positive linear I_θ vs. K trend; and (c) lower vertical GRF area-under-curve ($\int F_y dt$), with a constant-ramp down $\int F_y dt$ vs. K trend; and vice versa. Figure 15a shows predicted vs. actual propulsive impulses normalized to body mass, across all 5 values of K, the former calculated using model constants in Table 3. The average model-predicted values (0.71 ± 0.11 N-s/kg) were within 10% of the experimental values (0.64 ± 0.11 N-s/kg) across the full range of $K = 50$ to 250 Nm/rad. The absolute errors were 0.074 ± 0.067 N-s/kg across the full range of K, with lower values (0.043 ± 0.034 N-s/kg) for $K \leq 150$ N-m/rad and relatively higher errors (0.121 ± 0.93 N-s/kg) for $K \geq 200$ Nm/rad. Figure 15b shows 3D simulation of minimum stiffness (K_{min}) vs. desired propulsive impulse (P_d) vs. late stance duration (ΔT_{LS}), for arbitrarily chosen model constant set {$\Delta\theta_{LS} = -0.57$ rad, $\Delta\dot{\theta} = 0.28$ rad/s, $I_\theta = -0.14$ rad-s, $\int F_y dt = 68.2$ N-s}. As expected, the model correctly predicts that slower walking speeds (i.e., higher ΔT_{LS}) necessitates greater K to achieve a given P_d, and vice versa.

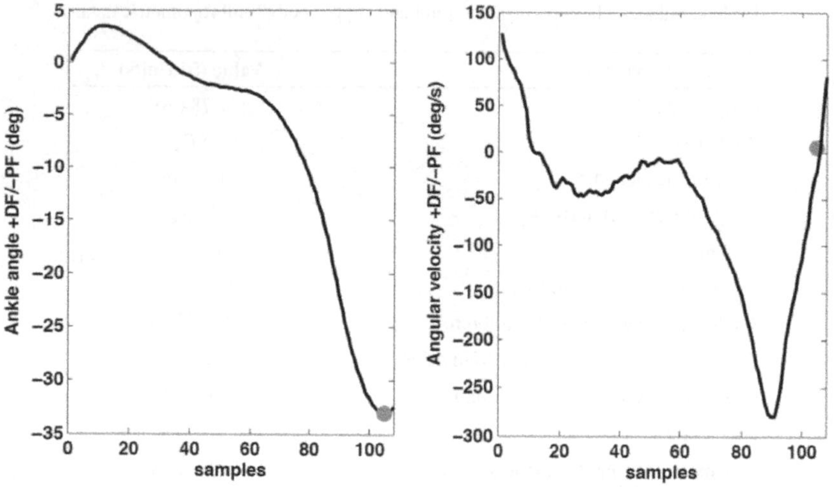

Figure 13. Exemplar ankle angle (*left*) and angular velocity (*right*) time series during unassisted (K = 0 Nm/rad) record-only walking. The traces are shown only for the mid-to-terminal stance period, i.e., t = 0 represents heel-off and filled circles represent toe-off. Four out of five model constants (ΔT_{LS}, $\Delta\theta_{LS}$, $\Delta\dot{\theta}_{LS}$, I_{θ}) are computed from these data.

Further validation was performed by comparing our results against existing literature. The effect of leg propulsion on walking speed and its relationship to muscle activity has been widely studied (e.g., Bowden et al. 2006, Turns et al. 2007, Neptune et al. 2001, Peterson et al. 2010). However, none of those investigations deployed robotics for PF gait therapy. Hence, a direct comparison cannot be made between findings from those studies and those reported here. Still, the predicted values of A-P positive impulse during the mid-to-terminal stance epoch reported here are much higher than those reported elsewhere, e.g., (Khanna et al. 2010) for healthy subjects (0.165 to 0.272 N-s/kg, n = 21). The primary reason for this is that unlike those studies, here we report impulses resulting from an actuated device. In addition, there are other differences: (1) Experimental conditions: for example, in (Peterson et al. 2010) control (able-bodied) subjects walked over split-belt treadmill without any robotic device vs. over ground with supplemental robotic assistance and unilateral robot mass loading condition in our study; and (2) Subject demographics: older nondisabled subjects (65.2 ± 9.6 yrs.) in (Peterson et al. 2010) vs. a young healthy (18 yr. old) subject in this study.

Proof-of-concept: Data from Stroke Survivor

The participant with stroke had pronounced foot drop (< 2° volitional dorsiflexion) making that a logical target for intervention. Training was conducted during 3xweekly visits with 48 hours between visits. On each visit, the session began with treadmill walking at self-selected speed (visit 1: 30 cm/s, visit 4: 36 cm/s) with the robot donned but not providing any assistance. This was followed by two 20-min trials of anklebot-assisted walking during which the anklebot provided swing assistance using a commanded dorsiflexion reference (Fig. 6c). To ensure safety, a support harness was worn, with easy access to an emergency stop switch, and two human "spotters" were present on either side. Footswitch voltage thresholds for each sub-event were verified and if needed, adjusted on each visit to

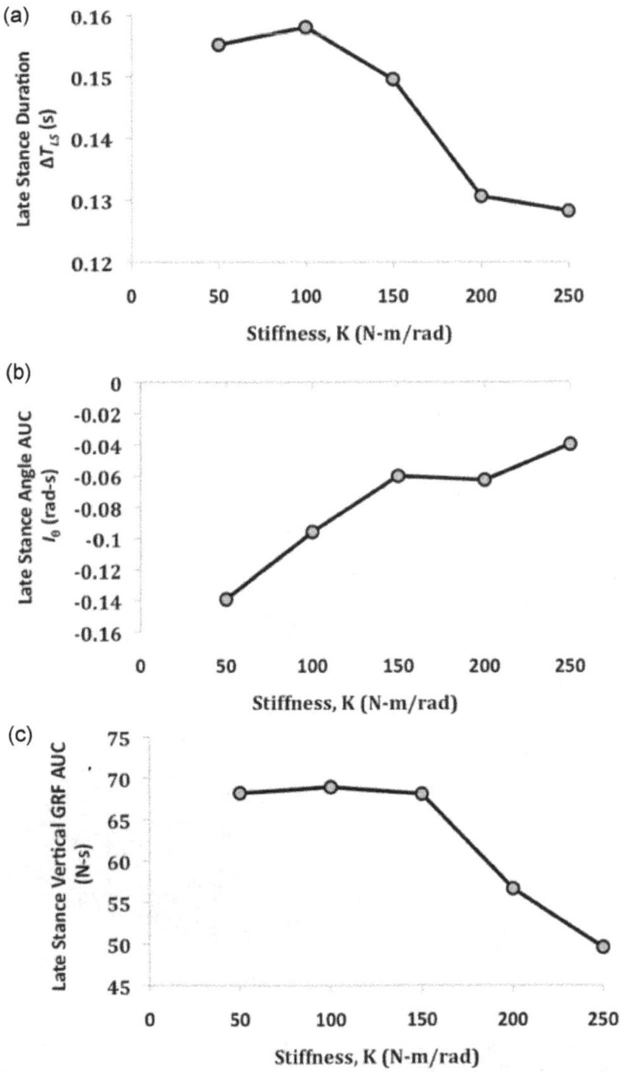

Figure 14. Variation of key model constants with stiffness, K: (a) Late stance duration (ΔT_{LS}); (b) Late stance angle area-under-curve (I_θ); and (c) Late stance vertical GRF area-under-curve $\int F_y dt$.

account for any inter-visit variability. The controller gains were set initially using Eq. (11) ($\gamma = 0.75$, $K_h = 50$ N-m/rad, $\theta_d = 20°$, $K_{min} = 150$ N-m/rad, $b = 1.5$ N-m-s/rad) and adjusted across visits based on participant feedback and clinician observations during assisted walking, as well as prior performance. We compared the peak swing angle during unassisted trials conducted on the first ("pre") visit and after 3 ("post") training visits (Fig. 16). There was a 200% increase in the pre-post peak swing angle (2.5 ± 1.1° vs. 7.6 ± 0.8°), and these gains were of similar magnitude after completion of training at 6 weeks (7 ± 0.5°) and notably, retained at 6-week follow-up (8 ± 0.8°) demonstrating initial efficacy of the approach.

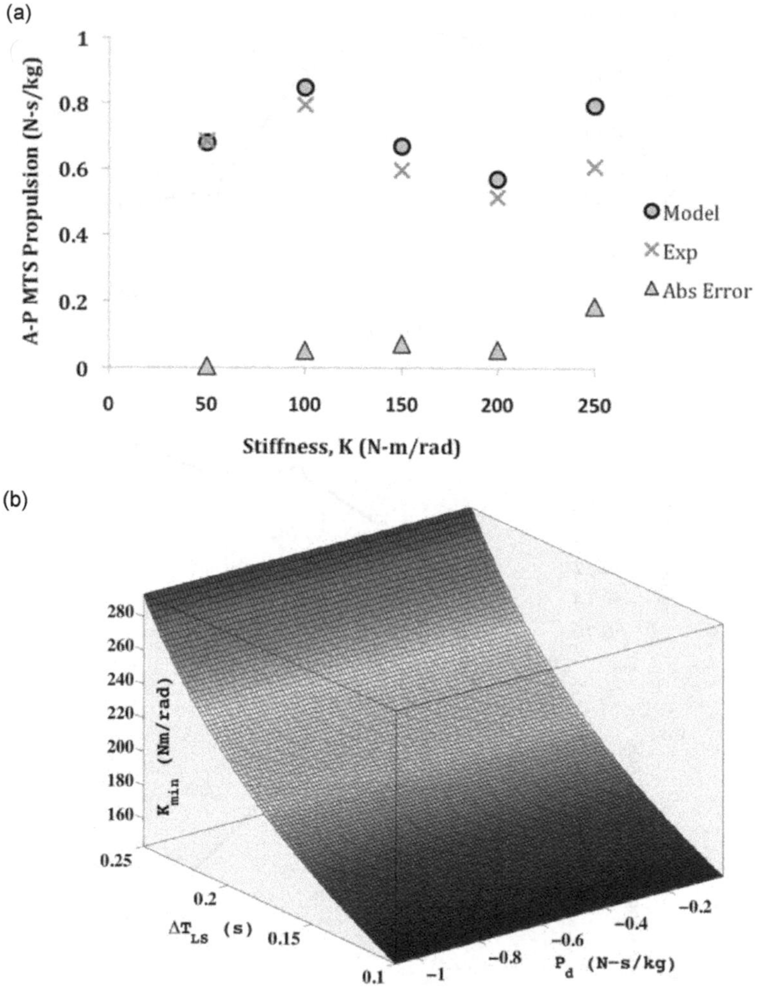

Figure 15. Model validation and simulations. (a) Comparison of model vs. actual A-P mid-to-terminal stance positive propulsive impulse for 5 values of stiffness, K. Also shown in the plot are the absolute values of the model vs. experiment residuals, which are "small" demonstrating accuracy of this first form model; and (b) 3D plot showing nonlinear relationship of minimum stiffness (K_{min}, Z-axis) as a function of desired propulsion (P_d, X-axis) and late stance duration (ΔT_{LS}, Y-axis).

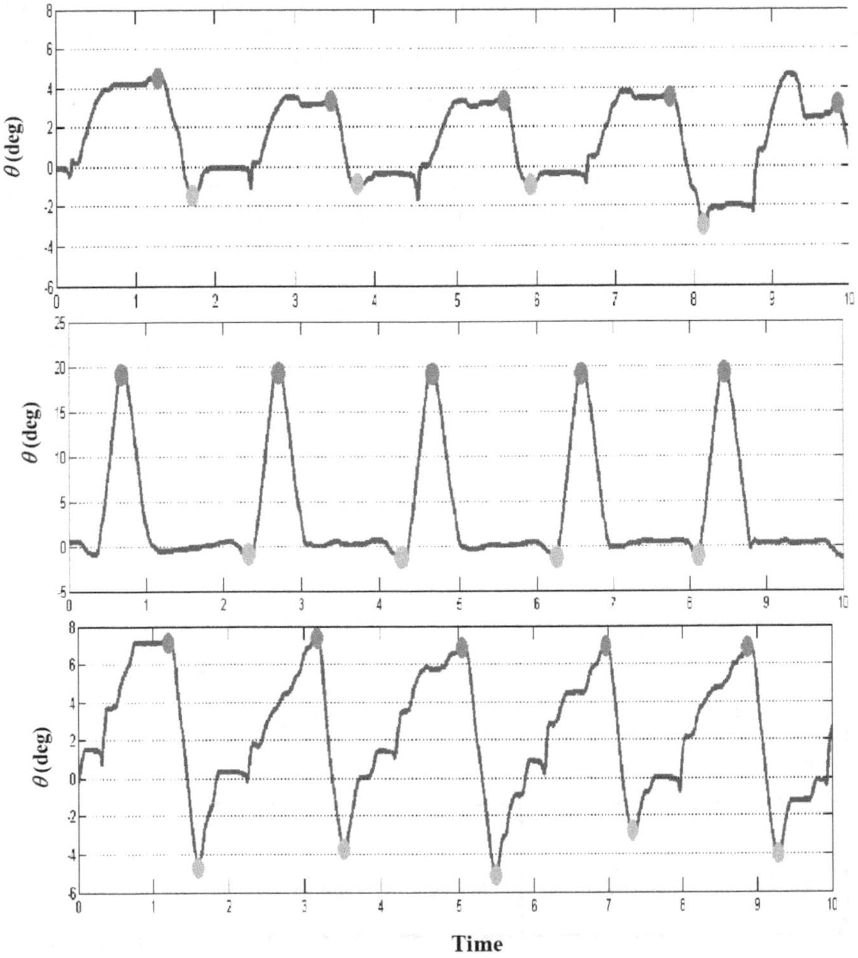

Figure 16. Paretic ankle angular data from the participant with stroke subject showing 5 cycles of treadmill gait on the first visit (*Top*) Without assistance, (*Middle*) With swing assistance, commencing at toe-off (circles). The peak swing angle was close to the desired value (20°) in each gait cycle during the assisted trials thus realizing the commanded reference; (*Bottom*) without assistance after 3 sessions. The peak swing angle (top circles) showed a marked increase after just 3 training sessions. Note difference in the *Y*-axis scale between assisted vs. unassisted trials.

Clinical and Rehabilitation Relevance

The models developed here are meant to predict reasonable initial values ("look-up" table) for appropriate robotic assistance for individual hemiparetic gait deficits during anklebot-assisted walking, i.e., they allow gait deficit-adjusted controller tuning. This is a major step toward implementing deficit-adjusted (i.e., where one deficit is more pronounced than another) customization of robotic therapy to individual hemiparetic gait profiles. Further, model may be used to systematically progress robotic support over an intervention concomitant with recovery, an approach consistent with our previous studies (e.g., Forrester et al. 2011, Roy et al. 2013).

In control terms, these models inform us of the controller gains (K and/or b) needed for deficit-adjusted robotic gait therapy to target an impairment specific to that phase. Using the swing, loading response, and push-off models we can set robot parameters to reasonable initial values based on deficit severity and type, i.e., minimum stiffness needed for desired swing clearance (Eq. (7), $K_{min,swing}$), the minimum damping needed for constraining loading response forces (Eq. (11), $b_{min,land}$), and the minimum stiffness needed for desired push-off propulsion (Inq. (16), K_{min}). This yields a centerpiece of our work: a deficit-based map in the controller space characterized by bounded or unbounded regions in the K-b plane that informs us of the robot (controller) parameters needed to target and alleviate one or more of these deficits during assisted walking (Fig. 17). Note that, in practice, the robot stiffness predicted by two of these models—enhanced swing clearance (Roy et al. 2013) and push-off propulsion, needs to be intersected with the achievable anklebot impedance ranges[4] (Roy et al. 2009) to prevent instability, i.e.,

$$K = [K_{\min}, \infty) \cap (0, K(b)_{\max,stab}) = [K_{\min}, K(b)_{\max,stab}) \tag{17}$$

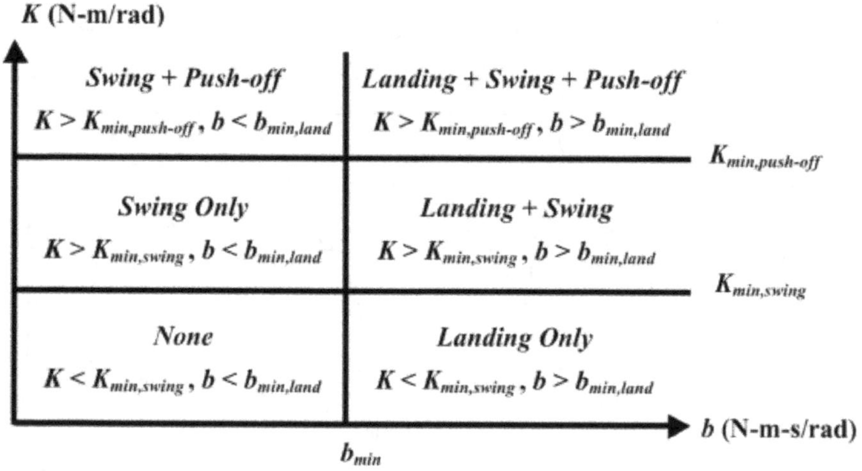

Figure 17. Deficit-based map in the controller parameter (K-b) plane. The map is an extension to that reported in (Roy et al. 2013) by incorporating the push-off model. Depending on the type of gait deficit being targeted, reasonable values of K and/or b can be chosen for desired performance.

CONCLUSIONS

Here we presented a sub-task controlled, deficit-adjusted approach for anklebot-assisted gait training. The approach used thus far represents a simple but versatile approach to calibrate and control the anklebot during assisted walking: (a) it can precisely time robotic support to hemiparetic gait in a way that accounts for dynamic step-to-step variability—a critical safety factor to prevent destabilization, while targeting different

[4] Characterized by K-b uncoupled stability curve (Roy et al. 2009) that is derived by measuring the highest K attainable at a given b before instability (persistent non-decaying oscillations) occurs ($K(b)_{max,stab}$), across a range of values of b.

deficits across patients to customize locomotor training. The latter is facilitated by sagittal-plane models capable of predicting a reasonable initial value for training inputs needed to parameterize anklebot outputs to individual hemiparetic deficits during assisted gait; (b) the unaffected (non-robot donning) foot can be used to establish sub-events when the affected (robot donning) foot is incapable of providing accurate information. We recognize, though, that additional information may be needed for this purpose—and hence have built the capacity to actuate the anklebot using the paretic knee angle information; and (c) the commanded kinematics can be shaped "on-the fly".

Initial validation tests verified that robotic support could be precisely and consistently timed to occurrence of sub-events, and that the models accurately predicted robotic support as per the specific nature of the deficit. Data from a chronic stroke survivor with drop foot showed that anklebot-assisted gait training progressively and significantly reduced foot drop after 6 weeks. We expect that over time, this approach will "teach" the central nervous system to take over from gradual withdrawal of robotic support in order to supplant the robot with volitional movements, i.e., avoid "learned nonuse" of the paretic ankle.

FUTURE DIRECTIONS

Preliminary Clinical Findings

Initial findings from our ongoing treadmill-based anklebot training study are very promising. In four chronic stroke subjects with dorsiflexion swing deficits[5] (peak swing: $-2.7° \pm 3.5°$, mid swing: $-3.8° \pm 3.1°$), six weeks (3x weekly) of treadmill-anklebot training targeted to swing dorsiflexion, led to robust and significant improvements in swing function (Peak swing angle: $+7.3° \pm 4.7°$, $\Delta = 370\%$; Mid swing angle: $+4.8° \pm 4.3°$, $\Delta = 226\%$). These gains were even greater at 6-week follow-up (Peak swing angle: $+8.4° \pm 6.7°$, Mid swing angle: $5.1° \pm 7.3°$). However, to our surprise, subjects also showed gains in plantar flexor push-off at 6 weeks (~3-fold increase in mid-to-terminal stance A-P propulsive impulse) and this was retained with even higher values at 6 weeks post-completion (9.8 N-s to 17.8 N-s), even though this effect was not targeted. Notably, all 4 patients discarded use of their ankle brace at follow-up and anecdotal evidence suggests better independent mobility function at home.

Expanding the Control Approach

Early testing of the anklebot has lent support to the idea of using a joint-specific modular approach as a therapeutic modality for hemiparetic stroke. Use of the seated approach has answered initial questions about the potential for augmenting traditional task-specific therapies, and may prove to be a valuable enhancement of early interventions by providing a platform to address underlying impairments in the realm of motor control. Additional work is needed to establish the optimal timing for modular robotics treatment across the spectrum of motor recovery after stroke, beginning with the inpatient

[5] Swing function was characterized by paretic peak dorsiflexion angle and mid swing angle during unassisted treadmill walking. Mid swing angle in particular, has special significance since the foot-floor local minima typically occurs at this time point. Hence, these measures are reasonable surrogates of foot-floor clearance or ground elevation during the swing phase.

rehabilitation hospital setting through the outpatient phase of clinical follow-up care. The modular anklebot offers a means to probe the effectiveness of early intensive robotic training for promoting neural plasticity associated with motor learning and whether this will increase the prospects for long term improvements in mobility and balance functions. The integration of the anklebot into actual task-oriented gait training, i.e., over ground is also evolving toward giving clinicians the ability to focus on specific paretic side deficits, such as foot drop, improper ankle-foot orientation at foot strike, and/or weak propulsion in late stance. In addition, unlike our upper-extremity adaptive algorithm (Krebs et al. 2003), the anklebot adaptive approach does not *auto*-modulate the controller gains or the commanded inputs in order to up- or down-regulate robotic outputs—e.g., amount or rate of assistance—to changes in performance—e.g., walking speed (presently adjusted manually based on therapists' observation and self-reported patient feedback). Refinements to our control approach will incorporate memory or performance-based auto-modulations to robotic outputs, both step-by-step and across training visits toward a more human-robot cooperative learning framework. In the future we anticipate the real-time integration of electroencephalography-monitored cognitive workload and cortical dynamics with robot-derived motor control measures as inputs to the anklebot adaptive controllers to cross a functionally meaningful threshold in the development of more effective robotics-based neurorehabilitation.

Other Applications

Other Neurologic Populations: Potential biomedical and human performance applications for adaptive control modular ankle exoskeletons go well beyond just hemiparetic stroke. Ankle exoskeletons with event triggered actuation could be customized to provide deficit-adjusted adaptive assistance across sub-movements of the gait cycle for a diversity of other chronic and rehabilitative neurological conditions that involve mobility disability. Approximately half of spinal cord injuries (SCI) are motor incomplete and could therapeutically benefit from human-robotics cooperative learning customized to their gait deficit profiles. Persons with more severe or motor complete SCI lesions could possibly leverage adaptive control ankle exoskeletons to enhance ground reaction forces in a more normal biomechanical manner. Adaptive control ankle exoskeletons could be considered in series with the more rigid velocity and/or position controlled multi-joint gait pattern generating exoskeletons in limited use for SCI. To our knowledge, no multi-joint exoskeleton currently utilized for SCI uses actuation at the ankle joint or adaptive control at the ankle or other joints to enhance or customize mobility tasks or balance. More than half of people with multiple sclerosis report falls within the prior 2 to 6 months, illustrating the need for more intelligent adaptive control exoskeletons or active orthotics engendering such engineering advances to enhance gait patterning and safety in the setting of neurological symptoms variably including spasticity, cerebellar, gait apraxia, and/or corticospinal weakness including foot drop. Similarly, individuals with idiopathic Parkinson's disease or the more common Parkinson's syndrome invariably develop gait and balance abnormalities that are often asymmetric, involve the ankle, and are known to benefit from sensorimotor cueing which could be provided in a step-by step customized fashion to account for their heterogeneity in gait patterning and freezing. While a detailed list of central nervous system conditions causing mobility disability is beyond the scope of this text, it can be generalized that the majority involve paresis, i.e., partial weakness and not plegia (complete paralysis), and are thus potential candidates for adaptive control robotics integrated with impedance control that can customize cooperative robotics to neurological needs and gait patterning.

Amputation Prostheses: The adaptive control approach may be utilized in the context of amputation prostheses that is designed to replace lost limbs in a patient, to help the patient recover mobility and sensory function. In particular, leg prostheses provide mechanical support, shock absorption, balance, and forward propulsion. Thus, the underlying approach may be utilized for example, in battery-powered motorized amputation prostheses to provide precise timing and appropriately scaled assistance. This has potential to advance the field of active amputation prostheses.

Training Healthy Joints: In (Roy et al. 2011) we demonstrated that when moved passively, the ankle is weakest when turning inward, stronger when tilting from side-side, and strongest when simply moving up-down, demonstrating highly anisotropic (e.g., direction dependent) passive mechanical impedance, in both healthy young and old, and chronic stroke survivors. These findings can be applied to train healthy people to exercise their ankles in specific ways that strengthen them and help reduce future injuries. Thus, the adaptive control approach presented here may be used for human performance augmentation of lower limb joints including the ankle using modular exoskeletons. The approach may also lead to mechanical "smart" footwear that provide scaled and timed resistance in lateral (e.g., side-side) foot motion for stability while providing timed assistance for up-down motion.

Regulating Foot Pressure and Ground Reaction Forces in Diabetic Neuropathy: One or more steps of the adaptive approach may be used in the context of regulating foot pressure and ground reaction forces in diabetic neuropathy. Approximately 9.3% of people have diabetes mellitus type 2, including 26% of individuals over 65 years of age. A substantial portion of these people will develop peripheral neuropathy that leads to reduced sensation, particularly in the toes and feet. This is followed by intrinsic foot muscle wasting and secondary orthopedic problems consisting of hammer toes, Charcot Joints, lateral toe deviations, and thinning of the metatarsal pads. All of these conditions, combined with the insensate foot, lead to foot ulcers, which produce a 50% five year survival curve, as they are only treatable with "static orthotics" to better distribute the foot pressure forces. The stance push-off movement model described here may be utilized to impart restorative torque on the subject's shoe and thus enable precise timing of impulse, or ground reaction forces that are seen by the toes and foot, thereby enabling a dynamic real time control of the pressures that lead to foot ulcers, and exacerbate ulcers. Additionally, the progression elements of the stance model can produce motor learning during ambulatory conditions, which affords inroads in the fields of podiatry and orthotics for the care of diabetic neuropathy, and other neuropathies such as peripheral arterial occlusive disease, yielding new therapies to reduce foot damage, prevent ulcers, and ultimately prevent amputations.

Motor Learning to Improve Outcomes for Podiatry, Orthopedics, and Prosthetics: Some elements of the approach may be utilized in the context of motor learning to improve outcomes for podiatry, orthopedics, and prosthetics. Selected post-operative care conditions in podiatry and orthopedics could optimize outcomes if ground reaction forces (impulse) were controlled for safety, and if progressive motor learning were optimized in the post-operative rehabilitative recovery period. The adaptive, deficit-adjusted control approach may potentially meet these needs. In the field of prosthetics, the ground reaction impulses are conducted up the prosthetic shank, and over many years, repetitive use and pounding can cause pain and complications at the stump. The stance push-off model may be utilized to produce a bio-inspired walking pattern utilizing the adaptive controller in a deficit severity adjusted manner, with machine learning to adapt the underlying prosthetic device to cushion the stump can be used to improve outcomes in prosthetics. For those with polytrauma, and subsequent tibialis anterior (e.g., swing phase deficit) or peroneal nerve damage with foot eversion weakness, or stance phase deficit due to lumbar 5 sacral 1 or sciatic trunk or incomplete tract injury, the modular deficit severity adjustable units can be adapted to serve as a task-oriented functional mobility therapeutic tool to extend the therapists capabilities. This would enable modelling for optimization of progression that would serve as a cumulative repository for assisting and informing the recovery of future similar polytrauma and orthopedic or mixed neurological-orthopedic cases.

REFERENCES

Andriacchi, T. P., J. A. Ogle and J. O. Galante. 1977. Walking speed as a basis for normal and abnormal gait measurements. J. Biomech., 10: 261–68.

Bosecker, C. J. and H. I. Krebs. 2009. MIT-Skywalker. pp. 542–9. *In*: Proc. IEEE Int. Conf. Rehab. Robotics, Kyoto, Japan.

Bowden, M. G., C. K. Balasubramanian, R. R. Neptune and S. A. Kautz. 2006. Anterior–posterior ground reaction forces as a measure of paretic leg contribution in hemiparetic walking. Stroke, 37: 872–76.

Burgar, C. G., P. S. Lum, A. M. Scremin, S. L. Garber, H. L. Van der Loos, D. Kenney and P. Shor. 2011. Robot-assisted upper limb therapy in acute rehabilitation setting following stroke: Department of Veterans Affairs multisite clinical trial. Journal of Rehabilitation Research and Development, 48: 445–458.

Dempster, W. T. and G. R. L. Gaughran. 1967. Properties of body segments based on size and weight. Am. J. Anatomy, 120: 33–54.

Dennis, M. S., K. M. Lo, M. McDowall and T. West. 2002. Fractures after stroke: Frequency, types, and associations. Stroke, 33(3): 728–734.

Ekkelenkamp, R., J. Veneman and J. Van der kooij. 2007. LOPES: a lower extremity powered exoskeleton. pp. 3132–33. *In*: Proc. IEEE Int. Conf. Robotics Automation, Roma, Italy.

Forrester, L. W., A. Roy, H. I. Krebs and R. F. Macko. 2011. Ankle training with a robotic device improves hemiparetic gait after a stroke. Neurorehabil. Neural. Repair, 25: 369–77.

Forrester, L. W., A. Roy, A. Krywonis, G. Kehs, H. I. Krebs and R. F. Macko. 2013. Modular ankle robotics training in early sub-acute stroke: A randomized controlled pilot study. Neurorehabil. Neural. Repair, 33: 85–97.

Forster, A. and J. Young. 1995. Incidence and consequences of falls due to stroke: A systematic inquiry. BMJ, 311: 83–6.

Heart disease and stroke statistics: American Heart Association. 2009. Available from: www.americanheart.org/statistics.

Hesse, S. and D. Uhlenbrock. 2000. A mechanized gait trainer for restoration of gait. J. Rehabil. Res. Dev., 37: 701–08.

Hornby, T. G., D. D. Campbell and J. H. Kahn. 2008. Enhanced gait-related improvements after therapist- versus robotic-assisted locomotor training in subjects with chronic stroke: a randomized controlled study. Stroke, 39: 1786–92.

Iqbal, K. and A. Roy. 2004. Stabilizing PID controllers for a single-link biomechanical model with position, velocity, and force feedback. J. Biomech. Eng., 126: 838–43.

Iqbal, K. and A. Roy. 2009. A novel theoretical framework for the dynamic stability analysis, movement control, and trajectory generation in a multi-segment biomechanical model. J. Biomech. Eng., 131: 011002.

Jezernik, S., G. Colombo, T. Keller, H. Frueh and M. Morari. 2003. Robotic orthosis Lokomat: A research and rehabilitation tool. Neuromod., 6: 108–115.

Kanis, J., A. Oden and O. Johnell. 2001. Acute and long-term increase in fracture risk after hospitalization for stroke. Stroke, 32(3): 702–706.

Khanna, I., A. Roy, M. M. Rodgers, H. I. Krebs, R. F. Macko and L. W. Forrester. 2010. Effects of unilateral robotic limb loading on gait characteristics in subjects with chronic stroke. J. Neuro. Eng. Rehab., pp. 7–23.

Krebs, H. I., J. J. Palazzolo, L. Dipietro, M. Ferraro, J. Krol, K. Rannekleiv, B. T. Volpe and N. Hogan. 2003. Rehabilitation robotics: performance-based progressive robot-assisted therapy. Autonomous Robots, Kluwer Academics, 15: 7–20.

Li, D. Y., A. Becker, K. A. Shorter, T. Bretl and E. T. Hsiao-Wecksler. 2011. Estimating system state during human walking with a powered ankle-foot orthosis. IEEE/ASME T Mech, 16: 835–44.

Lo, A. C., P. D. Guarino, L. G. Richards, J. K. Haselkorn, G. F. Wittenberg, D. G. Federman and P. Peduzzi. 2010. Robot-assisted therapy for long-term upper-limb impairment after stroke. New England Journal of Medicine, 362(19): 1772–1783.

Macko, R. F., G. V. Smith, C. L. Dobrovolny, J. D. Sorkin, A. P. Goldberg and K. H. Silver. 2005. Treadmill training improves fitness and ambulatory function in chronic stroke patients. Stroke, 36: 2206–11.

Neptune, R. R., S. A. Kautz and F. E. Zajac. 2001. Contributions of the individual ankle plantar flexors to support, forward progression and swing initiation during walking. J. Biomech., 34: 1387–98.

Perry, J., M. Garrett, J. K. Gronley and S. J. Mulroy. 1995. Classification of walking handicap in the stroke population. Stroke, 26: 982–89.

Peterson, C. L., J. Cheng, S. A. Kautz and R. R. Neptune. 2010. Leg extension is an important predictor of paretic leg propulsion in hemiparetic walking. Gait Posture, 32: 451–56.

Ramnemark, A., L. Nyberg, B. Borssen, T. Olsson and Y. Gustafson. 1998. Fractures after stroke. Osteoporosis International, 8(1): 92–95.

Roy, A., H. I. Krebs, D. J. Williams, C. T. Bever, L. W. Forrester, R. F. Macko and N. Hogan. 2009. Robot-aided neurorehabilitation: A novel robot for ankle rehabilitation. IEEE T Rob, 25: 569–82.

Roy, A., H. I. Krebs, C. T. Bever, L. W. Forrester, R. F. Macko and N. Hogan. 2011. Measurement of passive ankle stiffness in subjects with chronic hemiparesis using a novel ankle robot. J. Neurophysiol., 105: 2312–49.

Roy, A., L. W. Forrester and R. F. Macko. 2011. Short-term ankle motor performance with ankle robotics training in chronic hemiparetic stroke. J. Rehab. Res. Dev., 48: 417–30.

Roy, A., H. I. Krebs, J. E. Barton, R. F. Macko and L. W. Forrester. 2013. Anklebot-assisted locomotor training after stroke: A novel deficit-adjusted control approach. pp. 2167–2174. *In*: Proc. IEEE Int. Conf. Robotics and Automation (ICRA), Karlsruhe, Germany.

Schmidt, H. 2004. HapticWalker–A novel haptic device for walking simulation. pp. 60–7. *In*: Proc. EuroHaptics, Munich, Germany.

Turns, L. J., R. R. Neptune and S. A. Kautz. 2007. Relationships between muscle activity and anteroposterior ground reaction forces in hemiparetic walking. Arch. Phys. Med. Rehabil., 88: 1127–35.

APPENDIX

The following notations and definitions will be used throughout the development of the push-off propulsion model: For arbitrary segment angles, x_i, $y_i \in \mathfrak{R}$, we denote operators $S : \mathfrak{R} \rightarrow [0, 1]$, $C : \mathfrak{R} \rightarrow [0, 1]$ as $S_i = \sin x_i$, $C_i = \cos x_i$; $S_{x \pm y} = \sin(x \pm y)$; $\Delta x_{ij} = x_j - x_i$ $(i \neq j)$; $S'_i = \sin(x_i - x_i(0))$, $C'_i = \cos(x_i - x_i(0))$ where $x_i(0) = x_i(t)|_{t=0}$. In the two-segment model (Fig. 8), the body mass is assumed to be concentrated as a point mass (m_b) on the distal end of the linkage with length L. The foot including ankle joint, is a separate rigid body with mass (m_f) located at the foot COM; M is total mass (body mass m_b plus foot mass m_f); H is total height (body linkage length, L plus foot height, y_a); (x_a, y_a) is ankle position from forefoot; l_f is ankle distance from forefoot; r_f is foot COM distance from forefoot; c_a is foot COM horizontal position from ankle; I_f and I_a are foot moment of inertia about COM and about ankle respectively; I_b is body moment of inertia, so that body moment of inertia about ankle is $I_b + m_b r_b^2$; and subscripts "1" (or "a") and "2" (or "b") refer to ankle joint and body, respectively; and P_x^+ denotes mid-to-terminal stance A-P positive propulsive impulse. Other variables are defined in the sections below as development of the model evolves.

To derive a model for propulsion dynamics, we assume a body posture at the instant of terminal stance—i.e., the body COM is slightly anterior to the vertical (representing forward COM momentum) with the foot (except toe) above the base of support. We initially use segment angles from horizontal as generalized coordinates with x-axis directed toward posterior of the foot (Fig. 8), with θ_1 and θ_2 representing the ankle angle and body angle from horizontal, respectively; additionally, $\dot{\theta}_1 \approx \theta_1 - \theta_1(t = 0)$ represents the angle of foot COM from horizontal. Then, the kinetic (T) and potential (V) energies of the segments are given by:

$$T = \frac{1}{2}(I_f + m_f r_f^2 + m_b l_f^2)\dot{\theta}_1^2 + \frac{1}{2}(I_b + m_b r_b^2)\dot{\theta}_2^2 + m_b l_f r_b \dot{\theta}_1 \dot{\theta}_2 C_{12} \tag{1a}$$

$$V = [m_f r_f S_1' + m_b(l_f S_1 + r_b S_2)]g \tag{1b}$$

with $S_1 = \sin\theta_1$, $C_1 = \cos\theta_1$, $S_{12} = \sin\Delta\theta_{21}$, $C_{12} = \cos\Delta\theta_{21}$, $S'_1 = \sin(\theta_1 - \theta_1(t = 0))$ and $\Delta\theta_{21} = \theta_2 - \theta_1$ as defined by convention. Also, per this notation, $S_{x\pm y\pm z} = \sin(x \pm y \pm z)$ and $C_{x\pm y\pm z} = \cos(x \pm y \pm z)$. Using Lagrangian formulation, the dynamic equations of the model are obtained as:

$$\tau = M_1\ddot{\theta}_1 + M_{12}(\ddot{\theta}_2 C_{12} - \dot{\theta}_2^{\,2} S_{12}) + (m_f r_f C_1' + m_b l_f C_1)g \tag{2}$$

$$-\tau = M_2\ddot{\theta}_2 + M_{12}(\ddot{\theta}_1 C_{12} + \dot{\theta}_1^{\,2} S_{12}) + (m_b r_b C_2)g \tag{3}$$

where

$$M_1 = I_f + m_f r_f^2 + m_b l_f^{\,2} \tag{4}$$

$$M_2 = I_b + m_b r_b^2 \tag{5}$$

$$M_{12} = m_b l_f r_b. \tag{6}$$

Next, these equations are transformed into natural coordinates (θ, φ) via the following transformations:

$$\theta_1 = \theta + \varphi - 90^\circ, \theta_2 = \varphi + 90^\circ \tag{7}$$

$$\dot{\theta}_1 = \dot{\theta} + \dot{\varphi}, \dot{\theta}_2 = \dot{\varphi} \tag{8}$$

$$\ddot{\theta}_1 = \ddot{\theta} + \ddot{\varphi}, \ddot{\theta}_2 = \ddot{\varphi} \tag{9}$$

$$C_{\theta_1} = S_{\theta+\varphi}, S_{\theta_1} = -C_{\theta+\varphi}, C_{\theta_2} = -S_\varphi, S_{\theta_2} = C_\varphi. \tag{10}$$

The dynamics in Eqs. (2) and (3) can now be written as

$$\tau = M_1\ddot{\theta} + (M_1 - M_{12}C_\theta)\ddot{\varphi} - M_{12}\dot{\varphi}^2 S_\theta + (m_f r_f S_{\theta+\varphi-\psi} + m_b l_f S_{\theta+\varphi})g \tag{11}$$

$$-\tau = M_{12}C_\theta\ddot{\theta} + (M_2 - M_{12}C_\theta)\ddot{\varphi} + (\dot{\theta} + \dot{\varphi})^2 S_\theta - (m_b r_b S_\varphi)g. \tag{12}$$

To compute GRFs, we use d'Alembert's principle to obtain

$$F_x = m_f a_{f,x} + m_b a_{b,x} \tag{13}$$

$$F_y = (m_b + m_f)g + m_b a_{b,y} + m_f a_{f,y} \tag{14}$$

where the components of foot and body COM accelerations are computed using the following expressions

$$a_{f,x} = r_f(\ddot{\theta} + \ddot{\varphi})C_{\theta+\varphi-\psi} - r_f(\dot{\theta} + \dot{\varphi})^2 S_{\theta+\varphi-\psi} \tag{15}$$

$$a_{f,y} = r_f(\ddot{\theta} + \ddot{\varphi})S_{\theta+\varphi-\psi} + r_f(\dot{\theta} + \dot{\varphi})^2 C_{\theta+\varphi-\psi} \tag{16}$$

$$a_{m,x} = l_f(\ddot{\theta} + \ddot{\varphi})C_{\theta+\varphi} - l_f(\dot{\theta} + \dot{\varphi})^2 S_{\theta+\varphi} - r_b(\ddot{\varphi}C_\varphi - \dot{\varphi}^2 S_\varphi), \tag{17}$$

$$a_{m,y} = l_f(\ddot{\theta} + \ddot{\varphi})S_{\theta+\varphi} + l_f(\dot{\theta} + \dot{\varphi})^2 C_{\theta+\varphi} - r_b(\ddot{\varphi}S_\varphi + \dot{\varphi}^2 C_\varphi). \tag{18}$$

To relate the GRF to ankle torque, we consider the foot as it pivots around the ankle (Fig. 8) and its dynamic equation:

$$\tau_a = I_a(\ddot{\theta} + \ddot{\varphi}) + l_f(F_x C_{\theta+\varphi} + F_y S_{\theta+\varphi}) - m_f g c_a. \tag{19}$$

where $\tau_a = \tau_h + \tau_{PF}$ is the net torque at the ankle, comprised of the human (τ_h) and robot (τ_{PF}) torques. Using methods described previously (see Experiments), we "baseline subtract" the unassisted human torque ($\tau_{h,K=0}$), so that $\tau_a = \tau_a - \tau_{h,K=0} \approx \tau_{PF}$. Then, we swap τ_a in Eq. (19) with

$$\tau_{PF} = K(\theta^* - \theta) + b(\dot{\theta}^* - \dot{\theta}) \tag{20}$$

where $\theta^* = \theta^*_{PF}$ and $\dot{\theta}^* = 0$, to obtain

$$K\theta^*_{PF} - K\theta - b\dot{\theta} = I_a(\ddot{\theta} + \ddot{\varphi}) + l_f(F_x C_{\theta+\varphi} + F_y S_{\theta+\varphi}) - m_f g c_a. \tag{21}$$

We now links propulsion dynamics to robotic support. To do so, we first calculate the mid-to-terminal stance A-P propulsive impulse then integrate Eq. (21) from t_{HO} to t_{TO^-} and rearrange to obtain:

$$K(\theta^*_{PF}\Delta T_{LS} - I_\theta) = I_a(\dot{\theta} + \dot{\varphi})\Big|_{t_{HO}}^{t_{TO^-}} + l_f \int_{t_{HO}}^{t_{TO^-}} (F_x C_{\theta+\varphi} + F_y S_{\theta+\varphi})dt - b\Delta\theta_{LS} - m_f g c_a \Delta T_{LS}. \tag{22}$$

where $\Delta T_{LS}(s, K) = t_{TO^-} - t_{HO}$ is late stance duration (a proxy for gait speed) with $t_{TO^-} = \lim_{\delta \to 0+} t_{TO} - \delta$, $I_\theta(s,K) = \int_{HO}^{TO^-} \theta(t,K)dt$ is ankle angle vs. time area-under-curve, and $\Delta\theta_{LS}$ is the ankle angular displacement from t_{HO} to t_{TO^-}. The first bracketed RHS term in Eq. (22) may now be written as

$$(\dot{\theta} + \dot{\varphi})\Big|_{t_{HO}}^{t_{TO^-}} = \Delta\dot{\theta} + \Delta\dot{\varphi} \tag{23}$$

where $\Delta\dot{\theta}(s, K)$ and $\Delta\dot{\varphi}$ are difference in ankle and body angular velocities from heel-off to toe-off, respectively. Substituting Eq. (23) into (22) and rearranging, we obtain:

$$l_f \int_{t_{HO}}^{t_{TO^-}} (F_x C_{\theta+\varphi} + F_y S_{\theta+\varphi})dt = K(\theta^*_{PF}\Delta T_{LS} - I_\theta) - I_a(\Delta\dot{\theta} + \Delta\dot{\varphi}) + b\Delta\theta_{LS} + m_f g c_a \Delta T_{LS}. \tag{24}$$

Note that our desired output variable P_x^+ does not explicitly appear in LHS of Eq. (24). To proceed further, consider that

$$\left|F_x C_{\theta+\varphi}\right| \le \left|F_x\right|, \left|F_y S_{\theta+\varphi}\right| \le \left|F_y\right|, \forall t \in D = [\sigma_1, \sigma_2]$$

for arbitrary bounded domain $D = [\sigma_1, \sigma_2]$. Moreover, since $|F_y S_{\theta+\varphi}|$ and $|F_x C_{\theta+\varphi}|$ are both Riemann-integrable (bounded and continuous everywhere), we can use the following general inequality property:

$$\int_D |F_x C_{\theta+\varphi}|\, dt \le \int_D |F_x|\, dt = P_x^+ \tag{25}$$

$$\int_D | F_y S_{\theta+\varphi} | dt \le \int_D | F_y | dt. \tag{26}$$

Therefore, the integral term in LHS in Eq. (24) reduces to

$$\int_{t_{HO}}^{t_{TO^-}} (F_x C_{\theta+\varphi} + F_y S_{\theta+\varphi}) dt \le P_x^+ + \int_{t_{HO}}^{t_{TO^-}} | F_y | dt. \tag{27}$$

Using inequality (27) and Eq. (24), and rearranging, we obtain a nonlinear inequality between K and P_x^+, given by

$$P_x^+ \ge \{K(\theta_{PF}^* \Delta T_{LS} - I_\theta) - I_a(\Delta\dot{\theta} + \Delta\dot{\varphi}) + b\Delta\theta_{LS} \\ + m_f g c_a \Delta T_{LS} - l_f \int_{t_{HO}}^{t_{TO^-}} | F_y | dt\} / l_f. \tag{28}$$

The minimum stiffness, K_{min} needed to achieve $P_x^+ \ge P_d$ can be obtained by imposing the condition of Eq. (12) in inequality (28):

$$K(\theta_{PF}^* \Delta T_{LS} - I_\theta) - I_a(\Delta\dot{\theta} + \Delta\dot{\varphi}) + b\Delta\theta_{LS} + m_f g c_a \Delta T_{LS} \\ \ge l_f (P_d + \int_{t_{HO}}^{t_{TO^-}} | F_y | dt). \\ - \int_{HO}^{TO^-} (y_a R_x - x_a R_y) dt\} / (I_\theta^* - I_\theta) \tag{43}$$

10

Open Architecture High Value Added Robot Manufacturing Cells

Delbert Tesar

ABSTRACT

A major position paper was presented by (Tesar 1978) in the science magazine in 1978 outlining the increasing weakness of the "light" mechanicals (including the strengthening role of industrial robotics) in the U.S. with significant trade imbalance at that time (1977). Today, both heavy and light machinery show significant negative trade imbalances. Mechanical component trade did well in 1977 with considerable university research in sensors and operational software (more on this in Section Dynamic Manipulator Modelling). One decade later, Tesar presented a detailed forecast (Tesar 1989) for the fifth generation (the super robot) of industrial robots with emphasis on high value added light machining. The paper says that no light machining, no rapid product changeover, and no open architecture existed in 1989. Of the twelve major task objectives listed for the super robot, six (or 50%) are not met yet today. At that time, robot durability was 20,000 hours. Today, it is 100,000 hours. Unfortunately, force disturbances from machining create deflections 20x higher than the manipulator's repeatability. No in-depth metrology has been provided by industry, resulting in a grossly incomplete model. The forecast was that in two to three decades, such a model would be feasible with real-time processors far superior to those in 1989 (which is now validated) making real-time computation (5 to 10 m-sec.) of the model possible, i.e., to make the manipulator "electronically rigid", with feedforward predictive compensation in terms of 114 model parameters. The paper briefly discusses the importance of actuators with a nominal forecast on

Carol Cockrell Curran Chair in Engineering Director, Robotics Research Group, The University of Texas at Austin, 1 University Station, r9925, Austin, tx 78712.
E-mail: tesar@mail.utexas.edu

their required development. This need is outlined in Ref. (Tesar and Butler 1989) where in-depth descriptions were given for modularity in mechanical structures (serial, parallel and mixed). This leads to the need for highly-certified, high-performance actuators in minimum sets (Intel in computer chips) to open up the robot's architecture, enable rapid repairs and updates, reduce costs and permit a universal evolving operational software (Microsoft for computer operating software) and a competitive supply chain (Dell for personal computers). In 2013, Tolio et al. suggests a need for rapid product changes based on complex standard product shapes using standard tools with an analysis based on industrial engineering principles. Also in 2013, Neto et al. lays out the theme for precision machining under force disturbances in a Portugal workshop highlighting the Comet project funded by the European commission (more in Section Robot Machining Processes). Finally, Barnfather et al. (2016) provides a comparison between expensive machine tools and compliant, low accuracy robots used for large vessel machining employing manipulators in multiple locations.

Keywords: Open architecture, robotic manufacturing

SUMMARY

Justification

There are fundamentally two levels of products that require reprogrammable manufacturing cells. Both of these cell classes are equally important economically. The first involves precision machining tasks for high valued products (aircraft, automobiles, household appliances, etc.). These high valued tasks are yet to be satisfied by sufficient precision under force disturbances in quickly reprogrammable cells (in seconds).

The low valued tasks are intended to reduce human drudgery in highly repetitive/simple/low precision/low force "assembly" tasks of standardized parts whose quality is assured in carefully calibrated mass production. These low value added quickly reprogrammable cells require that their life cycle cost must be brought down considerably. This means that it must be easily programmed by a high-school level technician and also maintained by plug-and-play of a minimum set (few spares) of low-cost standardized modules.

Hence, both of these cell classes must be assembled on demand from highly-certified components and one universal/evolving operating system (for each class which permits 10 sec. (or less)) changeover to meet product parameter changes. The dominance of one-off/long-duration/high-cost production in car industry (see App. B) is a clear indication of lack of both of these classes of cells. The low-valued cells would operate in food/game/household (etc.) industries. Economically, food production in the U.S. accounts for 2x (or more) of the value of car industry (see App. B). On the other hand, aircraft industries still require numerous/expensive/large /unchanging precision jigs to make airframe parts.

Two Leading Cell Programs

The Robotics Research Group at UTexas (Tesar 1989) has been developing a tech base for dexterous (industrial serial robots) manufacturing systems for 30 years. The second, more recent, effort is by the Stuttgart Fraunhofer Institute in terms of the European Commission funded project COMET (Schneider 2013, Lehmann et al. 2013, Schneider et al. 2014). The RRG group performed extensive metrology on industrial robots over several years (1985–95), obtaining 120(+) of 150 necessary parameters to create a relatively complete dynamic model of the manipulator. Two limitations became apparent: the first was the latency of the computation (more than 10 seconds) and completely inadequate actuator technology (both of which will be shown to be feasible today). The COMET project shows similar limitations: in particular: sensor limitations, lack of sufficient actuator responsiveness, inadequate modeling parameter definition, and small work volume tasks to prevent backlash crossover in the actuator gear trains.

Development Strategy

The goal should be to develop two in-depth demonstrators of both classes of cells to enable industrial decision makers to commit to major investment in component and system technologies, as was done in micro-electronics and all scales of computers from 1970–90. These systems must be modular—made up of highly-certified actuators (sensors, controllers, power supplies, link structures, end-effectors, etc.) in minimum sets (to reduce cost of production and design) in a competitive supply chain to enable rapid maintenance and refreshment in a continuously evolving tech base as we now have for personal computers and social media. It is now feasible to create these essential demonstrators in the next five years.

Required Development

This development will require the dedicated commitment of a carefully assembled team of researchers over a wide range of technologies. The following list of technologies will summarize those basic topics for high value added manufacturing cells (the concentration of this paper). Doing so for high valued tasks should increase the market for industrial robots by 5x.

Generally, it is recognized that industrial robots are 50x less stiff than precision machine tools. They are also 20x (or more) less precise under machining force disturbances. Machine tools operate under critically measured model parameters (metrology) with relatively simple (mostly linear) operating software. Unfortunately, large work volume machine tools are very expensive. Hence, the allure of large work volume industrial robots which, today, are low accuracy (20x less than their repeatability) under force disturbance, represent very complex highly-coupled input subsystems to make their operating systems very complex, where no standard manipulator is supplied with an accurate set of model parameters based on in-depth metrology. Table 1 is a listing of the ten most important development tasks.

ROBOT MACHINING LITERATURE

Overview

A major position paper was presented by Tesar (Tesar 1978) in the *Science* magazine in 1978 outlining the increasing weakness of the "light" mechanicals (including the

Table 1. Listing of required development tasks for high value added manufacturing cells.

Development Task	Technical Issues	Goal
1. Stiff Manipulator Link	Two major links must be stiff, both in torsion and in bending. Torsion usually is the most demanding.	10x better than actuator compliance.
2. Actuator Torque Density	5 or 6 actuators must produce highly complex duty cycle torques (once/sec.) to reduce volume and manipulator weight.	10x better than the SOA (state-of-the-art).
3. Joint Bearings	Standard rolling element bearings are too large and too compliant. Use cross roller or grooved roller bearings in shortest force path structures to reduce weight.	5–10x better than SOA.
4. Exceptional Actuator Gear Trains	Recent development at RRG/UTexas confirms a 5x improvement in load capacity; 10x higher stiffness (no rolling elements in the load path); and, virtually no backlash.	5–10x better than SOA.
5. Actuator Responsiveness To Command	To rapidly compensate for force disturbances (deformations, inertia forces), the gear train should contain almost no inertia and be fundamentally linear.	5–10x better than SOA.
6. Link/System Metrology	Suppliers of precision industrial robots must provide in-depth parameter descriptions for compliance, mass, and actuator values for up to 150 parameters.	Virtually none now exists as standard specifications.
7. Internal Actuator Sensors	Essentially, 10 sensors are nec. to describe in real-time the nonlinear nature of electro-mechanical actuators in order to accurately describe their performance.	Only 2 or 3 sensors are now commonly available.
8. External System Sensors	Smaller/stiffer force torque sensors, high quality responsive machine vision may be combined to accurately describe end-effector motion in real-time (< 1 m-sec.)	Development must continue to improve these external sensors by 2 to 3x.
9. Fast Computational Model Representation	All nonlinear parameters may be described in m-sec. with a total system description in 5 to 10 msec. This is now possible due to low-cost, high computational density controllers, which enable large work volume.	Present models are very simplistic. Dramatic improvement is necessary.
10. Decision-Making Criteria	Complex industrial manipulators are highly nonlinear, whose perf. must be numerically assessed in terms of 50 to 100 competing criteria. Each criterion must be ranked by the operator to best meet the objectives of the precision tasks.	100 criteria have been described, 50 have been demonstrated in near real-time ($\approx$ 10 m-sec) at RRG, UTexas.

strengthening role of industrial robotics) in the U.S. with significant trade imbalance at that time (1977). Today, both heavy and light machinery show significant negative trade imbalances. Mechanical component trade did well in 1997 with considerable university research in sensors and operational software (more on this in Section Dynamic Manipulator Modelling).

One decade later, Tesar presented a detailed forecast (Tesar 1989) for the fifth generation (the Super Robot) of industrial robots with emphasis on high value added light machining. The paper says that no light machining, no rapid product changeover, and no open architecture existed in 1989. Of the twelve major task objectives listed for the super robot, six (or 50%) are not met yet today. A that time, a robot's durability was 20,000 hours. Today, it is 100,000 hours. Unfortunately, force disturbances from machining create deflections 20x higher than the manipulator's repeatability. No in-depth metrology was provided by industry, resulting in a grossly incomplete model. The forecast was that in two to three decades, such a model would be feasible with real-time processors far superior to those in 1989 (which is now validated) making real-time computation (5 to 10 m-sec.) of the model possible—i.e., to make the manipulator "electronically rigid", with feedforward predictive compensation in terms of 114 model parameters. The paper briefly discusses the importance of actuators with only a nominal forecast on their required development. This need was implied in Ref. (Tesar and Butler 1989) where in-depth descriptions were given for modularity in mechanical structures (serial, parallel, and mixed). This leads to the need for highly-certified, high-performance actuators in minimum sets (Intel in computer chips) to open up the robot's architecture, enable rapid repairs and updates, reduce costs, and permit a universal evolving operational software (Microsoft for computer operating software), and a competitive supply chain (DELL for personal computers). In 2013, Tolio et al. suggests a need for rapid product changes based on complex standard product shapes using standard tools with an analysis based on industrial engineering principles. Also in 2013, Neto et al. lays out the theme for precision machining under force disturbances in a Portugal workshop highlighting the COMET project funded by the European Commission (more in Section Robot Machining Processes). Finally, Barnfather et al. (2016) provides a comparison between expensive machine tools and compliant, low accuracy robots used for large vessel machining employing manipulators in multiple locations.

Dynamic Manipulator Modeling

The Robotics Research Group at UTexas, Austin has a long history in the development of an analytic description of the system-level parameters of serial robot manipulators. Reference (Thomas and Tesar 1982, Fresonke et al. 1988) provide generalized models for mass and compliance parameters for N DOF systems using both rotating and sliding joints. This included joint and link compliances, link mass values, and actuator torque demands. During the 1980–90 decade, normal computation would have been inadequate. Wander and Tesar 1991 made an effort to count all computations and allot these to multiple parallel processors, with nominal success. Behi and Tesar 1991 made an early effort to use modal analysis to measure local mass and compliance parameters, which work well on machine tools but proved inadequate for highly-coupled non-linear manipulators. Then, Wander et al. 1992 show what the performance requirements would be for airframe manufacturing (precision hole drilling and rivet placement).

The use of precision, expensive jigs remains dominant for this task today. In Ref. (Tisius et al. 2009), the UTexas program published a paper describing 25 years of work on manipulator performance criteria (100 described, 50 operational) to assess performance based on operator-selected and ranked criteria for a very wide range of applications (Tisius et al. 2008). At that time (2005), Tesar argued the case for open architecture manufacturing systems with emphasis on high-valued operations. A paper in process (Tesar 2015) outlines what is meant by machine intelligence. Based on the ranked criteria, the decision system uses very fast algebraic computations (in 5 to 10 m-sec.) which essentially linearizes the system over the very short motion increment making unsolvable continua mathematics unnecessary. Computational intelligence structures the decision (serial, parallel, or mixed) to further prioritize the decision algebra to minimize computational requirements to dramatically reduce decision uncertainty under human command.

Manipulator Compliance

Clearly, the robot manipulator deflects under force disturbances (50x more than machine tools). This was recognized by the RRG/UTexas program in 1985 by Thomas et al. where a "real time analysis" was performed for the stiffness distribution that would represent the lowest overall system compliance. In 1989, Hudgens and Tesar (Hudgens and Tesar 1992) presented some early results on a ten-year metrology effort at UTexas describing some compliance parameters identified on a standard industrial robot. In 2005, Zhang et al. studied deformation compensation for robot cleaning of aluminum castings to remove operators from this unhealthy task. The result was a 5x improvement (from 0.02" to 0.004") in position accuracy but only in a small work volume relative to the size of the manipulator. In 2012, Dumas et al. (Wander et al. 1992, Dumas et al. 2012) made a detailed analysis of the first three joints (the last three remained locked) of a 2.7 m Kuka KR-240-2 manipulator (which has a good repeatability of 0.05" but low accuracy); very detailed deflection plots due to the first three joints under load are given. Variations of 2 to 9% of joint compliances predict up to 80% of the EE deformation under a static load. Finally, a detailed paper (Slavkovic et al. 2013) out of Belgrade, presents an unnecessarily complex set of analytics (compared to that in Ref. (Thomas and Tesar 1982, Fresonke et al. 1988, Hudgens et al. 1991). They perform XYZ simulations for numerous work volume plots with measured compliance of 0.001" under a load of 20 lb. In a small work space, they simulate motion errors of 0.004" and force errors of 0.04".

Robot Manipulator Mass Content

Virtually all later papers for light machining robots have some concern for robot mass content. Nonetheless, the potential inertia forces are kept purposely low (relative to machining forces) by moving the end-effector very slowly. Such analyses were given in Ref. (Thomas et al. 1985), where the mass content was used to size the actuators under reasonable dynamic motions.

Manipulator Joint Deflection Sensing

The issue of joint deflection for light machining robots is basic for machining force deflection compensation since most of the end-effector deflection (about 70 to 80%) is due to joint compliance. In 2006, Krishnamoorthy and Tesar outlined a detailed collection of ten sensors necessary to adequately describe the internal performance of each of the N actuators in the manipulator. The two most critical (and expensive) are the torque and position sensors at the rotary output plate of the actuator. High resolution and rapid numerical assessments can now be envisioned at each actuator. Also, each actuator must be fully described in terms of a minimum of 40 embedded performance maps to fully describe their nonlinear nature and to do so in 1 milli-secs., or less. This full description can be used to predict response to command, torque levels, accuracy, endurance, remaining useful life, efficiency, etc. In 2011, Dumas et al. provided a minimal set of joint compliance measurements with attractive contour charts (and expected measurement noise). The results show very low deviation/errors in stiffness values after five tests.

Robot Manipulator Metrology

The RRG at UTexas set up a full metrology effort for an industrial robot in 1985–95 as described in Ref. (Hudgens et al. 1992) by Hudgens et al. This used careful machine vision end-effector measurement for a large number of poses to identify real geometric and real compliance (link and joint) parameters. The goal was to identify up to 150 as-built parameters using carefully measured joint command input parameters. It must be noted that metrology (which provides all model parameters) is not calibration which measures a few parameters of interest in a restricted work volume of the manipulator. Calibration corrects locally where metrology corrects globally.

Actuator Development

The reality of modular mechanical systems is that they may require actuator replacement either due to failure or reduced performance or due to upgrades. Doing so with inaccurate mechanical interfaces would destroy the parametric description of the manipulator gained by careful metrology. Shin and Tesar (Shin and Tesar 2004), provide a detailed interface of high positional accuracy and very high stiffness to enable rapid actuator replacement with nominal effects on the system model-operational software.

In Ref. (Krishnamoorthy and Tesar 2006) Krishnamoorthy and Tesar, describe the necessity of multiple low-cost sensors in each actuator of a careful inference by Bayesian correlation to "mathematically replace" the existence of the failed sensor. Finally, Richard et al. (2011) describe in an unpublished report how to make actuator modules with higher stiffness using flexure joints. These flexure joints show exceptional complexity, volume, and potential weight. They do, however, provide low backlash for small motions (and low stiffness).

The RRG group at UTexas has pursued actuator designs and prototyping since 1975, with real growth of 4 to 8 orders of magnitude in the last two decades. Much of this development for a wide class of low complexity, multi-speed, high resolution, high torque density, high stiffness, and high responsiveness remains privileged and is being documented in a full patent portfolio.

Robot Machining Processes

The opportunity to use the large work volume to perform precision machining tasks would revolutionize the industrial robot marketplace and should expand its value by up to 5x (today, the U.S. market holds steady at $1 bil/lyr.). One of the first requirements is to provide real-time (in 5 to 10 m-sec.) compensation for machining force-generated deflections. Given that capability (some outlined in the previous sections), it, then, becomes desirable to visually assist the operator in assessing expected performance of the manipulator throughout its work volume. This has been achieved by simulation in an unpublished report by Chang and Tesar (Chang and Tesar 2006) at UTexas in 2006.

The work by Pan et al., in 2006, shows that the depth of cut (DOC) and the width of cut (WOC) greatly influence the potential for tool chatter which causes destructive manipulator vibrations. Rather preliminary results for chatter control are presented for a very small work volume using an ABB IRB6000 large industrial robot. A realistic paper by De Vlieg (De Vlieg 2010) at Electro—impact shows the application of industrial robots in airframe assembly with an overall task accuracy of ± 0.03" making necessary an accuracy of ± 0.01" by the manipulator. An accuracy of ± 0.006" is considered very desirable. On line model-based programming is essential to reduce operator time demands. An unmeasured 10 ft. industrial robot would provide values of 0.08" to 0.16" over its work space. Using a first level of geometry/compliance metrology gets this down to 0.02" or 4 to 8x better, getting close to the system's repeatability. Two special demonstrations got these values down to ± 0.01" and ± 0.005". A second paper in 2011 by Electro—impact (Rodberg 2011) authored by Rodberg, shows that it is feasible to lay out large volume complex shapes made up of composite fiber layups.

A rather undocumented paper by Do et al. (Do et al. 2012), describes a lightweight dual arm 14 DOF manipulator system capable of generating 12 lb. force between the end-effectors. Chen and Dong (Chen and Dong 2012), looks at issues of path planning, stiffness modeling, and chatter-induced vibrations in an attractive survey paper with numerous references. They describe a 50" ABB manipulator as having 0.002" repeatability with a machining accuracy of 0.04", i.e., 20x worse in a rather small work volume. Some deflection compensation improves this accuracy by 10x.

Dogar et al. (Dogar et al. 2013), provide a rather general discussion of the assembly problem with no numerical documentation. Another paper of this class by Cheng and Chen (Cheng and Chen 2014) uses a quality 'adult' robot to teach a motion plan to a lower cost simpler robot using predictive analytic MARKOV learning processes. The process is a simple tube in a hole insertion with a necessary accuracy of 0.002". Simple $\Delta\chi$ movement is accurate to 0.02".

The next five papers (Schneider 2013, Lehmann et al. 2013, Schneider et al. 2014a, Karim and Verl 2014, Schneider et al. 2014b) deal with the important light machining project COMET at Stuttgart's Fraunhofer Institute and the University of Stuttgart. This appears to have been led by Ulrich Schneider (Schneider 2013, Schneider et al. 2014a) with overview given by C. Lehmann et al., expressing concerns for gearing nonlinearity and backlash. In Ref. (Schneider 2013) Schneider, describes the merits of a parallel piezo-micro-manipulator with a motion range of 0.02" to correct for small end-effector deflections. Unfortunately, this micro-manipulator will also generate force disturbances to generate deflections in the supporting manipulator.

Reference (Schneider et al. 2014a) is a detailed description of the COMET project. They report that end-effector accuracy is 0.04" relative to repeatability of 0.004", which is 10x better than that for most industrial manipulators. They also suggest that link deflections are small relative to actuator/joint deflections. They are deeply concerned about gear backlash, gear nonlinearity, gear inertia, and crossover shocks due to gear backlash (note that 20 years of work at UTexas have greatly reduced these actuator-related weaknesses). The COMET project was able to perform numerous tests using primarily add-on solutions without any significant physical modifications of the industrial robot manipulator obtaining a general improvement of 3 to 5x better machining accuracy in small work volumes. Because of the high gear inertia in the actuators, rapid force disturbances cannot be compensated. They recommend calibration within the desired work volume to cancel basic geometric errors. A useful actuator deflection chart shows a 2x higher stiffness on one side of a significant jump in the actuator output at the crossover highlighting the nonlinear nature and backlash of the existing gear trains. Rapid oscillations from programming shocks or from backlash shocks cannot be eliminated. This appears to restrict their demonstrations to rather small work volumes where oscillations are less likely. Reference (Karim and Verl 2014) gives a description of composite machining for auto components, aircraft components, and for wind turbine blades, where machining forces are quite low. They recommend the large Kuka 500-3, MT for these fully integrated work cells where the actuator gears are 2x stiffer than for most Kuka robots. They use a very high speed spindle with quick automatic tool changer to maximize up-time and cryogenic cooling to reduce tool wear. Reference (Schneider et al. 2014c) by Schneider gives a general description of the problem recommending an accuracy of 0.002" for many robot machining tasks. The paper discusses standard path planning issues with some effort on local calibration by trial and error optimization in a small work volume. Overall, the paper proposes a machining accuracy improvement of 10 to 30x over standard robot accuracy.

Wells and Walker, of Electro-impact (Wells and Walker 2014) describes using a Kuka robot manipulator for laser cutting (i.e., no cutting forces) in a large work volume with an accuracy of ± 0.015" which is 3x worse than its repeatability of ± 0.005". To do this, they do not use the last rotary actuator and note that too many long actuator torque tubes exist to provide high stiffness at the end-effectors. Finally, Barnfather et al. (Barnfather et al. 2016) of Sheffield, U.K., describe a general mathematics formulation to create a real-time model for industrial robots in the same manner that has been pursued for machine tools. No numerical results are given.

Operational Sensors

Giraud and Jouvencel (Giraud and Jouvencel 1995) gives an early (1995) elementary description of the need for sensor data management. Xiong and Svenson (Xiong and Svenson 2002) provide a long paper (2002) describing the use of predictive analytics (not real time) to perform data management from an array of sensors. In 2006, Kolba and Collins give a similar description for sensor data management using a discrimination-directed search to improve data resolution. In 2010, Ashok et al. (Krishnamoorthy et al. 2011, 2012) provide in-depth analytical tools for simultaneous sensor and process management in multi-input, multi-output (MIMO) systems with emphasis on near real time. This applies to both internal actuator sensor subsystems and to external manipulator guidance sensors. One of the principal objectives is to minimize data reduction errors.

Error Reduction Micro-manipulators

It is reasonable to presume that layered control can be used to have a small scale intelligent device which could rapidly compensate for errors in a large scale sluggish device such as a large robot manipulator. The RRG UTexas program performed in-depth design and modeling of a 6 DOF parallel micro-manipulator (Hudgens et al. 1991) for that purpose in 1988–91. The 6-legged structure used necked-down flexible joints driven by 6 small geared D.C. prime movers in a compact package with a motion range of 0.2" in a module that was 4" high and 6" in diameter. The program did not get to a demonstration stage because it could not be assured of creating a reverse disturbance on the supporting manipulator. Schneider et al. (Schneider et al. 2014c), continued their work on the micro-macro piezo manipulator in an extensive test facility with a measured accuracy under load of 0.004" in steel, which is very good (8x better than normal). The piezo device range is 0.004" to 0.007" with no backlash. The test is for a machining tool supported by the piezo module. Hence, it is not proven that the piezo module does not disturb the supporting manipulator to essentially cancel the benefits of the piezo module.

Real-time Machining Compensation

A series of detailed papers on real-time machine modeling and deflection compensation were presented by the RRG at UTexas (Tesar 1989, Hernandez et al. 1989, Hudgens et al. 1991, Tosunoglu et al. 1990) from 1989–91. This work showed the need for real-time computation in the super robot (Tesar 1989), the real-time computation to accurately represent force induced end-effector deflections due to joint compliances (Hernandez et al. 1989), the question of stability due to forced vibrations, stability in force control (Tosunoglu et al. 1990), and software controller design for flexible manipulators (Tosunoglu et al. 1990). Also, Tesar 2000 described how all this was relevant to products and systems at the human scale. Finally, Shabbir et al., used machine vision for end-effector error compensation to statistically reduce errors relative to commanded output. They claim a position improvement of 7x and a rotation improvement of 4 to 8x on an ABB (model IRB-4400) with a laser tracker system from Leica (At901-MR). This calibration procedure is not carried in real time to compensate for machining induced deflections. Reference (Tesar 2015) provides a decision structure based on machine intelligence (to locally linearize these highly coupled and nonlinear systems) and computational intelligence to structure the decision algebra to best benefit from the serial geometry of the serial manipulator. All of this means that rapid response to force disturbances is dramatically enhanced by this modern decision process for mechanical systems, benefitting by ever-more cost-effective and powerful computation-based controllers.

HIGH ACCURACY MACHINING CELLS

Concept of Intelligent Machines

The benefits of the massive technology associated with computers can now be expanded by changing the basis for machine control from analog and stability algorithms to criteria-based decision making such that task performance, condition-based maintenance, and fault tolerance become possible for complex production systems such as 40 DOF precision assembly cells for airframe manufacture. This generic approach would also apply to all intelligent machines such as aircraft, automobiles, harvesting equipment, medical equipment, etc. (Tesar 2015).

Most mechanical systems are based on an outgrown control paradigm associated with the criteria of stability and a few ancillary criteria such as overshoot, settling time, and steady state error. Not only are these criteria irrelevant to the critical operation of most high value added production systems, issues such as task performance (precision, force, obstacle avoidance, etc.), condition-based maintenance (when should a component be replaced to maintain system performance) and fault tolerance (operation even under a fault) cannot be addressed by this out-dated approach to control. The successful fly-by-wire approach used in fighter aircraft shows that criteria-based decision making not only works but that it is essential to generalize the architecture of production systems, make agile manufacturing feasible for high value added operations including advanced manufacturing cells, and to reduce life cycle cost.

Figure 1 describes what is meant by this new continuum of machine operation. Each operational concept (task performance, condition-based maintenance, and fault tolerance) is based on a "residual" (or difference) between a predicted model reference (based on a parametric description of how the "as built" machine should perform) with a sensor reference (based on actual parameters measured by distributed sensors with

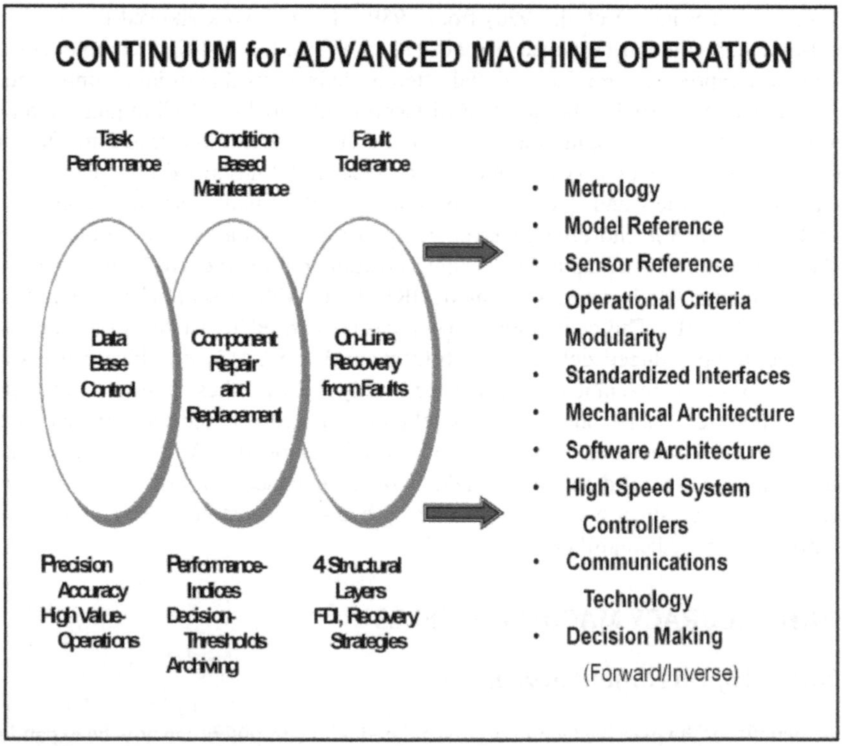

Figure 1. Continuum for advanced machine operation.

the system). This difference model then can be used by the decision making software to maximize performance, to identify faults and to recommend the best configuration to mask the fault, or to recommend the replacement of a component which is adversely affecting the system's performance. To obtain these benefits, a massive development of foundation technologies such as metrology, operational criteria, decision making, modularity, communications technology, etc. must be undertaken. Few of these foundation technologies now exist or are being developed or taught in our academic institutions. It is also necessary to mention that U.S. federal research funding for manufacturing has provided virtually no support for this revolutionary but essential technology. For example, it can be forecast that condition-based maintenance will be common to automobiles within this decade. Why not now also vigorously pursue this same technology for production systems and bring excitement back to our manufacturing industry and to the discipline of mechanical engineering.

Manufacturing Cell Development

Abstract

Manufacturing cells represent the full integration of all component and system technologies to form a self-contained system capable of process control, configuration management, rapid product change over, quick change out of cell modules for repair and tech mods, and to do so at reduced costs and enhanced performance (See Fig. 2).

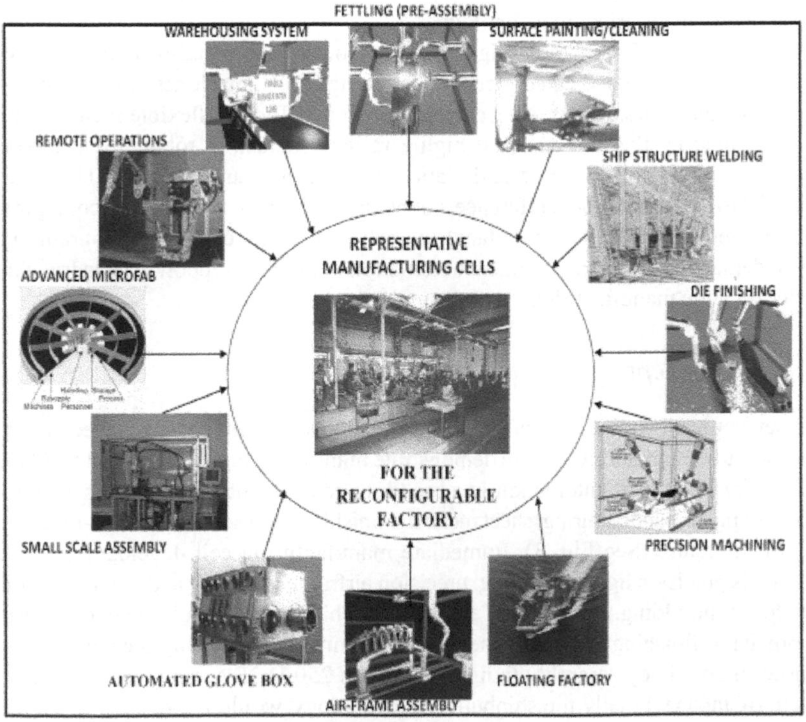

Figure 2. Representative manufacturing cells.

For example, an airframe assembly cell might contain 40 DOF involving two active assembly robots (7 DOF each), two supporting force robots (five DOF each), and four rigidized fixturing robots (four DOF each) all assembled on demand from standard modules (links, actuators, electronic controllers, operational software, etc.) exhibiting high performance (accuracy) and reduced life cycle cost (rapid module replacement and up-dates) and also moving towards condition-based maintenance (to permit operation by a nominally trained technician) (Tesar 1989, Lehmann et al. 2013, Schneider et al. 2014a).

Background

The need for flexible automation was demonstrated in 1995 when U.S. auto plants were strained because of market demand. The inability to surge rapidly to meet this demand has highlighted a weakness in our manufacturing technology. What is required is a national effort for Next Generation Manufacturing (NGM), similar to the COMET project in Europe (Schneider et al. 2014a). To do so will require a level of integration of diverse technologies far beyond that being utilized today. For example, the commonly implemented fly-by-wire operation of military aircraft rarely occurs in production systems. The need for intelligent automation has crystallized around the concept of "agile manufacturing" which essentially means the rapid response of the production system to market demands. This implies a top-to-bottom revolution in the "intelligence" associated with manufacturing systems allowing direct intervention either by an output from a market driven database or by the human operator whose skills have been honed by a supportive and friendly technology.

The concept of manufacturing cells came out of the metal-machining systems first assembled using dedicated precision machines (N.C. machines, automatic screw machines, conveyors, robots, etc.) during the 1960's. Today's flexible manufacturing is represented by the low cost and highly reliable industrial robot. It is, however, illustrative of an unresponsive and static approach to manufacturing. The present industrial robot has no model reference, cannot be rapidly interfaced to a reconfigurable and parametrically accurate manufacturing cell and cannot change its program in the cell on demand. This is not intelligent automation nor does it support what is frequently labeled as agile manufacturing.

Program Concept

This development outline will concentrate on all the component and system technologies necessary to design, fabricate, and demonstrate manufacturing cells from 14 to 40 DOF suitable for intelligent automation in a broad range of applications such as airframes, shipbuilding, remanufacturing, sheet metal die finishing, and advanced micro-electronics chip manufacture (See Fig. 3). Immediate manufacturing cell development for the program is precision light machining, precision airframe assembly and modular welding cells for shipbuilding. Elements of the light machining cell would be equally useful for precision finishing for auto panel dies. The airframe assembly technology could be used in auto body assembly (a first generation 250 DOF system has been built by Nissan of Japan). Finally the shipbuilding technology would prove equally useful in an advanced microchip fabrication facility or in fettling cells for product clean-up, painting, and polishing.

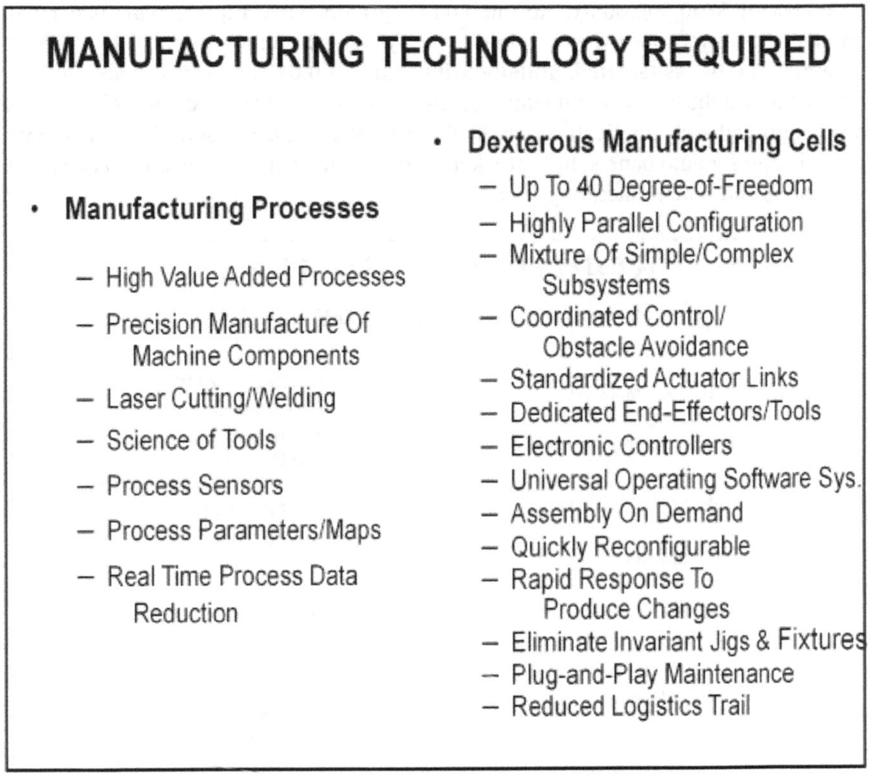
MANUFACTURING TECHNOLOGY REQUIRED
• Manufacturing Processes
– High Value Added Processes
– Precision Manufacture Of Machine Components
– Laser Cutting/Welding
– Science of Tools
– Process Sensors
– Process Parameters/Maps
– Real Time Process Data Reduction
• Dexterous Manufacturing Cells
– Up To 40 Degree-of-Freedom
– Highly Parallel Configuration
– Mixture Of Simple/Complex Subsystems
– Coordinated Control/ Obstacle Avoidance
– Standardized Actuator Links
– Dedicated End-Effectors/Tools
– Electronic Controllers
– Universal Operating Software Sys.
– Assembly On Demand
– Quickly Reconfigurable
– Rapid Response To Produce Changes
– Eliminate Invariant Jigs & Fixtures
– Plug-and-Play Maintenance
– Reduced Logistics Trail

Figure 3. Manufacturing technology required.

The Development of a Class of Modular, Dexterous Manipulator Structures For High Accuracy Manufacturing Cells

Program Objective

This paper will outline an advanced collection of component technologies (actuator modules, electronic controllers, operational software, standardized interfaces, etc.) to dramatically improve positional accuracy of robot manipulators under database control even under load or external disturbances. Demanding military applications remain unfulfilled because of the extraordinary cost of prototype development and evaluation. Hence, the overall deployed cost must come down (at least by a factor of 2) and the performance must go up by several orders of magnitude. For some industrial applications, a speed increase of 10 times is essential. For military and space applications, development costs must come down by a factor of 10. Unfortunately, almost all industrial robots cannot produce accurate motions from a computer command (positional errors of 0.1" are common) which means they can only be used for low valued, imprecise operations or have to be taught in place (a very expensive and demanding procedure) to repeat the same motion every cycle which means the system cannot adapt to any changes in either the product or the process. This reality is borne out by the fact that the auto industry spends 4x the cost of the robot to integrate it into their operations (at a cost of $800,000,000/yr. to one company alone (See Fig. 4). This paper will provide suggestions to not only improve the system's accuracy (for an 84" reach, an accuracy of 0.02" is targeted, initially, with a goal of 0.01" in the reasonable future), reduce the weight and cost of the manipulator, but also to generalize this technology to make it immediately applicable to 40 DOF manufacturing cells for airframe assembly, die finishing for auto panels, light stock machining, etc., with emphasis on batch mode operations and remanufacturing (See Fig. 5).

ROBOT INTEGRATION

(In Automotive Assembly Lines)

1. **EXPERIENCE AT MAJOR AUTO MAKER**
   - **Purchases 1000 Robots at a Time** (4000 Units Per Year)
   - **Three Suppliers**
   - **Low Cost** ($45,000 Each)
   - **High Reliability** (Up to 90,000 Hours)
   - **Great Smoothness**
2. **INTEGRATION PARAMETERS**
   - **Long Installation Time**
   - **No Absolute Accuracy**
   - **No Data Base Control** (Extensive Teaching In Place)
   - **Failure Means Robot Replacement** (Recalibration, Reteaching in Place)
     – 4X The Robot Cost ($200,000)
3. **FEASIBLE INTEGRATION GOALS**
   - **Cut Integration Costs by 50%**
   - **Savings Would Be $400,000,000 Per Year**
   - **Parametric Model For Accuracy** (0.01" req.)
   - **Develop Modular Technology** (Rapid Repairs and Tech Mods)

Figure 4. Robot integration.

Program Concept

A series of development efforts will be suggested for a structured design procedure for an optimum selection of link, actuator, controller, and software modules, etc., to best meet a given set of operational requirements. A full database on commercially available technology will be pursued. One initial reality from the database on gear trains is that the best is from Japan, costs 3x less and is 3x more accurate than that available in the U.S.[1] Another major requirement would be to develop a science of design, while also designing prototypes. This includes the design in small packages (actuator and link modules), standardized interfaces (electronic and structural), configuration management at all levels (gear trains, sensors, actuator scaling, distribution of parameters, etc.) to reduce the parameters faced by the designer (synthesis) and measurement techniques (metrology) to document as built (not only as designed). The goal is to create advanced component technologies (sensors, gear trains, interfaces, etc.), integrate them in a full

[1]Actuator work at UTexas now exceeds this actuator tech base by 2 to 4 orders of magnitude.

collection of modules with standardized interfaces which can be recombined to meet a broad range of precision, high performance tasks for both industrial and military applications. Even though the performance is very high, costs will go down because only a few distinct modules (minimum sets) will be sufficient to populate all conceivable systems (even 40 DOF cells for airframe assembly) and to integrate these modules on demand to meet the customer's requirements.

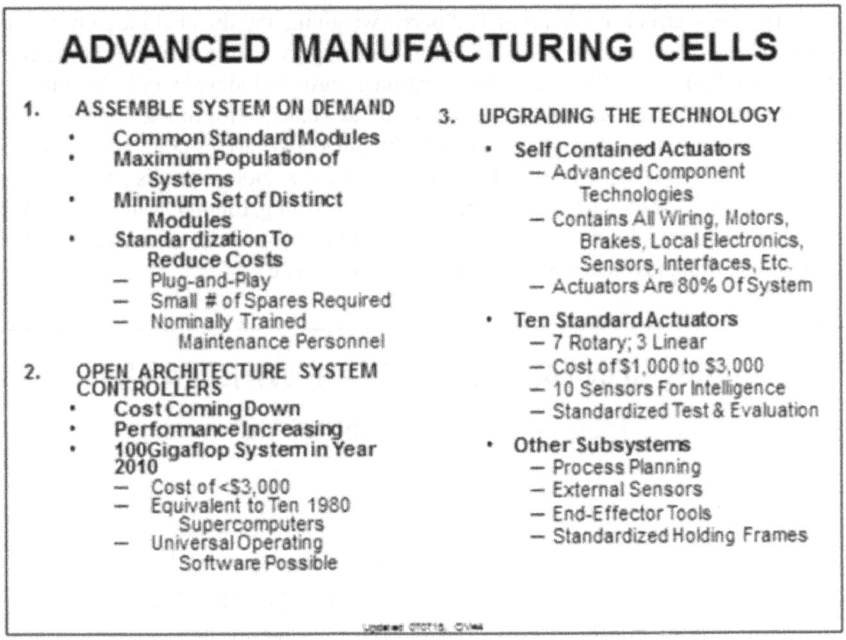
ADVANCED MANUFACTURING CELLS
1. ASSEMBLE SYSTEM ON DEMAND
• Common Standard Modules
• Maximum Population of Systems
• Minimum Set of Distinct Modules
• Standardization To Reduce Costs
– Plug-and-Play
– Small # of Spares Required
– Nominally Trained Maintenance Personnel
2. OPEN ARCHITECTURE SYSTEM CONTROLLERS
• Cost Coming Down
• Performance Increasing
• 100Gigaflop System in Year 2010
– Cost of <$3,000
– Equivalent to Ten 1980 Supercomputers
– Universal Operating Software Possible
3. UPGRADING THE TECHNOLOGY
• Self Contained Actuators
– Advanced Component Technologies
– Contains All Wiring, Motors, Brakes, Local Electronics, Sensors, Interfaces, Etc.
– Actuators Are 80% Of System
• Ten Standard Actuators
– 7 Rotary; 3 Linear
– Cost of $1,000 to $3,000
– 10 Sensors For Intelligence
– Standardized Test & Evaluation
• Other Subsystems
– Process Planning
– External Sensors
– End-Effector Tools
– Standardized Holding Frames

Figure 5. Advanced manufacturing cells.

PRESENT DEVELOPMENT ACTIVITY FOR MACHINING CELLS

Development of A Reprogrammable Assembly Cell for Airframe Manufacture

Objective

The goal of this paper is to create a precision assembly cell of 40+ Degrees-Of-Freedom (DOF) driven by advanced actuator modules under the control of criteria based performance system software where the requirements are established on-line by the operator, in keeping with a CAD data-base of the airframe component being assembled. The total assembly cell would be made up of a minimum number of distinct actuators (perhaps 5) to create a very large population of modular subsystems (3 DOF fixtures, 4 DOF force robots, 7 DOF dexterous positioning robots, etc.). This assembly cell and its operating system would itself be assembled on demand to meet the class of airframe assembly being considered. Overall, the objective is to reduce assembly cost (50% of the airframe cost) to 60% of its present level, allow rapid assembly set ups without specialized fixtures, and allow timely design changes with almost no additional cost.

Background

Large production runs of aircraft are a thing of the past. Fewer airframe companies will be asked to assemble fewer of any given airplane design and to be prepared to assemble several different airframes in the same facility. Knowledgeable airframe manufacturing experts suggest that 50% (or more) of the cost of the airframe is due to the assembly effort (based on people, space, and equipment resources required). For example, the F18 C/D Nose Barrel is made up of 120 parts (weighing 350 lbs.) held together by up to 1000 rivets. The Nose Barrel is 7 ft. long and roughly 3 ft. in diameter at the front and 4 ft. in diameter at the back. The machined parts including rivet holes must be accurate to ±0.01 inch. All holes must be round to +0.002" (for hi loks) and +0.005" (for rivets). The complexity of the final assembly looks like an egg-crate making access for riveting a critical issue late in the assembly process. Note that this same class of technology could be used to make a machining cell (drilling, routing, grinding, etc.) for light stock such as airframe skins, or a welding cell for ship structures or truck bodies, a remanufacturing task such as deriveting, etc. (See Fig. 6).

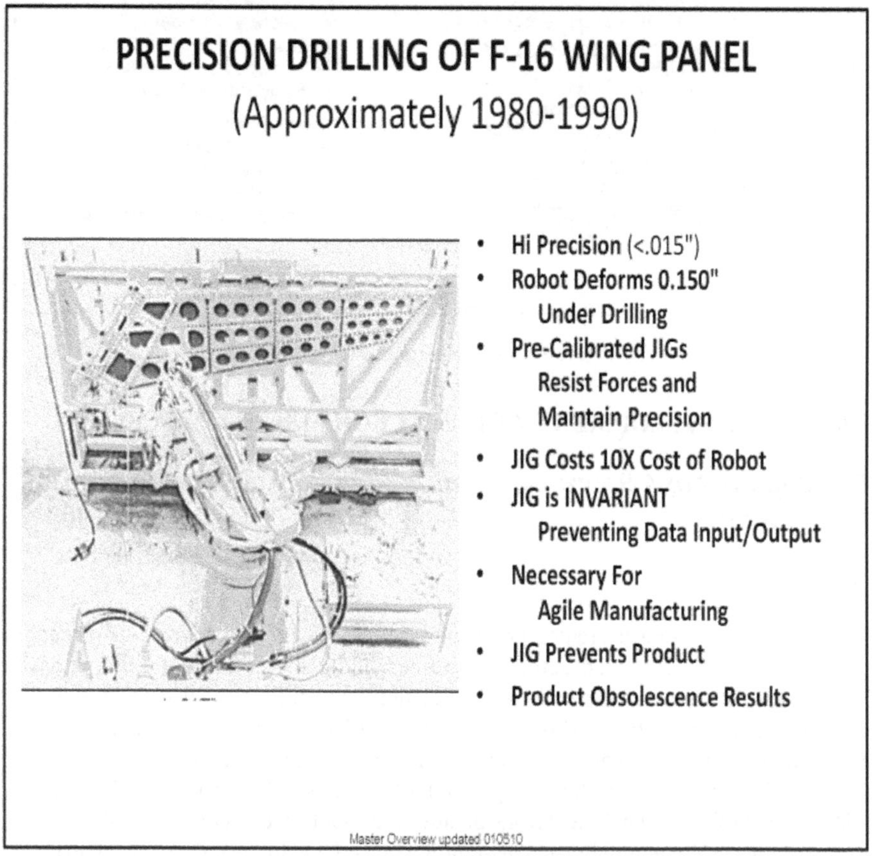
PRECISION DRILLING OF F-16 WING PANEL
(Approximately 1980-1990)
• Hi Precision (<.015")
• Robot Deforms 0.150" Under Drilling
• Pre-Calibrated JIGs Resist Forces and Maintain Precision
• JIG Costs 10X Cost of Robot
• JIG is INVARIANT Preventing Data Input/Output
• Necessary For Agile Manufacturing
• JIG Prevents Product
• Product Obsolescence Results
Master Overview updated 010510

Figure 6. Precision drilling of F-16 wing panel.

Suggested Development

It is now possible to develop the necessary spectrum of actuators (perhaps 5) to populate all 40 active joints in the assembly cell. Central to this facility would be two 7 DOF accuracy robots capable of creating working forces (of say 200 lb.) while maintaining accurate end-effector position to 0.01" in the work volume. An early version of this 7 DOF system was developed at The University of Texas under a DARPA grant. The generic actuator module uses the best component technologies (sensors, brakes, motors, gear trains, etc.) available today (some developed during this project). The 7 DOF positioning robot would place the parts in the work volume, put in tack rivets, and hold a riveting gun to perform the final assembly of the part. Because all of these parts deform under load, a 4 DOF force robot (2 would be required) would create an opposing force to that of the riveting device to prevent local deflection of the subassembly. The whole assembly would be held by up to six 3 DOF fixture robots holding the total assembly at various hard points to maintain its position in space. All 40 DOF would be controlled by a system controller whose operating software would be directly responsive to the CAD database of the assembly itself (See Figs. 7, 8).

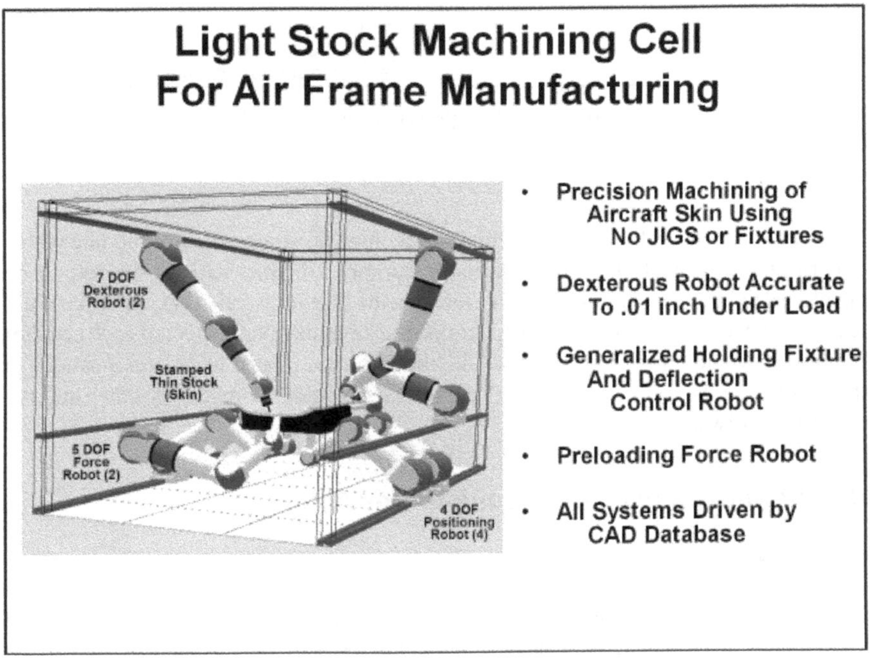
Light Stock Machining Cell
For Air Frame Manufacturing
7 DOF Dexterous Robot (2)
Stamped Thin Stock (Skin)
5 DOF Force Robot (2)
4 DOF Positioning Robot (4)
• Precision Machining of Aircraft Skin Using No JIGS or Fixtures
• Dexterous Robot Accurate To .01 inch Under Load
• Generalized Holding Fixture And Deflection Control Robot
• Preloading Force Robot
• All Systems Driven by CAD Database

Figure 7. Light stock machining cell.

CELL #4: ROTATING HOLDING FRAME WITH TWO FORCE AND TWO ACCURACY ROBOT PROTOTYPES

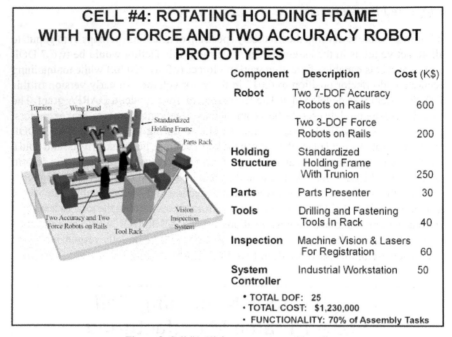

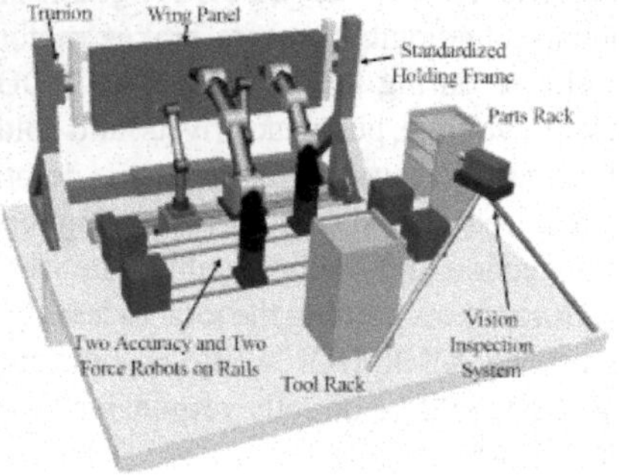

Component	Description	Cost (K$)
Robot	Two 7-DOF Accuracy Robots on Rails	600
	Two 3-DOF Force Robots on Rails	200
Holding Structure	Standardized Holding Frame With Trunion	250
Parts	Parts Presenter	30
Tools	Drilling and Fastening Tools In Rack	40
Inspection	Machine Vision & Lasers For Registration	60
System Controller	Industrial Workstation	50

- **TOTAL DOF: 25**
- **TOTAL COST: $1,230,000**
- **FUNCTIONALITY: 70% of Assembly Tasks**

Figure 8. Cell #4: High accuracy assembly cell.

Program Output

The principal deliverable would be a high force, high accuracy actuator module with a generic electronic controller for that actuator. Perhaps 5 distinct actuator sizes would be developed to fully populate all the active joints in the assembly cell. Also, a measurement technology would be put in place to determine the "as built" parameters of each actuator module. Its standardized interfaces would allow it to be placed in several locations in the cell where the on-board "as built" parameters would automatically enter into the operating software of the system.

Recent Data on Industrial Manipulator Accuracy

Objective

The goal is to use recent test data to evaluate the status of standard industrial robot manipulators in producing accurate positioning for precision tasks under load. Generally, it has long been known that robots are repeatable, but their absolute accuracy enabling computer commands to perform variable tasks is far below what is needed in many high valued industrial tasks. This recent data continues to support this established assessment.

Background

Reprogrammability of dexterous machines requires that they be absolutely accurate (to some acceptable tolerance) under computer command. Presently, standard industrial robot manipulators are repeatable only if all conditions of their operation (move sequence, temperature, load, speed, etc.) remain the same. This repeatability is remarkably high (approximately 0.001 inch), which strongly suggests high quality gear transmissions and joint bearings. But, their absolute accuracy (0.1 to 0.2 inch) under full load conditions will be 100 to 200x less than their inherent repeatability. This limitation is the principal barrier to the deployment of robots for complex sequential assembly which must be rapidly reconfigured to meet product design changes. Also, precision light machining tasks under load (drilling, routing, trimming, cutting, etc.) cannot be performed without the assistance of jigs and fixtures. To open up the factory to rapid product changeover, to frequent redesigns, to assembly with error recovery, to less dedicated open architecture systems (to reduce the threat of obsolescence), etc., then, robot manipulators must be made absolutely accurate. High valued operations are not now feasible, such that the return on investment for the robot supplier remains far too low. This is why U.S. sales of robots remain at the relatively low value of $1.0 billion/year.

Analysis of Recent Data

A recent study by Airbus UK[a] supplied the data presented in Figs. 9, 10 for four European industrial robots. The full load capacity of each manipulator was placed at its Tool Center Point (TCP) and that the robot was posed in an extended position for this measurement (worst case configuration). Also, numerous tests were run to reduce the influence of measurement error. A dynamic linear force was applied to the robot in five distinct configurations and the deflection/accuracy results were averaged. The manipulator accuracy deteriorates when the scale of the move (under load) goes from 0.004 inch up to 20 inches. Robot D does very poorly beyond 2 inches, while the other robots do quite well in this small move test (an accuracy of about 0.002 inch). The reference "desired" accuracy was 0.008 inch. Repeatability is 6 to 10x better than this number. Full envelope accuracy (no compensation) under load, however, yields results which were 7 to 33x above this desired accuracy. A restricted envelope with modest PID compensation improved these results to 5 to 14x above the desired accuracy level. Static deflection was 10 to 34x greater, while dynamic deflection was 15 to 32x greater than the desired value. Robot drift independently created a lack of accuracy that was 4 to 7x more than the desired level. But, the gear train backlash exceeded the desired accuracy by 4 to 10x. All of these results strongly support the long-standing contention that industrial robots cannot presently perform high valued functions in a rapidly changing product environment.

Proposed Development

Since 1985, The University of Texas has proposed the metrology of industrial robots. This means that every key as-built parameter (6 stiffness coefficients per link, 6 geometric parameters per link, 6 temperature expansion coefficients per link, etc.) must be identified in a reliable measurement procedure. This was done for a T3-776 in 1990 for 125 of its 146 parameters (stiffness, geometry, inertia). It was shown that

[a] Robot Capability Test and Development of Industrial Robot Positioning System for the Aerospace Industry, M. Summers, AeroTech, SAE, Dallas, TX, 2005.

STATUS OF INDUSTRIAL ROBOT POSITIONING ACCURACY

I. EXCELLENT ROBOT REPEATABILITY
- **At 0.001 inch**
  - Under Unchanging Conditions
  - Move Sequence
  - Load, Speed, Temperature
- **Excellent Mechanical Components**
  - Gear Trains, Bearings
  - Output Position Sensors

II. POOR ABSOLUTE ACCURACY
- **100 to 200x Above Repeatability**
  - From 0.1 inch to 0.2 inch
- **Systems Are Not Reprogrammable**
  - Performs Only Low Valued Tasks
  - Data Base Control Not Feasible
  - No Rapid Product Changeover

III. HIGH ACCURACY FOR HIGH VALUED TASKS
- **High Return On Investment**
  - Precision Light Machining
  - Drilling, Routing, Grinding, Trimming, etc.
  - No Jigs and Fixtures
  - Sequential Assembly With Error Recovery
- **Open Architecture Systems**
  - Reduced Impact of Obsolescence
  - Rapid Tech Mods
  - Diffusion of Technology
  - Plug-and-Play Repair
- **Basis For Enlarged Robot Market**
  - Today at $1.0 Billion/Year
  - Move To High Margin Applications
  - Aircraft Parts Preparation
  - Complex Sequential Assembly

Figure 9. Status of industrial robot positioning accuracy.

accuracy could be improved by 18x.[b] Today's laser measurement technology makes this system metrology more time efficient and cost effective. Because of the complexity of this system-level metrology effort, it is highly recommended to perform the required metrology at the individual link and actuator module level,[c] which would dramatically improve the parametric accuracy as well as permit an automated measurement strategy. To do so requires stiff and accurate standardized interfaces to assemble the modular manipulator structure while preserving the value of the parameters measured at the module level.[d] Hence, a pathway to accuracy in open architecture industrial robots is available. First, the inherent repeatability of today's industrial robot confirms the

[b] Static Robot Compliance and Metrology Procedures with Application to a Light Machining Robot, J. Hudgens and D. Tesar, UT Austin, Report to Texas Higher Education Coordinating Board for ATP under Project No. 003658-156 and DOE Grant No. DE-FG02-86NE37966, Aug. 1992.

[c] Robust Metrology Procedures for Modular Systems Using Indoor GPS Metrology System, D. Tesar and S. H. Kang, UT Austin, Report to DOE, December 2004.

[d] Analytic Integration of Tolerances in Designing Precision Interfaces for Modular Robotics, S. Shin and D. Tesar, UT Austin report to DOE under grant DE-FG04-04EW37966 and ATI subcontract, May 2004.

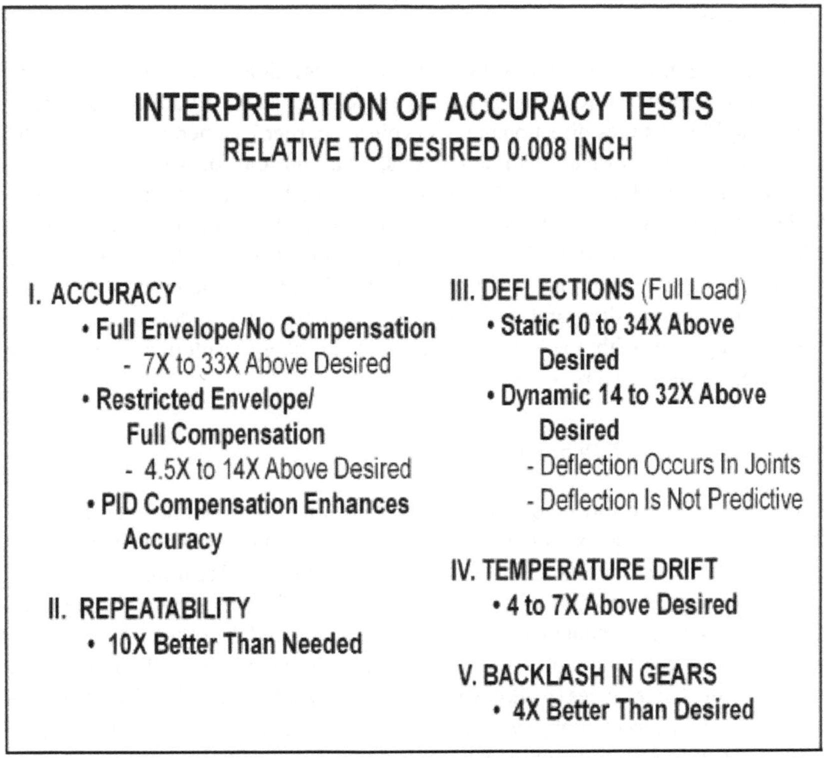
INTERPRETATION OF ACCURACY TESTS
RELATIVE TO DESIRED 0.008 INCH
I. ACCURACY
• Full Envelope/No Compensation
- 7X to 33X Above Desired
• Restricted Envelope/ Full Compensation
- 4.5X to 14X Above Desired
• PID Compensation Enhances Accuracy
II. REPEATABILITY
• 10X Better Than Needed
III. DEFLECTIONS (Full Load)
• Static 10 to 34X Above Desired
• Dynamic 14 to 32X Above Desired
- Deflection Occurs In Joints
- Deflection Is Not Predictive
IV. TEMPERATURE DRIFT
• 4 to 7X Above Desired
V. BACKLASH IN GEARS
• 4X Better Than Desired

Figure 10. Interpretation of accuracy tests.

quality of its mechanical structure. Real time compensating software (for the effects of loads, inertia forces, temperature, etc.) can, now, dramatically improve their accuracy to perhaps a factor of 2x their repeatability. Doing so would modernize much of our manufacturing and move us towards high valued product development, which the robot industry badly needs to expand its market beyond its present $1 billion level.

EUROPEAN COMET PROJECT[2] (IMPROVING ROBOTIC MACHINING ACCURACY)

Overview

COMET was a European project on using industrial robots to perform high valued precision machining tasks. A major development effort was pursued at the Fraunhofer Institute for Manufacturing, Stuttgart, Germany. The demonstrations show that sensor based real time deflection compensation can reduce cutting errors in steel by 3 to 5x. The principal limitations are actuator compliance and gearing backlash, neither of which were directly addressed in this demonstration project.

[2] Improving Robotic Machining Accuracy Through Experimental Error Investigation and Modular Compensation, U. Schneider et al. International Journal of Advanced Manufacturing Technology, Online: June 10, 2014.

Background

Precision machining represents significant force variations due to rapid changes in cutting depths (material removal rates) and especially when the cutting tool enters the work or rapidly changes direction at a workpiece "corner". Generally, robot joints are necessarily compliant (gear teeth deflections, gear and joint bearing deflections, and backlash—free motion in contact crossovers). Nonetheless, solving this "actuator" problem would enable a dramatic return on investment for robot manufacturers and their customers. Today, only 2% of industrial robots are used in processes such as cutting, milling, and grinding, and only for rather simple/lower-quality products made up of plastics, composites, and softer metals. If this accuracy under force disturbance could be fully dealt with, the high value tasks would likely result in a 5x expansion in the market for industrial robotics.

Fraunhofer COMET Project

This demonstration effort involved error compensation, both online and off line. The offline effort used programming tools based on more accurate kinematic and dynamic models of the industrial robot. The online compensation used real time sensing of the end-effector (cutter) position with corrective feedback through a rugged small motion 6 DOF parallel piezo-actuated module between the tool and the robot. This project also showed that a weak foundation floor could add to the robot's overall compliance and enable outside disturbances (walking, rolling lift trucks, etc.) to affect the resulting tool accuracy. Backlash is clearly present in the actuator torque crossover which represents an input shock to result in significant jumps as well as oscillations at the output tool. This class of error is virtually impossible to predict offline or compensate for online. Hence, quality design to eliminate actuator backlash is essential. Note that this was not dealt with in the COMET project. The measured backlash tool error of just the first axis of the Kuka KR125 represented an error of 0.008 to 0.022", which is still very large for precision work. Software control related errors were considered very small (≈ 0.001"). Force related errors could easily reach 0.01", most of which can be compensated for if tool motion is relatively slow. Unfortunately, tool related chatter is very destructive to precision operations using a flexible robot manipulator.

COMET Project Results

Figure 3 of the Fraunhofer paper shows the linear compliance and backlash for Joint 1 of the KR125. It is interesting that the forward stiffness is 2x that of the backward stiffness and a large jump (backlash) occurs at the crossover. Figures 6, 7 show dramatic differences in crossover shock and vibration frequencies when machining in the y direction (good) and in the x direction (poor). The dominant frequencies are 5.9 hz (x dir.) and 24 hz (y dir.), and the deflections are significantly higher when using the x direction. Using online model compensation with the piezo actuated small motion module, the accuracy was improved by 3 to 5x (from 0.025" down to 0.004") for steel. Unfortunately, the demonstration is for planar small motion range cuts.

This set of results cannot then be extrapolated for the full work volume of the robot, which would, then, show the effects of backlash and programming shocks which the limited demonstrations do not encompass. This limited set of results highlights the essential necessity for full parametric metrology (perhaps, up to 150 basic parameters for the full robot model) and a revolutionary class of high accuracy actuators (low compliance, no backlash, low inertia for high responsiveness, etc.). Repeatability of these robots is excellent. The ability to reprogram precision machining tasks on demand will not be possible until the actuator is also accurate to a level equivalent to 50% of the system's repeatability.

BASIC MACHINING TECHNOLOGY

Manufacturing Processes and Associated Tooling

Program Objective

The suggestion is to create an accurate parametric model of a finite number of processes (and their related tooling) to form the basis for digital control in an advanced manufacturing cell. These processes involve light machining tasks, assembly tasks, and some force controlled forming tasks.

Background

The heart of the intelligent automation concept is digital control of all functions in the manufacturing cell. This means that the core function (the manufacturing processes) must be understood in great detail and fully documented parametrically. Criteria which adequately and accurately represent the system must be developed. These criteria will take on a variety of forms and many will have synergistic effects. Criteria fusion will be implemented to combine all relevant parameters defining performance measures. This performance evaluation must be achieved in realtime (< 5 msec) so that the decision level of the cell can best allocate resources to match the needs of the process. To do so will involve extensive testing of the process under a wide range of conditions to confirm the analytical model.

Program Development

Ten or more processes should be studied. They would include: drilling, routing, grinding, deburring, laser cutting, laser welding, arc welding, riveting, bolt assembly, force fit assembly, hole punching, bending to fit, and pin alignment. In each case unique tooling would be developed, if necessary, to enhance the performance of the process in the cell; for example, force neutral grinding heads to not only reduce the force disturbances in the cell but also to enhance the precision in the grinding process. In each process, the optimum forces (in all 6 DOF), speeds, required supporting stiffness, etc. would be evaluated and documented parametrically. Based on these results a finite number of operational criteria (and associated norms) would be developed for each process to become the foundation for performance related control of the manufacturing cell. In most cases, advanced metrology would be utilized to directly establish these criteria. In some cases, motor current signature analysis and other diagnostic techniques would be implemented to fully characterize the operating criteria. Innovative approaches would be developed for operating parameter acquisition and distribution, such as non-contacting and wireless sensors, respectively.

A standardized approach at criteria development and criteria fusion would be established where appropriate to aid in modular controls implementation which is applicable to a wide variety of manufacturing processes. The resulting criteria fusion would be integrated with the total knowledge base and utilized in real-time control to achieve true intelligent automation. Combining the attributes of a modular controls architecture, and the precision resulting from real-time, knowledge based control would allow rapid change over from one product to another, thus capturing the essence of agile automation.

Sensor Based Manufacturing Process Control

Purpose

The goal here is to establish a criteria-based decision making procedure for maximum performance dependent on a residual (differences) between a predicted parametric model of the manufacturing process and its actual description derived from a suite of sensors. The required sensors would be developed to best meet the real-time parametric description of the spectrum of selected processes.

Background

Manufacturing cells will be increasingly driven by digital technology to ensure maximum performance while allowing rapid reconfigurability to match product design changes in response to market needs. This core reality of intelligent automation leads to a fundamental requirement that the manufacturing process be parametrically modeled and sensed in real-time. The difference of the modeled and sensed parameters creates the residual upon which the process control can occur. Each process description would provide a series of operational criteria (and norms) to form the basis for this control. Here the "residual" acts as the driver in these process criteria to make the final decision for control (in real-time).

Currently, most industrial robots (the precursors of a full manufacturing cell) are taught in place (offline) with minimal intervention from the factory computer database. Once programmed (at great cost of time), the robot structure then repeats this identical sequence of motions, meaning that any changes in the process or the physical environment go unnoticed. The addition of process sensors and a high level of process control can significantly enhance the operation of the robot (by preventing drift from the desired operation), by making it possible to rapidly change the basic parameters of the process, and to make the whole complex of fixturing, end-effectors, multiple processes, obstacle avoidance, process archiving, etc., possible for 40 DOF manufacturing cells.

One of the most challenging issues associated with the tasks enumerated above is the integration of multisensory data (or sensor fusion). The problem of multisensor integration involves the acquisition, fusion, and interpretation of information received from sensors for the purpose of enhancing the control of a manufacturing process. For example, information collected by one sensor may be combined with information collected by another sensor to generate knowledge that cannot be inferred from either sensor alone.

Suggested Development

Create the technology for 10 or more process sensors to completely define the actual condition of a selected set of manufacturing processes. These feasible process sensors might be:

temperature	light	range
6 DOF force	surface roughness	proximity
6 DOF deformation	sound	touch
position	vibration	slip
velocity	radiation	machine vision
acceleration	electromagnetic, etc.	etc.

The required set of sensors would be brought together to best measure each selected process. This sensor fusion is the integration of all real-time data into performance indices which can be compared with the desired (modeled) performance indices of the process. This comparison (residual) would then be used in the process controller subsystem to enhance the overall performance of the manufacturing cell. Finally, performance indicators would be made available to the human operator for oversight and for use as the basis for archiving the processes for lessons learned and other process redesign.

Required Information Technology

The opposite of fixed automation of the past 100 years is the production of precision products based on accurate process information available in real-time (< 5 milli-sec.) making it possible to adapt to changing conditions while the product is being made. This also means that quality assurance and product data archiving can be done while the product is being produced. To do so requires a full database under the control of the product designer (who responds to the customer) through a full information network in the factory or among several factories if they are making compatible components which make up a larger product at a downstream assembly plant (increasingly the norm in advanced industrial countries). This perspective on manufacturing technology has recently been termed “agile” manufacturing which is central to success in our future manufacturing activity. The technical sub-tasks in this case are:

Non-Destructive Testing and Inspection
Database (for Design and Operation)
Archiving of Product Data
Networking (Determinism and Fault Tolerance)
Task Planning Algorithms
Spectrum of Process Sensors (Force, Temperature, Position, Machine Vision, etc.)
Sensor Fusion Software Technology

One of the obstacles to such a modern approach to manufacturing through the use of an active process control is the dearth of intelligent sensors and actuators which are capable of monitoring manufacturing processes and changing them in response to measured parameters. There are three main aspects of information technology associated with agile manufacturing: (1) data acquisition through distributed, smart sensors, (2) the reduction, fusion, and processing of the signals into relevant operational information for the manufacturing process, and (3) data handling, transfer, multiplexing, storage, and retrieval from a design to market database.

Die Finishing: Example of Precision Machining

Objective

The design of die molds for sheet metal stamping is a complex art for which a lot of trial and error is required to perfect the final shape of the molded sheet. Molded sheet is used in a very wide range of products including autos, appliances, weapons, etc. It is suggested to combine recent advances in materials science (metal forming and springback analysis), robotics (to finish grinding the die mold under CAD control), and machine vision (to measure the final die shape) into a modern die finishing cell.

Demonstration Project

The suggested demonstration involves the rapid design, production, and evaluation of sheet metal panel dies, reducing the total design to delivery cycle time from 6 months to 2 weeks as an example of the gains possible in intelligent manufacturing. Panel dies are required to stamp out the final shapes from steel sheets. It is one of the longest lead time items for new product designs. The basic shape can be obtained by using NC machines down to an acceptable margin (say 0.020 inch). These complex shapes are now finished by hand by experienced (and well paid) craftsmen (who are now retiring). It is now proposed to combine the recent developments for "springback" with advanced controls, sensing and measurement, and path planning governed by the panel CAD database to finish grind to final specs (down to 0.005") by using specialized finishing tools on the end of a dexterous precision robot. Final touch up by craftsmen may still be necessary but the time commitment and cost would be dramatically reduced. This would be a self-contained system using the most advanced technologies at each stage of the die production (design, machining, finishing, evaluation and test, etc.). It would be truly "agile" in that the designer of the die would be in real-time control of the whole process using the very best computational tools (FEM, neural nets, CAD databases, etc.), the very best in production equipment (high accuracy robots, specialized finishing tools, process sensors), and advanced measurement technologies (vision metrology, laser calibration devices).

As the technology solidifies, the role of the craftsman would be raised to the level of inspection and final touch up (perhaps eliminating 95% of his present manual labor). On the other hand, the role of the die designer would be expanded to include real-time oversight in the rapid production and test of each die providing a much more cost effective technology and accurate lessons learned for future designs.

Suggested Development

The process modeling group would develop a very accurate modeling strategy to predict springback of the pressed panel as it is released from the die and, therefore, provide the designer a valuable tool to improve the "up-front" design of the die. Machine vision would be used as a measurement system to check the panel dimensions after release from the die. Based on this accurate modeling and measurement, lessons learned can be embedded in a neural network software package which can be implemented on a standard P.C. To complete the finishing cell, it becomes necessary to develop a high accuracy modular robot manipulator capable of performing the finish grinding of the die to an accuracy approaching 0.005" all under digital control based on CAD database information from the design of the die.

REQUIRED PRECISION MANIPULATOR DEVELOPMENT

Robot Manipulator Design Process

Objective

Robot manipulators of 6(+) DOF have an unusually complex task performance capacity. In addition, they may contain more than 100 significant physical parameters to define that capacity. It is suggested here to develop a task-dominated design procedure to best select the physical parameters of the manipulator with emphasis on the actuator parameters.

Background

In 1875, the field of mechanisms and machines got a boost in effectiveness from the architecture of linkages laid out by F. Releaux. This was, then, solidified with the science of kinematic synthesis which showed how to determine the link dimensions and, therefore, fix the input/output motion functions. Once these geometric dimensions were known, then link embodiments could be obtained to maintain force and deflection requirements, evaluate the effects of inertia forces at speed, determine the size of flywheels to maintain constant speed of operation, set up crankshafts to coordinate multiple mechanisms (as in a sewing machine), and them provide for the proper sizing of the power source (prime mover).

Today, a similar process is in place to determine the geometry of manipulators, their link lengths and how the joints are combined to yield a desired level of dexterity. The first step is to break down the task into a finite number of geometric sub-tasks (these could be straight line, circular, parabolic, point path trajectories joining local motions such as cusp geometries, inflections, etc. Note that a similar breakdown of the motion geometry is necessary for the three rotational motions of the end-effector). Link sizing questions are associated with dexterity margins, workspace coverage, obstacle avoidance, etc. Once a manipulator geometry is chosen, then the management of the actual motion depends on numerous operational criteria that can be determined in real time (See Fig. 11).

Task Performance

Once the task geometry has been determined and a suitable manipulator link/joint assembly chosen, then questions of structural integrity against task/inertia forces must be dealt with. The foremost problem is to determine the best distribution of the actuator parameters to be used to drive each of the joints of the manipulator. To achieve this most complex objective can at first appear overwhelming since there are so many choices. It is recommended here to work within limited task domains (say welding, deburring, transport, assembly, light machining, etc.). Then, within each task domain, define performance maps with control parameters (position, velocity, acceleration, force, power, etc.) and output parameters (efficiency, accuracy, elapsed time, durability, tool wear, etc.). Combine these task performance maps with representative geometric motion plans (say 10) described above to obtain an overall capability requirement for the desired manipulator.

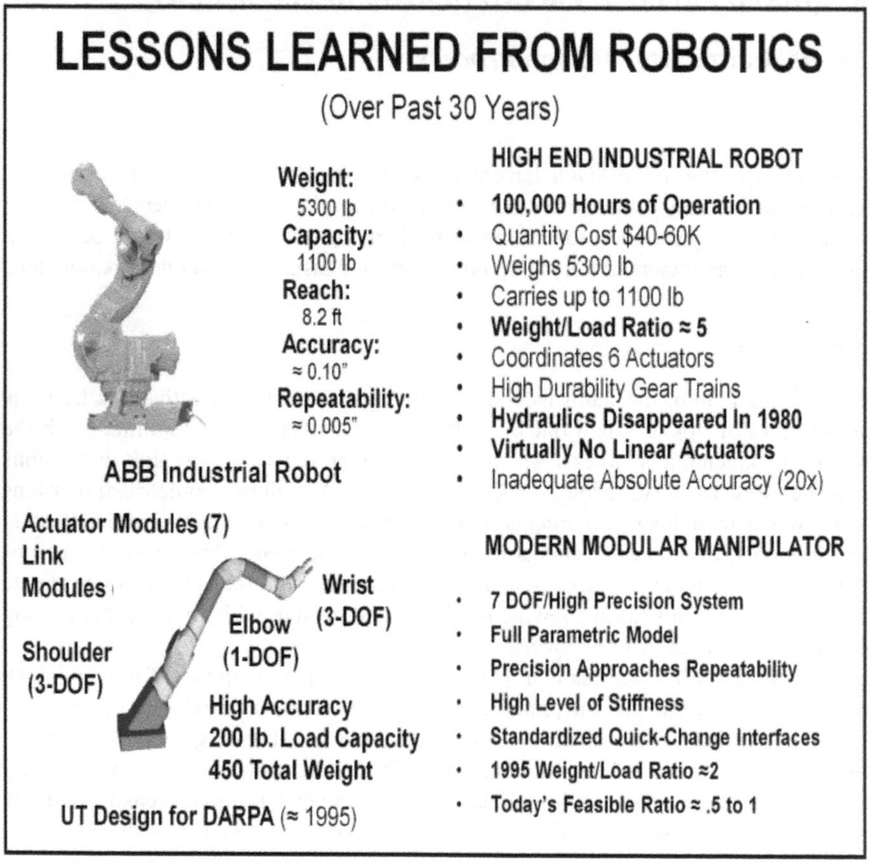

Figure 11. Lessons learned from robotics.

Actuator Parameter Distributions

The task performance maps and the end-effector motion plans fix the capacity requirements of the manipulator at its end-effector. Each actuator represents a finite number of performance parameters (lost motion, accuracy, stiffness, torque capacity, effective inertia, speed/acceleration limits, etc., all affected by internal parameters for the motor (speed, torque, acceleration, etc.) and gear train (stiffness, inertia, lost motion, reduction ratio, etc.). Using the geometric transforms (reduction ratio, geometric gain functions) between the actuators and the end-effector, performance capability envelopes can be created to explicitly demonstrate the impact of the local actuator parameters. (These envelopes would be generalizations of basic distribution rules for principal actuator parameters—gear reduction ratio, gear stiffness, lost motion, torque density, etc.). These performance capability envelopes become decision surfaces acting to give the designer a global perspective to visualize how his input design parameters affect the system's task performance envelope. Naturally, considerable uncertainty will be involved in the definition of these envelopes. Differencing these envelopes would provide the designer with norms to develop strategies to better select the necessary actuator parameters for the desired manipulator.

Ultimate Goal

The dominant components in all robot systems (mobile platforms, manipulators, surgical cells, manufacturing cells, etc.) are the actuators at the driven joints. For any application domain (auto assembly, airframe parts preparation, surgery, entertainment, battlefield systems, etc.), the ultimate question is what is the minimum set of actuators to satisfy the maximum population of systems required for that domain. Then, this minimum set would enable significant improvements in supply chain management (repair, re-supply, up-grades, certification of performance, condition-based maintenance, etc.) enabling these solutions to be assembled on demand (standardized interfaces, universal operating system software) by the user just as we now do for our personal computers. This is the basis for a next wave of technology for the mechanicals where the performance-to-cost ratio is always improving, Tesar 2015.

Actuators for Intelligent Machines at UTexas

Objective

The University of Texas has established a major program to develop advanced actuators to drive virtually all machines from 1 Degree-Of-Freedom (DOF) conveyers, transfer devices, parts elevators, etc., 2 and 3 DOF parts presenters, 6 to 10 DOF robot manipulators, through to 25 to 40 DOF manufacturing cells. These actuators will be made up of five classes and will represent a standard for investment. Each class will be composed of 10 or more standard sizes which will evolve under a continuous and concentrated design and testing effort to have the density of technology we now expect to see in our micro-electronics chips. The overall goal is to dramatically improve the performance of these actuators while reducing their costs and making it possible to assemble a wide range of dexterous, reprogrammable machines on demand.

Background

The primary improvement in electrical actuators over the past four decades has been the better use of materials (magnets, wire insulation) and electronic controllers (to provide responsive control of brushless D.C. motors having many poles). The University of Texas at Austin showed in 1988 that a concentrated design effort could improve the actuator performance by 200x, making a system that was not only frameless containing the bearing for the machine but it also was dual providing fault tolerance. Since that time, the program has made a dramatic improvement in brake design (more than 10x), gear transmission design (70x), communication wire reduction (12x), standardized interfaces (for quick change out), architecture (5 baseline rotary and 3 linear actuator designs), and showed how to embed the electronic controller within the actuator itself. This work has made an improvement of 4 orders of magnitude over the past two decades.

Program Activity

UTexas has built and tested a precision actuator module allowing the development of a precision modular manipulator capable of 0.01" accuracy over its work volume (its reach is 72") carrying a 50 lb. load at 50 inches per sec. We have built a baseline rotary actuator prototype (37x better performance than state-of-the-art). We have designed a switched reluctance motor (high torque, low velocity, no gear train) which is competitive with our baseline rotary actuators. Finally, we have designed a linear actuator (based on a sophisticated spindle screw drive) which shows that electrical systems can supersede hydraulics in many applications. We have developed four distinct test-beds to evaluate these actuators (endurance, condition-based maintenance, non-linear characterization, and parametric metrology) to ensure that our concepts and designs meet their objectives.

Suggested Development

We will further develop our baseline designs (where costs, performance, and high production quantities are involved) for both rotary and linear actuators. The development of our gear trains will continue in order to provide a generalized design procedure, establish a finite number of standard designs, provide multiple inputs (for layered or dual control) and to move towards their economical production. Parametric metrology will occur at the module level where all "as-built" parameters will be stored in the embedded electronic controller. We are setting up a 10 sensor environment in these actuators to ensure that they are not only intelligent (to enable enhanced performance, condition-based maintenance, and fault tolerance) but to enable us to create a full architecture for reconfigurable actuators of the future. Emphasis will be put on high precision postage stamp sized position and torque sensors. The electronic controller architecture must be dramatically expanded, based on performance map structured decision making. The concepts of multiple wiring circuits, layered control, multiple measurands for distinct physical phenomena will all be combined to make the actuator intelligent and expand its performance envelope.

Actuator Development for Light Machining Cell for Airframe Manufacture

Objective

Precision machining of light stock components (skins) of airframes is a high cost procedure which is not only inflexible (using rigid, invariant jigs), it represents a throwback to the use of templates and fixed automation in manufacturing. It is proposed to modernize this function by developing a data base driven light manufacturing cell (of up to 40 DOF) which is fully integrated and capable of precision operations even under large machining force variations. All 40 active joints would be driven by high accuracy—high load capacity actuator modules selected from a carefully designed and optimized set of 5 actuator sizes (i.e., a minimum set).

Background

Airframe manufacturing involves numerous light machining operations (i.e., grinding, drilling, routing, trimming, forming, press fit, rivet assembly, etc.). Today much of the accuracy of these operations is provided by expensive **jigs and fixtures**, which are unchangeable leading to extended implementation learning curves, **early airframe obsolescence**, and virtually **no data base control of the manufacturing process**. This lack of data base control and process feedback in real-time is illustrated by the information barrier depicted symbolically for the jigs and fixtures. In order to remove the barrier of these jigs and fixtures, the dynamic and process machining forces will generate deformations in the robot which must be compensated for in real-time (i.e., in less than 10 milli-sec.) by providing corrective commands to the system's actuators. However, contrary to the fly-by-wire operation of modern fighter aircraft, **existing robot manipulators operate with virtually no parametric description of their physical plant**. Because of the highly non-linear and coupled nature of multi-degree of freedom manipulators, this is obviously a major digital control problem requiring the highest order of software development and precision actuator technology. The University of Texas Robotics Research Group has studied this complex manufacturing operation since 1980. It established a major metrology laboratory for accurate measurements of all system parameters (mass, link dimensions, deformation coefficients, actuator properties, etc.) of which there maybe 140 to 160. For example, it has **reduced the positioning error of a Cincinnati Milacron T3-776 industrial robot by 18x from 0.370" to 0.020"**. It has demonstrated the feasibility of real-time machining force disturbance rejection by using a complete model of the robot to reject deformations caused by sudden changes in the cutting operation (i.e., going around a corner).

Proposed Development

It is necessary to develop actuator technology sufficient to populate a 40 DOF airframe component machining cell (such as stamped thin stock) to perform accurate drilling, routing, grinding, etc., by means of standard tools held by two 7 DOF dexterous manipulator robots. This cell would support the unmachined stamped part on a vacuum pad of the same shape. The vacuum pad would be held by four 4 DOF positioning robots. The machining force acting on the thin stock would be resisted by two 5 DOF force robots. This 40 DOF system would be controlled by an operating system where CAD part data governs the path planning of all tools and actuated systems (See Figs. 9, 12). The operation of such a cell implies the existence of a technology which remains incomplete today. Questions of high speed computational software, fusion of multiple performance criteria in real time, subsystems, etc., must still be met. Also, a full interactive and time efficient process for parametric metrology must be developed at the actuator (and link) module level to enable the operator to have a complete physical model of each precision robot deployed in the cell. The benefit to manufacturing would be enormous enabling **a**

high return on investment, quick changeover of the machining operation (no jigs) to prevent product obsolescence, and a **resurgence of mechanical technology** by creating a more aggressive and complete technology compatible with other technical sectors (i.e., telecommunications, computers, aircraft, etc.) (See Figs. 9, 12).

Suggested Result

It is suggested to develop a class of actuator (and link) modules with standardized interfaces which are capable of high accuracy under load suitable to populate 40 DOF light machining cell. These actuators would involve advanced component technologies (motors, brakes, gear trains, sensors, etc.) under a full parametric model (by a science of metrology at the modular level) to guarantee accurate operation in terms of a complete real time software based on a collection of operational criteria (torque saturation, temperature limits, deflection compensation, etc.). One of the additional goals would ultimately be to reduce life cycle costs (repair, tech mods, etc.) of the full-machining cell.

ROBOT METROLOGY

Emphasis on Intelligent Manufacturing and Automation

The present industrial robot represents a low cost ($50,000) and durable (100,000 hours of operation) system presently used for simple low-value added tasks (handling, gluing, painting, etc.) which has saturated its market at $1 bil/yr in the U.S. The Airbus study section shows that this technology is sufficiently repeatable (to a few thousandths of an inch), substantially insensitive to temperature changes, and no longer subject to concerns for free motion due to high backlash (or lost motion) in ever-improving

PROPOSED DEVELOPMENT TO IMPROVE INDUSTRIAL ROBOT ACCURACY

I. Measure "As Built" System Parameters
- **Manipulator Is A Complex System**
  - Excellent Bearings, Gear Trains
  - Nonlinear Stiffness Coefficients
  - Numerous Geometric Dimensions
  - Unpredictable Temperature Affects
- **Metrology Is Essential**
  - 6 Geometric, 6 Stiffness, and 6 Temperature Coefficients Per Link
  - Modern Laser Measurement System
  - Cost Effective/Time Efficient
  - Automated Data Reduction Feasible
  - Calibration Feasible At System Level

II. Modularity Improves Metrology
- **Assemble Systems On Demand**
  - Standardized Highly Certified Components
  - Links, Actuators, Tools, etc.
  - Standardized Accurate Interfaces
- **Mass Produce Modular Components**
  - Especially High Accuracy Actuators
  - Automate Their Metrology
  - Embed all "as Built" Parameters
  - Reduce Costs

III. Universal Operating System Software
- **UTexas OSCAR Language**
  - Runs All Modular Systems
  - Provides Real Time Compensation for Load/Temperature Induced Errors
  - Provides Basis for Optimized Motion Programming
- **Goal is to Enhance Overall Accuracy**
  - To 2x Inherent Repeatability
  - Enables Rapid Product Changeover
  - Enables Plug-and-Play Repair

Figure 12. Proposed development to improve industrial robot accuracy.

actuator based gear trains. This Airbus study strongly suggests that it is now feasible to make robotics sufficiently accurate (say 2x their repeatability), to perform high value added functions (precision drilling, cutting, grinding, broaching, etc.), and therefore dramatically expand its base market from $1 bil to $5 bil/yr. To do this, the architecture of the robot must become open so that it can be assembled from a minimum set of highly certified/standardized modules (actuators, interfaces, links, tools, controllers, etc.) on demand to make up versatile manufacturing cells of up to 40 DOF. Highly certified, in this case, means that these modules must be described parametrically in great detail, especially for their "as-built" geometry (to account for tolerance errors) and for compliance (to account for deformations under load). It is suggested to automate this data acquisition for this minimum set of modules to embed the data and its uncertainty in embedded chips in the modules, to automatically download this data to an existing commercial operating system software, OSCAR (See Section **Open Architecture Operating Systems**) (for open architecture manufacturing cells as Microsoft Windows is to personal computers), to perform high valued tasks (with varying force levels) with sufficient accuracy which are reprogrammable in seconds in order to revolutionize the concept of mass production systems (as might be envisioned for the automobile industry, airframes, kitchen appliances, etc.). This rapid reprogrammability means rapid product changeover, rapid repair and tech mods by quick-change module replacement, etc. just as we now do for the production of our personal computers. Accuracy requires full parametric awareness of all manufacturing cell modules in order to compensate for local inaccuracies and task induced deformations in real time. This parametric density can only be achieved by a new level of concentrated science for measurement/metrology.

The university of Texas has proposed open architecture systems since 1975, worked on their metrology science since 1985, developed actuators which are 4(+) orders better than the state-of-the-art, created a science for very accurate quick-change interfaces, developed a new science for intelligent actuators and tools based on performance maps and created a commercial system level software proven to reprogram complex cells (in this case, a 40 DOF surgical cell) in seconds.

Now, the task is to finish the module definition by automatically obtaining the "as-built" parameters for geometry and compliance of all mechanical manufacturing cell modules. Here, we suggest to again review the metrology literature in-depth, create a measurement science to minimize the required data, automate the data acquisition, establish levels of uncertainty in the data, engage the leading supplier of measurement systems (See Fig. 13), show how to commercialize this work as rapidly as possible, and to demonstrate to industry that given this parametric definition, an operating system (OSCAR) is available to operate any manufacturing cell made up of these modules by automatically downloading the data to OSCAR as the cell is being assembled.

This step should dramatically reduce the 4x cost penalty that now exists to make robots work in manufacturing cells, permit rapid module refreshment (just as we now do for computer chips) and rapid repair, set the stage for a $5 bil/yr market for industrial robots, create a new excitement for young scientists in manufacturing, reverse the trend in job losses in manufacturing, strengthen our $50 bil/yr actuator industry in the U.S. and compensate for the 50(+) year disincentive to our manufacturing machine industry as a result of the U.S. DoD offset program which is one of the key facets associated with our trade deficit of $300 bil/yr of manufactured goods.

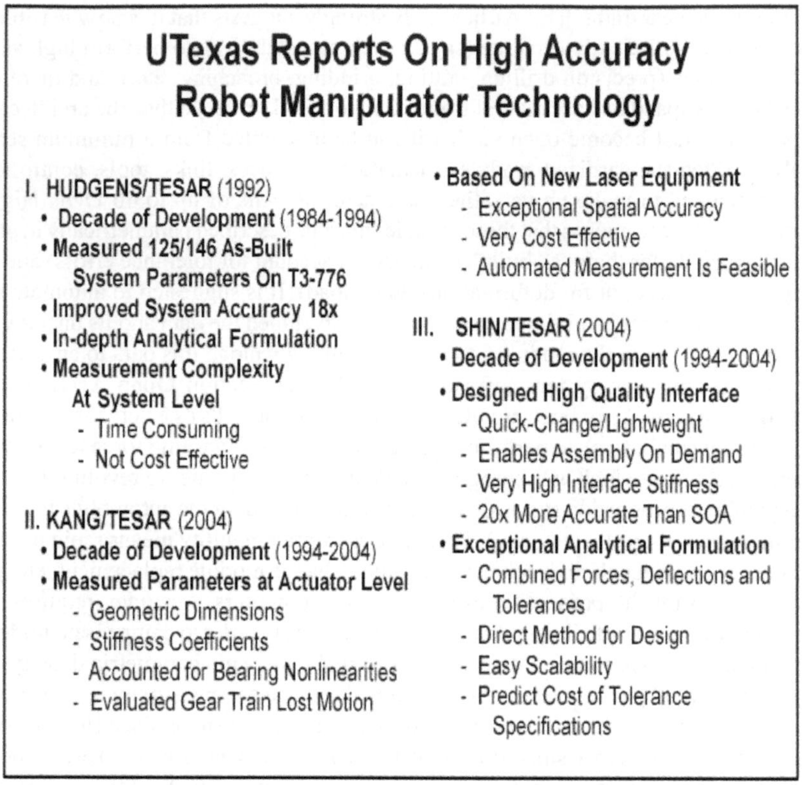
UTexas Reports On High Accuracy
Robot Manipulator Technology
I. HUDGENS/TESAR (1992)
• Decade of Development (1984-1994)
• Measured 125/146 As-Built System Parameters On T3-776
• Improved System Accuracy 18x
• In-depth Analytical Formulation
• Measurement Complexity At System Level
- Time Consuming
- Not Cost Effective
II. KANG/TESAR (2004)
• Decade of Development (1994-2004)
• Measured Parameters at Actuator Level
- Geometric Dimensions
- Stiffness Coefficients
- Accounted for Bearing Nonlinearities
- Evaluated Gear Train Lost Motion
• Based On New Laser Equipment
- Exceptional Spatial Accuracy
- Very Cost Effective
- Automated Measurement Is Feasible
III. SHIN/TESAR (2004)
• Decade of Development (1994-2004)
• Designed High Quality Interface
- Quick-Change/Lightweight
- Enables Assembly On Demand
- Very High Interface Stiffness
- 20x More Accurate Than SOA
• Exceptional Analytical Formulation
- Combined Forces, Deflections and Tolerances
- Direct Method for Design
- Easy Scalability
- Predict Cost of Tolerance Specifications

Figure 13. UTexas reports on high accuracy robot manipulator technology.

Metrology Test-Bed for Actuator Modules

Background

Robot metrology seeks to identify parameters that are not accurately known and that affect the performance of the robot under operation. These parameters include the geometric dimensions, compliance and mass properties of joints and links, and actuator control parameters.

Precision operations, such as airframe assembly, require positioning tolerances of 0.01 inch with loads in excess of 100 lb. Typical industrial robots, whose control software uses geometric parameters specified by their design values, can have accuracies as poor as 0.2 inch. Careful metrology techniques can be used to measure the actual geometric and stiffness parameters. This information can be used to increase the accuracy of a robot's tool positioning and compensate for deflections under load (an improvement of 20x has been achieved in the laboratory). The system metrology process for monolithic robots has proven to be difficult and time-consuming. Modular robot metrology (the metrology of robots that are composed of modules) is much more elementary, as a consequence of the reduced number of parameters in each module. An automated environment is needed to perform modular robot metrology in a cost-effective manner.

Objective

An automated metrology test environment must be able to measure three-dimensional coordinates with high accuracy. Lasers have become the standard for precision measurement. However, their implementation in a full 3D coordinate measurement system has been limited from a economic standpoint, costing as much as $500,000. The low-cost laser interferometer is inherently a linear displacement transducer that easily can be configured to measure distances. A coordinate measuring system using three lasers and the concept of trilateration (in which the length of the six sides of a tetrahedron are known and the coordinates with respect to one corner can be calculated) can be developed for less than 1/10 the cost of a laser tracker. A loading mechanism that applies independent forces and moments along and about each of three orthogonal axes is essential to simplify and standardize the loading procedure. The forces may be applied using an *xyz* frame with linear actuators that move along the three orthogonal directions of interest. Moments may be applied about the same axes by rotary actuators on a gimbals frame. The relationships between measured position, loads, and deflections can be used to accurately determine the 6x6 geometric parameters as well as the equivalent compliance parameters representing the stiffness of the module subsystem.

Suggested Development

It, then, becomes necessary to develop an automated metrology test-bed intended for the determination of geometric and compliance parameters of rotary actuator modules for intelligent machines (robots). It may also have the ability to measure backlash, stiction, input/output nonlinearity, and many other parameters requiring the application of loads and/or measurement of displacement. The measurement system is a relatively low cost alternative to the one-off systems available today, such as laser trackers. Three lasers (in a plane but not co-linear) will be focused on a single point (retroreflector) on the output plate of the module.

OPEN ARCHITECTURE OPERATING SYSTEMS

Intelligent Robot Systems and Control Software

Background

The goal of the Robotics Research Group (RRG) at The University of Texas at Austin is to develop open architecture mechanical systems that exhibit increased performance at lower costs through the development of standardized building blocks. To achieve this goal, the RRG has pursued three key areas of development. The first is the development of Standardized Actuators and their associated component technologies. The second thrust of the RRG is the development of operational software and analytics for manipulators. The third is the development of manufacturing workcells and the associated software to design, control and integrate such systems (Fig. 14).

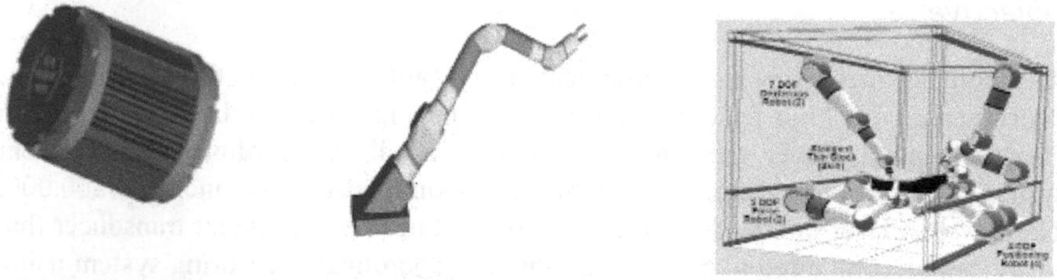

Figure 14. Open architecture robot components/systems.

1. Foundation Technology

The RRG has a 40 year history of analytical development in the field of robotic systems. The core of this development is in the establishment of generalized modeling of robotic systems in terms of kinematic, dynamic and compliance properties. Over the years, this has allowed us to analytically define independent performance criteria that can be used to optimize the control and design of robotic systems. In this area, over 100 different performance criteria have been developed, with 10 just in the area of obstacle avoidance. This basic development now acts as the foundation for our system-level research into six distinct areas.

2. System Design & Configuration Management

This research allows us to systematically design manipulator systems and processing cells from a limited set of standardized building blocks (actuators). Whereas, the traditional approach to such designs has been to treat actuators as black boxes, our approach involves using the actuator properties (gear ratio, motor inertia, etc.) actively in the design process. This research also addresses the configuration management issues faced due to the exponential explosion of choices when faced with modular systems (actuators, electronic controllers, end-effector tools, links, software, etc.).

3. Motion Planning and Obstacle Avoidance

Our research in motion planning focuses on the development of criteria that can be used to describe complex six dimensional paths in space while controlling the smoothness, accuracy, inflexion, etc. of the path. The eventual goal is to relate the path properties with the system capabilities and the constraints placed by the application environment (i.e., the process demands) to develop optimal motion plans. Obstacle avoidance is a key part of motion planning and we are developing first and second order criteria that are based on artificial potential forces. Obstacle avoidance is required for processing cells and can also greatly enhance the man machine interface by automatically guiding an operator towards obstacle free paths and/or towards specific targets.

There is an increasing use of mobile systems for remote operations, surveillance, bomb dismantlement, etc. Most of these systems are augmented by a manipulator for handling purposes, with the manipulator almost always treated as independent of the platform. The goal of this work is to scientifically study the interactions between a mobile system and the manipulator and to develop analytical techniques that improve the task performance of the complete system. Issues such as mobile system positioning versus manipulator accuracy, dynamics under high speed operation, etc. will be studied.

4. Decision Making & Performance

Manipulator systems are very nonlinear, uncertain, highly coupled, and implicit; thus requiring novel intelligent control methods that are different from conventional control approaches. As mentioned earlier, RRG's approach utilizes performance criteria (relative priorities set by human intervention) to better match manipulators capabilities with the task requirements. This research thread aims at developing techniques that allow the selection of appropriate criteria based on task constraints, methods to combine

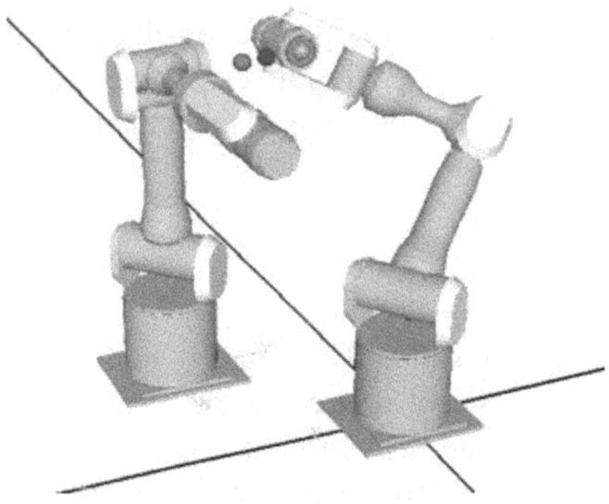

Figure 15. Framework for motion planning.

performance criteria, and conflict resolution strategies in the case of disjoint task goals, all while keeping the operator as the final arbiter.

Our decision making approach makes the assumption that the manipulator is redundant (i.e., it exhibits excess resources) and hence its performance can be improved by intelligently utilizing the redundancy. Unfortunately, there are few redundant systems commercially available and almost nonexistent in industrial use. One of our goals is to adapt our decision making system and develop new performance criteria that can be used to improve the performance of standard industrial systems (6 DOF or less). Our approach here will be to actively manage the task constraints faced by a manipulator and identify any underutilized resources that can be used to optimize the operation of the system.

5. Human Machine Interfaces

RRG's effort in this area has focused on the development of novel manual controllers and the actuators required for such a system. Additionally, we have developed a general software library for interfacing various categories of manual controllers into a telerobotic system. Realizing that a manual controller by itself is not a sufficient interface for an intelligent system, we plan to formally develop a generic human machine interface that

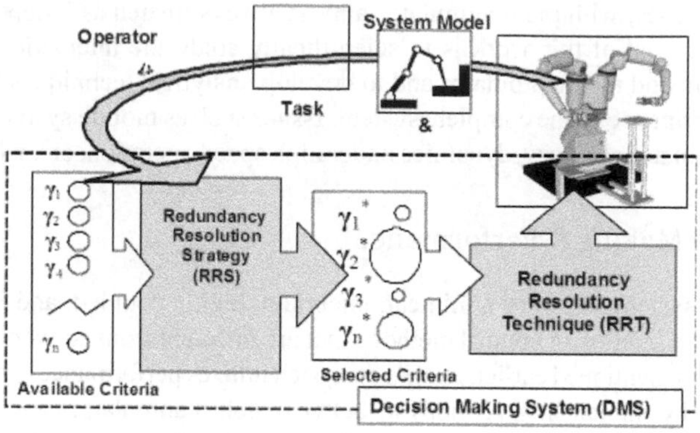

Figure 16. Criteria based decision making.

can support various input mechanisms such as a manual controller, GUI, voice, etc. This interface will have to be semantically complete and will demonstrate its functionality in a hierarchical fashion. An important goal of this work will be to illustrate the increased capability of our decision making resource management system in an intuitive and user friendly way.

6. End Effector Tools

The ability of a processing cell is directly related to its intelligent use of end-effector tools for specific tasks. This research addresses the generalized modeling of tools and how the tool model can be related to manipulator performance criteria. The goal is to provide the best match of the tool properties (usually invariant) to the full adaptability of all resources in the manipulator. As such, a procedure to formalize the parametric description of various tools is being developed.

7. Operational Software

All analytical activity at the RRG is currently embedded in its system operational software framework called OSCAR. OSCAR is an object-oriented library of C++ components that offer generalized kinematics, dynamics, performance criteria, obstacle avoidance, decision making and machine interfacing. This framework is based on well defined interfaces and allows easy substitution of OSCAR components with externally developed components. The generalized nature of OSCAR and the object oriented structure have led to reduced program development time by entry-level personnel and made possible the development of a universal processing cell controller.

While OSCAR provides the building blocks for intelligent machine software, the manner and mechanisms through which OSCAR components are composed remains manual and dependent on the software developer. This effort will develop formal operators (similar in concept to mathematical and logical operators such as +, –, and, or, etc.) that could then be used to construct OSCAR specific applications. These high level operators will lead to consistent programs with quantified performance and increased reliability. Key requirements that will drive this development will be integration, real-time capability, and machine independence.

Structured Decision-Making (SDM)

1. A Visualization Framework for Real-Time Decision-Making in a Multi-Input Multi-Output System, P. Ashok and D. Tesar, *IEEE Systems Journal*, Vol. 2, Issue 1, pp. 129–145, March 2008

Human beings have the capacity to make quick and accurate decisions when multiple objectives are involved provided they have access to all the relevant information. Accurate visual measures/decision surfaces (maps) are critical to the effectiveness of this process. This paper introduces a methodology that allows one to create a visual decision making interface for any multi-input multi-output (MIMO) system. In this case, the MIMO is thought of in the broadest sense to include battlefield operations, complex system design, and human support systems (rehabilitation). Our methodology starts with a Bayesian causal network approach to modeling the MIMO system. Various decision making scenarios in a typical MIMO system are presented. This is then followed by a description of the framework that allows for the presentation of the relevant scenario dependent data to the human decision maker (HDM). This presentation is in the form of 3-D surface plots called decision surfaces. Additional decision making tools (norms)

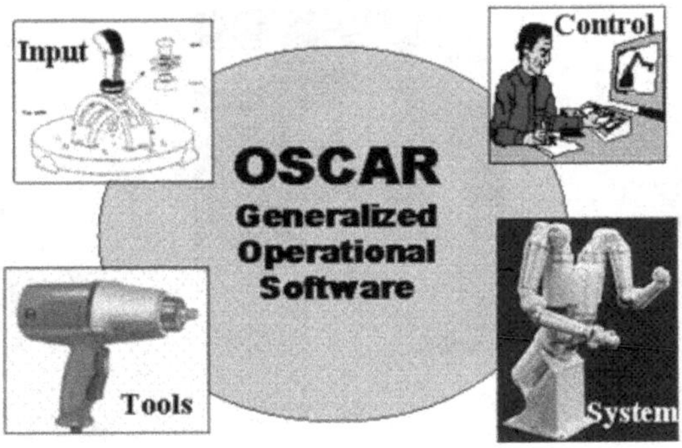

Figure 17. Conceptual basis for real time system software.

are then presented. These norms allow for single value numbers to be presented along with the decision surfaces to better aid the HDM. We then present some applications of the framework to representative MIMO systems. This methodology easily adapts to systems that grow bigger and also when two or more systems are combined to form a larger system (See Fig. 18).

2. The Need for a Performance Map Based Decision Theory Simultaneous Sensor & Process Fault Detection & Isolation in Multi Input Multi Output Systems, Part I: Theory, G. Krishnamoorthy, P. Ashok, and D. Tesar, Submitted to *IEEE Systems Journal*, in review, July 2001

This paper argues for the need for a new decision theory based on performance maps. This theory will complement both the optimization math theories and fuzzy logic math theory. Its principal purpose is to aid in decision making of complex nonlinear systems by involving humans and simultaneously ensuring that the math does not simplify the true nature of the system, thereby allowing full utilization of the system. The many advantages of such a theory based on performance maps is discussed in this paper.

3. Efficient Management of Information in Multi-Sensored Cyber Physical Systems, P. Ashok, G. Krishnamoorthy, and D. Tesar, *Journal of Sensors*, Vol. 2011, Art. ID 321709, 15 pp., Nov. 2011

Cyber Physical Systems (CPS) typically have a large number of sensors monitoring the various physical processes involved. Some sensor failures are inevitable and may sometimes have catastrophic effects. The relational nature of the diverse measurand sensor data can be very useful in detecting faulty sensors, monitoring the health of the system and in reducing false alarms. This paper provides procedures on how one may

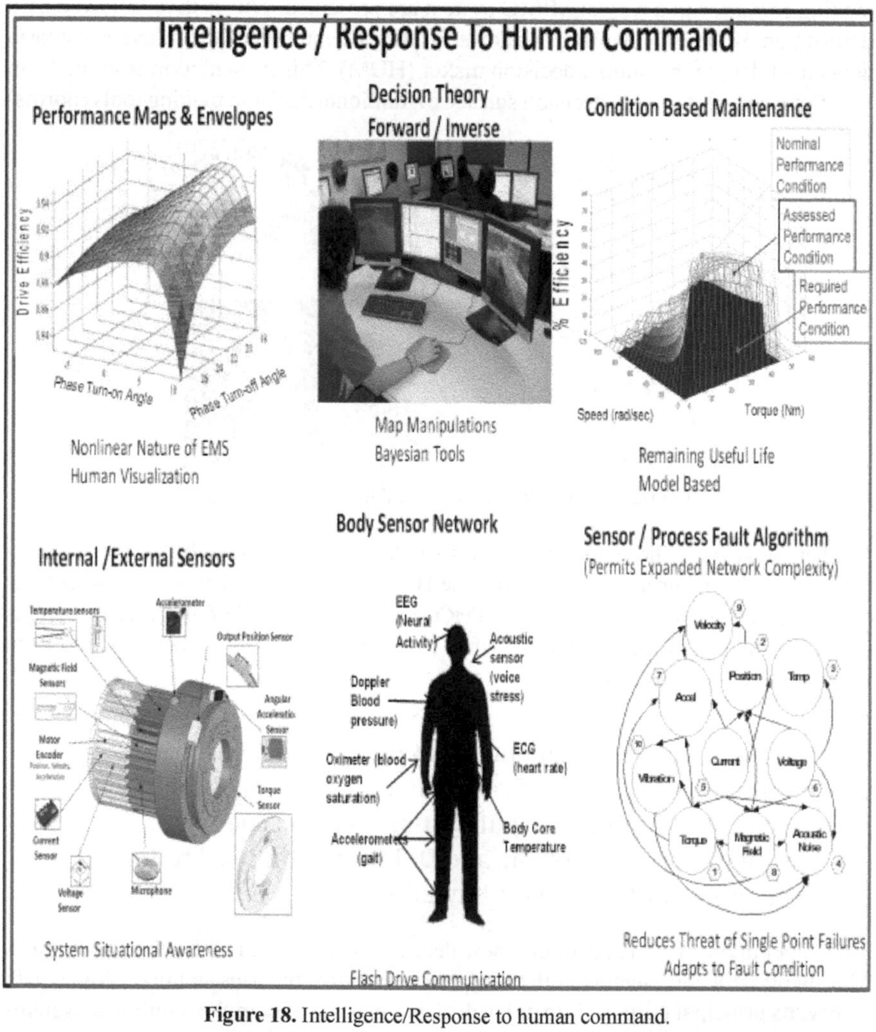

Figure 18. Intelligence/Response to human command.

integrate data from the various sensors, by careful design of a sensor data network and once such as network has been adopted, choices available in real time for enhancing the reliability, safety and performance of the overall system (See Figs. 19, 20).

4. Simultaneous Sensor and Process Fault Management in Multi Input Multi Output Systems Part I: Theory Simultaneous Sensor & Process Fault Detection & Isolation in Multi Input Multi Output Systems, Part II: Application Considerations, G. Krishnamoorthy, P. Ashok, and D. Tesar Submitted to *IEEE Systems Journal*, Nov. 2011

Dependable sensor data is vital in complex systems that rely on a suite of sensors for both control as well as condition monitoring purposes. With any unanticipated deviations in sensor values, the challenge is to determine if the anomalous behavior is the result of one or more flawed sensors or if it is indicative of a potentially more serious system level fault (i.e., avoid false alarms). This is part I of a two part paper that describes a

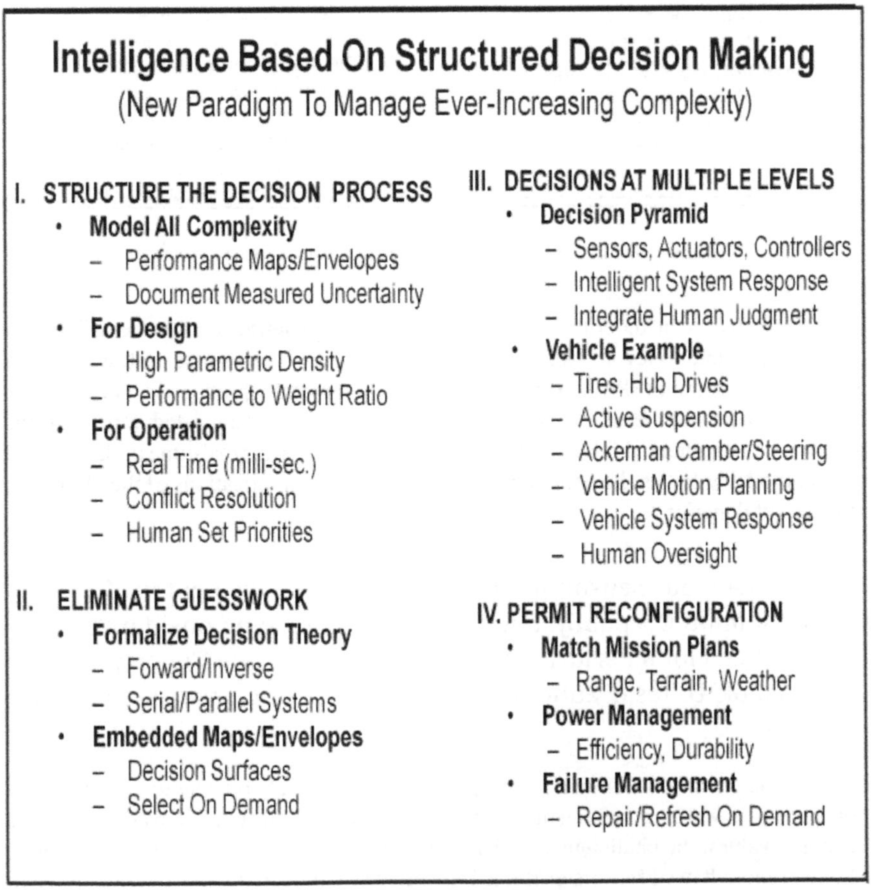

Figure 19. Intelligence based on structured decision making.

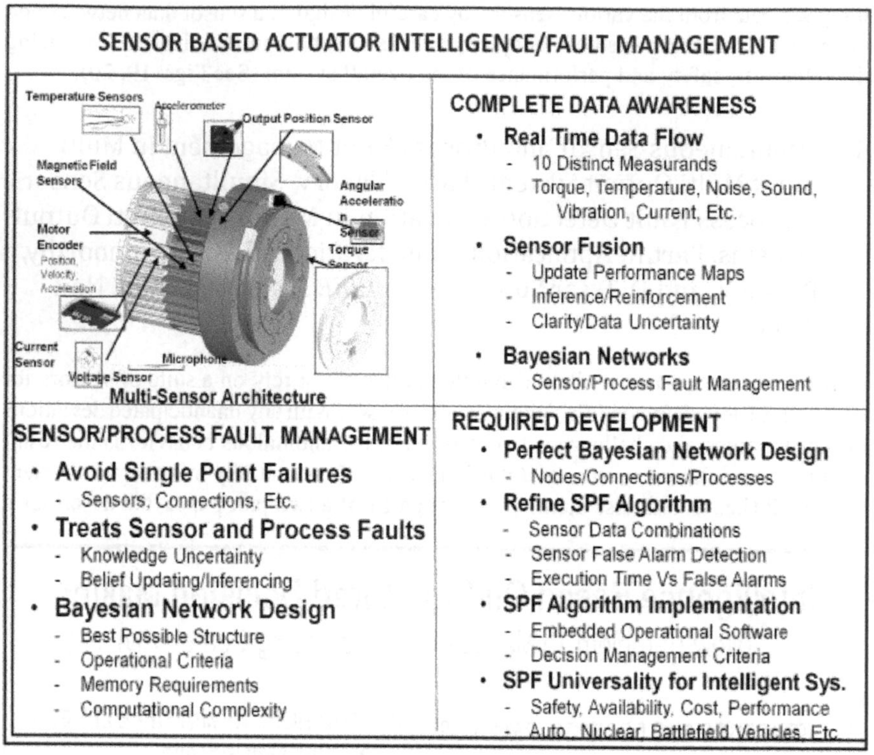

Figure 20. Sensor based actuator intelligence/fault management.

Bayesian network methodology to distinguish between sensor and process faults. A detailed literature review of existing methodologies is presented first, followed by the Sensor/Process Fault (SPF) algorithm. The limitations of the proposed algorithms are presented and some means to overcome these limitations are suggested. Part II of this paper discusses potential false alarms, an algorithm to allow for real-time updates of the system model based on the beliefs in the sensor data, and applications of the algorithm in different domains.

5. Simultaneous Sensor and Process Fault Management in Multi Input Multi Output Systems Part II: Application Considerations The Need for a Performance Map Based Decision Theory, P. Ashok, D. Tesar, Submitted to *IEEE Systems Journal*, recived Aug. 2010

Dependable sensor data is vital in complex systems that rely on a suite of sensors for both control as well as condition monitoring purposes. With any unanticipated deviations in sensor values, the challenge is to determine if the anomalous behavior is the result of one or more flawed sensors or if it is indicative of a potentially more serious system level fault (i.e., avoid false alarms). In Part I of this paper a novel methodology based on Bayesian networks was introduced to simultaneously detect both sensor and process faults. In this paper we discuss factors that affect the algorithm such as the effect of the Bayesian network structure, sensor characteristics and the effect of discretization. An algorithm that facilitates model refreshment is then introduced. This is followed by details of implementation of the algorithms on an Electro-Mechanical Actuator (EMA) test bed and brief review of the applicability of this algorithm.

ESSENTIAL RESEARCH ACTIVITY FOR ADVANCED MANUFACTURING CELLS

The suggested research would concentrate on the use of advanced computer technology to dramatically accelerate the development of a science of intelligent machines. The massive computational resources now available make it possible to treat the full parametric description of a much more general class of manufacturing cell structures, making it possible for the researcher to think much more openly and freely of generic hierarchical design and control strategies which should lead to a maximum level of productivity of new ideas and technology evaluated by complete simulations. This increased computational capacity will mean that the following can be addressed:

1. Open architecture

Future manufacturing cells will be composed of easily scaled structural modules (elbows, knuckles, shoulders, wrists, micromanipulator, mixed large and small control structures, etc.) to provide finite packages of proven technology to be rapidly assembled into generic intelligent machines.

2. Redundant structures

Serial and parallel machines have excess inputs for a very high level of dexterity and obstacle avoidance capability, but they require a correspondingly high level of decision-making intelligence to operate in real-time (for enhanced precision, load capacity, speed, etc.).

3. Optimal design

Initial success in the use of optimization techniques to the multi-parameter/multi criteria problem associated with robotics has led to improved distribution of actuator parameters. This computationally intensive effort must be expanded to include the full range of design parameters (link dimensions, mass distribution and deformation parameters, etc.).

4. Graphical simulation

In order to design or operate complex "robotic" structures, their full operational characteristics must be on display with great fidelity to the designer as well as to the machine operator. Training functions (as demonstrated by the Link aviation trainer) will be increasingly necessary to skill factory personnel when making complex interactive decisions associated with the operation of a network of systems (even some of which are outside the factory itself).

5. Operational software

Symbolic programming can now be applied to the complex analytical formulations required to completely describe the dynamic state of a robot and to form the basis for a generic operational language capable of off-line programming and disturbance rejection. Proven formulations can be embedded in dedicated processors to provide very high speed computations. This software can be written in an architecture that allows full extensibility and reuse.

6. Sensor fusion

The sensor subsystem must provide data on the operation of the intelligent machine and its interaction with the environment. System sensors include force, position, velocity, acceleration, actuator current and voltage, etc. Process sensors include range finding (to obstacles or targets), tactile, proximity, force, etc. Sensor fusion deals with the reduction of all this conflicting, incomplete, fuzzy, excess information into predictive state commands for the operation of the system.

7. Computer architecture

The top-down approach, made feasible by gigaflop personal computers, will make it possible to develop specialized computer hardware and software modules (arithmetics, array processors, etc.) uniquely suited to intelligent machines. In addition, a new generation of generic, modular, layered hardware controllers must be developed to match the modular and layered mechanical architecture that will be forthcoming. Recent developments suggest that distributed P.C.'s, in a carefully designed communications network, will dramatically reduce costs and increase flexibility and performance.

8. Resource management

This represents the real-time adjustment of the input parameters to best enhance the controllability of the fully nonlinear nature of intelligent machine structures. The dynamic model must include the first, second (and sometimes third) order geometric representations of the system in order to account for the full cross-coupling of mass, deformation, and drive system characteristics.

9. Decision making

As the system becomes generalized with more redundancy (extra DOF), more layers of control (for disturbance reduction), and more modularity (elbows, shoulders, etc.), it will become more non-deterministic. Hence, intensive integration of decision theory will be required to balance hundreds of (possibly conflicting) criteria in real-time in order to make the most efficient use of excess system resources, resulting in the best overall performance.

10. Light machining

The heart of the factory of the future will require inexpensive generic systems to perform precision light machining operations by direct computer control in order to have a maximum value-added benefit and response to the individual consumer. This requires a complete dynamic model implemented with feedforward compensation in real-time to make the system electronically rigid.

11. Metrology

Metrology is a semi-automatic means of identifying all significant parameters in an existing system. Unfortunately, to identify hundreds of parameters (146 of more for a 6 DOF robot manipulator) is an extremely demanding technical task. Limited parametric knowledge of the system reduces the potential for computer-based operation. It is now recognized that metrology must take place at the modular level, where the measured as built parameters are imbedded in the module and downloaded to the system controller, whenever the module is put into the system.

12. Man-machine interface

As the technology becomes more complex, a greater need (not less) will develop for a balanced control (or intervention) by man and machine. This will require a much higher level of machine intelligence (to treat multiple slaves, changes of scale, error filtering, reorientation, time lags, etc.) to obtain the full benefit of the technology for man.

13. Multi-arm systems

These parallel machines include dual arm robots, multi-fingered hands, walking machines, etc. which frequently require 3 to 4 times the number of prime mover commands in order to achieve the desired output (usually 6-DOF spatial motion of the output link or object). This move towards manufacturing cells becomes the most demanding problem for internal decision making and software development facing the intelligent machine research community.

14. Intelligent actuators

It is now timely to concentrate on the development of a full architecture for actuators based on a 10-sensor environment (position, torque, temperature, voltage, current, etc.), multiple input gear trains for layered control, reconfigurable circuit windings, fault tolerant brakes, clutches, etc. to achieve an aggressive building block for intelligent machines in the same manner that the computer chip does for the electronics industry. Five classes of these actuators have been identified (standard, high precision, high rigidity, backdrivable, and high speed), four of which have been designed and two of which have been built and tested.

15. Condition-based maintenance

It now becomes possible to develop a full parametric model for an intelligent actuator and therefore analytically describe its performance (envelope) under a wide range of conditions. Should this performance begin to degrade, overall performance indices would be used to alert the operator to either reduce the demands on this particular actuator (in the larger system operation) or to replace it (plug-and-play) with a new actuator at the earliest opportunity. This capability enables even a nominally trained technician to maintain rather complex (say up to 40 DOF) manufacturing cells.

16. Actuator testing

It becomes necessary to perform in-depth testing of these actuators. Four test-beds have been conceptualized, two have been built and two are being developed. The endurance test-bed essentially evaluates normal attributes such as reliability, saturation, temperature effects, cogging, etc. The condition-based maintenance test-bed evaluates partial failures and identifies their source in a structured fault tree. The nonlinear test-bed evaluates the actuators response to a wide range of non-linear dynamic loads (in order to obtain enhanced performance envelopes) and the metrology test-bed obtains all necessary geometric and compliance parameters for each actuator in order to analytically embed each actuator in the larger system.

17. Electronic controllers

System level controllers are becoming widely available at the gigaflop scale as a lowest commodity. Electronic controllers at the actuator level remain very modest based on an archaic analog concept of control. To support the 10 sensor environment, condition-based maintenance, and reconfiguration to avoid faults requires a revolution in electronic actuator controllers. In addition, they must be designed with standardized interfaces (to allow plug-and-play), be very compact (to be embedded in the actuator itself), and temperature tolerant (since there are numerous heat sources in the actuator module).

REFERENCES

Barnfather, J. D. et al. 2016. A performance evaluation methodology for robotic machine tools used in large volume manufacturing. Robotics and Computer-Integrated Manufacturing, 37: 49–56, February 2016.

Barnfather, J. D. et al. 2016. A performance evaluation methodology for robotic machine tools used in large volume manufacturing. Robotics and Computer Integrated Manufacturing, 37: 49–56.

Behi, F. and D. Tesar. 1991. Parametric identification for industrial manipulators using experimental modal analysis. IEEE Transactions on Robotics and Automation, 7(5): 642–652, October 1991.

Chang, K. and D. Tesar. 2006. Physical modelling of tools necessary for robot manipulation. Major Report, Robotics Research Group, UTexas, Austin, May 2006.

Chen, Y. and F. Dong. 2012. Robot machining: Recent development and future research issues. International Journal of Advanced Manufacturing Technology, online August 12, 2012 (Springerlink.com).

Cheng, H. and A. Chen. 2014. Adult robot enabled learning process in high precision assembly automation. Journal of Manufacturing Science and Engineering, 136: 1–10, April 2014.

De Vlieg, R. 2010. Expanding the use of robotics in air frame assembly via accurate robot technology. SAE International, 2010-01-1846.

Do, H. M. et al. 2012. Dual arm robot for packaging and assembling IT products. 8th IEEE International Conference on Automation Science and Engineering, August 20–24, 2012, Seoul, South Korea.

Dogar, M. et al. 2013. Towards coordinated precision assembly with robot teams. Computer Science and Artificial Intelligence Lab, MIT (mdogar@mail.mit.edu), 2013.

Dumas, C. et al. 2011. Joint stiffness identification of six-revolute industrial robots. Robotics and Computer Integrated Manufacturing, 27: 881–888, 2011.

Dumas, C. et al. 2012. Joint stiffness identification of industrial serial robots. Robotica, 30: 649–659.

Fresonke, D., E. Hernandez and D. Tesar. 1988. Deflection prediction for serial manipulators. 1988 IEEE J. Robotics & Automation, 1: 482–487, April 1988.

Giraud, C. and B. Jouvencel. 1995. Sensor selection: A geometric approach. IEEE 0-8186-7108-4/95, 1995, pp. 535–560.

Han, C. S. and A. E. Traver. 1989. Using CAD/CAM in the design of a stewart platform type of micromanipulator. Proceedings ASME 1989 Design Automation Conference, Anaheim, CA, July 1989, pp. 251–256.

Han, C. S., A. E. Traver and D. Tesar. 1989. Optimum design of a 6-DOF fully parallel micromanipulator for enhanced robot accuracy. Proceedings ASME 1989 Design Automation Conference, Montreal, Canada, September 1989, pp. 17–20.

Han, C. S., J. C. Hudgens, A. E. Traver and D. Tesar. Modelling, synthesis, analysis and design of high resolution micromanipulator to enhance robot accuracy. Proceedings of the 12th International Conference on Soil Mechanics and Foundation Engineering, Rio de Janeiro, Brazil, August 1989. Also, Proceedings IROS '91, IEEE/RSJ International Workshop on Intelligent Robots and Systems '91, Osaka, Japan, 2: 1157–1162, November 1991.

Hernandez, E., R. Sreedhar and D. Tesar. 1989. Computational requirements for the design and control of the fifth generation robot. Mechanical Engineering Systems, September 1989.

Hudgens, J. and D. Tesar. 1988. A fully parallel six degree-of-freedom micromanipulator: Kinematic analysis and dynamic model. Proceedings 20th ASME Mechanisms Conference, pp. 29–37, September 1988.

Hudgens, J. C., E. Hernandez and D. Tesar. 1991. A compliance parameter estimation method for serial manipulators. Applications of Modelling and Identification to Improve Machine Performance, DSC-Vol. 29, ASME 1991, New York, pp. 15–23, November 1991.

Hudgens, J. C., E. Hernandez and D. Tesar. 1991. A compliance parameter estimation method for serial manipulators. Applications of Modelling and Identification to Improve Machine Performance, DSC-Vol. 29, ASME, 1991, New York, pp. 15–23, December 1991.

Hudgens, J. C. and D. Tesar. 1991. Analysis of a fully parallel six degree-of-freedom micromanipulator. Proceedings 1991 Fifth International Conference on Advanced Robotics, Pisa, Italy, 1: 814–820, June 1991.

Hudgens, J. C., E. Hernandez and D. Tesar. 1991. A compliance parameter estimation method for serial manipulators. Applications of Modelling and Identification to Improve Machine Performance, DSC-Vol. 29 ASME 1991, New York, pp. 15–23, Dec. 1991.

Hudgens, J. and D. Tesar. 1992. Static robot compliance and metrology procedures with application to a light machining robot. The University of Texas at Austin, Report to Texas Higher Education Coordinating Board for Advanced Technology Program under Project No. 003658-156 and U.S. Department of Energy under Grant No. DE-FG02-86NE37966, August 1992.

Karim, A. and A. Verl. 2014. Challenges and obstacles in robot machining, ieeexplore.ieee.org, University of Stuttgart, Germany, 2014.

Kolba, M. P. and L. M. Collins. 2006. Information—Theoretic sensor management for multimodal sensing, 0-7803-9510-7, IEEE, 2006.

Krishnamoorthy, G. and D. Tesar. 2006. A full sensor fault tolerance for intelligent electromechanical actuators. Proceedings Electric Machines Technology Symposium, American Society of Naval Engineers (ASNE), Philadelphia, PA, May 22–24, 2006.

Krishnamoorthy, G. and D. Tesar. 2006. Intelligence for a full architecture of electromechanical actuators and drive systems, ACT 2006, Bremen, Germany, June 14–16, 2006.

Krishnamoorthy, G. et al. 2011. Guidelines for managing sensors in cyber physical systems with multiple sensors. Journal of Sensors, Vol. 2111, Art. ID 321709, 2011, p. 15.

Krishnamoorthy, G. et al. 2012. Simultaneous sensor and process fault detection and isolation in multi-input/multi-output systems, Part I: Theory; Part II: Application Considerations. IEEE Journal, ID GPT-10/29/2010, December 2012.

Lehmann, C. et al. 2013. Machining with industrial robots: The COMET project approach, WRSM 2013, CCIS 371, pp. 27–36.

Neto, P. and A. P. Moriera. 2013. Robotics in smart manufacturing. International Workshop, WRSM 2013, Porto Portugal, June, pp. 26–28.

Pan, Z. et al. 2006. Chatter analysis of robotic machining process. Journal of Materials Processing, Technology, 173: 301–309.

Richard et al. 2011. Concept of Modular Flexure-based Mechanisms for Ultra-high Precision Robot Design. (muracelle.richard@epfl.ch); Ecole Polytechnic, Lausanne, Switzerland, April 19, 2011.

Rodberg, T. 2011. One piece AFP spar manufacture. SAE International, 2011-01-2592.

Schneider, U. 2013. Compensation of errors in robot machining with a parallel 3-D-Piezo compensation mechanism. 46th CIRP Conference on Manufacturing Systems, 2013, proceeds, CIRP, pp. 305–310.

Schneider, U. et al. 2014a. Improving robotic machining accuracy through experimental error investigation and modular compensation. International Journal of Advanced Manufacturing Technology, online, June 10, 2014.

Schneider, U. et al. 2014b. Combining holistic programming with kinematic parameter optimization for robot machining. Conference ISR Robotik, 2014.

Schneider, U. et al. 2014c. Integrated approach to robotic machining with macro/micro-actuation. Robotics and Computer Integrated Manufacturing, 30: 636–697.

Shabbir, K. M. and G. Yang. 2010. Geometrical approach for online error compensation of industrial manipulators. Conference on Advanced Intelligent Mechatronics, Montreal, Canada, July 6–9, 2010.

Shin, S. H. and D. Tesar. 2004. Analytical method for designing modular robot interfaces with high connection accuracy. Proceedings, IEEE Conference On Robotics and Automation, New Orleans, LA, April 26–May 1, 2004.

Slavkovic, N. R. et al. 2013. Cartesian compliance identification and analysis of an articulated machining robot. FME Transactions, 2013, pp. 83–95.

Tesar, D. 1978. Mission-oriented research for light machinery. Science, Sept. 8, 1978, 201: 880–887.

Tesar, D. 1989. Thirty-year forecast: The concept of a fifth generation of robotics—The super robot. ASME Trans., Manufacturing Review, 2(1): 16–25, March 1989.

Tesar, D. and M. S. Butler. 1989. A generalized modular architecture for robot structures. ASME Trans., Manufacturing Review, 2(2): 91–118, June 1989.

Tesar, D. 1989. Thirty-year forecast: The concept of a fifth generation of robotics—The super robot. ASME Transactions on Manufacturing Review, 2(1): 16–25, March 1989.

Tesar, D. 2002. Human scale intelligent mechanical systems. Journal of Engineering & Automation Problems, Moscow Center for Scientific and Technical Information, Moscow, Russia, Vol. 4, 2002.

Tesar, D. 2005. Open architecture manufacturing systems. Plenary Presentation, Conference on Innovation in Manufacturing Processes, Seoul, Korea, September 29, 2005.

Tesar, D. 2015. The next wave of technology, Taylor and Francis Group, UK, DOI: 10 1080/10798587.2015.1118202, Two-page Summary, Mechanical Engineering, September 2015, No. 09, Vol. 137, Dynamic Systems and Control, pp. 1–2.

Thomas, M. and D. Tesar. 1982. Dynamic modelling of serial manipulator arms. ASME Journal of Dynamic Systems, Measurement and Control, 104: 218–228, September 1982.

Thomas, M., H. C. Yuan-Chou and D. Tesar. 1985. Optimal actuator stiffness distribution for robotic manipulators based on dynamic criteria. IEEE International Conference on Robotics & Automation, St. Louis, Missouri, pp. 275–281, March 25–28, 1985.

Thomas, M., H. C. Yuan-Chou and D. Tesar. 1985. Optimal actuator sizing for robotic manipulators based on local dynamic criteria. ASME Journal of Mechanisms, Transmissions and Automation in Design, 107: 163–169, June 1985.

Tisius, M., M. Pryor, C. Kapoor and D. Tesar. 2009. An empirical approach to performance criteria for manipulation. ASME Journal of Mechanisms and Robotics, 1: 1–12, August 2009.

Tolio, T. and M. Urgo. 2013. Design of flexible transfer lines: A case-based reconfiguration cost assessment. Journal of Manufacturing Systems, 32: 325–334.

Tosunoglu, S., S. Lin and D. Tesar. 1990. A controller design for compliant manipulators modelled with elastic links and joints. Proceedings of the 29th IEEE Conference on Decision and Control, Honolulu, Hawaii, Vol. 3, 1936–1942, December 1990.

Wander, J. and D. Tesar. 1991. Development of robot deflection compensation for improved machining accuracy. The University of Texas at Austin, Report to DOE under Grant No. DE-FG02-86NE37966 and Texas Higher Education Coordinating Board for Advanced Technology Program under Grant No. 4679, August 1991.

Wander, J., J. Hudgens, M. Aalund and D. Tesar. 1992. Precision routing by an industrial robot using real-time deflection compensation. Proceedings of Manufacturing International 1992, ASME, New York, pp. 101–114, March 1992.

Wells, D. and A. Walker. 2014. Integrating ultrasonic cutting with high accuracy robotic automatic fiber placement for production flexibility. SAMPE Tech, Seattle, WA June 2014.

Xiong, N. and P. Svensson. 2002. Multi-sensor management for information fusion issues and approaches. Information Fusion, Volume 3, 2002, pp. 165–186.

Zhang and Hui et al. 2005. Machining with flexible manipulator: Toward improving robotic machining performance. International Conference on Advanced Intelligent Mechatronics, Monterrey, CA, U.S.A., July 24–28, 2005.

APPENDIX A

BRIEF ASSESSMENT OF THE 2013 ROBOTICS ROADMAP

D. Tesar, UTexas, Austin, May 1, 2013

Concern: The robotics literature and the 2013 roadmap is/has become biased towards external sensor based autonomy and away from physical task performance that is enhanced by using human operator visualization. Intelligence is a universal need but it must exist for all technologies to create balance and to maximize performance while reducing cost. To accelerate this development requires a fully open architecture based on a minimum set of highly certified actuators (just as the chip is the driver of computers and social media) of high performance and low cost.

Background: Joe Engelberger sparked a wave of interest in multi-purpose machines (robotics) which is a great teacher to expand intelligence for all machines to maximize human choice. This generality forces a move towards open architecture to assemble (and reconfigure) machines on demand to best meet this increasing choice spectrum. Ten major reports (aircraft, ships, human rehabilitation, manufacturing cells, structured decision-making, space robots, battlefield robots, battlefield vehicles, commercial vehicles and freight trains) all show that the intelligent actuator is the core component in all these systems to respond to human command, to achieve high level task performance, at reduced cost. Four of these reports (listing attached) provide 600 pages of detailed science and development strategies on robotics which supersedes a great deal of the content of the 2013 Robotics Roadmap.

Comments On Roadmap: The roadmap has four primary sections: manufacturing, human rehabilitation, space and service robots, and military field robots. In manufacturing, it is important to note that the real U.S. market for industrial robots is $1 bil/yr and all are imported (To get this to $5 bil/yr will require a revolution in robotics capable of high value functions). Yes, Europe, Korea, Japan all have large R&D programs for robotics while the U.S. has almost no focused development. The 2013 roadmap provides almost no science or critical technology guidance for manufacturing. They mention that robots must adapt rapidly to product design changes (which means high accuracy actuators, extensive cell component metrology, quick change high accuracy interfaces, real-time decision loops (in 5 milliseconds), operator task visualization, excess internal and external sensors, fault management, etc.). They discuss the need for complex hands which could never be used for high value tasks. Instead, there is a tendency to recommend unproductive ideas such as nano-robots, humanoids, passive actuator compliance and full dependence on external sensor feedback and self-learning. They do mention the need for high performance actuators (with no guidance on the needed science), the creation of a needed educational robot system, and accurate end-effector positioning using dedicated sensors.

The section on rehabilitation and assist technology has very helpful suggestions on the balance between the system and individual, the need to reduce costs, to move the technologies to the home, to enable the disabled to work, etc. There is no doubt that this technology spectrum could create a win-win for human well-being and economic benefits. But the reality is that assist technology must provide morning to night functional support (there are at least 9 basic functions). Cost will come down only if we open up the architecture which means a minimum set of 6 low-cost/low weight actuators to enable a clinician to use a configuration manager to match each patient's real needs. A full body orthosis may require 24 actuators of 5 to 6 distinct sizes which can then be produced in a responsive supply chain at ever-increasing performance to cost ratios. Additive manufacturing can make the required "body sockets" on demand embedding a wide array of interface body sensors. This system may require the careful data reduction from 30 to 50 sensors (some in undergarments) in 10 to 30 milliseconds to coordinate the position, velocity, acceleration and force in up to 10 actuators. This cannot be done using brain wave signals. All of this can be operated by a domain specific operating SFW system (with 3rd party apps) that is constantly upgraded just as Microsoft's Windows is. Note that the roadmap process on p. 60 does not highlight the benefits of open architecture, minimum set of standardized actuators, or domain specific SFW.

The roadmap sections on service and space robotics are very disappointing. The space robot section is structured as a program plan to NASA headquarters with little guidance on required component and system technology development. They also make extraordinary claims of creating technologies for dozens of earth based fields mentioned elsewhere in the roadmap. They never mention open architecture which could reduce flyweight by 20x for long duration site missions. For service robots, almost all the concentration is on external sensors and autonomous operation. They mention that ROS is a freely available operating system. Several such SFW systems already exist elsewhere. They tend to emphasize small-scale systems when the dominant need is at the human scale. They do suggest a need for energy-efficient actuators with no suggestions for the required science. Again the idea of passive compliance as a key development topic is overstated.

The section on military robots (with emphasis on field applications) is a careful and balanced presentation. They clearly state their requirements (mobility, task performance, survivability, lower cost, low weight, easy maintenance, etc.). Here there is a clear call for open architecture, standardized components (commonality), domain specific software, reduced soldier decision burdening; a balanced need for intelligent actuators, internal and external sensors, data recovery from lost sensors, adaption to task uncertainty, prioritizing of sensor data and quicker decision-making. They want one soldier to be able to simultaneously operate several robots and they give a useful listing of critical development needs (especially for EOD). This implies the need for a family of specialized tools (not hands) automatically selected from a library to perform complex disassembly tasks.

Technology Overview: Many attributes of open architecture and intelligent actuators has been given above. All mechanical systems are highly nonlinear and their complexity will continue to increase. The need for power dense, high performance, low-cost actuators is clear. These then require performance maps for 20 to 40 operational criteria to be combined into envelopes to create decision surfaces to structure the decision process to enable latencies of 5 milliseconds or less. This requires a careful actuator metrology. It then requires an expansion of the architecture of actuators (duality for position, velocity, acceleration, torque—especially for fault tolerance; layered control at two or more scales, miniaturization, etc.). Sensors are becoming remarkably low-cost enabling their wide/distributed use (internally/externally) throughout the system. This provides excess sensor data (enabling inferences of data from a lost sensor) which supports feed forward decision-making at the subsystem level of 5 milliseconds and at the system level (without solving a set of highly non-linear complex differential equations) in 20 milliseconds. This leads to the concept of forward (for serial structures) and inverse (for parallel structures) decisions to enable the treatment of systems of hundreds of components in 100 milliseconds. All of this suggests that it is now possible to create systems with virtually no single point failures, use of CBM to provide for timely component replacement, archive lessons learned, and provide for the application of sensor/process/fault management operating systems. This is built using distinct/tested operating criteria, domain specific software, human goal setting, task planning; all oversighted by operator visualization.

The 2013 robotics roadmap frequently highlights non-core or tangential ideas (expensive complex hands, humanoids, passive compliance, brain signal control, etc.) and diminishes much of the technology critical to the mechanical portion of the system (multi-function actuators, reduced single point failures, forward/inverse decision-making, standardized interfaces, component minimum sets, etc.) and unfortunately it gives young engineers and scientists little guidance on what they should concentrate on to meet real needs. We need to do much better.

University of Texas Major Reports

1. A 230-page assessment of the weakness of U.S. manufacturing with emphasis on open architecture using standardized high accuracy actuators for high value added functions in the Future of Open Architecture and Reconfigurable Manufacturing Cells to provide rapid response to product revisions to quickly follow market demands.
2. A 90-page description on human assist (orthotics) technology under review by the VA associate administrator to meet the needs of 10 million disabled in the U.S. in Long Term Development Plan for Mobile Assist/Orthotic Rehabilitation for the Disabled.
3. Cost and weight are pressing problems for space robotics, which will soon be impacted by long-term availability and the need for refreshment for lunar base type facilities in a 165-page Open Architecture Development of Space Robotics.
4. A 220-page in-depth tech base study: Development Roadmap for Battlefield Robots (Emphasis on Open Architecture) to show how modularity can rapidly move cost-effective/life-saving robots forward.

APPENDIX B

Economic Impact of the Next Wave of Technology (NWT)[3] (Informal Assessment)

D. Tesar, UTexas, Austin, July 23, 2015

Overview: U.S. production of manufactured goods in 2015 will be $5.8 trillion, with 50% durable (computers, vehicles, machinery) and 50% non-durable (food, petroleum, chemicals, paper, plastics). The NWT tech base is central to the design, manufacture, and utilization of this full spectrum of products, many that operate under human command. This brief is intended to highlight the impact on these products of the ten technical categories described in the full NWT paper.[4]

Background: The U.S. is a major producer of manufactured goods as represented by the projected 2015 production levels:

Durable (Bil $)		Non-Durable (Bil $)	
Machinery	400*	Food/Beverages	934*
Computers/Electronics	350	Textiles/Mills	73*
Electricals	126*	Paper/Printing	282
Vehicles	609*	Petroleum	617
Aircraft	232*	Chemicals	762
Ships	34*	Plastics/Rubber	229

It is estimated that the NWT tech base directly supports $1 trillion of the economic sector represented by the starred items. It is recommended that decision makers recognize the magnitude of this opportunity to invest industrial and federal R&D resources to strengthen the general field of mechanical engineering and its neighboring disciplines.

Tech Base Prioritization: The included matrix chart lists the averaged rankings of the ten categories in the NWT in terms of 13 economically weighted applications. The result is given in 3 tiers:

Top Tier		2nd Tier		3rd Tier	
Availability	10.0	Command/Response	7.67	Man & Machine	6.95
Computational Intel.	9.36	Sensors	7.37	Visualization	6.76
Product Relevance	9.32	Machine Intel.	7.52	Actuators	6.16
Actuator Intelligence	8.99				

Most of the key applications involve high-end machine systems where maximum availability is a high priority (orthotics, vehicles, aircraft, battlefield systems, etc.). The same systems greatly benefit from quality computational intelligence (including surgical cells, battlefield systems, human visualization, and decision-making).

[3] Supported by 4 Expert Reviews (attached).

[4] Submitted to the IEEE Systems Journal.

Clearly, to respond to changing physical task requirements; actuators and actuator intelligence become essential components. None of the NWT ten categories are unimportant. This tabulation is intended to prioritize those categories that need the most long-term attention to revitalize a large economic sector in U.S. manufacturing.

Balance Among All Technologies: The central issue raised by the Next Wave of Technology (NWT) paper is the essential balance among all technologies in order to have a vibrant development and production tech base for all products having significant economic importance. Recently, it has become popular by futurists to discuss the "singularity" where computers (artificial intelligence) will overtake all other reasoning capabilities represented by mankind. Or, to describe robotics as the ultimate warfare technology, or to describe food (agriculture) production as fully automated to no longer need human labor, or to describe factories without lights, etc. This may be entertaining, but it does not meet the real design, production, and operational needs of our present and near-term products that sustain our economy. For example, the U.S. market for industrial robots today is $1 bil/year, whereas that for vehicles exceeds $600 bil/year.

A recent article[5] by the present author shows the required balance of all technologies to ensure a continued improvement of vehicle performance (safety, efficiency, drivability, availability, life cycle cost, refreshability, etc.). Autonomy and fuel efficiency has become popular development topics. The news media believe that this means more computer technology, more decision-making software, and less emphasis on the mechanical components. Just the opposite is true. Autonomy for a cross-country truck will require fast acting intelligent actuators to respond to command in 10 milli-sec., or less. For example, at 70 mph, 10 milli-sec. represents a travel length of 1 ft. Loss of traction (in poor weather) may be responded to in this 10 milli-sec. time frame. Today's outdated driveline (engine, transmission, clutches, torque tubes, differentials, etc.) can easily require 2 sec. (2000 milli-sec. or a travel distance of 200 ft.) to react to a traction loss or a rapid maneuver command. Hence, the singular effort for autonomy (system command to the outside environment) will have a modest role, where fuel efficiency, safety, response to maneuvers, etc. are involved.

The three central technologies that must come into balance are the mechanicals, electricals, and software structures. Those products that diminish the mechanicals will soon be superseded by those that bring the mechanicals back into balance with the other two technologies. The listing in this paper of $1 trillion in the U.S. product spectrum represents the economic importance of this recommendation. This is the main theme of the Next Wave of Technology as reviewed by 4 expert technologists.

[5] Open Architecture Vehicles of the Future, IFToMM Theory of Mechanisms and Machines, July 15, 2015, pp. 107–127.

Impact On Major Applications From The Next Wave of Technology

Application/ Eco. Import.	Rele-vance	Mach. Intel.	Comp Intel	Sen-sors	Man & Mach.	Visual-ization	Com Resp	Actu-ators	Avail-ability	Act. Intel.
MFG. Cells (10)	9	5	8	7	3	2	8	5	7	7
Orthotics, (5) Exoskeletons	8	10	10	7	10	7	10	10	10	7
Surg. Cells (2)	8	7	10	8	10	10	8	7	8	9
Commercial Vehicles (10)	7	7	8	8	5	3	4	5	10	7
Freight Trains (7)	6	5	5	5	4	3	4	3	8	5
Oil/Gas Ops(7)	5	4	4	4	2	2	2	3	8	8
Ship Ops. (4)	5	3	4	4	3	3	3	3	4	4
Aircraft (4)	3	4	4	4	6	3	5	4	10	8
Space Ops. (2)	3	3	7	5	2	2	2	2	10	7
Battle 'bots (4)	7	7	10	7	4	5	7	7	8	7
Battle Veh. (4)	10	10	10	10	7	7	8	8	10	10
Human Vis. (5)	7	5	8	4	10	10	7	4	3	7
Decision Mk(7)	10	7	10	3	10	10	10	5	7	10
Normalized	9.32	7.52	9.36	7.37	6.95	6.76	7.67	6.18	10.0	8.99

11

The Adaptive Control Algorithm for Manipulators with Joint Flexibility

*Krzysztof Kozłowski** and *Piotr Sauer*

ABSTRACT

Two control problems, present in this chapter are sent point and trajectory tracking control for robots with elastic joints. Firstly, discuss a simple non-adaptive case of the set point and illustrated by simulation results. Secondly, present a new adaptive tracking control scheme for robots with elastic joint. The proposed algorithm is the control algorithm extension proposed by Loria and Ortega. It assumes that link and motor positions are available for measurements and model of manipulator with elastic joints contain dynamic friction components on both link and motor sides. Construction of this algorithm is based on Lyapunov stability theory. In this chapter, shown is the proof of semiglobal asymptotic stability of the proposed system. The presented algorithm was tested and verified on experimental set-up. The experimental set-up consists of manipulator with two degree of freedom and control system based on card with signal processor DS1102. Two DC motors with harmonics gear were used as the actuators of manipulator. These gears are a source of flexibility in the joints. Software implementation of the proposed control algorithm is described using fast prototyping methodology in MATLAB and SIMULINK environment.

Keywords: Joint flexibility, set point control, trajectory control, adaptive control, feedback linearization, Lyapunov stability theory

Chair of Control and Systems Engineering, Poznań University of Technology, ul. Piotrowo 3a, 60-965 Poznań, Poland.
E-mail: piotr.sauer@put.poznan.pl
* Corresponding author: krzysztof.kozłowski@put.poznan.pl

INTRODUCTION

Many current manipulators are equipped with elastic motion transmission elements as harmonic drives, transmission belts and long shafts. These elements introduce elastic deformations at the joints. Such deformations are regarded as a source of problems, especially when accurate trajectory tracking of high sensitivity to end-effector forces is mandatory. It is shown that the joint elasticity should be taken into account in the modeling of a manipulator and designing of a control algorithm with the help of highly nonlinear equations. The modeling of flexible-joint robots is far more complex than that of rigid-joint robots. Two different models are used in the modeling of the flexible joint robots: complete and reduced models (De Luca and Tomei 1996). The reduced model was proposed by Spong (1987). It differs from the complete model in the assumption that the kinetic energy of the rotors is determined only by their own rotation (Ott 2008). One of the assumptions to build a complete model is that elasticity of a gear is modeled as a linear spring located between the rotor and the subsequent link (Ott 2008). Readman and Belanger (1990) presented a precise model based on Lagrange dynamics without simplifying assumptions. Very often a manipulator with flexible joints contains uncertain elements, i.e., elements which are not exactly known (for example system parameters which are difficult to measure with certainty, disturbance inputs which cannot be precisely predicted, nonlinear elements which are difficult to characterize exactly). If dynamic parameters are not exactly known, adaptive control laws must be designed to guarantee stabilization or tracking. Control of a system with uncertain elements has been discussed in the robotics literature (Barany and Colbaugh 1998, Corless 1989, Leitmann 1981). Spong proposed an adaptive controller for flexible-joint robots by using the singular perturbation formulation (Spong 1989, 1995). Khorasani proposed an adaptive controller using the concept of integral manifolds for n-link flexible-joint robots (Khorasani 1992). Madoński et al. (2014) presented an application of a special case of an Active Disturbance Rejection Controller (ADRC) for controlling a flexible joint manipulator designed as a rehabilitation device. Using the ADRC approach, the modeling uncertainty in the system is partially decoupled from the system, which increases the robustness of the whole control framework against both internal and external disturbances. Sauer and Kozlowski (2007) presented an adaptive tracking controller for a manipulator with only a revolute joint in which link and motor positions are available for measurements.

In this chapter, a new adaptive tracking controller that is an extension of the one proposed by Loria and Ortega (1995) is presented. The construction of this controller is based on Lyapunov theory (Barany and Colbaugh 1998, Colbaugh et al. 1994, Corless 1990, Slotine and Li 1991). The algorithm presented assumes that link and motor positions are available for measurements and the model of the manipulator with elastic joints contain dynamic friction components on both link and motor sides. In this chapter, the proof of semiglobal asymptotic stability of the proposed system is shown. This proof shows that the domain of attraction can be arbitrarily enlarged by increasing the gain.

The chapter is organized as follows. In Section **Nonadaptive Set Point Control of a Two Link Robot** a simple non-adaptive case of the set point is discussed. A more general mathematical description of the robot model, control algorithm and the stability properties of the adaptive tracking controller are described in Section **Adaptive Control Algorithm**. It also describes an experimental set-up and implemetation of the control algorithm. The experimental set-up consists of two links and a control system based on a card with signal processor DSP1102. Finally, concluding remarks are given.

NONADAPTIVE SET POINT CONTROL OF A TWO LINK ROBOT

In this section a simple example of one link robot with joint flexibility is considered assuming that all parameters of this mechanical structure are exactly known. This example is discussed in (Isidori 1995) and it is considered here because stabilization to equilibrium point discussed in reference (Isidori 1995) is not fully correct. The mechanical structure of the robot is depicted in Fig. 1.

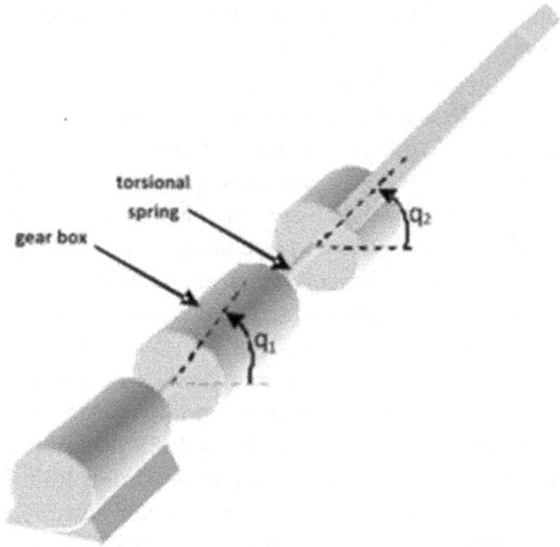

Figure 1. Mechanical structure of a simple one link robot with joint flexibility.

The system presented in Fig. 1 is described in terms of generalized coordinates, one variable q_1 represents angular position of the motor and is measured with respect to a fixed coordinate system shown in Fig. 1. It can be seen that the actuator shaft is connected directly to the gearbox that is linked via a torsional spring to the link. This is a simple one-link arm. Its joint variable is denoted by q_2 and stands for angular displacement of the arm measured again with respect to a fixed coordinate system. Elasticity constant of the spring here represents elastic coupling between the gearbox and link, and it is denoted by constant parameter *K*. Denoted by J_1 and F_1, inertia constant of the motor and viscous friction coefficient actuator equation can be written in the following form:

$$J_1\ddot{q}_1 + F_1\dot{q}_1 + \frac{K}{N}\left(\frac{q_1}{N} - q_2\right) = \tau \tag{1}$$

where τ represents torque generated by the motor. On the other hand, the link equation can be written as follows:

$$J_2\ddot{q}_2 + F_2\dot{q}_2 + K\left(q_2 - \frac{q_1}{N}\right) + mgd\cos q_2 = 0. \tag{2}$$

In the last equation J_2 and F_2 stand for inertia and velocity coefficient constants, respectively, m is the mass of the link, d represents the position of the center of gravity of the link, N stands for gear ratio and g is gravity constant. Note that Eq. (1) given here differs from the one presented in Isidori's book, namely the third term given in Eq. (1) has a minus sign when compared to the one given in this reference book. It is due to the fact that we consider this term in comparison to Eq. (2) as being on a different side of the torsional spring. This reasoning comes from a physical interpretation of the mechanical system.

Consequently, we follow analysis discussed in reference (Isidori 1995) and apply an apparatus of the differential geometry. Following this analysis one can choose the state vector as $\boldsymbol{x} = [q_1, q_2, \dot{q}_1, \dot{q}_2]$ consisting of angular positions and their velocities. Equations (1) and (2), taking into account the state vector with input $u = \tau$ can be written in the following form:

$$\dot{\mathbf{x}} = \mathbf{f}(\mathbf{x}) + \mathbf{g}(\mathbf{x})u$$

where

$$\mathbf{f}(\mathbf{x}) = \begin{bmatrix} x_3 \\ x_4 \\ -\dfrac{K}{J_1 N^2} x_1 + \dfrac{K}{J_1 N^2} x_2 - \dfrac{F_1}{J_1} x_3 \\ \dfrac{K}{J_2 N} x_1 - \dfrac{K}{J_2} x_2 - \dfrac{mgd}{J_2} \cos x_2 - \dfrac{F_2}{J_2} x_4 \end{bmatrix}, \quad \mathbf{g}(\mathbf{x}) = \begin{bmatrix} 0 \\ 0 \\ \dfrac{1}{J_1} \\ 0 \end{bmatrix}. \tag{3}$$

This system is under-actuated and has one control signal, namely the torque generated by the actuator. It is assumed that all parameters of the system given by Eq. (3) are known and one can apply one of the control algorithms originated in the differential geometry. It would be natural to select an output of the system as the angular position of the link q_2. Therefore one can write the output function as:

$$y = h(\mathbf{x}) = x_2. \tag{4}$$

Calculating the first Lie derivative of this output function results in

$$L_{\mathbf{f}} h(\mathbf{x}) = \frac{\partial h}{\partial \mathbf{x}} \mathbf{f}(\mathbf{x}) = \begin{bmatrix} 0 & 1 & 0 & 0 \end{bmatrix} \left(\mathbf{f}(\mathbf{x}) + \mathbf{g}(\mathbf{x}) u \right) = f_2(\mathbf{x}) = x_4$$

Here $\mathbf{f}(\mathbf{x})$ is represented as a vector $\mathbf{f}(\mathbf{x}) = [f_1(\mathbf{x})\ f_2(\mathbf{x})\ f_3(\mathbf{x})\ f_4(\mathbf{x})]^T$

Consequently, higher order Lie derivatives are:

$$L_{\mathbf{f}}^2 h(\mathbf{x}) = \frac{\partial}{\partial \mathbf{x}} \left[L_{\mathbf{f}} h(\mathbf{x}) \right] \mathbf{f}(\mathbf{x}) = f_4(\mathbf{x})$$

$$L_{\mathbf{f}}^3 h(\mathbf{x}) = \frac{\partial f_4}{\partial x_1} x_3 + \frac{\partial f_4}{\partial x_2} x_4 + \frac{\partial f_4}{\partial x_4} f_4(\mathbf{x})$$

due to the fact that $f_4(\mathbf{x})$ does not depend on x_3. From the previous calculations it also follows that

$$L_{\mathbf{g}}h(\mathbf{x}) = L_{\mathbf{g}}L_{\mathbf{f}}h(\mathbf{x}) = L_{\mathbf{g}}L_{\mathbf{f}}^2h(\mathbf{x}) = 0$$

and

$$L_{\mathbf{g}}L_{\mathbf{f}}^3h(\mathbf{x}) = \frac{\partial L_{\mathbf{f}}^3h(\mathbf{x})}{\partial x_3}\frac{1}{J_1} = \frac{\partial f_4}{\partial x_1}\frac{1}{J_1} = \frac{K}{J_1J_2N}.$$

The considered system has relative degree $r = 4 = n$ at each point x^0 in the state space, here the dimension of the system $n = 4$. Thus one can conclude that the system can be exactly linearized via state feedback and coordinate transformation around any point x^0 of state space. The linearizing coordinates are defined as follows:

$$z_1 = h(\mathbf{x}),\ z_2 = L_{\mathbf{f}}h(\mathbf{x}),\ z_3 = L_{\mathbf{f}}^2h(\mathbf{x}),\ z_4 = L_{\mathbf{f}}^3h(\mathbf{x}). \tag{5}$$

Taking into account Eq. (4), it is easy to notice that

$$h(\mathbf{x}) = y, \quad L_{\mathbf{f}}h(\mathbf{x}) = \frac{dy}{dt}, \quad L_{\mathbf{f}}^2h(\mathbf{x}) = \frac{d^2y}{dt^2}, \quad L_{\mathbf{f}}^3h(\mathbf{x}) = \frac{d^3y}{dt^3}.$$

Also note that output function and its time derivatives up to the third order have nice physical interpretation, namely they represent angular position, velocity, acceleration and jerk of the link with respect to the fixed coordinate frame.

The linearizing feedback is

$$u = \frac{-L_{\mathbf{f}}^4h(\mathbf{x}) + v}{L_{\mathbf{g}}L_{\mathbf{f}}^3h(\mathbf{x})} \tag{6}$$

and v has the following form:

$$v = c_0h(\mathbf{x}) + c_1L_{\mathbf{f}}h(\mathbf{x}) + c_2L_{\mathbf{f}}^2h(\mathbf{x}) + c_3L_{\mathbf{f}}^3h(\mathbf{x}).$$

Taking into account the previously defined linearizing coordinates, the linear system in z coordinates can be written as follows:

$$\begin{bmatrix} \dot{z}_1 \\ \dot{z}_2 \\ \dot{z}_3 \\ \dot{z}_4 \end{bmatrix} = \begin{bmatrix} 0 & 1 & 0 & 0 \\ 0 & 0 & 1 & 0 \\ 0 & 0 & 0 & 1 \\ c_0 & c_1 & c_2 & c_3 \end{bmatrix} \begin{bmatrix} z_1 \\ z_2 \\ z_3 \\ z_4 \end{bmatrix}.$$

Now it is easy to choose coefficients c_0, c_1, c_2, c_3 such as the constant matrix above has eigenvalues with negative values making the system asymptotically stable. As a consequence in view of the transformation given by Eq. (5) leads the original coordinates in x to be asymptotically stable to the desired value in x coordinates. The last equation can be written in a matrix form as:

$$\dot{\mathbf{z}} = \mathbf{A}\mathbf{z}. \tag{7}$$

Based on Eq. (7) one can calculate

$$
(s\mathbf{I}-\mathbf{A})^{-1}=\frac{1}{\Delta}\begin{bmatrix} s^2(s-c_3)-c_1-c_2s & s(s-c_3)-c_2 & s-c_3 & 1 \\ c_0 & s^2(s-c_3)-c_2s & s(s-c_3) & s \\ sc_0 & c_0+c_1s & s^2(s-c_3) & -s^2 \\ c_0s^2 & c_0s+c_1s^2 & s^2c_2+c_1s+c_0 & s^3 \end{bmatrix}
$$

where Δ stands for the determinate of the characteristic equation and

$\Delta = s^4 - s^3c_3 - s^2c_2 - sc_1 + c_0$.

Selecting proper coefficient c_0, c_1, c_2, c_3 as discussed above guarantees that the system (6) is asymptotically stable and in the steady state all variables z tend to zero. Taking into account Eq. (6) and assuming for the sake of simplicity that all mechanical parameters are unit elements the transformation (5) results in the following equations:

$$
\begin{aligned}
z_1 &= x_2 \\
z_2 &= x_4 \\
z_3 &= x_1 - x_2 - g\cos x_2 - x_4 \\
z_4 &= x_3 + gx_4\sin x_2 - x_1 - x_2 + g\cos x_2
\end{aligned}
\quad \text{or } \mathbf{z} = \mathbf{\Phi}(\mathbf{x}). \tag{8}
$$

Note that if x^0 is an equilibrium point of the original system (3) then $\mathbf{f}(\mathbf{x}^0) = \mathbf{0}$. However, it is immediately seen that a state of this type cannot be an equilibrium for the vector field $\mathbf{f}(\mathbf{x})$, because the constraints $\mathbf{f}(\mathbf{x}^0) = \mathbf{0}$ and $x_2^0 = 0$ are not compatible. Note that $x_2^0 = 0$ is equivalent by definition that the output function $h(\mathbf{x}^0) = x_2^0 = 0$.

Equating vector $\mathbf{f}(\mathbf{x})$ to zero with the assumption of unit elements of the mechanical elements results in

$$
x_1^0 = \frac{\pi}{2},\quad x_2^0 = \frac{\pi}{2},\quad x_3^0 = 0,\quad x_4^0 = 0.
$$

Now making use of the control law given by Eq. (6) results in $x_2^0 = 0$, $x_4^0 = 0$, $x_1^0 = g$, and $x_3^0 = 0$. Comparing both results it is evident that both conditions discussed above are not compatible. In this case, one may try to order a point of this type equilibrium by means of feedback. This condition can be established for some real number c such as the following condition is satisfied:

$$
\mathbf{f}(\mathbf{x}^0) + \mathbf{g}(\mathbf{x}^0)c = \mathbf{0}. \tag{9}
$$

Taking into account unit elements of the mechanical parameters Eq. (9) and assuming that $x_2^0 = 0$ results in:

$$
\begin{aligned}
x_3^0 &= 0 \\
x_4^0 &= 0 \\
x_1^0 + c &= 0 \\
x_1^0 - g &= 0
\end{aligned}
$$

that can be uniquely solved for c and x_1^c namely $x_1^0 = g$ and $c = -g$. The control law proposed by Isidor is as follows:

$$u = \frac{-L_{\mathbf{f}}^4 h(\mathbf{x}) + v}{L_{\mathbf{g}} L_{\mathbf{f}}^3 h(\mathbf{x})} + c. \tag{10}$$

As a consequence, the linearized system will be defined around the point $\mathbf{z} = \mathbf{0}$. Adding constant c to the right side of Eq. (6) as proposed in Eq. (10) is not correct in general because it does not guarantee the linearization of the original system. This is due do the fact that because in the last equation of (7) the term $\frac{c}{L_{\mathbf{g}} L_{\mathbf{f}}^3 h(\mathbf{x})}$ will appear (in general it should be $\frac{c}{L_{\mathbf{g}} L_{\mathbf{f}}^{n-1} h(\mathbf{x})}$ in case when the system has a relative degree equal to n). In this particular example $L_{\mathbf{g}} L_{\mathbf{f}}^3 h(\mathbf{x})$ is constant but in general it can be a nonlinear term, therefore this control law cannot linearize the system. The correct version of the control law should be written as follows

$$u = \frac{-L_{\mathbf{f}}^4 h(\mathbf{x}) + v + c}{L_{\mathbf{g}} L_{\mathbf{f}}^3 h(\mathbf{x})}. \tag{11}$$

As a consequence in the last equation of the system (7) constant term appears as c. Taking into account the general solution the steady state of the system (7) with control law (11) can be calculated as follows

$$\lim_{s \to 0} \frac{1}{\Delta} \begin{bmatrix} 1 \\ s \\ -s^2 \\ s^3 \end{bmatrix} c.$$

Taking into account the analytical expression of the determinant in the considered case one can calculate that

$$\lim_{t \to \infty} z_1(t) = -\frac{c}{c_0}, \ \lim_{t \to \infty} z_2(t) = 0, \ \lim_{t \to \infty} z_3(t) = 0 \quad \textit{and} \quad \lim_{t \to \infty} z_4(t) = 0 .$$

As can be seen, the steady state of the first variable z_1^0 can be changed and it is equal to $-\frac{c}{c_0}$. Now when this is taken into account, Eq. (8) takes the following form in the steady state:

$$\begin{aligned}
z_1^0 &= -\frac{c}{c_0} = x_2^0 \\
z_2^0 &= x_4^0 \\
z_3^0 &= x_1^0 - x_2^0 - g\cos x_2^0 - x_1^0 \\
z_4^0 &= x_3^0 + g x_4^0 \sin x_2^0 - x_1^0 + x_2^0 + g\cos x_2^0.
\end{aligned} \tag{12}$$

From Eq. (12) it can be seen that the linearized system is not necessarily defined around the point $z^0 = 0$. In other words one cannot set up an arbitrary value to the angular displacement x_2^0 applying to the nonlinear system (3) control law defined by Eq. (11).

The theoretical considerations were verified by simulation but since only the steady state is of interest it can be calculated easily by hand as done here.

ADAPTIVE CONTROL ALGORITHM

Mathematical Description of an Adaptive Tracking Controller

In general the system considered here can be described by the following equations

$$\mathbf{M}_l(\mathbf{q}_l)\ddot{\mathbf{q}}_l + \mathbf{C}_l(\mathbf{q}_l,\dot{\mathbf{q}}_l)\dot{\mathbf{q}}_l + \mathbf{F}_l + \mathbf{G}_l(\mathbf{q}_l) = \mathbf{K}(\mathbf{N}^{-1}\mathbf{q}_m - \mathbf{q}_l) \tag{13}$$

$$\mathbf{J}_m\ddot{\mathbf{q}}_m + \mathbf{N}^{-1}\mathbf{K}(\mathbf{q}_m\mathbf{N}^{-1} - \mathbf{q}_l) + \mathbf{F}_m = \mathbf{u}, \tag{14}$$

where $\mathbf{M}_l(\mathbf{q}_l)$ is a $n \times n$ link inertia matrix, $\mathbf{C}_l(\mathbf{q}_l, \dot{\mathbf{q}}_l)$ is a $n \times n$ matrix containing the centripetal and Coriolis terms, $\mathbf{G}_l(\mathbf{q}_l)$ is a $n \times 1$ vector containing the gravity terms. Friction forces are represented by vectors $\mathbf{F}_l(n \times 1)$ for links and $\mathbf{F}_m(n \times 1)$ for actuators. Vector $\mathbf{q}_l$ $(n \times 1)$ represents the link displacements, and $\mathbf{q}_m$ is a $n \times 1$ vector of the actuator displacements. $\mathbf{J}_m$ is a $n \times n$ positive definite constant diagonal actuator inertia matrix. $\mathbf{N}$ is a positive–definite constant diagonal $n \times n$ matrix containing the gear ratios for each harmonic drive and $\mathbf{K}$ is a $n \times n$ diagonal matrix of the torsional stiffness. Finally, $\mathbf{u}$ is a $n \times 1$ vector of input torque control. The system described by Eqs. (13), (14) follows the description given in reference (Tian and Goldenberg 1995) assuming that the joint torques can be measured. Note that Eqs. (13) and (14) are generalization of Eqs. (1) and (2) developed for non-adaptive case.

This model was defined assuming that:

- matrix $\mathbf{M}_l(\mathbf{q}_l)$ is positive definite, while matrix which has the form $\mathbf{N} = \dot{\mathbf{M}}_l(\mathbf{q}_l) - 2\mathbf{C}_l(\mathbf{q}_l, \dot{\mathbf{q}}_l)$, is a skew-symmetric matrix,
- we can write $\mathbf{C}(\mathbf{x},\mathbf{y})\mathbf{z} = \mathbf{C}(\mathbf{x},\mathbf{z})\mathbf{y}$ for any vector $\mathbf{z} \in R^n$, and $\mathbf{C}(\mathbf{x},\mathbf{y})$ is bounded for the sake of $\mathbf{x}$ and linear in $\mathbf{y}$, and $\|\mathbf{C}(\mathbf{x},\mathbf{y})\| \leq k_c \|\mathbf{y}\|$, $k_c > 0$, where $\|\cdot\|$ denotes Euclidean norm,
- the manipulator works in horizontal plane only (therefore gravity term can be neglected).

Consider a manipulator with two degrees of freedom presented in Fig. 2. In this particular example we have:

- Cartesian coordinates x_1, y_1, and x_2, y_2 are described as follows (compare Fig. 2):

$$x_1 = l_{C_1} sin(q_{l1}), \tag{15}$$

$$y_1 = l_{C_1} cos(q_{l1}), \tag{16}$$

$$x_2 = l_1 sin(q_{l1}) + l_{C_2} sin(q_{l1} + q_{l2}), \tag{17}$$

$$y_2 = l_1 cos(q_{l1}) + l_{C_2} cos(q_{l1} + q_{l2}). \tag{18}$$

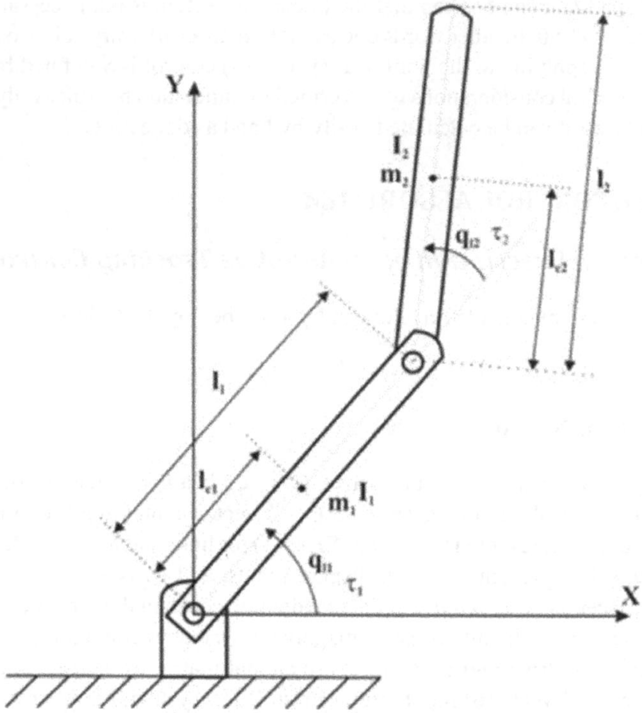

Figure 2. A scheme of a manipulator with two degrees of freedom.

- inertia matrix

$$\mathbf{M}_l = \begin{bmatrix} I_{l1} + I_{l2} + m_1 l_{C_1}^2 + m_2[l_1^2 + l_{C_2}^2 + 2l_1 l_{C_2} cos(q_{l2})] & I_{l2} + m_2 l_{C_2}^2 + m_2 l_1 l_{C_2} cos(q_{l2}) \\ m_2 l_{C_2}^2 + m_2 l_1 l_{C_2} cos(q_{l2}) + I_{l2} & m_2 l_{C_2}^2 + I_{l2} \end{bmatrix} \quad (19)$$

- matrix containing the centripetal and Coriolis terms

$$\mathbf{C}_l = \begin{bmatrix} -2m_2 l_1 l_{C_2} sin(q_{l2})\dot{q}_{l2} & -m_2 l_1 l_{C_2} sin(q_{l2})\dot{q}_{l2} \\ m_2 l_1 l_{C_2} sin(q_{l2})\dot{q}_{l1} & 0 \end{bmatrix}, \quad (20)$$

- matrix containing the gravity terms, which in our case is

$$\mathbf{G}_l = \begin{bmatrix} 0 \\ 0 \end{bmatrix}, \quad (21)$$

- vector containing friction forces for links

$$\mathbf{F}_l = \begin{bmatrix} F_{ld1}\dot{q}_{l1} + F_{ls1} sgn(\dot{q}_{l1}) \\ F_{ld2}\dot{q}_{l2} + F_{ls2} sgn(\dot{q}_{l2}) \end{bmatrix}, \quad (22)$$

- vector containing friction forces for actuators

$$\mathbf{F}_m = \begin{bmatrix} F_{md1}\dot{q}_{m1} + F_{ms1}sgn(\dot{q}_{m1}) \\ F_{md2}\dot{q}_{m2} + F_{ms2}sgn(\dot{q}_{m2}) \end{bmatrix}, \tag{23}$$

- stiffness matrix

$$\mathbf{K} = \begin{bmatrix} K_1 & 0 \\ 0 & K_2 \end{bmatrix}, \tag{24}$$

- gear ratio matrix

$$\mathbf{N} = \begin{bmatrix} N_1 & 0 \\ 0 & N_2 \end{bmatrix}, \tag{25}$$

- actuators inertia matrix

$$\mathbf{J}_m = \begin{bmatrix} J_{m1} & 0 \\ 0 & J_{m2} \end{bmatrix}. \tag{26}$$

Equations (13) and (14) can be rewritten in the following form

$$\begin{aligned} &\mathbf{M}_l\ddot{\mathbf{q}}_l + \mathbf{C}_l\dot{\mathbf{q}}_l + \mathbf{G}_l + \mathbf{F}_l = \mathbf{K}\left(\mathbf{N}^{-1}\mathbf{q}_m - \mathbf{q}_l\right), \\ &\mathbf{k}_m^{-1}\left[\mathbf{J}_m\ddot{\mathbf{q}}_m + \mathbf{F}_m + \mathbf{N}^{-1}\mathbf{K}\left(\mathbf{N}^{-1}\mathbf{q}_m - \mathbf{q}_l\right)\right] = \mathbf{i}, \end{aligned} \tag{27}$$

where $\mathbf{k}_m$ is a diagonal matrix (in our case (2×2)), which contains electromotive force constants (SEM), and $\mathbf{i}$ is a vector (in our case (2×1)) which contains currents. A set of Eqs. (13) and (14) is similar to that presented in references (Loria and Ortega 1995, Tian and Goldenberg 1995) with the extensions described above. The tracking error is defined as:

$$\tilde{\mathbf{q}}_l = \mathbf{q}_l - \mathbf{q}_{ld}, \tag{28}$$

$$\tilde{\mathbf{q}}_m = \mathbf{q}_m - \mathbf{q}_{md}, \tag{29}$$

where $\mathbf{q}_{ld}$ and $\mathbf{q}_{md}$ denote $n \times 1$ vectors of the desired link and motor trajectories, respectively. It is also assumed that all $\mathbf{q}_{ld} \in C^4$, and $\|\mathbf{q}_{ld}\|, \|\dot{\mathbf{q}}_{ld}\|, \ddot{\mathbf{q}}_{ld}\| \le B_d$.

The new adaptive version of the control law is derived from the Lyapunov stability theory (Kozłowski and Sauer 1998). The adaptation law for links can be described in the following form:

$$\dot{\tilde{\mathbf{a}}}_l = -\eta\hat{\mathbf{a}}_l + \varepsilon\mathbf{\Gamma}_l^{-1}\mathbf{Y}_l^T(\tilde{\mathbf{v}}_l + \tilde{\mathbf{q}}_l). \tag{30}$$

and the adaptation law for actuators:

$$\dot{\tilde{\mathbf{a}}}_m = -\eta\hat{\mathbf{a}}_m + \varepsilon\mathbf{\Gamma}_m^{-1}\mathbf{Y}_m^T(\tilde{\mathbf{v}}_m + \tilde{\mathbf{q}}_m), \tag{31}$$

where $\mathbf{Y}_l$ and $\mathbf{Y}_m$ are appropriate matrices (regressor factors) associated with links and motors, respectively. In Eqs. (30) and (31) ε is a small positive constant, $\boldsymbol{\Gamma}_l$, $\boldsymbol{\Gamma}_m$ are positive definite constant gain matrices and $\tilde{\mathbf{a}}_l(\tilde{\mathbf{a}}_m)$ is an estimate error $\tilde{\mathbf{a}}_l = \mathbf{a}_l - \hat{\mathbf{a}}_l(\tilde{\mathbf{a}}_m = \mathbf{a}_m - \hat{\mathbf{a}}_m)$ between parameter $\mathbf{a}_l(\mathbf{a}_m)$ and its estimate $\hat{\mathbf{a}}_l(\hat{\mathbf{a}}_m)$. The adaptive version of control algorithms for the system given by Eqs. (13) and (14) can be written in the following form:

$$\mathbf{u} = \hat{\mathbf{J}}_m\ddot{\mathbf{q}}_{md} + \hat{\mathbf{N}}^{-1}\hat{\mathbf{K}}(\hat{\mathbf{N}}^{-1}\mathbf{q}_{md} - \mathbf{q}_{ld}) + \hat{\mathbf{F}}_m(\dot{\mathbf{q}}_{md}) - \mathbf{K}_{pm}\tilde{\mathbf{q}}_m - \mathbf{K}_{dm}\tilde{\mathbf{v}}_m \tag{32}$$

where $\ddot{\mathbf{q}}_{md}$ is calculated by performing the differentiation operation twice to the following equation:

$$\mathbf{q}_{md} = \hat{\mathbf{N}}\hat{\mathbf{K}}^{-1}\left[\hat{\mathbf{M}}_l(\mathbf{q}_l)\ddot{\mathbf{q}}_{ld} + \hat{\mathbf{C}}_l(\mathbf{q}_l,\dot{\mathbf{q}}_{ld}) + \hat{\mathbf{G}}_l(\mathbf{q}_l) + \hat{\mathbf{F}}_l(\dot{\mathbf{q}}_{ld}) - \mathbf{K}_{pl}\tilde{\mathbf{q}}_l - \mathbf{K}_{dl}\tilde{\mathbf{v}}_l\right] + \hat{\mathbf{N}}\mathbf{q}_{ld} \tag{33}$$

where $\tilde{v}_j$ for $j = l, m$ is calculated from Eq. (34).

$$\tilde{\mathbf{v}}_j = diag\left\{\frac{b_{ij}s}{s + a_{ij}}\right\}\tilde{\mathbf{q}}_j, \quad j = l, m, \tag{34}$$

where $\tilde{v}_l$ and $\tilde{v}_m$ denote vectors of the filtered position error for the link and actuator, respectively.

In (32) and (33) " ~ " denotes an appropriate estimate. Equations (32) and (33) can be presented in general form:

$$\mathbf{q}_{md} = \mathbf{Y}_l(\mathbf{q}_{ld},\dot{\mathbf{q}}_{ld},\ddot{\mathbf{q}}_{ld})\tilde{\mathbf{a}}_l - \mathbf{K}_{pl}\tilde{\mathbf{q}}_l - \mathbf{K}_d\tilde{\mathbf{v}}_l,$$

$$\mathbf{u} = \mathbf{Y}_m(\mathbf{q}_{md},\dot{\mathbf{q}}_{md},\ddot{\mathbf{q}}_{md})\tilde{\mathbf{a}}_m - \mathbf{K}_{pm}\tilde{\mathbf{q}}_m - \mathbf{K}_{dm}\tilde{\mathbf{v}}_m. \tag{35}$$

Adaptive gain coefficients for position control of the motor and current control for the first link and second link can be defined, respectively:

$$\boldsymbol{\Gamma}_l = diag\left[\gamma_{M_l} \quad \gamma_N \quad \gamma_{F_{ls}} \quad \gamma_{F_{ld}}\right], \tag{36}$$

$$\boldsymbol{\Gamma}_m = diag\left[\gamma_{J_m} \quad \gamma_K \quad \gamma_{F_{ms}} \quad \gamma_{F_{md}}\right]. \tag{37}$$

Matrices $\mathbf{Y}_m$, $\mathbf{Y}_l$, $\tilde{\mathbf{a}}_m$, $\tilde{\mathbf{a}}_l$ have the following form:

$$\mathbf{Y}_m(\mathbf{q}_{md},\dot{\mathbf{q}}_{md},\ddot{\mathbf{q}}_{md}) = \left[\ddot{\mathbf{q}}_{md} \quad \left(\frac{\mathbf{q}_{md}}{\hat{\mathbf{N}}} - \mathbf{q}_{ld}\right) \quad sgn(\dot{\mathbf{q}}_{md}) \quad \dot{\mathbf{q}}_{md}\right], \tag{38}$$

$$\mathbf{Y}_l(\mathbf{q}_{ld},\dot{\mathbf{q}}_{ld},\ddot{\mathbf{q}}_{ld}) = \left[\ddot{\mathbf{q}}_{ld} \quad \mathbf{q}_{ld} \quad sgn(\dot{\mathbf{q}}_{ld}) \quad \dot{\mathbf{q}}_{ld}\right], \tag{39}$$

$$\tilde{\mathbf{a}}_m^T = \left[\tilde{\mathbf{J}}_m \quad \tilde{\mathbf{K}} \quad \tilde{\mathbf{F}}_{ms} \quad \tilde{\mathbf{F}}_{md}\right], \tag{40}$$

$$\tilde{\mathbf{a}}_l^T = \left[\tilde{\mathbf{M}}_l \quad \tilde{\mathbf{N}} \quad \tilde{\mathbf{F}}_{ls} \quad \tilde{\mathbf{F}}_{ld}\right]. \tag{41}$$

The Stability Properties of the Adaptive Tracking Controller

The stability properties of the adaptive tracking controller, which was presented in Section Mathematical Description of an Adaptive Tracking Controller, can be derived using the following theorem (Corless 1990).

Theorem 1. Consider the dynamic system $\dot{x}_1 = f_1(x_1, x_2, t)$, $\dot{x}_2 = f_2(x_1, x_2, t)$. Let $V(x_1, x_2, t)$ be a Lyapunov function for the system with the properties:

$$\mu_1 \| x_1 \|^2 + \mu_2 \| x_2 \|^2 \leq V \leq \mu_3 \| x_1 \|^2 + \mu_4 \| x_2 \|^2$$
$$\dot{V} \leq -\mu_5 \| x_1 \|^2 - \mu_6 \| x_2 \|^2 + \varepsilon$$

where ε and μ_i are positive scalar constants. Define $\sigma_1 = max(\mu_3/\mu_5, \mu_4/\mu_6)$ and $r_i = (\sigma_1 \varepsilon/\mu_i)^{1/2}$ for $i = 1, 2$. Then for any initial state $x_1(0)$, $x_2(0)$ the model will evolve so that $x_1(t)$, $x_2(t)$ are uniformly bounded and converge exponentially to the closed balls B_{r_1}, B_{r_2}, respectively, where

$$B_{r_1} = \{x_i : \| x_i \| \leq r_i\}.$$

The control law given by Eq. (32) is semiglobally asymptotically stable because the domain of atraction can be arbitrarily enlarged by increasing the gain of the controller. First we consider the case in which we neglect flexibility ($K \rightarrow \infty$) (Loria and Ortega 1995, Nicosia and Tomei 1994). In this case the model given by Eqs. (13) and (14) is reduced to

$$\mathbf{D}_l(\mathbf{q}_l)\ddot{\mathbf{q}}_l + \mathbf{C}_l(\mathbf{q}_l, \dot{\mathbf{q}}_l)\dot{\mathbf{q}}_l + \mathbf{G}(q_l) + \mathbf{F} = \mathbf{u} \quad (42)$$

where $\mathbf{D}_l(\mathbf{q}_l) = \mathbf{M}_l(\mathbf{q}_l) + \mathbf{N}^2 \mathbf{J}_m$ and $\mathbf{F} = \mathbf{F}_l(\dot{\mathbf{q}}_l) + \mathbf{N}\mathbf{F}_m(\dot{\mathbf{q}}_m)$.

It is assumed that $\mathbf{F}$ is a combined $n \times 1$ vector containing the dynamic friction terms associated with link bearings and actuators. For Eq. (42) the following adaptive control law will be proposed:

$$\mathbf{u} = \hat{\mathbf{D}}_l(\mathbf{q}_l)\ddot{\mathbf{q}}_{ld} + \hat{\mathbf{C}}(\mathbf{q}_l, \dot{\mathbf{q}}_{ld})\dot{\mathbf{q}}_{ld} + \hat{\mathbf{G}}(\mathbf{q}_l) + \hat{\mathbf{F}}(\dot{\mathbf{q}}_{ld}) - \mathbf{K}_{pl}\tilde{\mathbf{q}}_l - \mathbf{K}_{dl}\tilde{\mathbf{v}}_l. \quad (43)$$

First the following Lyapunov function candidate will be chosen:

$$V(\tilde{\mathbf{q}}_l, \dot{\tilde{\mathbf{q}}}_l, \tilde{\mathbf{v}}_l, t) = \frac{1}{2}\dot{\tilde{\mathbf{q}}}_l^T \mathbf{D}_l \dot{\tilde{\mathbf{q}}}_l + \frac{1}{2}\tilde{\mathbf{q}}_l^T \mathbf{K}_{pl} \tilde{\mathbf{q}}_l + \frac{1}{2}\tilde{\mathbf{v}}_l \mathbf{K}_{dl} \mathbf{B}_l^{-1} \tilde{\mathbf{v}}_l + \varepsilon \tilde{\mathbf{q}}_l^T \mathbf{D}_l \dot{\tilde{\mathbf{q}}}_l + \varepsilon \tilde{\mathbf{v}}_l^T \mathbf{D}_l \dot{\tilde{\mathbf{q}}}_l + \frac{1}{2}\tilde{\mathbf{a}}_l^T \mathbf{\Gamma}_l \tilde{\mathbf{a}}_l \quad (44)$$

where $\mathbf{\Gamma}_l$ is a symmetric positive–definite matrix, ε is a constant value coefficient. The function $V(t)$ in (44) can be bounded in the following manner:

$$\mathbf{z}^T \mathbf{Q}_1 \mathbf{z} + \frac{1}{2}\mathbf{\Gamma}_{lm} \| \tilde{\mathbf{a}}_l \|^2 \leq V(t) \leq \mathbf{z}^T \mathbf{Q}_2 \mathbf{z} + \frac{1}{2}\mathbf{\Gamma}_{lM} \| \tilde{\mathbf{a}}_l \|^2 . \quad (45)$$

where vector $\mathbf{z}$ is defined as $\mathbf{z} = [\tilde{\mathbf{q}}^T, \dot{\tilde{\mathbf{q}}}^T, \tilde{\mathbf{v}}^T]^T$, and matrices $\mathbf{Q}_1$, $\mathbf{Q}_2$:

$$\mathbf{Q}_1 = \begin{bmatrix} \frac{1}{2}k_{plm} & \frac{\varepsilon}{2}d_{lm} & 0 \\ \frac{\varepsilon}{2}d_{lm} & \frac{1}{2}d_{lm} & -\frac{\varepsilon}{2}d_{lm} \\ 0 & -\frac{\varepsilon}{2}d_{lm} & \frac{1}{2}k_{dlm}b_{lM}^{-1} \end{bmatrix},$$

$$\mathbf{Q}_2 = \begin{bmatrix} \frac{1}{2}k_{plM} & -\frac{\varepsilon}{2}d_{lM} & 0 \\ -\frac{\varepsilon}{2}d_{lM} & \frac{1}{2}d_{lM} & -\frac{\varepsilon}{2}d_{lM} \\ 0 & -\frac{\varepsilon}{2}d_{lM} & \frac{1}{2}k_{dlM}b_{m}^{-1} \end{bmatrix}.$$

The matrices $\mathbf{Q}_1$ and $\mathbf{Q}_2$ are positive definite, respectively, if

$$\varepsilon < \sqrt{\frac{k_{plm}}{d_{lm}}}, \varepsilon < \sqrt{\frac{k_{plm}k_{dlm}b_{lM}^{-1}}{d_{lm}(k_{plm} + k_{dlm}b_{lM}^{-1})}},$$

$$\varepsilon < \sqrt{\frac{k_{plM}}{d_{lM}}}, \varepsilon < \sqrt{\frac{k_{plM}k_{dlM}b_{lm}^{-1}}{d_{lM}(k_{plM} + k_{dlM}b_{lm}^{-1})}}. \tag{46}$$

As a consequence, the Lyapunov function (45) can be bounded as

$$\lambda_m(\mathbf{Q}_1)\|\mathbf{z}\|^2 + \frac{1}{2}\mathbf{\Gamma}_{lm}\|\mathbf{a}_l\|^2 \le V(t) \le \lambda_M(\mathbf{Q}_2)\|\mathbf{z}\|^2 + \frac{1}{2}\mathbf{\Gamma}_{lM}\|\mathbf{a}_l\|^2. \tag{47}$$

where $\lambda_m(\mathbf{Q}_1)$, $\lambda_M(\mathbf{Q}_2)$ denote the smallest and the biggest eigenvalues of matrices $\mathbf{Q}_1$ and $\mathbf{Q}_2$, respectively.

Taking the first derivative of the Lyapunov function, one can get

$$\dot{V} = \dot{\tilde{\mathbf{q}}}_l^T \mathbf{D}\ddot{\tilde{\mathbf{q}}}_l + \frac{1}{2}\dot{\tilde{\mathbf{q}}}_l^T \dot{\mathbf{D}}\dot{\tilde{\mathbf{q}}}_l + \tilde{\mathbf{q}}_l^T \mathbf{K}_{pl}\dot{\tilde{\mathbf{q}}}_l + \tilde{\mathbf{v}}_l^T \mathbf{K}_{dl}\mathbf{B}_l^{-1}\dot{\tilde{\mathbf{v}}} + \varepsilon\dot{\tilde{\mathbf{q}}}_l^T \mathbf{D}\dot{\tilde{\mathbf{q}}}_l + \varepsilon\tilde{\mathbf{q}}_l^T \dot{\mathbf{D}}\dot{\tilde{\mathbf{q}}}_l + \varepsilon\tilde{\mathbf{q}}_l^T \mathbf{D}\ddot{\tilde{\mathbf{q}}}_l^T +$$

$$+ \varepsilon\dot{\tilde{\mathbf{v}}}_l^T \mathbf{D}\dot{\tilde{\mathbf{q}}}_l + \varepsilon\tilde{\mathbf{v}}_l^T \dot{\mathbf{D}}\dot{\tilde{\mathbf{q}}}_l + \varepsilon\tilde{\mathbf{v}}_l^T \mathbf{D}\ddot{\tilde{\mathbf{q}}}_l + \tilde{\mathbf{a}}_l^T \mathbf{\Gamma}_l \dot{\tilde{\mathbf{a}}}_l. \tag{48}$$

Taking into account the skew symmetric property of the matrix $\dot{\mathbf{D}}(\mathbf{q}_l) - 2\mathbf{C}(\mathbf{q}_l, \dot{\mathbf{q}}_l)$ and Eqs. (34), (42) and (43) the last equation can be rewritten in the following form:

$$\dot{V} = -\dot{\tilde{\mathbf{q}}}_l^T \mathbf{F}(\dot{\mathbf{q}}_l) + \dot{\tilde{\mathbf{q}}}_l^T \mathbf{F}(\dot{\mathbf{q}}_{ld}) - \varepsilon\tilde{\mathbf{q}}_l^T \mathbf{F}(\dot{\mathbf{q}}_l) + \varepsilon\tilde{\mathbf{q}}_l^T \mathbf{F}(\dot{\mathbf{q}}_{ld}) - \varepsilon\tilde{\mathbf{v}}_l^T \mathbf{F}(\dot{\mathbf{q}}_l) +$$

$$+ \varepsilon\tilde{\mathbf{v}}_l^T \mathbf{F}(\dot{\mathbf{q}}_{ld}) + \tilde{\mathbf{a}}_l^T \mathbf{\Gamma}\dot{\tilde{\mathbf{a}}}_l - \varepsilon\tilde{\mathbf{q}}_l^T \mathbf{Y}_l\tilde{\mathbf{a}}_l - \varepsilon\tilde{\mathbf{v}}_l^T \mathbf{Y}_l\tilde{\mathbf{a}}_l - \dot{\tilde{\mathbf{q}}}_l^T \mathbf{Y}_l\tilde{\mathbf{a}}_l +$$

$$-\tilde{\mathbf{v}}_l^T\mathbf{K}_{dl}\mathbf{B}_l^{-1}\mathbf{A}_l\tilde{\mathbf{v}}_l-\dot{\tilde{\mathbf{q}}}_l^T\mathbf{C}_d\dot{\tilde{\mathbf{q}}}_l+\varepsilon\dot{\tilde{\mathbf{q}}}_l^T\mathbf{D}\dot{\tilde{\mathbf{q}}}_l+\varepsilon\tilde{\mathbf{q}}_l^T\left(\mathbf{C}_d^T-\mathbf{C}_d\right)\dot{\tilde{\mathbf{q}}}_l+$$
$$+\varepsilon\tilde{\mathbf{q}}_l^T\left(\mathbf{C}^T-\mathbf{C}_d^T\right)\dot{\tilde{\mathbf{q}}}_l-\varepsilon\tilde{\mathbf{q}}_l^T\mathbf{K}_{pl}\tilde{\mathbf{q}}_l-\varepsilon\tilde{\mathbf{q}}_l^T\mathbf{K}_{dl}\tilde{\mathbf{v}}_l+$$
$$-\varepsilon\tilde{\mathbf{v}}_l^T\mathbf{A}_l^T\mathbf{D}\dot{\tilde{\mathbf{q}}}_l+\varepsilon\dot{\tilde{\mathbf{q}}}_l\mathbf{B}_l^T\mathbf{D}\dot{\tilde{\mathbf{q}}}_l+\varepsilon\tilde{\mathbf{v}}_l^T\left(\mathbf{C}^T-\mathbf{C}_d^T\right)\dot{\tilde{\mathbf{q}}}_l+$$
$$+\varepsilon\tilde{\mathbf{v}}_l^T\left(\mathbf{C}_d^T-\mathbf{C}_d\right)\dot{\tilde{\mathbf{q}}}_l-\varepsilon\tilde{\mathbf{v}}_l^T\mathbf{K}_{pl}\tilde{\mathbf{q}}_l-\varepsilon\tilde{\mathbf{v}}_l^T\mathbf{K}_{dl}\tilde{\mathbf{v}}_l. \tag{49}$$

where $\mathbf{C}$, $\mathbf{C}_d$ denote $\mathbf{C}(\mathbf{q}_l, \dot{\mathbf{q}}_l)$, $\mathbf{C}(\mathbf{q}_l, \dot{\mathbf{q}}_{ld})$, respectively.

Next, upper bounds for some expressions are calculated. First take $\varepsilon\tilde{\mathbf{q}}_l^T(\mathbf{C}^T-\mathbf{C}_d^T)\dot{\tilde{\mathbf{q}}}_l$, the following calculations can be obtained.

$$\varepsilon\tilde{\mathbf{q}}_l^T\left(\mathbf{C}^T-\mathbf{C}_d^T\right)\dot{\tilde{\mathbf{q}}}_l=\varepsilon\left[\tilde{\mathbf{q}}_l^T\mathbf{C}^T(\mathbf{q}_l,\dot{\mathbf{q}}_l)\dot{\tilde{\mathbf{q}}}_l-\tilde{\mathbf{q}}_l^T\mathbf{C}^T(\mathbf{q}_l,\dot{\mathbf{q}}_{ld})\dot{\tilde{\mathbf{q}}}_l\right]=$$
$$=\varepsilon\left[\dot{\tilde{\mathbf{q}}}^T\mathbf{C}(\mathbf{q},\dot{\mathbf{q}})\tilde{\mathbf{q}}-\dot{\tilde{\mathbf{q}}}^T\mathbf{C}(\mathbf{q},\dot{\mathbf{q}})\tilde{\mathbf{q}}\right]=\varepsilon\left[\dot{\tilde{\mathbf{q}}}^T\mathbf{C}(\mathbf{q},\tilde{\mathbf{q}})\dot{\mathbf{q}}+\right.$$
$$\left.-\dot{\tilde{\mathbf{q}}}_l^T\mathbf{C}(\mathbf{q}_l,\tilde{\mathbf{q}}_l)\dot{\mathbf{q}}_{ld}\right]=\varepsilon\dot{\tilde{\mathbf{q}}}_l^T\mathbf{C}(\mathbf{q}_l,\tilde{\mathbf{q}}_l)\left(\dot{\mathbf{q}}_l-\dot{\mathbf{q}}_{ld}\right)=\varepsilon\dot{\tilde{\mathbf{q}}}_l^T\mathbf{C}(\mathbf{q}_l,\tilde{\mathbf{q}}_l)\dot{\tilde{\mathbf{q}}}_l=$$
$$=\varepsilon\dot{\tilde{\mathbf{q}}}_l^T\mathbf{C}(\mathbf{q}_l,\dot{\tilde{\mathbf{q}}}_l)\tilde{\mathbf{q}}_l\le\varepsilon k_c\|\dot{\tilde{\mathbf{q}}}_l\|^2\|\tilde{\mathbf{q}}_l\|. \tag{50}$$

The upper bound for $\varepsilon\tilde{\mathbf{v}}_l^T(\mathbf{C}^T-\mathbf{C}_d^T)\dot{\tilde{\mathbf{q}}}_l$ can be evaluated similarly.

To perform further calculations for the sake of simplicity the following model of friction will be used:

$$\mathbf{F}(\dot{\mathbf{q}}_l)=\mathbf{F}_c sgn(\dot{\mathbf{q}}_l)+\mathbf{F}_v\dot{\mathbf{q}}_l$$
$$\mathbf{F}(\dot{\mathbf{q}}_{ld})=\mathbf{F}_c sgn(\dot{\mathbf{q}}_{ld})+\mathbf{F}_v\dot{\mathbf{q}}_{ld}. \tag{51}$$

where $\mathbf{F}_v$ denotes a diagonal matrix consisting of viscous friction coefficients associated with the links, and $\mathbf{F}_c$ denotes static friction. Next, the following bounds can be established for the terms associated with the friction coefficients which appear in Eq. (49) taking into account Eq. (51)

$$-\dot{\tilde{\mathbf{q}}}_l^T\mathbf{F}(\dot{\mathbf{q}}_l)+\dot{\tilde{\mathbf{q}}}_l^T\mathbf{F}(\dot{\mathbf{q}}_{dl})\ \le\ -F_{vm}\|\dot{\tilde{\mathbf{q}}}\|^2$$
$$-\varepsilon\tilde{\mathbf{q}}_l^T\mathbf{F}(\dot{\mathbf{q}}_l)+\varepsilon\tilde{\mathbf{q}}_l^T\mathbf{F}(\dot{\mathbf{q}}_{dl})\ \le\ 2\varepsilon F_{cM}\|\tilde{\mathbf{q}}_l\|+\varepsilon F_{vM}\|\tilde{\mathbf{q}}_l\|\|\dot{\tilde{\mathbf{q}}}_l\|$$
$$\varepsilon\tilde{\mathbf{v}}_l^T\mathbf{F}(\dot{\mathbf{q}}_l)-\varepsilon\tilde{\mathbf{v}}_l^T\mathbf{F}(\dot{\mathbf{q}}_{dl})\ \le\ 2\varepsilon F_{cM}\|\tilde{\mathbf{v}}_l\|+\varepsilon F_{vM}\|\tilde{\mathbf{v}}_l\|\|\dot{\tilde{\mathbf{q}}}_l\|$$

where F_{vm} denotes the smallest eigenvalue of the matrix $\mathbf{F}_v$ and F_{cM} denotes the biggest eigenvalue of the matrix $\mathbf{F}_c$. By virtue of some properties we can establish the following bounds for the other terms in Eq. (49) which do not include elements with vector $\tilde{\mathbf{a}}$:

$$-\tilde{\mathbf{v}}_l^T\mathbf{K}_{dl}\mathbf{B}_l^{-1}\mathbf{A}_l\tilde{\mathbf{v}}_l\le-k_{dlm}b_{lM}^{-1}a_{lm}\|\tilde{\mathbf{v}}_l\|^2$$
$$-\dot{\tilde{\mathbf{q}}}_l^T\mathbf{C}_d\dot{\tilde{\mathbf{q}}}_l\le k_cB_d\|\dot{\tilde{\mathbf{q}}}_l\|^2$$
$$\varepsilon\dot{\tilde{\mathbf{q}}}_l^T\mathbf{D}\dot{\tilde{\mathbf{q}}}_l\ \le\ \varepsilon d_M\|\dot{\tilde{\mathbf{q}}}_l\|^2$$
$$\varepsilon\tilde{\mathbf{q}}_l^T\left(\mathbf{C}_d^T-\mathbf{C}_d\right)\dot{\tilde{\mathbf{q}}}_l\ \le\ 2\varepsilon k_cB_d\|\tilde{\mathbf{q}}_l\|\|\dot{\tilde{\mathbf{q}}}_l\|$$
$$\varepsilon\tilde{\mathbf{q}}_l^T\left(\mathbf{C}^T-\mathbf{C}_d^T\right)\dot{\tilde{\mathbf{q}}}_l\ \le\ \varepsilon k_c\|\dot{\tilde{\mathbf{q}}}_l\|^2\|\tilde{\mathbf{q}}_l\|$$

$$-\varepsilon\tilde{\mathbf{q}}_l^T\mathbf{K}_{pl}\tilde{\mathbf{q}}_l \leq -\varepsilon k_{plm}\|\tilde{\mathbf{q}}_l\|^2$$
$$-\varepsilon\tilde{\mathbf{q}}_l\mathbf{K}_{dl}\tilde{\mathbf{v}}_l \leq \varepsilon k_{dlM}\|\tilde{\mathbf{q}}_l\|\|\tilde{\mathbf{v}}_l\|$$
$$\varepsilon\tilde{\mathbf{v}}_l\mathbf{A}_l^T\mathbf{D}\dot{\tilde{\mathbf{q}}}_l \leq a_{lM}d_M\|\tilde{\mathbf{v}}_l\|\|\dot{\tilde{\mathbf{q}}}_l\|$$
$$-\varepsilon\dot{\tilde{\mathbf{q}}}_l^T\mathbf{B}_l^T\mathbf{D}\dot{\tilde{\mathbf{q}}}_l \leq -\varepsilon b_{lm}d_m\|\dot{\tilde{\mathbf{q}}}_l\|^2$$
$$-\varepsilon\tilde{\mathbf{v}}_l^T\left(\mathbf{C}^T-\mathbf{C}_d^T\right)\dot{\tilde{\mathbf{q}}}_l \leq \varepsilon k_c\|\dot{\tilde{\mathbf{q}}}_l\|^2\|\tilde{\mathbf{v}}_l\|$$
$$-\varepsilon\tilde{\mathbf{v}}_l^T\left(\mathbf{C}_d^T-\mathbf{C}_d\right)\dot{\tilde{\mathbf{q}}}_l \leq 2k_c\varepsilon B_d\|\tilde{\mathbf{v}}_l\|\|\dot{\tilde{\mathbf{q}}}_l\|$$
$$\varepsilon\tilde{\mathbf{v}}_l^T\mathbf{K}_{pl}\tilde{\mathbf{q}}_l \leq \varepsilon k_{plM}\|\tilde{\mathbf{v}}_l\|\|\tilde{\mathbf{q}}_l\|$$
$$\varepsilon\tilde{\mathbf{v}}_l^T\mathbf{K}_{dl}\tilde{\mathbf{v}}_l \leq \varepsilon k_{dlM}\|\tilde{\mathbf{v}}_l\|^2$$

where $\mathbf{C}(\mathbf{q},\dot{\mathbf{q}}) \leq k_c\|\dot{\mathbf{q}}\|$ and $k_c > 0$. The above inequalities are useful in evaluating the upper bound of the time derivative of the Lyapunov function given by Eq. (49). All constants with index M denote the biggest eigenvalue of an appropriate matrix. The elements of the matrices are denoted by corresponding small letters and are self-explanatory. Then the following adaptive law can be proposed:

$$\dot{\tilde{\mathbf{a}}}_l = -\eta\hat{\mathbf{a}}_l + \varepsilon\mathbf{\Gamma}_l^{-1}\mathbf{Y}_l^T(\tilde{\mathbf{v}}_l + \tilde{\mathbf{q}}_l). \tag{52}$$

When it and the previously evaluated upper bounds are taken into account, the time derivative of the proposed Lyapunov function can be evaluated as follows:

$$\dot{V}(t) \leq -\frac{\varepsilon}{2}\begin{bmatrix}\|\tilde{\mathbf{q}}_l\|\\\|\tilde{\mathbf{v}}_l\|\end{bmatrix}^T \overbrace{\begin{bmatrix}\frac{1}{2}k_{plm} & -k_{plM}-k_{dlM}\\ -k_{plM}-k_{dlM} & \frac{1}{2\varepsilon}k_{dlm}b_{lM}^{-1}a_{lm}\end{bmatrix}}^{\mathbf{Q}_1}\begin{bmatrix}\|\tilde{\mathbf{q}}_l\|\\\|\tilde{\mathbf{v}}_l\|\end{bmatrix}+$$

$$-\frac{\varepsilon}{2}\begin{bmatrix}\|\tilde{\mathbf{v}}_l\|\\\|\dot{\tilde{\mathbf{q}}}_l\|\end{bmatrix}^T \overbrace{\begin{bmatrix}\frac{1}{2\varepsilon}k_{dlm}b_{lM}^{-1}a_{lm} & -2k_cB_d-a_{lM}d_{lM}-F_{vM}\\ -2k_cB_d-a_{lM}d_{lM}-F_{vM} & \frac{1}{\beta}d_{lM}b_{lM}\end{bmatrix}}^{\mathbf{Q}_2}\begin{bmatrix}\|\tilde{\mathbf{v}}_l\|\\\|\dot{\tilde{\mathbf{q}}}_l\|\end{bmatrix}+$$

$$-\frac{\varepsilon}{2}\begin{bmatrix}\|\tilde{\mathbf{q}}_l\|\\\|\dot{\tilde{\mathbf{q}}}_l\|\end{bmatrix}^T \overbrace{\begin{bmatrix}\frac{1}{2}k_{plm} & -2k_cB_d-F_{vM}\\ -2k_cB_d-F_{vM} & \frac{1}{\beta}b_{lM}d_{lM}-\frac{2\alpha k_cB_d}{\varepsilon}\end{bmatrix}}^{\mathbf{Q}_3}\begin{bmatrix}\|\tilde{\mathbf{q}}_l\|\\\|\dot{\tilde{\mathbf{q}}}_l\|\end{bmatrix}+$$

$$-\left[\overbrace{\left[-\frac{\beta+2}{\beta}b_{lM}d_{lM}\varepsilon-\gamma k_cB_d+F_{vm}\right]}^{\lambda_1}+\varepsilon\overbrace{\left[\frac{1}{\beta}b_{lM}d_{lM}-d_{lM}\right]}^{\lambda_2}\right]\|\dot{\tilde{\mathbf{q}}}_l\|^2+$$

$$-\overbrace{\left[\frac{1}{2}k_{dlm}b_{lM}^{-1}a_{lm}+\frac{3}{2}\varepsilon k_{dlm}\right]}^{\lambda_3}\|\tilde{\mathbf{v}}_l\|^2+\varepsilon k_c(\|\tilde{\mathbf{v}}_l\|+\|\tilde{\mathbf{q}}_l\|)\|\dot{\tilde{\mathbf{q}}}_l\|^2+$$

$$-\tilde{\mathbf{a}}_l^T\mathbf{\Gamma}\,\eta\hat{\mathbf{a}}_l-\dot{\tilde{\mathbf{q}}}_l^T\mathbf{Y}_l\tilde{\mathbf{a}}_l+2\varepsilon F_{cM}(\|\tilde{\mathbf{q}}_l\|+\|\tilde{\mathbf{v}}_l\|)-\frac{\varepsilon}{2}k_{plm}\|\tilde{\mathbf{q}}_l\|^2-\frac{\varepsilon}{2}k_{dlm}\|\tilde{\mathbf{v}}_l\|^2. \tag{53}$$

where $\alpha+\gamma=1$ and $\alpha, \gamma, \beta>0$.

The condition for the positive definiteness of the matrix $\mathbf{Q}_1$ results in upper bound for ε (recall that $\varepsilon>0$):

$$\varepsilon<\frac{k_{plm}k_{dlm}a_{lm}}{4b_{lM}(k_{plM}+k_{dlM})^2}. \tag{54}$$

For the matrix $\mathbf{Q}_2$ the positive definiteness condition can be written in the following form:

$$\varepsilon<\frac{k_{dlm}b_{lm}a_{lm}d_{lM}}{2\beta(2k_cB_d+a_{lM}d_{lM}+F_{vM})^2}. \tag{55}$$

Then the positive definiteness of the matrix $\mathbf{Q}_3$ is guaranteed when two conditions are satisfied and derived from

$$\frac{1}{2\beta}k_{plm}b_{lM}d_{lM}-\left(2k_cB_d+F_{vM}\right)^2-\frac{1}{2\beta}k_{plm}\frac{2\alpha k_cB_d}{\varepsilon}>0$$

$$\frac{1}{2\beta}k_{plm}b_{lM}d_{lM}-\left(2k_cB_d+F_{vM}\right)^2=C_2>0$$

$$k_{plm}>\frac{2\beta\left(2k_cB_d+F_{vM}\right)^2}{d_{lM}b_{lM}} \tag{56}$$

$$\varepsilon>\frac{k_{plm}k_cB_d\alpha}{C_2\beta}. \tag{57}$$

Note that the positive definiteness of constant λ_1 imposes a lower bound on ε, i.e.,

$$\varepsilon<\frac{\beta}{\beta+2}\frac{F_{vm}-k_cB_d\gamma}{b_{lM}d_{lM}}. \tag{58}$$

In the case when $F_{vm}-k_cB_d\gamma>0$:

$$\gamma<\frac{F_{vm}}{k_cB_d} \tag{59}$$

λ_2 is positive under the following condition

$$b_{lM} > \beta. \tag{60}$$

Finally, λ_3 is always positive. Then define matrix $\mathbf{Q}$ as:

$$\mathbf{Q} = \begin{bmatrix} \varepsilon k_{plm} & -\frac{\varepsilon}{2}F_{vM} - \varepsilon k_c B_d & -\frac{\varepsilon}{2}k_{dlM} - \frac{\varepsilon}{2}k_{plM} \\ -\frac{\varepsilon}{2}F_{vM} - \varepsilon k_c B_d & F_{vm} - k_c B_d(\alpha+\gamma) - \varepsilon d_{lM} + d_{lm}b_{lm} & -\frac{\varepsilon}{2}F_{vM} - \varepsilon k_c B_d - \frac{\varepsilon}{2}a_{lM}d_{lM} \\ -\frac{\varepsilon}{2}k_{dlM} - \frac{\varepsilon}{2}k_{plM} & -\frac{\varepsilon}{2}F_{vM} - \varepsilon k_c B_d - \frac{\varepsilon}{2}a_{lM}d_{lM} & k_{dlm}b_{lM}^{-1}a_{lm} - \varepsilon k_{dlM} \end{bmatrix} \tag{61}$$

and note that $\mathbf{Q}$ is positive-definite if conditions (54)–(60) are satisfied. Further evaluation of the time derivative of Lyapunov function (53) is as follows:

$$\dot{V}(t) \leq -\mathbf{z}^T\mathbf{Q}\mathbf{z} + \varepsilon k_c(\|\tilde{\mathbf{v}}_l\| + \|\tilde{\mathbf{q}}_l\|)\|\dot{\tilde{\mathbf{q}}}_l\|^2 - \eta\tilde{\mathbf{a}}\mathbf{\Gamma}_l^T{}_l\hat{\mathbf{a}}_l - \dot{\tilde{\mathbf{q}}}_l^T\mathbf{Y}_l\tilde{\mathbf{a}}_l - \frac{\varepsilon}{2}k_{plm}\|\tilde{\mathbf{q}}_l\|^2 - \frac{\varepsilon}{2}k_{dlm}\|\tilde{\mathbf{v}}_l\|^2 +$$

$$+2\varepsilon F_{cM}(\|\tilde{\mathbf{q}}_l\| + \|\tilde{\mathbf{v}}_l\|). \tag{62}$$

Introducing bounds of the regresion function $\mathbf{Y}_l$, the expression: $-\tilde{\mathbf{a}}_l^T\mathbf{\Gamma}\eta\hat{\mathbf{a}}_l - \dot{\tilde{\mathbf{q}}}_l^T \mathbf{Y}\tilde{\mathbf{a}}_l$ is a scalar function.

Due to the fact that the matrix $\mathbf{Y}_l$ is a rectangular matrix ($\mathbf{Y}_l \in \mathbf{R}^{m\times n}$), the Frobenius norm should be applied:

$$\mathbf{A} \in \mathbf{R}^{m\times n} \Rightarrow \|\mathbf{A}\|_F = \sqrt{\sum_{i=1}^{m}\sum_{j=1}^{n}|a_{ij}|^2}. \tag{63}$$

As a consequence, all elements of regresion function $\mathbf{Y}_l(\mathbf{q}_l, \dot{\mathbf{q}}_{ld}, \ddot{\mathbf{q}}_{ld})$ are bounded as

$$-\dot{\tilde{\mathbf{q}}}_l^T\mathbf{Y}_l\tilde{\mathbf{a}}_l \leq \|\mathbf{Y}_l\|_F\|\dot{\tilde{\mathbf{q}}}_l\|\|\tilde{\mathbf{a}}_l\| = C_1\|\dot{\tilde{\mathbf{q}}}_l\|\|\tilde{\mathbf{a}}_l\|. \tag{64}$$

Taking into account Eq. (64), the upper bound of $-\tilde{\mathbf{a}}_l^T\mathbf{\Gamma}\eta\hat{\mathbf{a}}_l - \dot{\tilde{\mathbf{q}}}_l^T\mathbf{Y}\tilde{\mathbf{a}}_l$ can be calculated as follows:

$$-\eta\tilde{\mathbf{a}}_l^T\mathbf{\Gamma}_l\hat{\mathbf{a}}_l - \dot{\tilde{\mathbf{q}}}_l^T\mathbf{Y}_l\tilde{\mathbf{a}}_l = -\eta\tilde{\mathbf{a}}_l^T\mathbf{\Gamma}_l\tilde{\mathbf{a}}_l - \eta\tilde{\mathbf{a}}_l^T\mathbf{\Gamma}_l\mathbf{a}_l - \dot{\tilde{\mathbf{q}}}_l^T\mathbf{Y}_l\tilde{\mathbf{a}}_l \leq -\eta\Gamma_{lm}\|\tilde{\mathbf{a}}_l\|^2 +$$

$$+\eta\Gamma_{lM}\|\tilde{\mathbf{a}}_l\|\|\mathbf{a}_l\|+C_1\|\dot{\tilde{\mathbf{q}}}_l\|\|\tilde{\mathbf{a}}_l\|=-\frac{1}{2}\eta\Gamma_{lm}\|\tilde{\mathbf{a}}_l\|^2+C_1\|\dot{\tilde{\mathbf{q}}}_l\|\|\tilde{\mathbf{a}}_l\|+$$

$$-\left(\sqrt{\frac{\eta\Gamma_{lm}}{2}}\|\tilde{\mathbf{a}}_l\|-\frac{\eta\Gamma_{lM}}{\sqrt{2\eta\Gamma_{lm}}}\|\mathbf{a}_l\|\right)^2+\frac{1}{2}\frac{\eta^2\Gamma_{lM}^2}{\Gamma_{lm}\eta}\|\mathbf{a}_l\|^2\leq-\left(\frac{1}{2}\sqrt{\eta\Gamma_{lm}}\|\tilde{\mathbf{a}}_l\|+\right.$$

$$\left.-\frac{C_1}{\sqrt{\eta\Gamma_{lm}}}\|\dot{\tilde{\mathbf{q}}}_l\|\right)^2+\frac{C_1^2}{\eta\Gamma_{lm}}\|\dot{\tilde{\mathbf{q}}}_l\|^2-\frac{1}{4}\eta\Gamma_{lm}\|\tilde{\mathbf{a}}_l\|^2+\frac{1}{2}\frac{\eta\Gamma_{lM}^2}{\Gamma_{lm}}\|\mathbf{a}_l\|^2\leq$$

$$\frac{C_1^2}{\eta\Gamma_{lm}}\|\dot{\tilde{\mathbf{q}}}_l\|^2-\frac{1}{4}\eta\Gamma_{lm}\|\tilde{\mathbf{a}}_l\|^2+\varepsilon_1. \tag{65}$$

Assuming that the parameter $\mathbf{a}_l$ is bounded ($\|\mathbf{a}_l\|\leq\bar{a}$), one can write:

$$\varepsilon_1=\frac{1}{2}\frac{\eta\Gamma_{lM}^2}{\Gamma_{lm}}\|\mathbf{a}_l\|^2>0.$$

Next the upper bund of the time derivative of Lyapunov function can be calculated as:

$$\dot{V}(t)\leq-\mathbf{z}^T\mathbf{Q}\mathbf{z}+\varepsilon k_c\|\tilde{\mathbf{q}}\|\|\dot{\tilde{\mathbf{q}}}\|^2+\varepsilon k_c\|\tilde{\mathbf{v}}\|\|\dot{\tilde{\mathbf{q}}}\|^2+$$

$$+\frac{C_1^2}{\eta\Gamma_{lm}}\|\dot{\tilde{\mathbf{q}}}_l\|^2-\frac{1}{4}\eta\Gamma_{lm}\|\tilde{\mathbf{a}}_l\|^2+\varepsilon_1-\frac{\varepsilon}{2}k_{plm}\|\tilde{\mathbf{q}}_l\|^2-\frac{\varepsilon}{2}k_{dlm}\|\tilde{\mathbf{v}}_l\|^2+$$

$$+2\varepsilon F_{cM}(\|\tilde{\mathbf{q}}_l\|+\|\tilde{\mathbf{v}}_l\|). \tag{66}$$

Including term $\frac{C_1^2}{\eta\Gamma_{lm}}\|\dot{\tilde{\mathbf{q}}}_l\|^2$ in matrix $\mathbf{Q}$ one can define matrix $\mathbf{Q}'$ as:

$$\mathbf{Q}'=\begin{bmatrix}\varepsilon k_{plm} & -\frac{\varepsilon}{2}F_{vM}-\varepsilon k_c B_d & -\frac{\varepsilon}{2}k_{dlM}-\frac{\varepsilon}{2}k_{plM}\\ -\frac{\varepsilon}{2}F_{vM}-\varepsilon k_c B_d & F_{vm}-k_c B_d(\alpha+\gamma)-2\varepsilon d_{lM}+d_{lM}b_{lM}-\frac{C_1^2}{\eta\Gamma_{lm}} & -\frac{\varepsilon}{2}F_{vM}-\varepsilon k_c B_d-\frac{\varepsilon}{2}a_{lM}d_{lM}\\ -\frac{\varepsilon}{2}k_{dlM}-\frac{\varepsilon}{2}k_{plM} & -\frac{\varepsilon}{2}F_{vM}-\varepsilon k_c B_d-\frac{\varepsilon}{2}a_{lM}d_{lM} & k_{dlm}b_{lM}^{-1}a_{lm}-\varepsilon k_{dlm}\end{bmatrix}. \tag{67}$$

Taking into account Eq. (67), the time derivative of the Lyapunov function is bounded as:

$$\dot{V}(t) \le -\mathbf{z}^T\mathbf{Q}'\mathbf{z} - \frac{1}{4}\eta\Gamma_{lm} \| \tilde{\mathbf{a}}_l \|^2 + \varepsilon_1 + k_c \| \tilde{\mathbf{q}} \| \| \dot{\tilde{\mathbf{q}}} \|^2 + \varepsilon k_c \| \tilde{\mathbf{v}} \| \| \dot{\tilde{\mathbf{q}}} \|^2 +$$

$$-\frac{\varepsilon}{2}k_{plm} \| \tilde{\mathbf{q}}_l \|^2 - \frac{\varepsilon}{2}k_{dlm} \| \tilde{\mathbf{v}}_l \|^2 + 2\varepsilon F_{cM}(\| \tilde{\mathbf{q}}_l \| + \| \tilde{\mathbf{v}}_l \|). \tag{68}$$

The expression $2\varepsilon F_{cM}\left(\| \tilde{\mathbf{q}}_l \| + \| \tilde{\mathbf{v}}_l \|\right) - \frac{\varepsilon}{2}k_{plm} \| \tilde{\mathbf{q}}_l \|^2 - \frac{\varepsilon}{2}k_{dlm} \| \tilde{\mathbf{v}}_l \|^2$ can be recalculated as follows:

$$2\varepsilon F_{cM}\left(\| \tilde{\mathbf{q}}_l \| + \| \tilde{\mathbf{v}}_l \|\right) - \frac{\varepsilon}{2}k_{plm} \| \tilde{\mathbf{q}}_l \|^2 - \frac{\varepsilon}{2}k_{dlm} \| \tilde{\mathbf{v}}_l \|^2 =$$

$$= -\frac{\varepsilon}{2}k_{plm} \| \tilde{\mathbf{q}}_l \|^2 + 2\varepsilon F_{cM} \| \tilde{\mathbf{q}}_l \| - \frac{\varepsilon}{2}k_{dlm} \| \tilde{\mathbf{v}}_l \|^2 + 2\varepsilon F_{cM} \| \tilde{\mathbf{v}}_l \| =$$

$$-\frac{\varepsilon}{2}\left(\sqrt{k_{plm}} \| \tilde{\mathbf{q}}_l \| - \frac{2F_{cM}}{\sqrt{k_{plm}}}\right)^2 + \frac{2\varepsilon F_{cM}^2}{k_{plm}} - \frac{\varepsilon}{2}\left(\sqrt{k_{dlM}} \| \tilde{\mathbf{v}}_l \| - \frac{2F_{cM}}{\sqrt{k_{dlm}}}\right)^2$$

$$+\frac{2\varepsilon F_{cM}^2}{k_{dlm}} \le 2\varepsilon F_{cM}^2 \frac{k_{plm} + k_{dlm}}{k_{plm}k_{dlm}}. \tag{69}$$

As a result, the time derivative of the Lyapunov function can be bounded as:

$$\dot{V}(t) \le -\mathbf{z}^T\mathbf{Q}\mathbf{z} + \varepsilon k_c\left(\| \tilde{\mathbf{v}}_l \| + \| \tilde{\mathbf{q}}_l \|\right) \| \dot{\tilde{\mathbf{q}}}_l \|^2 - \frac{1}{4}\eta\Gamma_{lm} \| \tilde{\mathbf{a}}_l \|^2 + \varepsilon_1 + \varepsilon_2, \tag{70}$$

where ε_2 was defined as:

$$\varepsilon_2 = 2\varepsilon F_{cM}^2 \frac{k_{dlm} + k_{plm}}{k_{dlm}k_{plm}}. \tag{71}$$

The second component of Eq. (71) can be estimated as follows:

$$\begin{aligned} &\varepsilon k_c\left(\| \tilde{\mathbf{v}}_l \| + \| \tilde{\mathbf{q}}_l \|\right) \| \dot{\tilde{\mathbf{q}}}_l \|^2 = \varepsilon k_c \| \tilde{\mathbf{v}}_l \| \| \dot{\tilde{\mathbf{q}}}_l \| + \varepsilon k_c \| \tilde{\mathbf{q}}_l \| \| \dot{\tilde{\mathbf{q}}}_l = \\ &= \frac{\varepsilon k_c}{2}\left(2 \| \tilde{\mathbf{v}}_l \| \| \dot{\tilde{\mathbf{q}}}_l \| + 2 \| \tilde{\mathbf{q}}_l \| \| \dot{\tilde{\mathbf{q}}}_l \|\right) \| \dot{\tilde{\mathbf{q}}}_l \| \le \\ &\le \frac{\varepsilon k_c}{2}\left(\| \tilde{\mathbf{v}}_l \|^2 + \| \dot{\tilde{\mathbf{q}}}_l \|^2\right) \| \dot{\tilde{\mathbf{q}}}_l \| \le \varepsilon k_c \| \dot{\tilde{\mathbf{q}}}_l \| \| \mathbf{z} \|^2 . \end{aligned} \tag{72}$$

Now consider the lower bound of the Lapunov function:

$$V(t) \ge \frac{1}{2}d_{lm} \| \dot{\tilde{\mathbf{q}}}_l \|^2 + \frac{1}{2}k_{plm} \| \tilde{\mathbf{q}}_l \|^2 + \frac{1}{2}k_{dlm}b_{lM} \| \tilde{\mathbf{v}}_l \|^2 - \varepsilon d_{lm} \| \tilde{\mathbf{q}}_l \| \| \dot{\tilde{\mathbf{q}}}_l \| +$$

$$-\varepsilon d_{lM} \| \tilde{\mathbf{v}}_l \| \| \dot{\tilde{\mathbf{q}}}_l \| + \frac{1}{2} \| \tilde{\mathbf{a}}_l \|^2 \Gamma_m. \tag{73}$$

where

$$\frac{1}{2}k_{plm}\|\tilde{\mathbf{q}}_l\|^2 - \varepsilon d_{lm}\|\tilde{\mathbf{q}}_l\|\|\dot{\tilde{\mathbf{q}}}_l\| = \frac{1}{2}k_{plm}\left[\|\tilde{\mathbf{q}}_l\|^2 - 2\varepsilon\frac{d_{lm}}{k_{plm}}\|\tilde{\mathbf{q}}_l\|\|\dot{\tilde{\mathbf{q}}}_l\|\right] =$$

$$= \frac{1}{2}k_{plm}\left(\|\tilde{\mathbf{q}}_l\| - \|\dot{\tilde{\mathbf{q}}}_l\|\varepsilon\frac{d_{lm}}{k_{plm}}\right)^2 + \tag{74}$$

$$-\|\dot{\tilde{\mathbf{q}}}_l\|^2\frac{\varepsilon^2 d_{lm}^2}{k_{plm}^2}\frac{1}{2}.$$

$$\frac{1}{2}k_{dlm}b_{lM}\|\tilde{\mathbf{v}}_l\|^2 - \varepsilon d_{lm}\|\tilde{\mathbf{v}}_l\|\|\dot{\tilde{\mathbf{q}}}_l\| = \frac{1}{2}k_{dlm}b_{lM}\left(\|\tilde{\mathbf{v}}_l\|^2 - \frac{2\varepsilon}{k_{dlm}b_{lM}}\|\tilde{\mathbf{v}}_l\|\|\dot{\tilde{\mathbf{q}}}_l\|\right) =$$

$$= \frac{1}{2}k_{dlm}b_{lM}\left(\|\tilde{\mathbf{v}}_l\| - \frac{\varepsilon d_{lm}}{k_{dlm}B_{lM}}\|\dot{\tilde{\mathbf{q}}}_l\|\right)^2 +$$

$$-\frac{\varepsilon^2 d_{lm}}{2k_{dlm}b_{lM}}\|\dot{\tilde{\mathbf{q}}}_l\|^2. \tag{75}$$

Consequently, the lower bound of the Lyapunov function (73) can be calculated as follows:

$$\begin{aligned}
V(t) \geq\ & \frac{1}{2}d_{lm}\|\dot{\tilde{\mathbf{q}}}_l\| + \frac{1}{2}k_{plm}\left(\|\tilde{\mathbf{q}}_l\| - \|\dot{\tilde{\mathbf{q}}}_l\frac{\varepsilon d_{lm}}{k_{plm}}\right)^2 - \|\dot{\tilde{\mathbf{q}}}_l\|\frac{\varepsilon^2 d_{lm}^2}{k_{plm}}\frac{1}{2} + \\
& + \frac{1}{2}k_{dlm}b_{lM}^{-1}\left(\|\tilde{\mathbf{v}}_l - \frac{\varepsilon d_{lm}}{k_{dlm}b_{lM}}\|\dot{\tilde{\mathbf{q}}}_l\|\right)^2 - \frac{\varepsilon^2 d_{lm}^2}{2k_{dlm}b_{lM}} + \frac{1}{2}\|\tilde{\mathbf{a}}_l\|\Gamma_{lm} \geq \\
& \geq \left(\frac{1}{2}d_{lm} - \frac{\varepsilon^2 d_{lm}^2}{2k_{plm}} - \frac{\varepsilon^2 d_{lm}^2}{k_{dlm}b_{lM}}\right)\|\dot{\tilde{\mathbf{q}}}_l\| + \frac{1}{2}\|\tilde{\mathbf{a}}_l\|^2\Gamma_{lm} \geq \\
& \geq \left(\frac{1}{2}d_{lm} - \frac{\varepsilon^2 d_{lm}^2}{2k_{plm}} - \frac{\varepsilon^2 d_{lm}^2}{k_{dlm}b_{lM}}\right)\|\dot{\tilde{\mathbf{q}}}_l\| \geq \xi_1\|\dot{\tilde{\mathbf{q}}}_l\|.
\end{aligned} \tag{76}$$

Note that if the condition

$$\varepsilon^2 < \frac{k_{plm}b_{lM}}{d_{lm} + d_{lm}b_{lM}}, \tag{77}$$

is true, the parameter ξ_1, which is defined as:

$$\xi_1 = k_{plm}b_{lM} - \varepsilon^2 d_{lm}b_{lM} - \varepsilon^2 d_{lM}, \tag{78}$$

is always positive. Therefore one can write:

$$\|\dot{\mathbf{q}}\| \leq \frac{\sqrt{V}}{\sqrt{\xi_1}}. \tag{79}$$

From (65), (69), (72), (79), and matrix $\mathbf{Q}'$, $\dot{V}$ can be bounded as follows:

$$\dot{V}(t) \leq -\left(\lambda_m(\mathbf{Q}') - \varepsilon k_c \frac{\sqrt{V}}{\sqrt{\xi}}\right)\|\mathbf{z}\|^2 - \frac{1}{4}\eta\Gamma_{lm}\|\tilde{\mathbf{a}}_l\|^2 + \varepsilon_3. \tag{80}$$

The time derivative of the Lyapunov function considered here is of the form at Theorem 1 assuming that:

$$\mu_1 = \lambda_m(\mathbf{Q}_1) = \frac{1}{2}k_{dlm}b_{lM}^{-1},$$

$$\mu_2 = \frac{1}{2}\Gamma_{lm},$$

$$\mu_3 = \lambda_M(\mathbf{Q}_2),$$

$$\mu_4 = \frac{1}{2}\Gamma_{lM},$$

$$\mu_5 = \lambda_m(\mathbf{Q}') - \varepsilon k_c \frac{\sqrt{V}}{\sqrt{\xi}},$$

$$\mu_6 = \frac{1}{4}\eta\Gamma_{lm},$$

$$\varepsilon_3 = \varepsilon_1 + \varepsilon_2 = \frac{1}{2}\frac{\Gamma_{lM}^2\eta}{\Gamma_{lm}}\|\mathbf{a}_l\|^2 + 2\varepsilon F_{cM}^2\frac{k_{dlm}+k_{plm}}{k_{dlm}k_{plm}} =$$

$$= \frac{4F_{cM}^2\varepsilon\Gamma_{lm}(k_{dlm}+k_{plm}) + \eta\Gamma_{lM}^2 k_{dlm}k_{plm}\|\mathbf{a}_l\|^2}{2\Gamma_{lm}k_{dlm}k_{plm}}. \tag{81}$$

One can set the parameter ε_3 to a small value, if parameters ε, and η are small. Assuming that:

$$k_{plm} = k,$$

$$k_{dlm} = \sigma k \tag{82}$$

the parameter ε_3 can be written in the following form:

$$\varepsilon_3 = \frac{4F_{cM}^2\varepsilon\Gamma_{lm}(1+\sigma) + \eta\sigma\Gamma_{lM}^2 k\|\mathbf{a}_l\|^2}{\sigma\Gamma_{lm}k} \tag{83}$$

where σ is a positive constant. As a consequence $V(t)$ and $\dot{V}(t)$ satisfy Theorem 1 under the following definitions:

$$\sigma_1 = max\left\{\frac{\lambda_M(\mathbf{Q}_2)}{\left(\lambda_m(\mathbf{Q}) - \varepsilon k_c \frac{\sqrt{V}}{\sqrt{\xi}}\right)}, \frac{2\Gamma_{lM}}{\eta\Gamma_{lm}}\right\} \tag{84}$$

and radii

$$r_1 = \sqrt{\frac{2\sigma_1(16F_{cM}^2\varepsilon\Gamma_{lm}(1+\sigma)+\eta\sigma\Gamma_{lM}^2 k \|\mathbf{a}_l\|^2)}{\sigma^2 k^2 b_{lM}^{-1}\Gamma_{lm}}},$$

$$r_2 = \sqrt{\frac{2\sigma_1(16F_{cM}^2\varepsilon\Gamma_{lm}(1+\sigma)+\eta\sigma\Gamma_{lM}^2 k \|\mathbf{a}_l\|^2)}{\sigma\Gamma_{lm}^2 k}}. \tag{85}$$

Based on Theorem 1 $\mathbf{z}$ and $\tilde{\mathbf{a}}$ are uniformly bounded (which implies that $\tilde{\mathbf{q}}_l$, $\dot{\tilde{\mathbf{q}}}_l$ and $\tilde{\mathbf{v}}_l$ are uniformly bounded too) and converge exponentially to the closed balls B_{r_1}, B_{r_2}, respectively. The radius of the balls can be decreased by increasing coefficient k or increasing the parameter Γ_{lm}.

The adaptive version of the control algorithms for the system, given by Eqs. (13) and (14) is the same, assuming that Eq. (42) is written in the following form:

$$\boldsymbol{D}\ddot{\tilde{\mathbf{q}}} + [\boldsymbol{C}_1 + \boldsymbol{C}_d]\dot{\tilde{\mathbf{q}}} + \mathbf{Y}_1(\mathbf{q}_l, \dot{\mathbf{q}}_{ld}, \ddot{\mathbf{q}}_{ld}, \dot{\mathbf{q}}_{md}, \ddot{\mathbf{q}}_{md})\tilde{\mathbf{a}} + \boldsymbol{K}_p\tilde{\mathbf{q}} + \boldsymbol{K}_d\tilde{\mathbf{v}} = 0 \tag{86}$$

$$\dot{\tilde{\mathbf{q}}} = -A\tilde{\mathbf{v}} + B\dot{\tilde{\mathbf{q}}} \tag{87}$$

where $\boldsymbol{D} = blockdiag[\mathbf{M}_l, \mathbf{J}_d]$, $\tilde{\mathbf{q}} = [\tilde{\mathbf{q}}_l^T, \tilde{\mathbf{q}}_m^T]^T$, $\tilde{\mathbf{v}} = [\tilde{\mathbf{v}}_l^T, \tilde{\mathbf{v}}_m^T]^T$, $\boldsymbol{K}_d = blockdiag[\mathbf{K}_{dl}, \mathbf{K}_{dm}]$,

$A = blockdiag\ \{\mathbf{B}_1, \mathbf{B}_2\}$ and $\boldsymbol{C}_1 = \begin{bmatrix} \mathbf{C}(q_l, \dot{q}_l) & \mathbf{0} \\ \mathbf{0} & \mathbf{0} \end{bmatrix}$, $\boldsymbol{C}_d = \begin{bmatrix} \mathbf{C}(q_l, \dot{q}_{ld}) & \mathbf{0} \\ \mathbf{0} & \mathbf{0} \end{bmatrix}$,

$$\boldsymbol{K}_p = \begin{bmatrix} \mathbf{K}_{pl} + \mathbf{K} & -\mathbf{K} \\ -\mathbf{K} & \mathbf{K}_{pm} + \mathbf{K} \end{bmatrix}, \mathbf{Y}_1 = \begin{bmatrix} \mathbf{Y}_l(\mathbf{q}_l, \dot{\mathbf{q}}_{ld}, \ddot{\mathbf{q}}_{ld}) & \mathbf{0} \\ \mathbf{0} & \mathbf{Y}_m(\dot{\mathbf{q}}_{md}, \ddot{\mathbf{q}}_{md}) \end{bmatrix}$$

where $\mathbf{Y}_l$ and $\mathbf{Y}_m$ are appropriate matrices associated with the links and motors, respectively.

Now, consider the Lyapunov function candidate

$$V(t) = \frac{1}{2}\dot{\tilde{\mathbf{q}}}^T \boldsymbol{D}\dot{\tilde{\mathbf{q}}}^T + \frac{1}{2}\tilde{\mathbf{q}}^T \boldsymbol{K}_p\tilde{\mathbf{q}} + \frac{1}{2}\tilde{\mathbf{v}}^T \boldsymbol{K}_d B^{-1}\tilde{\mathbf{v}} + \varepsilon\tilde{\mathbf{q}}^T \boldsymbol{D}\dot{\tilde{\mathbf{q}}} - \varepsilon\tilde{\mathbf{v}}^T \boldsymbol{D}\dot{\tilde{\mathbf{q}}} + \frac{1}{2}\tilde{\mathbf{a}}^T \boldsymbol{\Gamma}\tilde{\mathbf{a}} \tag{88}$$

and note that (86) is similar to (42), as well as (88) is similar to (44). The main difference is that $\boldsymbol{K}_p$ is no longer diagonal but is positive–definite if $\mathbf{K}_{pl}$ and $\mathbf{K}_{pm}$ are positive–definite. Note also that the adaptation law for the motor parameters has the following form

$$\dot{\tilde{\mathbf{a}}}_m = -\eta\tilde{\mathbf{a}}_m + \varepsilon\boldsymbol{\Gamma}_m^{-1}\mathbf{Y}_m^T(\tilde{\mathbf{v}}_m + \tilde{\mathbf{q}}_m) \tag{89}$$

where $\boldsymbol{\Gamma}_m$ is a positive–definite matrix.

Example—The Control System With A Signal Processor For A Manipulator With Two Degrees Of Freedom

The Control Scheme Of A Manipulator With Two Degrees

The experimental set-up (Fig. 3) was designed based on a manipulator with two links with a different cross-section and stiffness. It is a SCARA manipulator. The links of the manipulator were made of an alloy of light metals. Each link has a cross-section of a rectangular shape. The manipulator consists of

- two DC motors (PRM 76-05TR) with the following parameters: nominal torque: 0.55 Nm, maximum torque: 2.6 Nm, maximum speed: 4000 rpm, maximum current: 19A, maximum voltage: 60V,
- two harmonic-drives (HDU C2) with the gear ratio 1:158,
- belt transmission with a gear ratio 1:1,
- encoders MOZ 30/500/5/K with a resolution 500 counts per revolution and voltage supply +5V,
- two tachometric generators with a constant voltage 3.3V/1000 revolutions,
- accelerometers,
- safety cut-off switch (TCST 2000).

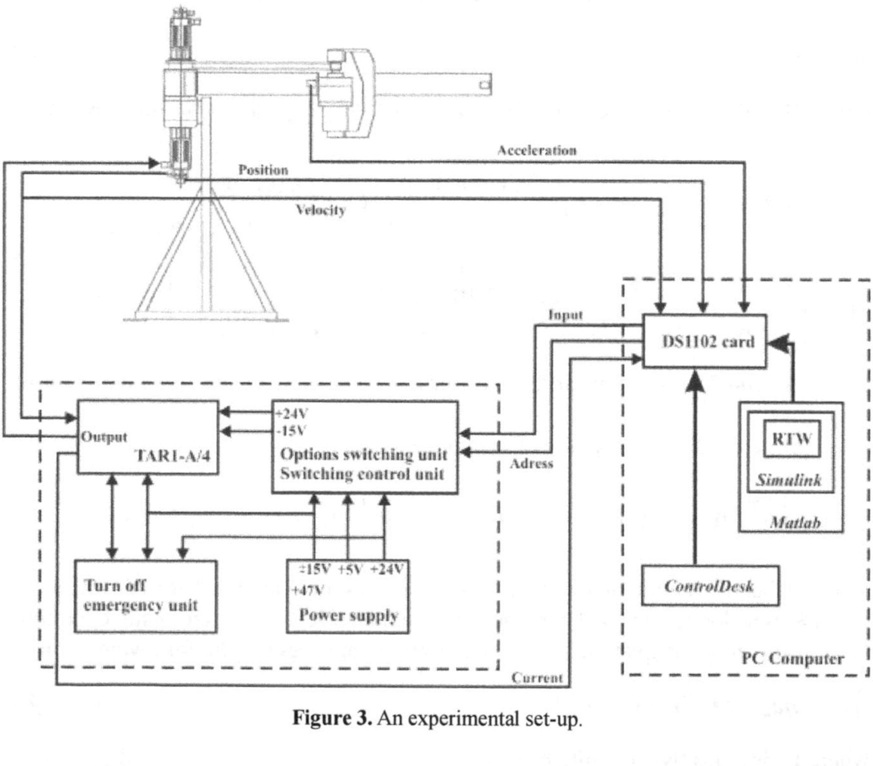

Figure 3. An experimental set-up.

The control system consists of the following elements:

- **Control unit TAR 1-A/4**. This is a power module which operates DC motor directly. The system TAR1-A/4 is a solid-state speed controller of DC motors with a maximum voltage of 50V and a maximum current of 15A. This module was used in control systems of the IRp6 robots. This control unit can be divided into the following functional blocks:
  - motor speed controller implemented in three operational amplifiers: the first amplifier protects the motor against the influence of the voltage drop at engine speed, the second amplifier compares the voltage proportional to motor speed with a predetermined voltage (speed setpoint), the third realizes a PI controller,
  - motor current PI controller, output signal from this controllers controls the modulator,
  - the tringular voltage generator—output signal is a tringular waveform voltage with a frequency of 4kHz,
  - modulator—converts the voltage signal from output of the current controller to pulse with adjustable filling,
  - class-D amplifier—amplifies the output pulses from the current controller to control the power transistors,
  - solid-state power amplifier—circuit of H-bridge,
  - overcurrent protection—motor overload protection.
- **Options switching unit**. This module was built in order to reconfigure the control unit. It allows performing the following functions:
  - unblocking the controller,
  - changing the voltage polarization,
  - switching the desired signal to the different loops of the controller, namely:
    - to the velocity controller loop,
    - to the current controller loop,
    - to the power amplifier loop.
- **A microprocessor controller unit**. This unit protects the switching module from incorrect configuration set by the operator, and sets up control signals necessary to perform non-failure operation of the unit described in the previous point. The controller consists of the following blocks:
  - block of manual dialing,
  - block select options using a PC,
  - executive module,
  - blockade.
- **Supply system**, which consists of a power supply of control units and power supply. The power supply delivers unstabilized voltage (50V) to the TAR1-A/4 unit.
- **DS1102 signal processor card (dSPACE Company)**, it is the most important element of the control system. This card has been built on the basis of TMS320C31 digital signal processor from Texas Instruments. It is equipped with:

- four analog-to-digital converters (ADC)—transducers which have a resolution of 12-bits and 16-bits,
- four 16-bit digital-to-analog converters (DAC),
- 16-bit input-output system,
- two interfaces of rotary encoders.

- **Turn off emergency unit**, which protects the manipulator against unspecified behavior. When the manipulator reaches the end position or the user presses the RESET button then be shut off motor voltage.
- **Measurement module**. This unit consists of the following elements: rate generators, encoders and accelerometers. Accelerometers (mounted at the end of each link) are connected to DSP card directly. The converter does not work in RS422 standard but in TTL standard open collector therefore an appropriate interface for the encoders was designed.

Implementation of the proposed control algorithms (which are presented in Section Mathematical Description of an Adaptive Tracking Controller) are described using fast prototyping methodology. The algorithms were implemented with the help of MATLAB environment with SIMULINK. To use the program implemented in SIMULINK to control the device using the DSP card, the MATLAB RTI toolbox (Real Time Interface) was employed. The RTI toolbox enables the use of SIMULINK blocks that represent the DS1102 card. The implemented algorithm uses the following blocks of the RTI toolbox (Fig. 7):

- DS1102DAC—digital-to-analog converters,
- DS1102ENC_POS—a block for cooperation with encoders which enable the measurement position relative to the zero position,
- DS1102IN – represents a 16-bit input DSP card.

Initially, simulation results were carried out, which initially verified the proposed control algorithm (Fig. 4).

First, model of the manipulator with two links (in Fig. 4 the block titled Object) was implemented in SIMULINK and subdivided into two subsystems:

- linear part of the system,
- non-linear part of the system added as feedback to the linear subsystem.

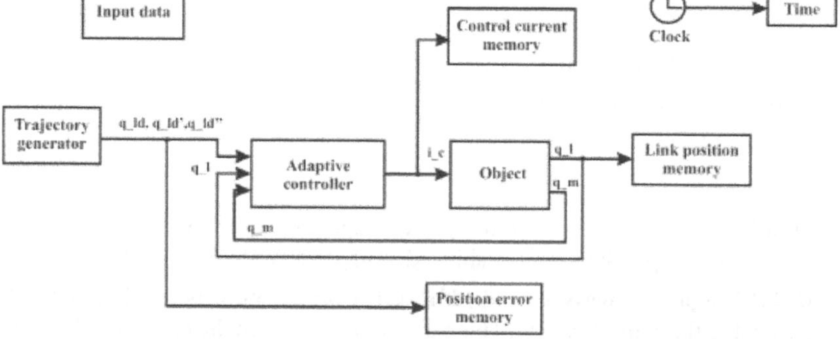

Figure 4. Scheme of adaptive control algorithm for simulation research.

The linear elements were described by the transfer function $G_l(s)$, with input signal as torque, and output signal as link position. The transfer function for the link is described by the following equations:

- first link

$$G_{l1}(s) = \frac{Q_{l1}(s)}{U_1(s)} = \frac{1}{\left[I_{l1} + d_2^2 m_2 + d_1^2(m_1 + m_2)\right]s^2 + F_{ld1}s + K_1}, \tag{90}$$

- second link

$$G_{l2}(s) = \frac{Q_{l2}(s)}{U_2(s)} = \frac{1}{\left[I_{l2} + d_2^2 m_2\right]s^2 + F_{ld2}s + K_2}, \tag{91}$$

where K_1, K_2 are stiffness coefficients for the first and second link, respectively.

The transfer function of the motor has the following form (where input signal is the current i, and output signal is the motor position q_m):

$$G_{mj}(s) = \frac{Q_{mj}(s)}{I\ (s)} = \frac{k_{mj}}{J_{mj}s\ \ + F_{mdj}s + \text{—}} \tag{92}$$

where $j = 1,2$.

Expressions $P_1,\ldots, P_7$ describe nonlinear elements in the following form:

- inertia of the first link

$$P_1 = -2d_1 d_2 m_2 cos(q_{l2})\ddot{q}_1 - \left(d_2^2 m_2 + d_1 d_2 m_2 cos(q_{l2})\right)\ddot{q}_2, \tag{93}$$

- inertia of the second link

$$P_2 = \left(d_2^2 m_2 + d_1 d_2 m_2 cos(q_{l2})\right)\ddot{q}_1, \tag{94}$$

- centripetal and Coriolis terms of the first link

$$P_3 = m_2 d_1 d_2 sin(q_{l2})\dot{q}_2^2 + 2m_2 d_1 d_2 sin(q_{l2})\dot{q}_1\dot{q}_2, \tag{95}$$

- centripetal and Coriolis terms of the second link

$$P_4 = -m_2 d_1 d_2 sin(q_{l2})\dot{q}_2^2, \tag{96}$$

- static friction term of the first and second link, respectively

$$P_5 = -F_{lsj}sgn(\dot{q}_j), \quad j = 1,2, \tag{97}$$

- static friction term of the first and second motor, respectively

$$P_6 = \frac{-F_{lsj}sgn(\dot{q}_j)}{k_{mj}}, \quad j = 1,2, \tag{98}$$

- gear and stiffness terms of the first and second link, respectively

$$P_7 = \frac{K_j}{k_{mj} N_j} q_j, \quad j = 1,2. \tag{99}$$

The implementation scheme of the manipulator with two links is shown in Fig. 5 The first and second links are implemented similarly (compare Fig. 6).

Next the adaptive control algorithm for the manipulator with two degrees of freedom was implemented. The following block diagram of control scheme for a manipulator with joint flexibility is proposed, see Fig. 8.

Next, this simulation program was transformed into a program to control of the real manipulator. For this purpose, the RTW toolbox (Real Time Workshop) and RTI toolbox were used. In the implemented control algorithm (Fig. 4), a block representing the mathematical description of the manipulator (object model) was replaced by the respective blocks of the DSP card. These blocks allow cooperation of this program with the real device. In place of the block **Object** (Fig. 4) a block of digital-to-analog converters (block DS1102DAC which allows controlling motors) were introduced and a block for cooperation with encoders (block DS1102ENC_POS which enables measuring the position of the manipulator links). Figure 7 shows a block diagram of an adaptive control algorithm of the manipulator implemented in SIMULINK.

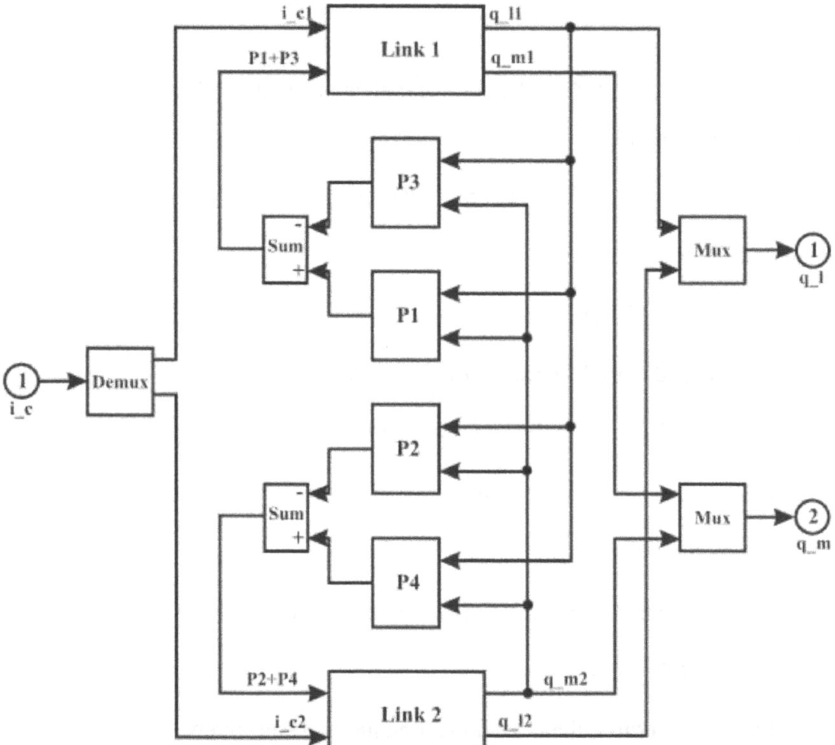

Figure 5. SIMULINK implementation of a two degrees of freedom manipulator with elastic joints (P1—Inertia of the link 1, P2—Inertia of the link 2, P3—Centripetal and Coriolis terms for link 1, P4—Centripetal and Coriolis terms for link 2).

The block named **Adaptive controller** includes implementation of the control algorithm which was described in Section Mathematical Description of an Adaptive Tracking Controller (Fig. 8). This block consists of the following elements:

- **Motor position controller**—estimates of link parameters: inertia, static and viscous friction are calculated.
- **Link position controller**—estimates of motor parameters are calculated.
- **Gear estimate,**
- **Stiffness estimate,**
- **Link (motor) position filter**—block performing differentiation operation on link (motor) error position.

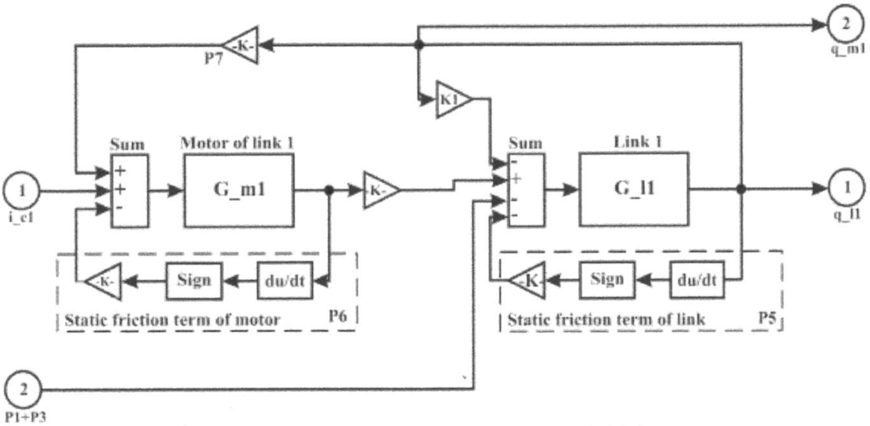

Figure 6. A block diagram for link 1 (P5—Static friction term of link 1, P6—Static friction term of motor 1, P7—Gear and stiffness term).

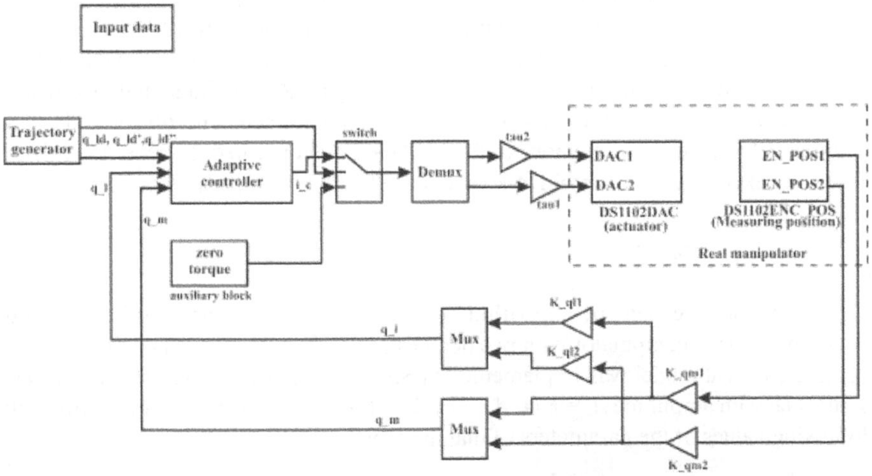

Figure 7. Scheme of the adaptive control algorithm for the real manipulator (tau1, tau2 adjust the control signals to the ranges of analog-to-digital converters, K_qm1, K_qm2—converting signals from the encoders to position signal in radians).

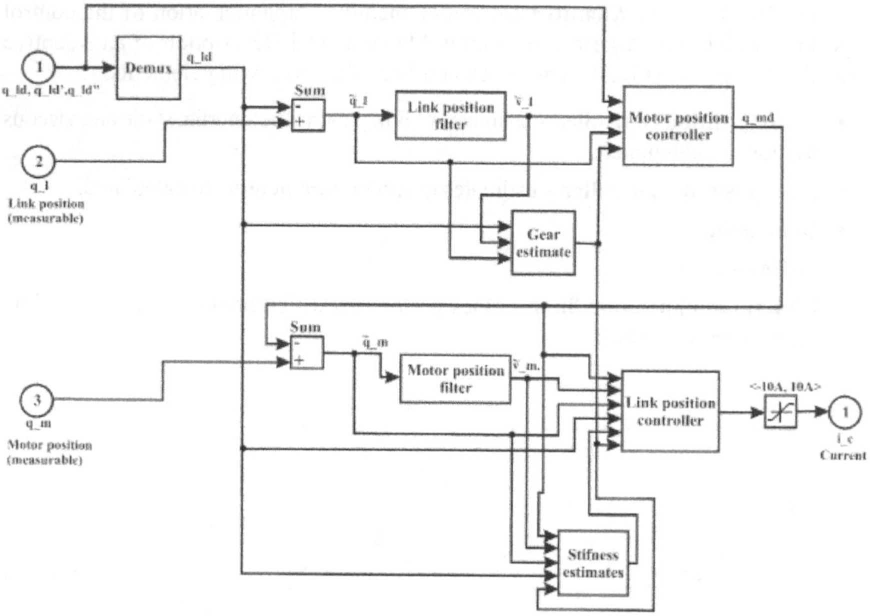

Figure 8. SIMULINK implementation of adaptive control algorithm.

The parameter estimates were calculated on the basis of the equation:

$$\dot{\hat{\mathbf{a}}}_{m,l} = -\eta \hat{\mathbf{a}}_{m,l} + \varepsilon \mathbf{\Gamma}_{m,l}^{-1} \mathbf{Y}_{m,l}^{T} (\tilde{\mathbf{v}}_{m,l} + \tilde{\mathbf{q}}_{m,l}). \tag{100}$$

In Fig. 9, a block diagram of a trajectory generator is presented, which allows setting two trajectories. The reference trajectory is a trajectory of the sine at a set amplitude and frequency, and zero trajectory, which keeps the manipulator in the position in which the internal coordinates are equal to zero. Choosing the type of trajectory is done with the START button, which is connected to the input of the DSP card. DS1102IN_C16 block represents a 16 bit of input of the DSP card, which is connected to the START button.

The implemented program was compiled using an RTI package and as a result of this action, C program and its object code were obtained. The object code was automatically loaded into memory of the DSP card and executed. In order to register date from experiments, Control Desk program (dSPACE Company) was used.

Experimental Results

The new adaptive control algorithm was tested carrying out simulations and experiments on a manipulator with two degrees of freedom. For the sake of simulation its mathematical model was implemented in SIMULINK. The desired trajectory was sinusoidal with amplitude $A_1 = \pi/4$, $A_2 = -\pi/4$ with full cycle during $T_1 = T_2 = 16s$. The following values of the parameters of manipulator were identified:

- links masses: $m_1 = m_2 = 3\ kg$,
- SEM: $k_{m1} = k_{m2} = 0.128\ \frac{Nm}{A}$,

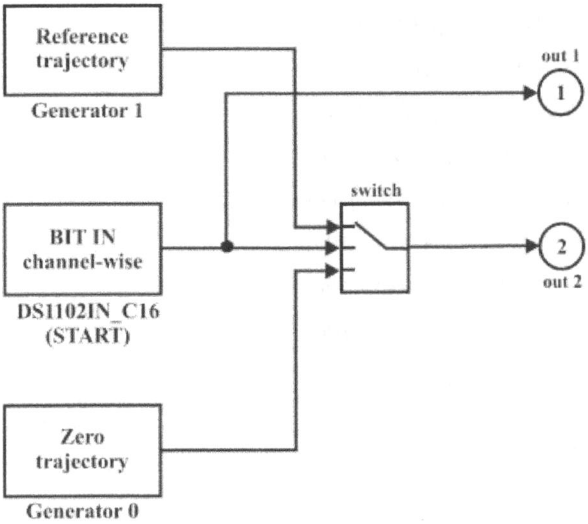

Figure 9. Scheme of a trajectory generator.

- joint stiffness: $K_1 = K_2 = 4000 \dfrac{Nm}{rad}$,
- motor inertias: $J_{m1} = J_{m2} = 6e-6\ kgm^2$,
- link inertias: $I_{l1} = I_{l2} = 0.27\ kgm^2$,
- distance of the load from the axis of rotation: $l_{c1} = l_{c2} = 0.31\ m$.

It was assumed that the purpose of control is achieved when the tracking trajectory error is smaller than 1%. On the basis of wide examinations carried out the following parameters of adaptive controllers were selected (the units are in SI and are omitted):

- PD controller

$$\begin{aligned}&K_{pl1}=0.75, \quad K_{dl1}=0.75, \quad K_{pl2}=0.2,\\&K_{dl2}=0.2, \quad K_{pm1}=0.75, \quad K_{dm1}=0.2,\\&K_{pm2}=0.2, \quad K_{dm2}=0.2.\end{aligned} \tag{101}$$

- Filter parameters

$$\begin{aligned}&a_{l1}=35, \quad a_{m1}=35, \quad a_{l2}=25, \quad a_{m2}=25,\\&b_{l1}=4, \quad b_{m1}=4, \quad b_{l2}=5, \quad b_{m2}=5,\end{aligned} \tag{102}$$

- Initial value of the estimates

$$\begin{aligned}&\hat{M}_{1a}(0)=0.0045, \quad \hat{M}_{2a}(0)=0.005,\\&\hat{M}_{1b}(0)=0.0051, \quad \hat{M}_{2b}(0)=0.005,\\&\hat{M}_{1c}(0)=0.0046, \quad \hat{M}_{2c}(0)=0.005,\\&\hat{M}_{1d}(0)=0.0048, \quad \hat{C}_{2}(0)=0.001,\end{aligned} \tag{103}$$

$$\begin{array}{ll}
\hat{C}_{1a}(0)=0.001, & \hat{F}_{ld2}(0)=6.3e- \\
\hat{C}_{1b}(0)=0.001, & \hat{F}_{ls2}(0)=3.3e- \\
\hat{F}_{ld1}(0)=6.3e-3, & \hat{J}_{m2}(0)=6.4e- \\
\hat{F}_{ls1}(0)=3.3e-2, & \hat{F}_{md2}(0)=3.9e- \\
\hat{J}_{m1}(0)=6.4e-5, & \hat{F}_{ms2}(0)=3.1e- \\
\hat{F}_{md1}(0)=3.9e-3, & \hat{K}_2(0)=105, \\
\hat{F}_{ms1}(0)=3.1e-2, & \hat{N}_2(0)=100, \\
\hat{K}_1(0)=105, & \hat{N}_1(0)=100.
\end{array}$$

- Adaptive gain coefficient

$$\begin{array}{ll}
\eta_{l1}=0.5 & \eta_{m1}=0.5 \\
\eta_{l2}=0.5 & \eta_{m2}=0.5 \\
\gamma_{M1a}=100, & \gamma_{M2a}=50, \\
\gamma_{M1b}=6, & \gamma_{M2b}=50, \\
\gamma_{M1c}=300, & \gamma_{M2c}=50, \\
\gamma_{M1d}=400, & \gamma_{C2}=100, \\
\gamma_{C1a}=500, & \gamma_{Fld2}=5e6, \\
\gamma_{C1b}=100, & \gamma_{Fls2}=1e2, \\
\gamma_{Fld1}=500, & \gamma_{Jm2}=3.5e7, \\
\gamma_{Fls1}=1e4, & \gamma_{Fmd2}=1.5e6, \\
\gamma_{Jm1}=5e8, & \gamma_{Fms2}=2e6, \\
\gamma_{Fmd1}=1e6, & \gamma_{K2}=0.01, \\
\gamma_{Fms1}=2e6, & \\
\varepsilon_{l1}=60, & \varepsilon_{l2}=60, \\
\varepsilon_{m1}=800, & \varepsilon_{m2}=100,
\end{array} \tag{104}$$

The experiment consists of tracking the desired link and motor trajectories. The large parameter values of $\gamma_{Fls1}, \gamma_{Fmd1}, \gamma_{Fms1}, \gamma_{Fld2}, \gamma_{Fls2}, \gamma_{Fmd2}, \gamma_{Fms2}, \gamma_{Jm1}, \gamma_{Jm2}$ are questionable but they result from many experiments carried out and from the fact that the initial estimates of friction and moment of inertia were very small (Eq. (103)). The stability analysis carried out in Section The Stability Properties of The Adaptive Tracking Controller, shows that $\tilde{\mathbf{q}}_l, \dot{\tilde{\mathbf{q}}}_l, \tilde{\mathbf{q}}_m, \dot{\tilde{\mathbf{q}}}_m, \tilde{\mathbf{v}}_1, \tilde{\mathbf{v}}_m$ and $\tilde{\mathbf{a}}$ are uniformly bounded and converge exponentially to the closed balls B_{r_1}, B_{r_2}, respectively. The radii of these balls depend on gains of the PD controller $\mathbf{K}_1, \mathbf{K}_m$ and eigenvalues of matrices $\mathbf{F}_c$, $\mathbf{Q}$, $\mathbf{\Gamma}_1, \mathbf{\Gamma}_m$. When gains of the PD controller $\mathbf{K}_1, \mathbf{K}_m$ or parameters Γ_{lm}, Γ_{mm} are increased, the radii of the balls will decrease.

In the program ControlDesk one can observe changes of the generalized positions and output signals. In Figs. 10, 11, 12, experimental results are presented. Control of the manipulator went smoothly with soft undulation, which results from mechanism flexibility. The robot obtained the desired position in the defined time (3 sec) without

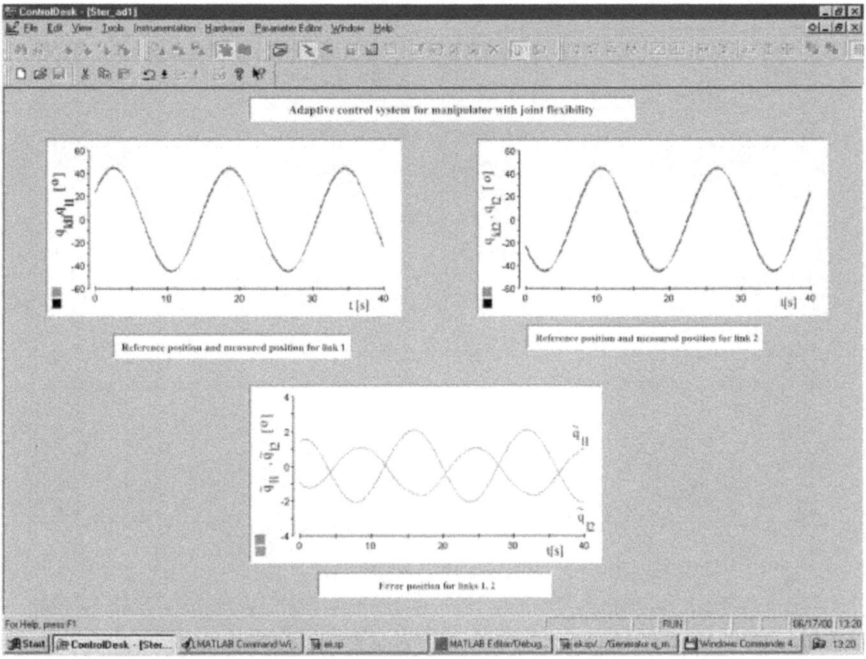

Figure 10. Position links and error link position for $k_{pl1} = 0.75$, $k_{pm1} = 0.75$, $k_{pl2} = 0.2$, $k_{pm2} = 0.2$.

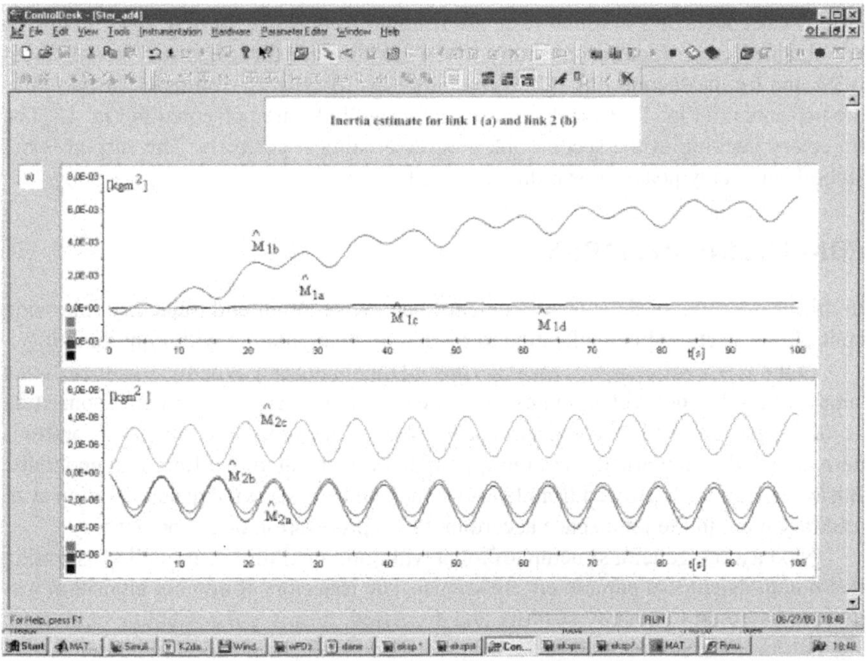

Figure 11. Inertia estimates for $k_{pl1} = 1.1$, $k_{pm1} = 0.9$, $k_{pl2} = 1.1$, $k_{pm2} = 0.95$.

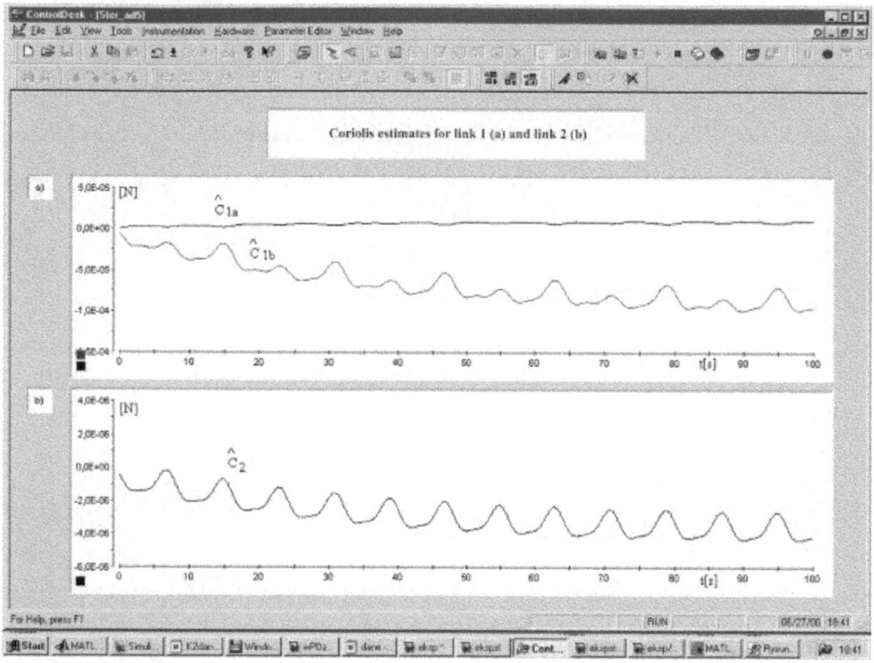

Figure 12. Coriolis estimates for $k_{pl1} = 0.75$, $k_{pm1} = 0.75$, $k_{pl2} = 0.2$, $k_{pm2} = 0.2$.

jerks. Peaks occurring in initial response resulted from static friction. In experiments we obtained error tracking trajectory in the steady state for the first link was about 0.5% and for the second link 0.2% (compare Fig. 10). Inertia estimates for two links are presented in Fig. 11. Coriolis estimates for two links are presented in Fig. 12. The trajectory tracking of the second link progressed almost perfectly. The current is not smooth and many peaks appear due to static friction.

CONCLUDING REMARKS

In this chapter two basic control problems, namely set point and trajectory following control, are discussed in application to rigid body manipulators with joint flexibility.

In the very beginning it is assumed that all kinematic and dynamic parameters of a simple one link robot with joint flexibility are known. It is a typical structure consisting of motor, gear box and torsional spring connected to one link robot. Set point control is derived based on linearization of the system that is originated in differential geometry. Basic algorithm is borrowed from Isidori's book and it is shown that stabilization at an arbitrary point in the joint space according to his proposition does not work.

Next a more general structure of robot with joint flexibility is modelled assuming that not all dynamical parameters are known. The trajectory following algorithm was derived based on Lyapunov stability and discussed, thus it is the adaptive case. Both simulation and experimental results are presented. In all cases experiments are carried out on set-up built at the Chair of Control and Systems Engineering.

REFERENCES

Barany, E. and R. Colbaugh. 1998. Global stabilization of uncertain mechanical system with bounded controls. Proceedings of the American Control Conference. Philadelphia, Pensylvania, pp. 2212–2216.

Colbaugh, R., K. Glass and E. Barany. Adaptive output stabilization of manipulators. Proceedings of the 33rd Conference on Decision and Control, pp. 1296–1302.

Corless, M. 1989. Tracking controllers for uncertain systems: Application to a Manutec R3 robot. Journal of Dynamic Systems Measurement and Control, 111: 609–618.

Corless, M. 1990. Guarenteed rates of exponential convergence for uncertain systems. Journal of Optimization Theory and Applications, 64(3): 481–494.

De Luca, A. and P. Tomei. 1996. Elastic joints. pp. 179–218. *In*: C. C. de Wit, B. Siciliano and G. Bastin (eds.). Theory of Robot Control. Springer, London.

Isidori, A. 1995. Nonlinear Control Systems, Springer Verlag.

Khorasani, K. 1992. Adaptive control of flexible-joint robots. IEEE Trans. on Robotics and Automation, 8(2): 250–267.

Kozłowski, K. and P. Sauer. 1998. On adaptive tracing control of rigid and flexible joint robots. Proceedings of the Fifth International Symposium on Methods and Models in Automation and Robotics, pp. 897–903.

Loria, A. and R. Ortega. 1995. On tracking control of rigid and flexible joint robots. Applied Mathematics and Computer Science, 5(2): 329–341.

Leitmann, G. 1981. On the efficacy of nonlinear control in uncertain linear systems. Journal of Dynamic Systems, Measurement and Control, 102: 95–102.

Madoński, R., M. Kordasz and P. Sauer. 2014. Application of a disturbance-rejection controller for robotic-enhanced limb rehabilitation trainings. ISA Transaction The Journal of Automation, 53: 899–908.

Nicosia, S. and P. Tomei. 1994. On the control of flexible joint robots by dynamic output feedback. Preprints of the Fourth IFAC Symposium on Robot Control, pp. 717–722.

Ott, Ch. 2008. Cartesian impedance control of redundant and flexible-joint robots. Springer Tracts in Advanced Robotics, Vol. 49, Springer-Verlag.

Readman, M. C. and P. R. Belanger. 1990. Analysis and control of a flexible joint robot. Proc., IEEE Conf. Decis. Contr. Honolulu. HI, USA, 4: 2551–2559.

Sauer, P. and K. Kozłowski. 2007. A new control algorithm for manipulators with joint flexibility, robot motion and control. Lecture Notes in Control and Information Sciences. Springer-Verlag London Limited 360: 119–136.

Slotine, J. and W. Li. 1991. Applied Nonlinear Control, Prentice Hall, Englewood Cliffs, NJ.

Spong, M. W. 1987. Modelling and control of elastic joint robots. Transactions of the ASME. Journal of Dynamic Systems, Measurement and Control, 109: 310–319.

Spong, M. W. 1989. Adaptive control of flexible-joint manipulators. Systems and Control Letters, 13: 15–21.

Spong, M. W. 1995. Adaptive control of flexible-joint manipulators: Comments on two papers. Automatica, 31(4): 585–590.

Tian, L. and A. A. Goldenberg. 1995. Robust adaptive control of flexible joint robots with joint torque feedback. Proceedings of the IEEE International Conference on Robotics and Automation, pp. 1229–1234.

12

Unification of Bipedal Robotic Walking using Quadratic Program-based Control Lyapunov Function

Applications to Regulation of ZMP and Angular Momentum

Pilwon Hur, Kenneth Chao* and *Victor Christian Paredes Cauna*

ABSTRACT

This chapter presents how robotic bipedal walking control can unify locomotion, manipulation, and force-based tasks into a single framework via quadratic programs utilizing control Lyapunov functions. We introduce two common examples where the unification can be applied: (1) ZMP-based pattern generation and locomotion, and (2) nonlinear dynamics with push recovery. Specifically, for ZMP-based pattern generation and locomotion, the proposed controller unifies both model predictive control-based center of mass planning and locally exponentially stabilizating control Lyapunov functions. For nonlinear dynamics with push recovery, the proposed controller unifies exponentially stabilizating control Lyapunov functions with centroidal angular momentum regulators. In both examples, unification of multiple control tasks into a single framework via quadratic program has benefits in terms of feasibility of constraints, simultaneous treatment of control objectives, and robustness and extensibility of this approach to various applications. In both examples, the end result was a single quadratic program that can dynamically balance all of these disparate objectives.

Department of Mechanical Engineering, Texas A&M University, College Station, TX, USA.
* Corresponding author: pilwonhur@tamu.edu

This approach can be implemented as a single framework and allow for more holistic implementation of controllers on bipedal robotic walking.

Keywords: Locomotion, control lyapunov function, quadratic program, angular momentum, unification, model predictive control

INTRODUCTION

Control of bipedal robotic walking is one of the main research interests in robotic literature. Among various approaches, two control methods have been used predominantly: (1) zero-moment point (ZMP)-based control (Kajita et al. 2003a, Kuindersma et al. 2014, Stephens and Atkeson 2010, Wieber 2006) and (2) nonlinear control methods that can enable dynamic walking (Ames 2014, Grizzle and Westervelt 2008, Morris and Grizzle 2005). Let's briefly look at both ZMP and nonlinear dynamic methods, respectively. ZMP is a point where influence of all generalized forces (e.g., forces, torques) acting on the foot of a robot can be replaced by one single force (Vukobratovic and Borovac 2004). ZMP-based methods usually include a simplified linear inverted pendulum model (LIPM) with the robot's center of mass (COM) and its ZMP. ZMP-based methods are simple enough that the algorithms can be implemented and computed in real time. Even though the ZMP-based methods can provide the dynamic balance all the time by keeping the ZMP within the support polygon under the foot, the resulting walking on the full-dimensional robot is usually limited to a quasi-static behavior. Nonetheless, various implementations of ZMP-based methods have been proposed and realized: kinematics-based control (Kajita et al. 2003a), force/torque-based control (Herzog et al. 2014, Kajita et al. 2003b, Macchietto et al. 2009) and whole body control with manipulation (Saab et al. 2013) to name a few. Experimental success of ZMP-based methods has motivated further attempts to unify ZMP-based methods with nonlinear control methods into a single-unified framework.

On the other end of the spectrum, several nonlinear control methods have been proposed to enable dynamic walking for bipedal robots. McGeer (1990) showed that exploiting the dynamics of a simple robotic system can achieve more agile and human-like walking even without any controls compared to ZMP-based methods. Several nonlinear control methods could fully utilize the benefits of the robot dynamics via Poincare analysis for the stability of the hybrid dynamical system. Efforts have been made for deriving proper initial conditions for passive robots, or driving a robot actively to the region of attraction within an invariant set where a stable periodic orbit exists. Examples of nonlinear control include direct trajectory optimization (Posa and Tedrake 2013), hybrid zero dynamics (Grizzle and Westervelt 2008), human-inspired control with partial hybrid zero dynamics (Ames 2014, Zhao et al. 2014). Even though these methods utilizing full-order dynamics of the system can provide natural benefits of human-like and energy-efficient robotic walking, generating gaits patterns that can simultaneously consider whole-body manipulation and robust walking in the presence of mild external perturbations is still a challenging control problem due to the high complexity of the dynamics and uncertainties of the environment.

In both methods, whether it is ZMP-based or nonlinear control methods, there are several criteria that need to be considered simultaneously when generating control laws. These criteria include stability, posture specification (e.g., trajectory tracking for each joint), other walking performance (e.g., control efforts, angular momentum, torque saturation, ZMP constraints). Kuindersma et al. (2014) proposed a quadratic program that combines MPC of walking pattern generation and locomotion control of real-time implementation. Ames and Powell (2013) introduced a unifying method of robotic locomotion and manipulation using control Lyapunov function (CLF) and quadratic programming (QP). The CLF-QP method was later experimentally realized with human-inspired walking control and MPC for full dynamics of robot (Powell et al. 2015). This unification method (CLF-QP) could combine the aforementioned various criteria into a single unified and implementable framework. The CLF-QP framework could dynamically balance all of various disjoint controller designs through weighted inequality constraints, which provide a holistic implementation of controllers on the robotic system. Even though some drawbacks on QP problems including real-time implementation and possible infeasibility may exist, the benefits of CLF-QP should not be undermined.

In our paper, we will demonstrate how CLF-QP framework can be used to unify locomotion controller design with two examples: (1) controller design using ZMP-based method and (2) controller design using nonlinear control method for push recovery. The structure of this paper is as follows. In Section **Control Lyapunov Function-Based Quadratic Program**, we will introduce and summarize the mathematical framework of CLF-QP that will be used in the two examples in Sections **Controlling Robot Locomotion With Zmp Constraints** and **Unification of Controller Design Using Nonlinear Control Methods for Push Recovery**, respectively. Section **Controlling Robot Locomotion With ZMP Constraints** will handle the unification of ZMP-based control using CLF-QP. ZMP constraints, linear inverted pendulum model (LIPM) for planning of COM trajectory, and a model predictive control (MPC) approach for pattern generation of LIPM will also be introduced. Section **Unification of Controller Design Using Nonlinear Control Methods for Push Recovery** will illustrate the unification of controller design using nonlinear control methods for push recovery. Specifically, human-inspired control, partial hybrid zero dynamics, and angular momentum regulation will be introduced. Finally, conclusion and future directions will be discussed in Section **Conclusion**.

CONTROL LYAPUNOV FUNCTION-BASED QUADRATIC PROGRAM

In this section, we summarize the framework of control Lyapunov function-based quadratic program (CLF-QP). This framework will be used in Sections **Controlling Robot Locomotion With Zmp Constraints** and **Unification of Controller Design Using Nonlinear Control Methods for Push Recovery** as a tool for unifying various control methodologies. Readers who want to know the detailed theoretical framework for CLF-QP are advised to refer to Ames and Powell (2013).

We assume that the equations of motion for a robot can be represented by the following form of the Euler-Lagrange equations:

$$D(q)\ddot{q} + H(q,\dot{q}) = Bu, \tag{1}$$

where $q \in Q \subset \mathbb{R}^{n\times 1}$, Q is the configuration space of a robot with degrees of freedom, D is the inertia matrix, H is a vector containing the Coriolis and gravity terms, and $B \in \mathbb{R}^{n\times n}$ is the torque map that determines which torque inputs actuate the robot. By reformulating (1), the following affine control system in the form of ODE can be obtained:

$$\dot{x} = f(x) + g(x)u. \tag{2}$$

where $x = (q,\dot{q}) \in TQ \subset \mathbb{R}^{2n}$, and TQ represents the tangent bundle of Q.

Since robotic walking involves discrete impacts, the discrete behavior in conjunction with the continuous dynamics can be formulated in terms of hybrid systems:

$$\mathcal{HC} = (\mathcal{D}, \mathcal{U}, \mathcal{S}, \Delta, (f, g)), \tag{3}$$

where $\mathcal{D} \subset TQ \subset \mathbb{R}^{2n}$ is the domain of a smooth submanifold of the continuous dynamics, $\mathcal{U} \subset \mathbb{R}^n$ is the set of all admissible control, $\mathcal{S}$ is a proper subset of $\mathcal{D}$ and is called the guard or switching surface at which discrete changes in the dynamics happen, Δ is a smooth map called the reset map and determines the discrete changes in dynamics, and (f, g) is the affine control system governing the continuous dynamics. The details of reset map can be found in Ames (2012).

Suppose that the nonlinear robot dynamics in the form of (2) is associated with an output y_2 of relative degree 2. Then, the input/output relation can be expressed as follows:

$$\ddot{y}_2(q) = L_f^2 y_2(q) + L_g L_f y_2(q) u = L_f + Au, \tag{4}$$

where L_f is the *Lie* derivative and A denotes the decoupling matrix which is always invertible since outputs can be chosen mutually exclusive (Zhao et al. 2014). If (2) is associated with both output y_1 of relative degree 1 and output y_2 of relative degree 2, the input/output relation can be expressed as follows:

$$\begin{bmatrix} \dot{y}_1(q,\dot{q}) \\ \ddot{y}_2(q) \end{bmatrix} = \begin{bmatrix} L_f y_1(q,\dot{q}) \\ L_f^2 y_2(q) \end{bmatrix} + \begin{bmatrix} L_g y_1(q,\dot{q}) \\ L_g L_f y_2(q) \end{bmatrix} u = L_f + Au, \tag{5}$$

The control law of feedback linearization for (4) or (5) is then given as follows:

$$u = A^{-1}(-L_f + \mu) \tag{6}$$

Applying the feedback law in (6) back to (4) or (5) will result in $\ddot{y}_2(q) = \mu$ or $[\dot{y}_1(q, \dot{q}),\ \ddot{y}_2(q)]^T = \mu$, respectively. Designing appropriate control μ can drive the outputs y_1 and y_2 to zero exponentially. For, $\eta \in H$ where H is a controlled space with appropriate dimension, defining the output coordinate $\eta = [y_2, \dot{y}_2]^T$ or $\eta = [y_1, y_2, \dot{y}_2]^T$ for (4) or (5) yields the following dynamics:

$$\dot{\eta} = F\eta + G\mu \tag{7}$$

where

$$F = \begin{bmatrix} 0 & I \\ 0 & 0 \end{bmatrix}, G = \begin{bmatrix} 0 \\ I \end{bmatrix} \quad \text{for dynamics (4)}$$

or (8)

$$F = \begin{bmatrix} 0 & 0 \\ 0 & I \\ 0 & 0 \end{bmatrix}, G = \begin{bmatrix} 1 & 0 \\ 0 & 0 \\ 0 & I \end{bmatrix} \quad \text{for dynamics (5)}$$

Please note that the original affine nonlinear system (2) was turned into a linear system (7) as a result of input/output feedback linearization (Isidori 1995, Sastry 1999).

However, input/output linearization also produces an uncontrolled nonlinear subsystem such that the original system (2) can be decomposed as follows:

$$\dot{\eta} = F\eta + G\mu$$
$$\dot{z} = Q(\eta, z) \tag{9}$$

where $z \in Z$ are the uncontrolled states, η are the controlled (or output) states as defined before, and a vector field Q is assumed to be locally Lipschitz continuous. With some appropriate control u (i.e., μ from (6)), η, becomes zero and the uncontrolled nonlinear dynamics (9) becomes $\dot{z} = Q(0, z)$, which is called a zero dynamics. In this case, the manifold z becomes invariant.

To ensure the stability of the continuous dynamics (9) (or equivalently (2)), for ***exponentially stabilizing control Lyapunov function*** (ES-CLF) $V: H \rightarrow \mathbb{R}$ which is a continuously differentiable function, there exist c_1, c_2, and c_3 are some positive real numbers such that (Ames et al. 2014, Blanchini and Miani 2008),

$$c_1\|\eta\|^2 \leq V(\eta) \leq c_2\|\eta\|^2 \tag{10}$$

$$\inf_{\mu \in U}[L_f V(\eta, z) + L_g V(\eta, z)\mu + c_3 V(\eta)] \leq 0 \tag{11}$$

Note that if there are no positive real c_1, c_2, and c_3, exponential stability of (9) is not guaranteed. When $V(\eta)$ is an ES-CLF, the solution of (9) satisfies the following:

$$\|\eta\| \leq \sqrt{\frac{c_2}{c_1}} e^{-\frac{c_3}{2}t} \|\eta(0)\|. \tag{12}$$

However, the definition of ES-CLF is not enough to handle the stability of the hybrid dynamics (3). That is, the convergence rate of ES-CLF is not fact enough to handle the hybrid dynamics. Therefore, ES-CLF is extended to have sufficiently rapid convergence rate to overcome the repulsion due to impact from the discrete dynamics. For ***rapidly exponentially stabilizing control Lyapunov function*** (RES-CLF), a one-parameter family of continuously differentiable functions $V_\epsilon: H \rightarrow \mathbb{R}$, there exist c_1, c_2, and c_3 are some positive real numbers such that

$$c_1\|\eta\|^2 \leq V_\epsilon(\eta) \leq c_2\|\eta\|^2 \tag{13}$$

$$\inf_{\mu \in U}[L_f V_\epsilon(\eta, z) + L_g V_\epsilon(\eta, z)\mu + \frac{c_3}{\epsilon} V_\epsilon(\eta)] \leq 0 \tag{14}$$

for all $0 \leq \epsilon \leq 1$, where c_1, c_2, and c_3 are some positive real numbers that are related to the convergence rate to the origin. When u_ϵ is any Lipschitz continuous feedback control law and $V_\epsilon(\eta)$ is an RES-CLF, the (13) and (14) imply that the solution of (9) satisfies the following:

$$V_\epsilon(\eta(t)) \leq e^{-\frac{c_3}{\epsilon}t} V_\epsilon(\eta(0)) \tag{15}$$

$$\|\eta\| \leq \frac{1}{\epsilon}\sqrt{\frac{c_2}{c_1}} e^{-\frac{c_3}{2\epsilon}t} \|\eta(0)\|. \tag{16}$$

Therefore, the rate of exponential convergence can be directly controlled by ϵ (Ames et al. 2014). Finally, the following optimization problem can be configured such that control effort can be minimized using quadratic program (QP):

$$u^* = \arg\min_{\mu} \mu^T \mu \tag{17}$$

$$L_f V_\epsilon(\eta, z) + L_g V_\epsilon(\eta, z)\mu + \frac{c_3}{\epsilon} V_\epsilon(\eta) \leq 0 \tag{18}$$

Then, we can eventually come to the following QP-based CLF formulation in terms of actual input u.

$$u^* = \arg\min_u u^T A^T A u + 2L_f^T A u \tag{19}$$

$$L_f V_\epsilon(\eta, z) + \frac{c_3}{\epsilon} V_\epsilon(\eta) + L_g V_\epsilon(\eta, z)\,(L_f + Au) \leq 0 \tag{20}$$

It should also be noted that the advantage of using QP based controller is that various constrains which are affine in the input can be imposed in addition to the stability constraints as opposed to min-norm control or LQR. In the next section, the other constrains that will be used for ZMP-based walking will be briefly introduced.

CONTROLLING ROBOT LOCOMOTION WITH ZMP CONSTRAINTS

To control robotic locomotion using ZMP methods, we can consider two methods which will be briefed here: (1) nonlinear robot control with ZMP constraints using RES-CLF and (2) linear inverted pendulum model for COM trajectory generation using MPC. And finally, later in this section, we will introduce how these two methods can be unified. The details can be found in (Chao et al. 2016).

Nonlinear Robot Control with ZMP Constraints

We assume that all movements of the robot happen only in the sagittal plane. Then, the ZMP constrains for dynamic balance can be described in terms of the ground reaction force (GRF), as shown in the following:

$$x_z^- \leq x_z\left(=-\frac{\tau_y}{F_z}\right) \leq x_z^+ \tag{21}$$

where x_z^- and x_z^+ are the lower and upper ends of support polygon, x_z is the ZMP position, τ_y is an ankle torque in the sagittal plane and F_z is the vertical GRF. To achieve robotic bipedal locomotion while satisfying the constraint of (21) and minimizing the control effort, CLF-QP with RES-CLF that was introduced in the previous section can be used. CLF-QP can then be formulated in the following way:

$$\bar{u}^* = \arg\min_{\bar{u}} \bar{u}^T H_{CLF}\, \bar{u} + f_{CLF}^T\, \bar{u} \tag{22}$$

$$\text{s.t. } \dot{V}(x) \leq -\epsilon V(x) \tag{23}$$

$$x_z^- \leq -\frac{\tau_y}{F_z} \leq x_z^+. \tag{24}$$

where H_{CLF} is the quadratic objective function, f_{CLF} is the linear objective function, (23) is RES-CLF condition, (24) is ZMP constraint and $x = [q\ \dot{q}]^T$ and $\bar{u}$ is the control input from the extended robot dynamics (extended from Eq. (1)) $D(q)\,\ddot{q} + H(q, \dot{q}) = [B\ J_h^T]\begin{bmatrix} u \\ F \end{bmatrix} \equiv \bar{B}\bar{u}$, where J_h is Jacobian matrix of the contact constraint $h(q)$, and F is the ground reaction force vector.

However, even though the inequality (24) is satisfied, the robot can still enter the states for which there is no feasible solution to the ZMP constraints (24). In the next subsection, we will introduce another method using linear inverted pendulum model and MPC.

Linear Inverted Pendulum Model for COM Trajectory Generation

To simplify the ZMP tracking, one can generate a COM trajectory with the linear inverted pendulum model (LIPM) that tracks the desired ZMP trajectory (or walking pattern generation). For this pattern generation, model predictive control (MPC) has been used in the literature. Since LIPM assumes a constant COM height, we have the following relation:

$$\ddot{x}_c = \frac{g}{z_0}(x_c - x_z) \equiv \omega^2 (x_c - x_z) \tag{25}$$

where z_0 is the constant COM height, x_c is the horizontal component of the COM, x_z is the ZMP. The continuous equation (25) is then discretized to implement MPC as following:

$$x_{t+1} = \begin{bmatrix} 1 & \Delta T & 0 \\ \omega^2 \Delta T & 1 & -\omega^2 \Delta T \\ 0 & 0 & 1 \end{bmatrix} x_t + \begin{bmatrix} 0 \\ 0 \\ \Delta T \end{bmatrix} u_t \tag{26}$$

where $x_t = [x_{ct}\ \dot{x}_{ct}\ x_{zt}]^T$, and ΔT is the sampling time. If N time step horizon is assumed, we have the following linear equation:

$$\bar{X} = \bar{A} x_{t_0} + \bar{B}\bar{U} \tag{27}$$

where $\bar{X} = [x_{t_0+1}^T \ldots x_{t_0+N}^T]^T$ and $\bar{U} = [u_{t_0+1}^T \ldots u_{t_0+N}^T]^T$ are the sequence of states and control inputs, respectively, and $\bar{A}$ and $\bar{B}$ are derived from (26). The predictive control can be computed from the following optimization problem:

$$\bar{U}^* = \arg\min_{\bar{U}} \bar{U}^T H_p \bar{U} + f_P^T \bar{U} \tag{27}$$

$$\text{s.t. } C\bar{U} \leq \bar{d} \tag{28}$$

where ZMP constraint is expressed in the linear constraint (28). The details of constraints and objective function will be explained in the next subsection where the unification will be introduced. Even though the pattern generation for LIPM via MPC has advantage of real-time implementation, the oversimplification of LIPM may generate control sequence that may not be feasible for the full nonlinear dynamics. Also, the ZMP trajectory generated for LIPM may not result in a feasible ZMP trajectory for the full nonlinear dynamics. To overcome these issues for both nonlinear robot control with ZMP constraints and LIPM with MPC, we will introduce the unification of both methods in the next subsection.

Unification of Pattern Generation and Locomotion via CLF-QP

Setup for Nonlinear CLF-QP

In this subsection, we will combine nonlinear CLF-QP with LIPM-MPC. The setup of CLF-QP for ZMP-based walking was already introduced in Section **Control Lyapunov Function-Based Quadratic Program**. For ZMP-based walking, we need relative degree two output function (4) for the input/output linearization. The RES-CLF for the linearized dynamics (7) is used as the following form (Ames et al. 2014),

$$V(x) = V(\eta) = \eta^T P \eta \tag{29}$$

where P can be obtained from the solution of the continuous time algebraic Riccati equation (CARE):

$$F^T P + PF - PGG^T P + Q = 0 \tag{30}$$

where $Q = Q^T > 0$, and $P = P^T > 0$. As mentioned in Section **Control Lyapunov Function-Based Quadratic Program**, exponentially stabilizing η with the convergence rate $\epsilon > 0$, η must satisfy the following condition:

$$\dot{V}(\eta) = L_f V(\eta) + L_g V(\eta) u \leq -\epsilon V(\eta) \tag{31}$$

where

$$L_f V(\eta) = \eta^T (F^T P + PF)\eta$$
$$L_g V(\eta) = 2\eta^T PG \tag{32}$$

In this setup, we assume the following for the output functions: (1) the height of COM is set to initial height, $y_{z_{COM}} = z_{COM}(q_0)$, (2) torso angle is fixed with respect to the inertial frame, $y_{\theta_{torso}}(q) = 0$, (3) during the single support phase, the orientation of swing foot, $y_{\theta_{foot}}(q)$, is zero (i.e., horizontal to the ground), and the horizontal and vertical positions of swing foot $y_{x_{foot}}(q)$ and $y_{z_{foot}}(q)$ are smooth polynomial functions where the velocities and accelerations at the beginning and at the end of swing phase are zero.

Here, we introduce all constraints used for CLF-QP. For exponential stability (31), we relaxed the constraint to improve the feasibility when there are multiple active constraints. The relaxation is penalized in the objective function.

$$\dot{V}(\eta) = L_f V(\eta) + L_g V(\eta) u \leq -\epsilon V(\eta) + \delta \tag{33}$$

Torque for each motor can exert was limited such that $|u| \leq u_{max}$. For ZMP constraint, we used (21). We also made sure that normal force at the foot is always positive, $F_z \geq 0$. The contact constraints are imposed in the equation of motion as follows (Murray et al. 1994),

$$\begin{aligned} \ddot{q} = J_h^T \left[\left(J_h D(q)^{-1} J_h^T \right)^{-1} - I \right] \times \left(J_h D(q)^{-1} H(q, \dot{q}) - J_h^T \dot{J}_h \dot{q} \right) \\ - J_h^T \left[\left(J_h D(q)^{-1} J_h^T \right)^{-1} - I \right] J_h D(q)^{-1} u \end{aligned} \tag{34}$$

where $h(q)$ is the holonomic constraint of contact points, J_h is the Jacobian matrix of $h(q)$. The detailed derivation of (34) can be found in (Chao et al. 2016).

The objective function for nonlinear CLF-QP is similar to (19). One modification to (19) is that minimization for relaxation is added with some weighting as following:

$$u^* = \arg\min_{\bar{u}} \bar{u}^T A^T A \bar{u} + 2\, L_f^T A \bar{u} + p\delta^2 \tag{35}$$

Where $\bar{u}$ is defined the same as in (22), and p is the weighting factor for the relaxation δ.

Setup for LIPM via MPC

Since robotic walking alternates single support (SS) and double support (DS) phases, the target duration of MPC horizon has N discrete points where $N = (T_{SS} + T_{DS})/\Delta T$, and T_{SS}, T_{DS}, ΔT are single support time, double support time, and MPC sampling time, respectively. Therefore, the horizon naturally consists of three phases of either SS-DS-SS or DS-SS-DS depending on the current phase. For the LIPM-MPC constraints, one of the goals is to enforce the COM to reach the position x_c^{goal} at the end of the trajectory. Also, ZMP over the horizon has to remain within the support polygon. The details can be found in (Chao et al. 2016, Stephens and Atkeson 2010, Stephens 2011, Wieber 2006).

The objective function for MPC is to minimize the control effort, while achieving ZMP trajectory tracking and driving the COM position to the desired location for the next step. The objective function is given as follows:

$$\bar{U}^* = \arg\min_{\bar{U}^*} \omega_1 \bar{U}^T \bar{U} + \omega_2 |\bar{X}_z - \bar{X}_z^{goal}|^2 \tag{36}$$

where ω_1, ω_2 are the weighting factors, $\bar{X}_z^{goal}$ is the desired ZMP trajectory, and x_c^{goal} is the desired COM terminal position at $t = t_0 + N$.

Unified QP combining Pattern Generation and ZMP-based Walking Control

The QP which combines both pattern generation and ZMP-based locomotion with RES-CLF is given as follows:

$$\begin{aligned}
&\underset{\bar{u}^*,\bar{U}^*,\delta^*}{\arg\min} \frac{1}{2}\begin{bmatrix}\bar{u}\\ \bar{U}\\ \delta\end{bmatrix}^T \begin{bmatrix} H_{CLF} & 0 & 0\\ 0 & H_p & 0\\ 0 & 0 & p\end{bmatrix}\begin{bmatrix}\bar{u}\\ \bar{U}\\ \delta\end{bmatrix} + \begin{bmatrix} f_{CLF}\\ f_p\\ 0\end{bmatrix}^T\begin{bmatrix}\bar{u}\\ \bar{U}\\ \delta\end{bmatrix}\\
&\text{s.t.} \begin{bmatrix} A_{eq,CLF} & 0 & 0\\ 0 & A_{eq,p} & 0\end{bmatrix}\begin{bmatrix}\bar{u}\\ \bar{U}\\ \delta\end{bmatrix} = \begin{bmatrix} b_{eq,CLF}\\ b_{eq,p}\end{bmatrix}\\
&\quad\;\; \begin{bmatrix} A_{iq,CLF} & 0 & -1\\ 0 & A_{iq,p} & 0\end{bmatrix}\begin{bmatrix}\bar{u}\\ \bar{U}\\ \delta\end{bmatrix} \le \begin{bmatrix} b_{iq,CLF}\\ b_{iq,p}\end{bmatrix}\\
&\quad\;\; (L_f^2 x_c + L_f L_g x_c \bar{u})\frac{z_0}{g} - x_c = -x_z
\end{aligned} \tag{37}$$

where the last equality constraint is from (25). As a synthesis constraint, it aims to equate the COM acceleration, $\ddot{x}_c$, in full dynamics to the $\ddot{x}_c$ in LIPM so that the pattern generation will plan the COM with the acceleration feedback from full dynamics. By solving the QP for each time step, the instantaneous torque input for ZMP-based locomotion considering both output tracking and COM planning on-the-fly can be derived.

Simulation Result and Future Direction

The unified controller introduced in the previous subsection is implemented on the model of AMBER 3[1] which is a human-sized, planar, and fully actuated bipedal robot with 6 degrees of freedom. For the simulation, the following parameters were implemented (Table 1).

With these parameters, the unified controller was implemented combining nonlinear CLF-QP and LIPM-MPC using MATLAB (v2015a, MathWorks, Natick, MA). The results are shown in Figs. 1, 2, and 3. As the figures show, tracking performance seems satisfactory. However, compared to the controller solving LIPM-QP and CLF-QP separately, several adjustments of the unified QP controller has to be made to ensure

Table 1. Important simulation parameters.

Parameter	Value	Parameter	Value
T_{ss}	2s	T_{Ds}	1s
MPC sampling time ΔT	0.1s	Length of MPC horizon	3s
Step Length	10 cm	Stride Height	5 cm

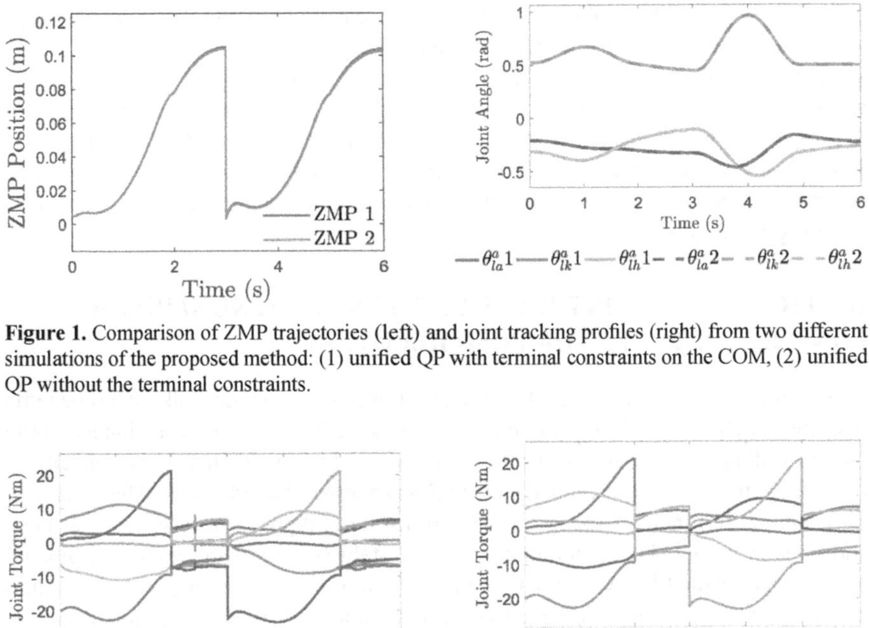

Figure 1. Comparison of ZMP trajectories (left) and joint tracking profiles (right) from two different simulations of the proposed method: (1) unified QP with terminal constraints on the COM, (2) unified QP without the terminal constraints.

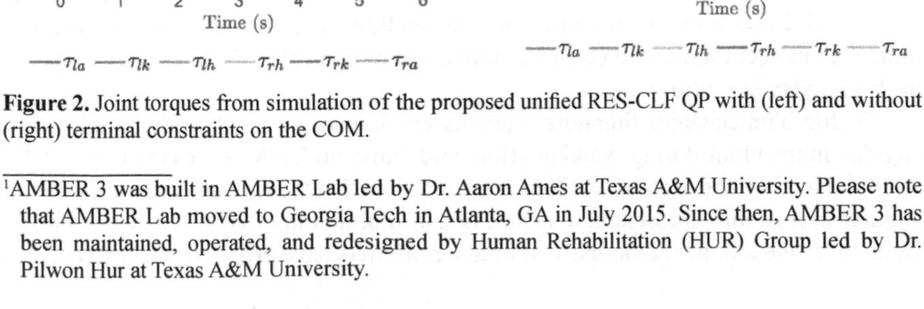

Figure 2. Joint torques from simulation of the proposed unified RES-CLF QP with (left) and without (right) terminal constraints on the COM.

[1]AMBER 3 was built in AMBER Lab led by Dr. Aaron Ames at Texas A&M University. Please note that AMBER Lab moved to Georgia Tech in Atlanta, GA in July 2015. Since then, AMBER 3 has been maintained, operated, and redesigned by Human Rehabilitation (HUR) Group led by Dr. Pilwon Hur at Texas A&M University.

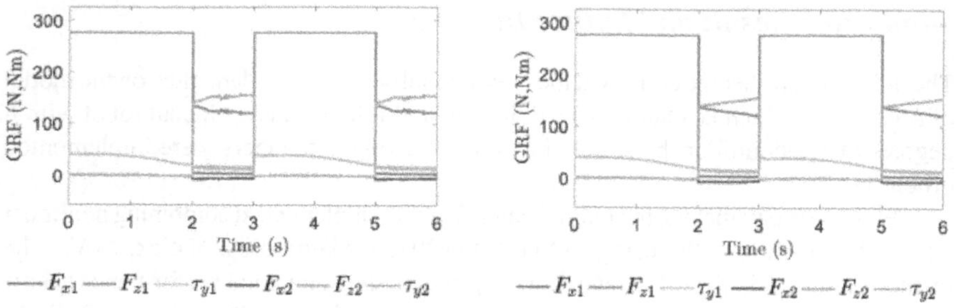

Figure 3. Ground reaction forces from simulation using the proposed unified RES-CLF QP with (left) and without (right) terminal constraints on the COM.

that the system would not be over-constrained. The first change was to remove x_c from the output vector in the unified framework because the resolved control input also minimized the cost function in unified pattern generation. Dropping of x_c could provide greater flexibility for integration with other tasks. Second, the COM terminal constraints were removed in the unified framework, since this would make the system over-constrained and the control input would lose continuity and cause chattering as shown in Figs. 2, and 3, although the derived joint trajectories and ZMP patterns (Fig. 1) are similar. Lastly, the direct ZMP feedback for updating x_{t_0} in real-time COM planning would easily cause the resolved x_c to diverge in the beginning of single support due to the impact in the full dynamics. As a result, only partial state feedback of the COM position and velocity from the full dynamics were applied. Future implementations will reason about hybrid system elements, such as impact equations, for a more general pattern generation scheme.

UNIFICATION OF CONTROLLER DESIGN USING NONLINEAR CONTROL METHODS FOR PUSH RECOVERY

As was mentioned in the Introduction section, nonlinear dynamic walking has benefits over the ZMP-based walking in terms of agileness, efficiency, and resemblance of the human walking. However, it lacks the robustness which is defined as the capacity to react to or the ability to keep balance against external perturbations. The causes of disturbances in robotic walking can be various, e.g., slippery surface, external push and uneven terrain. The importance of robust stabilization of the perturbed bipedal walking gait cannot be overemphasized in the situations where the gait stability and task performance of rescue robots, space robots, and any other service robots can be easily threatened via unexpected disturbances. However, despite these importance, robust stabilization of the perturbed bipedal walking is still a challenging research topic. In this section, we will consider unification of controller design via CLF-QP for push-recovery scenario.

In the biomechanics literature, humans are known to regulate the whole body angular momentum during walking (Herr and Popovic 2008, Popovic et al. 2004a, 2004b). Neptune and McGowan (2011) also reported that the regulation of whole-body angular momentum is essential to restoring and maintaining dynamic balance during walking. In the robotics community, several researchers have proposed control schemes exploiting angular momentum to recover balance of robots from external perturbations. Kajita et al. (2003b) presented a controller that regulated whole-body linear and angular momenta to the desired momenta. Goswami (2004) proposed to regulate the rate of angular momentum to zero in order to maintain stability.

As mentioned in Section **Control Lyapunov Function-Based Quadratic Program**, CLF-QP can be used to generate optimal controller in real time with guaranteed exponential stability. Furthermore, CLF-QP combined with the human-inspired control from the human walking data could provably result in stable and human-like robotic walking (Ames 2014). In this section, we introduce a control framework that unifies CLF-QP with human-inspired control and regulation of centroidal angular momentum, which we call CLF-K-QP. In the following subsections, we will show that the inclusion of centroidal angular momentum regulation in the CLF-QP with human-inspired control can robustly reject disturbance due to external perturbations while successfully tracking joint trajectories.

Angular Momentum in Bipedal Walking

In this subsection we briefly present angular momentum calculation. Detailed explanation can be found in (Orin and Goswami 2008). For each link *i*, linear and angular momenta about its COM can be computed as follows:

$$h_i = \begin{bmatrix} l_i \\ k_i \end{bmatrix} = \begin{bmatrix} m_i v_{c_i} \\ I_i^{COM} \omega_i \end{bmatrix} \tag{38}$$

where v_{c_i} and ω_i are linear velocity of link *i* COM and angular velocity of link *i*, respectively. m_i and I_i^{COM} are mass and inertia tensor with respect to the link *i* COM, respectively. Translating each momentum of link *i* to the origin of the local reference frame, with a distance from the origin to its center of mass as $\vec{p}_{c_i}$, and expressing the cross product using a skew symmetric matrix, $S(a)b = a \times b$, the momentum can be derived in matrix form as shown in the following equation:

$$h_i^O = \begin{bmatrix} l_i^O \\ k_i^O \end{bmatrix} = \begin{bmatrix} m_i I_{3\times 3} & m_i S(\vec{p}_{c_i})^T \\ \hat{I}_i & m_i S(\vec{p}_{c_i}) \end{bmatrix} \begin{bmatrix} \omega_i \\ v_i \end{bmatrix} = A_i^O \vec{V}_i \tag{39}$$

where $\vec{V}_i$ is the system velocity which contains ω_i and v_i of the link *i*, and $\hat{I}_i = I_i^{COM} + m_i S(\vec{p}_{c_i}) S(\vec{p}_{c_i})^T$.

Then, the collection of momentum for all links of the system can be given as follows:

$$h^O = \begin{bmatrix} h_1^O \\ \vdots \\ h_n^O \end{bmatrix} = diag[A_1^O \quad \dots \quad A_n^O]\vec{V} = A^O \vec{V} \tag{40}$$

where

$$\vec{V} = [(\vec{V}_1)^T \cdots (\vec{V}_n)^T]^T \tag{41}$$

The velocity at the origin of the local coordinate system can be computed using a Jacobian with respect to the link's reference frame as follows:

$$\vec{V}_i = J_i\dot{q} \tag{42}$$

where q is the joint angle vector, and J_i is the Jacobian of the reference frame on link i. Combining (41) with (42) can yield the following equation:

$$\vec{V} = [(J_1)^T \dots (J_n)^T]^T\,\dot{q} = J\dot{q} \tag{43}$$

Then, both (40) and (43) leads to:

$$h^O = A^O\,J\dot{q} = H^O\dot{q} \tag{44}$$

By projecting each momentum onto the COM, we get the following equation:

$$h^G = \sum_{i=1}^{n} X_i^{COM}\,h_i^O = \sum_{i=1}^{n} X_i^{COM}\,H_i^O\,\dot{q} \tag{45}$$

$$h^G = [X_1^{COM} \quad \dots \quad X_n^{COM}]H_i^O\,\dot{q} = X^{COM}\,H_i^O\,\dot{q} = A^G\,\dot{q} \tag{46}$$

CLF-QP and Human-Inspired Control

Robot Dynamics

Figure 4 gives the coordinate system and definition of joint angles: $\theta = [\theta_{sa}, \theta_{sk}, \theta_{sh}, \theta_{nsh}, \theta_{nsk}, \theta_{nsa}]^T$. The equation of motion for the robot using Euler-Lagrange formula is given as follows:

$$D(\theta)\,\ddot{\theta} + H(\theta, \dot{\theta}) = Bu \tag{47}$$

where $D(\theta) \in \mathbb{R}^{6\times6}$ is the inertial matrix and $H(\theta, \dot{\theta}) \in \mathbb{R}^{6\times1}$ contains the terms resulting from the Coriolis effect and the gravity vector. The torque map $B = I_{6\times6}$ under the assumption that the robot is fully-actuated and the control input, u, is the vector of torque inputs. With the notation $x = [\theta; \dot{\theta}]$, we assumed to have the following affine control system:

$$\dot{x} = f(x) + g(x)u \tag{48}$$

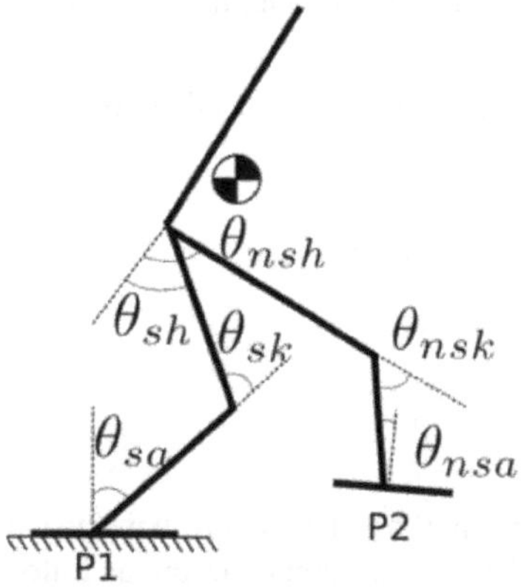

Figure 4. Flat footed robot model.

Human-Inspired Outputs

With the goal of achieving human-like robotic walking, we need to drive the actual robot output $y^a(\theta)$ to the desired human output $y^d(t, \alpha)$ that are represented by the canonical walking function (CWF):

$$y_{CWF} = e^{\alpha_4 t}(\alpha_1 \cos(\alpha_2 t) + \alpha_3 \sin(\alpha_2 t)) + \alpha_5 \tag{49}$$

Instead of using experimental human walking data directly, we are interested in seeking a low dimensional representation of human walking. It is known that hip position increases linearly with time whereas other human walking data can be fitted by CWF in the form as in (49) (Ames 2014, Sinnet et al. 2014). For our model (see Fig. 4), a total of 6 outputs are of interest for the fully-actuated 7-link bipedal robot. Therefore, we introduce human-inspired outputs as follows:

$$y(\theta, \dot{\theta}, \alpha) = \begin{bmatrix} y_1(\theta, \dot{\theta}, \alpha) \\ y_2(\theta, \alpha) \end{bmatrix} = \begin{bmatrix} y_1^a(\theta, \dot{\theta}) - v_{hip} \\ y_2^a(\theta) - y_2^d(\rho(\theta), \alpha) \end{bmatrix} \tag{50}$$

where $y_1(\theta, \dot{\theta}, \alpha)$ is the relative degree one output which is the difference between the actual hip velocity $y_1^a(\theta, \dot{\theta})$ and the desired hip velocity v_{hip}. The vector $y_2(\theta, \alpha)$ contains the 5 relative degree two human-inspired outputs which are the differences between the actual outputs $y_2^a(\theta)$ and the desired output $y_2^d(\rho(\theta), \alpha)$. Based on the linearity of hip position over time, $\rho(\theta)$ is utilized as a time parameterization of the given human walking gait.

Utilizing these outputs, the human-inspired controller can be utilized to drive both $y_1 \to 0$ and $y_2 \to 0$ in a provably exponentially stable fashion for the continuous dynamics. However, the robot may be "thrown-off" from the designed trajectory when impacts occur. This motivates the introduction of the partial hybrid zero dynamics (PHZD) constraints aiming to yield a parameter set α that ensures the tracking of relative degree two outputs will remain invariant even after the impacts. In particular, with the partial zero dynamics (PZD) surface defined as:

$$\mathbf{PZ}_\alpha = \{(\theta, \dot{\theta}) \in TQ : y_2(\theta, \alpha) = 0, L_f y_2(\theta, \alpha) = 0\} \tag{51}$$

The PHZD constraint can be explicitly stated as:

$$\Delta_R(S_R \cap \mathbf{PZ}_\alpha) \subset \mathbf{PZ}_\alpha \tag{52}$$

where Δ_R and S_R are the reset map and switching surface of the robot model, respectively.

Human-Inspired Control

With the human-inspired output functions defined in the previous subsubsection, we can perform input/output linearization. The details of this process are already introduced in Section **Control Lyapunov Function-Based Quadratic Program**. Since we have both relative degree one output and relative degree two output vectors, the input/output relation is the same as (5) and the feedback control is the same as (6). The linearized equation (7) has Lyapunov function defined as in (29) that is related to the solution of CARE (30). The Lyapunov function (29) is a RES-CLF once it satisfies the constraint (14). Finally, CLF-QP is defined with the cost function (17) or (19).

Unification of CLF-QP and Angular Momentum Regulator

In the previous subsection, we have already defined CLF-QP. In this subsection, angular momentum regulator will be combined via a QP. That is, the new controller (CLF-K-QP) adds a Lyapunov constraint to track the desired values of centroidal angular momentum as another constraint in the QP. As derived before, the centroidal angular momentum at the COM is expressed as follows:

$$h^G = A^G\dot{q} \tag{53}$$

Since the centroidal angular momentum is of interest in this application, the linear momentum will be discarded from now on.

$$k^G = [0_{3\times3}\ I_{3\times3}]h^G = Kx \tag{54}$$

As usual, the dynamics of the system is expressed as in (48). To track a desired angular momentum R, it is possible to construct a Lyapunov function based on the actual angular momentum as follows:

$$V_k = (k^G - R)^T(k^G - R) = (Kx - R)^T(Kx - R) \tag{55}$$

To ensure exponential convergence, the following condition must be met.

$$L_fV_k(x) + L_gV_k(x)u \leq -\frac{\gamma_k}{\epsilon_k}V_k(x) \tag{56}$$

where γ_k and ϵ_k are both positive and determine the rate of convergence. γ_k is a constant while ϵ_k can be changed by the control designer. A smaller ϵ_k would increase the convergence rate but it may affect feasibility of the optimization.

Since the system is already fully-actuated, we relax the tracking of CWF with the insertion of δ_1 and δ_2 as shown in (57, 58, 59) to compensate for disturbances in angular momentum at the cost of losing tracking performance. Therefore, an optimal u^* that tracks the CWF and the centroidal angular momentum is given by:

$$u^* = \arg\min_u u^TA^TAu + 2L_f^TAu + p_1\delta_1^2 + p_2\delta_2^2 \tag{57}$$

s.t.

$$\varphi_0(q,\dot{q}) + \varphi_1(q,\dot{q})(Au + L_f) \leq \delta_1 \tag{58}$$

$$\varphi_0^k(q,\dot{q}) + \varphi_1^k(q,\dot{q})^T u \leq \delta_2 \tag{59}$$

where

$$\begin{aligned}\varphi_0(\eta) &= L_fV(\eta) = \frac{\gamma}{\varepsilon}V(\eta)\\ \varphi_1(\eta) &= L_gV(\eta)^T\end{aligned} \tag{59}$$

$$\varphi_0^k = L_fV_k(x) + \frac{\gamma_k}{\epsilon_k}V_k(x) \tag{60}$$

$$\varphi_1^k = L_gV_k(x)^T \tag{61}$$

and δ_1 and δ_2 are positive slack variables inserted into constraints (58) and (59) to relax the conditions on tracking and angular momentum regulation in order to find feasible solution.

To compute the desired centroidal angular momentum R, it is possible to compute the ideal states from the PHZD surface which represents a lower dimensional presentation of the system. These ideal states can be plugged into (46) to compute R which will then be plugged into the optimization shown in (59).

Simulation Result and Future Direction

In the simulation, we assumed that the robotic model is perturbed by an external and impulsive force $F_{ext} = 700N$ horizontally at the torso in the direction of movement for $t = 0.05s$ and was applied when $\rho(\theta) = 0.21$. The proposed controller CLF-K-QP is compared with CLF-QP. In order to have specific metrics of performance we use the root-mean-square (RMS) values to compare the tracking performance for the different quantities we are interested in.

The results are shown in Figs. 5, 6 and Table 2. The principal differences found during an external push include torso angle, angular momentum and hip velocity (Fig. 6). After the disturbance, all of torso angle, centroidal angular momentum and hip velocity suffered from a significant deviations from their nominal trajectories. The CLF-QP controller tried to recover joint trajectory tracking for the torso back to its desired value faster compared to CLF-K-QP. However, CLF-K-QP did not immediately drive the torso angle to its desired trajectory as shown in Fig. 6a. As expected, CLF-K-QP performed better in tracking of angular momentum to its desired value compared to CLF-QP (Fig. 6b). Interestingly, CLF-K-QP showed better performance in hip velocity

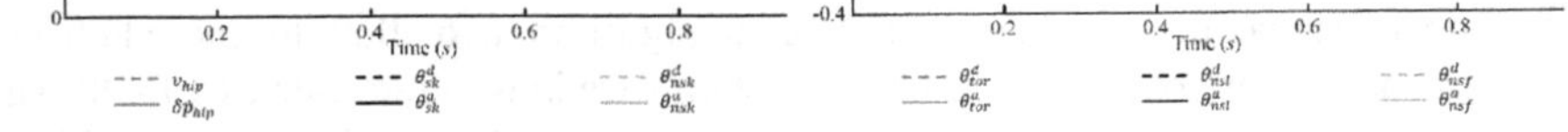

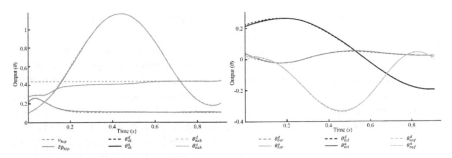

Figure 5. Outputs of unperturbed system. Note that superscript d represents desired trajectory whereas superscript a represents actual trajectory.

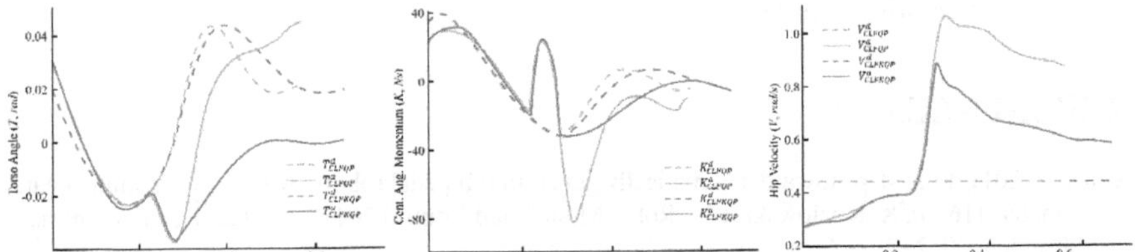

Figure 6. Torso angle evolution for one perturbed step. Note that superscript d represents desired trajectory whereas superscript a represents actual trajectory.

Table 2. Controller comparison.

RMS Values	CLF-K-QP			CLF-QP		
	Angular Momentum (Ns)	Hip Velocity (m/s)	Torso Angle (rad)	Angular Momentum (Ns)	Hip Velocity (m/s)	Torso Angle (rad)
Push	12.919	0.289	0.008	24.965	0.380	0.021
Norm	3.496	0.064	0.004	5.225	0.075	0.005

compared to CLF-QP (Fig. 6c). When perturbation as applied, hip velocity increased drastically for both controllers. However, CLF-K-QP outperformed CLF-QP in that CLF-K-QP had smaller peak hip velocity and kept the hip velocity significantly close to the desired hip velocity compared to CLF-QP. What is more inspiring is that even though CLF-QP was better at tracking torso angle, hip velocity tracking became worse. Based on these observations, we may summarize the following findings: (1) when the system is perturbed, CLF-K-QP put more effort in regulating angular momentum, (2) CLF-K-QP relaxes tracking of torso angle, and (3) the system can maintain its walking speed more robustly with CLF-K-QP. These three observations may not seem to work independently. For example, relaxed torso angle due to CLF-K-QP may have counteracted the undesired and excessive angular momentum, and eventually the system could maintain the walking speed better, whereas CLF-QP made the system stiffer. This may be similar to the natural behavior of human beings how protract their torso and outstretch their arms out to counteract the debilitating angular moments.

CONCLUSION

This chapter presented how robotic bipedal walking control can use the framework developed for locomotion and manipulation to unify ZMP-based tasks, force-based tasks, and angular momentum regulation into the same single framework via quadratic programs utilizing control Lyapunov functions. We introduced two possible examples where the unification can be applied: (1) ZMP-based pattern generation and locomotion, and (2) nonlinear dynamics with push recovery. In both examples, the end result was a single quadratic program that can dynamically balance all of these disparate objectives. This approach can be implemented as a single algorithm and allow for more holistic implementation of controllers on bipedal robotic walking. Practically, CLF-QP has been implemented in real-time to experimentally achieve 2D bipedal robotic walking. However, the speed of the QP depends on the feasibility of the constraints, so future study investigating the relation between feasibility of constraints and QP computation will be an interesting problem.

REFERENCES

Ames, A. 2011. First steps toward automatically generating bipedal robotic walking from human data. pp. 89–116. *In*: K. Kozłowski (ed.). Robot Motion and Control 2011, Vol. 422. London: Springer London, 2012.

Ames, A. and M. Powell. 2013. Towards the unification of locomotion and manipulation through control lyapunov functions and quadratic programs. pp. 219–240. *In*: D. C. Tarraf (ed.). Control of Cyber-Physical Systems. Heidelberg: Springer International Publishing, 2013.

Ames, A. 2014. Human-inspired control of bipedal walking robots. IEEE Trans. Automat. Contr., 59(5): 1115–1130, May 2014.
Ames, A., K. Galloway, K. Sreenath and J. Grizzle. 2014. Rapidly exponentially stabilizing control lyapunov functions and hybrid zero dynamics. IEEE Trans. Automat. Contr., 59(4): 876–891, 2014.
Blanchini, F. and S. Miani. 2008. Set-Theoretic Methods in Control. Basel: Birkhäuser, 2008.
Chao, K., M. Powell, A. Ames and P. Hur. 2016. Unification of locomotion pattern generation and control lyapunov function-based quadratic programs. pp. 3910–3915. *In*: Boston, M.A. American Control Conference, 2016.
Goswami, A. and V. Kallem. 2004. Rate of change of angular momentum and balance maintenance of biped robots. pp. 3785–3790. *In*: IEEE International Conference on Robotics and Automation, 2004. Proceedings. ICRA '04, 2004, Vol. 4.
Grizzle, J. W. and E. R. Westervelt. 2008. Hybrid zero dynamics of planar bipedal walking. pp. 223–237. *In*: Analysis and Design of Nonlinear Control Systems, Springer, 2008.
Herr, H. and M. Popovic. 2008. Angular momentum in human walking. J. Exp. Biol., 211(Pt 4): 467–81, Feb. 2008.
Herzog, A., L. Righetti, F. Grimminger, P. Pastor and S. Schaal. 2014. Balancing experiments on a torque-controlled humanoid with hierarchical inverse dynamics. pp. 981–988. *In*: 2014 IEEE/RSJ International Conference on Intelligent Robots and Systems, 2014.
Isidori, A. 1995. Nonlinear Control Systems, 3rd ed. London: Springer, 1995.
Kajita, S., F. Kanehiro, K. Kaneko, K. Fujiwara, K. Harada, K. Yokoi and H. Hirukawa. 2003a. Resolved momentum control: humanoid motion planning based on the linear and angular momentum. pp. 1644–1650. *In*: Proceedings 2003 IEEE/RSJ International Conference on Intelligent Robots and Systems (IROS 2003) (Cat. No. 03CH37453), 2003, Vol. 2.
Kajita, S., F. Kanehiro, K. Kaneko, K. Fujiwara, K. Harada, K. Yokoi and H. Hirukawa. 2003b. Biped walking pattern generation by using preview control of zero-moment point. pp. 1620–1626. *In*: 2003 IEEE International Conference on Robotics and Automation (Cat. No. 03CH37422), 2003, Vol. 2.
Kuindersma, S., F. Permenter and R. Tedrake. 2014. An efficiently solvable quadratic program for stabilizing dynamic locomotion. pp. 2589–2594. *In*: 2014 IEEE International Conference on Robotics and Automation (ICRA), 2014.
Macchietto, A., V. Zordan and C. R. Shelton. 2009. Momentum control for balance. *In*: ACM SIGGRAPH 2009 papers on - SIGGRAPH '09, 2009, 28(3): 1.
McGeer, T. 1990. Passive dynamic walking. Int. J. Rob. Res., 9(2): 62–82, April 1990.
Morris, B. and J. W. Grizzle. 2005. A restricted poincaré map for determining exponentially stable periodic orbits in systems with impulse effects: Application to bipedal robots. pp. 4199–4206. *In*: Proceedings of the 44th IEEE Conference on Decision and Control, 2005.
Murray, R. M., Z. Li and S. S. Sastry. 1994. A Mathematical Introduction to Robotic Manipulation, 1st ed. Boca Raton, 1994.
Neptune, R. R. and C. P. McGowan. 2011. Muscle contributions to whole-body sagittal plane angular momentum during walking. J. Biomech., 44(1): 6–12, Jan. 2011.
Orin, D. E. and A. Goswami. 2008. Centroidal momentum matrix of a humanoid robot: Structure and properties. pp. 653–659. *In*: 2008 IEEE/RSJ International Conference on Intelligent Robots and Systems, 2008.
Popovic, M., A. Hofmann and H. Herr. 2004a. Zero spin angular momentum control: definition and applicability. *In*: 4th IEEE/RAS International Conference on Humanoid Robots, 2004, 1: 478–493.
Popovic, M., A. Hofmann and H. Herr. 2004b. Angular momentum regulation during human walking: biomechanics and control. *In*: IEEE International Conference on Robotics and Automation, 2004. Proceedings. ICRA '04, 2004, 3: 2405–2411.
Posa, M. and R. Tedrake. 2013. Direct trajectory optimization of rigid body dynamical systems through contact. pp. 527–542. *In*: E. Frazzoli, T. Lozano-Perez, N. Roy and D. Rus (eds.). Algorithmic Foundations of Robotics X. Vol. 86. Berlin, Heidelberg: Springer Berlin Heidelberg, 2013.
Powell, M., E. Cousineau and A. Ames. 2015. Model predictive control of underactuated bipedal robotic walking. pp. 5121–5126. *In*: 2015 IEEE International Conference on Robotics and Automation (ICRA), 2015.

Saab, L., O. E. Ramos, F. Keith, N. Mansard, P. Soueres and J. -Y. Fourquet. 2013. Dynamic whole-body motion generation under rigid contacts and other unilateral constraints. IEEE Trans. Robot., 29(2): 346–362, April 2013.

Sastry, S. 1999. Nonlinear Systems—Analysis, Stability, and Control. New York: Springer, 1999.

Sinnet, R., S. Jiang and A. Ames. 2014. A human-inspired framework for bipedal robotic walking design. Int. J. Biomechatronics Biomed. Robot., Feb. 2014.

Stephens, B. J. and C. G. Atkeson. 2010. Push recovery by stepping for humanoid robots with force controlled joints. pp. 52–59. *In*: 2010 10th IEEE-RAS International Conference on Humanoid Robots, 2010.

Stephens, B. 2011. Push Recovery Control for Force-Controlled Humanoid Robots, Carnegie Mellon University, 2011.

Vukobratovic, M. and B. Borovac. 2004. ZERO-MOMENT POINT—THIRTY FIVE YEARS OF ITS LIFE. Int. J. Humanoid Robot., 01(01): 157–173, March 2004.

Wieber, P. 2006. Trajectory free linear model predictive control for stable walking in the presence of strong perturbations. pp. 137–142. *In*: 2006 6th IEEE-RAS International Conference on Humanoid Robots, 2006.

Zhao, H., M. Powell and A. Ames. 2014. Human-inspired motion primitives and transitions for bipedal robotic locomotion in diverse terrain. Optim. Control Appl. Methods, 35(6): 730–755, Nov. 2014.

Zhao, H., W. -L. Ma, A. Ames and M. Zeagler. 2014. Human-inspired multi-contact locomotion with AMBER2. pp. 199–210. *In*: 2014 ACM/IEEE International Conference on Cyber-Physical Systems (ICCPS), 2014.

13

Robust Adaptive Nonlinear Control for Robotic Manipulators with Flexible Joints

P. Krishnamurthy,[1] *F. Khorrami*[1,*] and *Z. Wang*[2]

ABSTRACT

The robust adaptive control problem for a multi-link robotic manipulator with flexible joints is considered in this chapter. A general uncertain dynamic model of a robotic manipulator with joint flexibility is considered including uncertainties in joint stiffness, Coriolis and centrifugal terms, friction, gravity, load torques and disturbances, and actuator inertia. Only the inertia matrix of the rigid links of the robotic manipulator is assumed to be known. A robust adaptive dynamic controller is designed through vector backstepping with adaptations for the joint stiffness matrix, actuator inertia matrix, and a lumped uncertain constant representing the other uncertainties in the system. It is shown that the designed dynamic controller provides global stabilization and practical tracking of reference trajectories specified for each of the joint variables. The efficacy of the designed dynamic controller is demonstrated through simulation studies on a two-link manipulator with joint flexibility.

Keywords: Flexible joint, robotic manipulator, robust adaptive control

[1] Control/Robotics Research Laboratory (CRRL), Dept. of Elec. and Comp. Engg., NYU Polytechnic School of Engineering, Six Metrotech Center, Brooklyn, NY 11201, USA.

[2] Toshiba America Electronic Components, Inc., San Jose, CA, USA.

* Corresponding author

INTRODUCTION

Joint flexibility frequently occurs in robotic manipulators due to underlying effects such as elasticity in the gears, shaft wind-up, and bearing deformations. Hence, the dynamic modeling and control of manipulators with flexible joints has been studied in the literature through a variety of approaches (Khorasani and Kokotovic 1985, Luca et al. 1985, Spong 1987, Spong et al. 1987, Spong 1989, Dawson et al. 1991, Lozano and Brogliato 1992, Nicosia and Tomei 1992, Jankowski and Brussel 1993, Readman 1994, Qu 1995, Jain and Khorrami 1998, Wang and Khorrami 2000, Park and Cho 2003, Chien and Huang 2007). In parallel with the specific application to robotic manipulators with flexible joints, the development of nonlinear control designs for various classes of nonlinear systems has been heavily studied in the literature over the last few decades (e.g., (Krstic et al. 1995, Khalil 1996, Sepulchre et al. 1997, Isidori 1999, Khorrami et al. 2003, Krishnamurthy and Khorrami 2003, Krishnamurthy and Khorrami 2008)) to provide a rich toolbox of control design methodologies applicable to various practically relevant nonlinear dynamic systems.

Drawing from the literature on various control design methodologies, several control approaches have been developed for flexible joint manipulators as briefly summarized below. A dynamic feedback linearization method was proposed in (Luca et al. 1985). Singular perturbation based control designs were developed in (Khorasani and Kokotovic 1985, Spong et al. 1987, Spong 1989, Readman 1994). An adaptive controller based on passivity properties was presented in (Lozano and Brogliato 1992) and an integrator backstepping based methodology was proposed in (Nicosia and Tomei 1992). While approaches such as (Lozano and Brogliato 1992, Nicosia and Tomei 1992) required the nonlinearities multiplying the unknown parameters in the manipulator dynamics to be known, robust controllers were developed in (Dawson et al. 1991, Qu 1995) to alleviate the difficulty in explicitly modeling the manipulator dynamics and to reject possibly unknown disturbances. An inverse dynamics based controller was developed in (Jankowski and Brussel 1993). A robust adaptive backstepping based control design was proposed in (Jain and Khorrami 1998) and extended in (Wang and Khorrami 2000) reducing overparametrization and allowing more general nonlinear uncertainties and disturbances. A fuzzy model reference based controller was proposed in (Park and Cho 2003) and a function approximation based adaptive controller for electrically driven flexible joint manipulators was developed in (Chien and Huang 2007).

A general uncertain dynamic model of a multi-link robotic manipulator with flexible joints is considered in this paper and a robust adaptive backstepping based controller is developed. The uncertain dynamic model is allowed to include both parametric and functional uncertainties (as general nonlinear function bounds) and disturbance inputs. The designed dynamic controller is robust to uncertainties in joint stiffness, Coriolis and centrifugal terms, friction, gravity, load torques and disturbances, and actuator inertia. Only the inertia matrix corresponding to the rigid links of the robotic manipulator is assumed to be known. The designed controller achieves practical tracking of reference trajectories specified for each of the joint variables, i.e., the tracking errors can be made arbitrarily small through appropriate choice of the control parameters. The controller includes adaptations for the joint stiffness matrix and actuator inertia matrix as well as an additional scalar adaptation for a lumped uncertain constant representing all other uncertainties in the dynamic system resulting in a dynamic controller of order $(2n + 1)$ for an n-link manipulator. While the basic design approach is similar to our prior work in

(Jain and Khorrami 1998, Wang and Khorrami 2000), the control design here addresses a more general model of the uncertain terms in the system dynamics. While (Jain and Khorrami 1998) considered polynomial bounds on uncertain terms and required the actuator inertia matrix to be known, the design here addresses general nonlinear bounds and allows uncertainty in the actuator inertia matrix. Also, the design here addresses uncertain state-dependent disturbance torques on both the robot link side and actuator side in a more general structure than (Jain and Khorrami 1998, Wang and Khorrami 2000). Furthermore, the proposed adaptive control design results in a lower dynamic order $(2n + 1)$ than in (Wang and Khorrami 2000) where a dynamic controller of order $(4n + 1)$ was proposed. The considered dynamic model of a multi-link manipulator and the control design problem addressed here are described in Section **Problem Formulation.** The proposed robust adaptive control design is presented in Section **Control Design**. Simulation studies for a two-link flexible joint manipulator are given in Section **Simulation Studies**.

PROBLEM FORMULATION

The notations utilized throughout this paper are summarized below before describing the specific control design problem considered here.

Notations: $\mathcal{R}$, $\mathcal{R}^+$, and $\mathcal{R}^k$ denote the set of real numbers, the set of non-negative real numbers, and the set of real k-dimensional column vectors, respectively. The notation $\max(a_1, \ldots, a_n)$ indicates the largest value among the numbers $a_1, \ldots, a_n$; the notation $\min(a_1, \ldots, a_n)$ similarly indicates the smallest value among the given arguments. With k being an integer, I_k denotes an identity matrix of dimension $k \times k$. With $\eta_1, \ldots, \eta_k$ being real numbers, $\mathrm{diag}(\eta_1, \ldots, \eta_k)$ denotes the $k \times k$ diagonal matrix with the ith diagonal element being η_i. $|a|$ denotes the Euclidean norm of a vector a. If a is a scalar, $|a|$ denotes its absolute value. If a is a matrix, $|a|$ denotes the Euclidean norm of the vector obtained by vertically stacking all the columns of a (i.e., Frobenius norm). If P is a symmetric positive-definite matrix, then $\lambda_{max}(P)$ denotes its maximum eigenvalue and $\lambda_{min}(P)$ denotes its minimum eigenvalue. A continuous function $\alpha : [0, a) \rightarrow [0, \infty)$ is said to belong to class K if it is strictly increasing and $\alpha(0) = 0$. It is said to belong to class K_∞ if furthermore $a = \infty$ and $\alpha(r) \rightarrow \infty$ as $r \rightarrow \infty$.

The model of a flexible joint manipulator under the assumption that the actuator motions can be modeled as torsion springs (Spong 1987) can be written as:

$$\ddot{q}_1 = -D^{-1}(q_1)\{C(q_1, \dot{q}_1)\dot{q}_1 + g(q_1) + \tau_{l1} + K(q_1 - q_2)\}$$
$$\ddot{q}_2 = J^{-1}\{K(q_1 - q_2) + \tau_{l2} + u\} \tag{1}$$

where q_1 and q_2 are the $n \times 1$ vectors of the link and actuator angles, respectively. $D(q_1)$ is the $n \times n$ symmetric positive definite inertia matrix corresponding to the rigid links of the manipulator. $J \triangleq \mathrm{diag}\{J_1, \ldots, J_n\}$ is an $n \times n$ constant diagonal matrix modeling the actuator inertia. $C(q_1, \dot{q}_1)$ is an $n \times n$ matrix that models Coriolis and centrifugal terms. $g(q_1)$ represents the gravity terms. $K \triangleq \mathrm{diag}\{K_1, \ldots, K_n\}$ is an $n \times n$ diagonal matrix of spring constants. τ_{l1} and τ_{l2} are $n \times 1$ time-varying and state-dependent vectors that include various uncertain torques (e.g., friction, disturbances, etc.) acting on the manipulator links side and actuators side, respectively. In general, the friction torques can be modeled through static or dynamic characterizations such as in (Sankaranarayanan and Khorrami 1997). Only the link inertia matrix $D(q_1)$ is assumed to be known. $C(q_1, \dot{q}_1)$, $g(q_1)$, K, J, τ_{l1}, and τ_{l2} are allowed to be uncertain. u is the $n \times 1$ vector of applied input torques.

Defining

$$\tau_{n1} = C(q_1, \dot{q}_1)\dot{q}_1 + g(q_1) + \tau_{l1}, \tag{2}$$

the dynamics of q_1 can be written as

$$\ddot{q}_1 = -D^{-1}(q_1)\{\tau_{n1} + K(q_1 - q_2)\}. \tag{3}$$

Based on physical considerations (e.g., characterizations of friction and disturbance effects, physical manipulator limits, etc.), the uncertain torque vectors τ_{n1} and τ_{l2} can be bounded as summarized in the assumption below.

Assumption A1. τ_{n1} and τ_{l2} can be bounded as

$$|\tau_{n1}|^2 \leq \theta_1[\psi_{11}(|q_1|^2) + \psi_{12}(|\dot{q}_1|^2) + \psi_1] \tag{4}$$

$$|\tau_{l2}|^2 \leq \theta_2[\psi_{21}(|q_2 - q_1|^2) + \psi_{22}(|\dot{q}_2|^2) + \psi_{23}(|\dot{q}_1|^2) + \psi_{24}(|q_2|^2) + \psi_2] \tag{5}$$

where θ_1, θ_2, ψ_1, and ψ_2 are unknown non-negative constants and ψ_{11}, ψ_{12}, ψ_{21}, ψ_{22}, ψ_{23}, and ψ_{24} are known class $\mathcal{K}_\infty$ functions.

Remark 1. Given any class $\mathcal{K}$ function $\Phi(s)$, it is possible to find a class $\mathcal{K}_\infty$ function $\tilde{\Phi}(s) = O[s]$ as $s \to 0^+$, and a positive constant, σ, to satisfy $\Phi(s) \leq \tilde{\Phi}(s) + \sigma$ for all $s \geq 0$. This can be easily seen by defining, for example, $\sigma = \sup_{s \leq 1} \Phi(s)$ and

$$\tilde{\Phi}(s) = \begin{cases} \Phi(1)s & \text{for } s \leq 1 \\ \Phi(s) + (s-1) & \text{for } s > 1. \end{cases} \tag{6}$$

Hence, without loss of generality, the Assumption A1 can be replaced with the equivalent Assumption A1' given below.

Assumption A1'. τ_n and τ_{l2} can be bounded as

$$|\tau_{n1}|^2 \leq \theta_1[\rho_{11}(|q_1|^2)|q_1|^2 + \rho_{12}(|\dot{q}_1|^2)|\dot{q}_1|^2 + \kappa_1] \tag{7}$$

$$\begin{aligned}|\tau_{l2}|^2 \leq \theta_2[&\rho_{21}(|q_2 - q_1|^2)|q_2 - q_1|^2 + \rho_{22}(|\dot{q}_2|^2)|\dot{q}_2|^2 + \rho_{23}(|\dot{q}_1|^2)|\dot{q}_1|^2 \\ &+ \rho_{24}(|q_2|^2)|q_2|^2 + \kappa_2]\end{aligned} \tag{8}$$

where θ_1, θ_2, κ_1, and κ_2 are unknown non-negative constants and ρ_{11}, ρ_{12}, ρ_{21}, ρ_{22}, ρ_{23}, and ρ_{24} are known class $\mathcal{K}_\infty$ functions.

Defining

$$x_1 = q_1, x_2 = \dot{q}_1, x_3 = q_2, x_4 = \dot{q}_2, \tag{9}$$

the dynamics (1) can be written as

$$\begin{aligned}\dot{x}_1 &= x_2 \\ \dot{x}_2 &= -D^{-1}(x_1)\{\tau_{n1} + K(x_1 - x_3)\} \\ \dot{x}_3 &= x_4 \\ \dot{x}_4 &= J^{-1}\{K(x_1 - x_3) + \tau_{l2} + u\}.\end{aligned} \tag{10}$$

The control objective is defined as practical tracking of a reference trajectory q_r specified for q_1, i.e., we seek to design a dynamic control law for u utilizing measurement of x_1, x_2, x_3, and x_4 to make the tracking error $|x_1 - q_r|$ converge to $[-\epsilon_e, \epsilon_e]$ for any arbitrary prespecified ϵ_e while keeping all closed-loop signals bounded. $q_r(t)$ is assumed to be a given thrice-differentiable bounded reference signal with bounded derivatives up to order three. The ith time derivative of q_r is denoted as $q_r^{(i)}$, i.e., $q_r^{(0)} = q_r$, $q_r^{(1)} = \dot{q}_r$, etc.

CONTROL DESIGN

The backstepping design proceeds by considering lower-dimensional subsystems and designing *virtual* control inputs (or equivalently, state transformations). The virtual control inputs in the first, second, and third steps are x_2, x_3, and x_4, respectively. In the fourth step, the control input u appears and can be designed.

Step 1: Defining the tracking error

$$z_1 = x_1 - q_r, \tag{11}$$

we get

$$\dot{z}_1 = x_2 - \dot{q}_r. \tag{12}$$

Defining the virtual control law x_2^* for x_2 as

$$x_2^* = -G_1 z_1 + \dot{q}_r \tag{13}$$

with G_1 being an $n \times n$ matrix and defining the Lyapunov function

$$V_1 = z_1^T P_1 z_1 \tag{14}$$

with P_1 being any $n \times n$ symmetric positive definite matrix, we get

$$\dot{V}_1 = -z_1^T [P_1 G_1 + G_1^T P_1] z_1 + 2 z_1^T P_1 z_2 \tag{15}$$

where

$$z_2 = x_2 - x_2^* = x_2 + G_1 z_1 - \dot{q}_r. \tag{16}$$

Picking G_1 such that

$$P_1 G_1 + G_1^T P_1 \geq v_1 I_n \tag{17}$$

with v_1 being a positive constant, (15) reduces to

$$\dot{V}_1 \leq -v_1 |z_1|^2 + 2 z_1^T P_1 z_2. \tag{18}$$

Step 2: Defining the Lyapunov function

$$V_2 = V_1 + z_2^T P_2 z_2 \tag{19}$$

with P_2 being any $n \times n$ symmetric positive definite matrix, we get

$$\dot{V}_2 \leq -v_1 |z_1|^2 + 2 z_1^T P_1 z_2 - 2 z_2^T P_2 D^{-1}(x_1) \tau_{n1} + 2 z_2^T P_2 D^{-1}(x_1) K (x_3 - x_1)$$

$$+ 2 z_2^T P_2 G_1 [z_2 - G_1 z_1] - 2 z_2^T P_2 \ddot{q}_r \tag{20}$$

Since $x_2 = z_2 - G_1 z_1 + \dot{q}_r$, we have $x_2^2 \leq 3(|z_2|^2 + |G_1 z_1|^2 + |\dot{q}_r|^2)$. Since $\rho_{12}(|x_2|^2)|x_2|^2$ is a class $\mathcal{K}_\infty$ function of $|x_2|^2$, we can write

$$\rho_{12}(|x_2|^2)|x_2|^2 \leq 9|z_2|^2\rho_{12}(9|z_2|^2) + 9|G_1 z_1|^2\rho_{12}(9|G_1 z_1|^2) + 9|\dot{q}_r|^2\rho_{12}(9|\dot{q}_r|^2). \quad (21)$$

Hence, using (7) and (21),

$$|2z_2^T P_2 D^{-1}(x_1)\tau_{n1}| \leq \frac{1}{8}v_1|z_1|^2 + c_2|z_2|^2 + \theta_1\mu_1(x_1, q_r, z_2)z_2^T P_2 D^{-2}(x_1)P_2 z_2 + d_1 \quad (22)$$

where c_2 is any positive constant and

$$\mu_1(x_1, q_r, z_2) = \frac{32}{v_1}[\rho_{11}(|x_1|^2) + 9|G_1|^2\rho_{12}(9|G_1 z_1|^2)] + 2 + \frac{9}{c_2}\rho_{12}(9|z_2|^2) \quad (23)$$

$$d_1 = \kappa_1 + \frac{1}{8}v_1|q_r|^2 + 9|\dot{q}_r|^2\rho_{12}(9|\dot{q}_r|^2). \quad (24)$$

Note that $\mu_1(x_1, q_r, z_2)$ is a known function of its arguments and d_1 is a time-varying signal that is upper bounded by an unknown non-negative constant.

Using (20) and (22), we get

$$\dot{V}_2 \leq -\frac{7}{8}v_1|z_1|^2 + 2z_1^T P_1 z_2 + 2z_2^T P_2 D^{-1}(x_1)K(x_3 - x_1) + 2z_2^T P_2 G_1[z_2 - G_1 z_1] - 2z_2^T P_2 \ddot{q}_r + c_2|z_2|^2 + \theta_1\mu_1(x_1, q_r, z_2)z_2^T P_2 D^{-2}(x_1)P_2 z_2 + d_1. \quad (25)$$

The virtual control law for x_3 is defined as

$$x_3^* = x_1 - \hat{K}D(x_1)\alpha_1(z_1, z_2, \ddot{q}_r) - \frac{1}{2}\hat{\beta}\mu_1(x_1, q_r, z_2)\hat{K}D^{-1}(x_1)P_2 z_2 \quad (26)$$

where

$$\alpha_1(z_1, z_2, \ddot{q}_r) = G_2 z_2 + \frac{1}{2}c_2 P_2^{-1}z_2 + P_2^{-1}P_1 z_1 + G_1 z_2 + \frac{8}{v_1}G_1^4 P_2 z_2 - \ddot{q}_r, \quad (27)$$

G_2 is an $n \times n$ matrix, $\hat{K}$ is an $n \times n$ diagonal matrix, and $\hat{\beta}$ is a scalar. The matrix $\hat{K} \triangleq \text{diag}\{\hat{K}_1, \ldots, \hat{K}_n\}$ is an adaptation estimate for the uncertain diagonal matrix K^{-1}. The scalar $\hat{\beta}$ is an adaptation estimate for a lumped uncertain non-negative constant β^* that will be defined further below.

G_2 is picked such that

$$P_2 G_2 + G_2^T P_2 \geq v_2 I_n \quad (28)$$

with v_2 being a positive constant.

Defining

$$z_3 = x_3 - x_3^* \quad (29)$$

and using (25) and (26), we get

$$\begin{aligned}\dot{V}_2 \leq & -\frac{13}{16}\nu_1|z_1|^2 - \nu_2|z_2|^2 + 2z_2^T P_2 D^{-1}(x_1)K z_3 \\ & - (\hat{\beta} - \theta_1)\tau_1 - 2z_2^T P_2 D^{-1}(x_1)K(\hat{K} - K^{-1})\left\{D(x_1)\alpha_1(z_1, z_2, \ddot{q}_r)\right. \\ & \left. + \frac{1}{2}\hat{\beta}\mu_1(x_1, q_r, z_2)D^{-1}(x_1)P_2 z_2\right\} + d_1 \end{aligned} \tag{30}$$

where τ_1 denotes[1]

$$\tau_1 = \mu_1(x_1, q_r, z_2)z_2^T P_2 D^{-2}(x_1)P_2 z_2. \tag{31}$$

Since K and $\hat{K}$ are diagonal matrices, (30) can be rewritten as

$$\begin{aligned}\dot{V}_2 \leq & -\frac{13}{16}\nu_1|z_1|^2 - \nu_2|z_2|^2 + 2z_2^T P_2 D^{-1}(x_1)K z_3 \\ & - (\hat{\beta} - \theta_1)\tau_1 - \sum_{i=1}^{n} K_i\left(\hat{K}_i - \frac{1}{K_i}\right)\gamma_i + d_1 \end{aligned} \tag{32}$$

where γ_i, $i = 1, \ldots, n$, denote

$$\gamma_i = e_i^T \tilde{\gamma} e_i,\ i = 1, \ldots, n \tag{33}$$

where $e_i \in \mathcal{R}^n$ is the ith unit vector, i.e., with i^{th} element being 1 and zeros elsewhere, and $\tilde{\gamma}$ denotes

$$\begin{aligned}\tilde{\gamma} = 2 & \left\{D(x_1)\alpha_1(z_1, z_2, \ddot{q}_r)\right. \\ & \left. + \frac{1}{2}\hat{\beta}\mu_1(x_1, q_r, z_2)D^{-1}(x_1)P_2 z_2\right\} z_2^T P_2 D^{-1}(x_1). \end{aligned} \tag{34}$$

Designing the adaptation dynamics for $\hat{K}_i$, $i = 1, \ldots, n$, as

$$\dot{\hat{K}}_i = -\sigma_{K_i}\hat{K}_i + \frac{1}{c_{K_i}}\gamma_i,\ i = 1, \ldots, n \tag{35}$$

with σ_{K_i}, $i = 1, \ldots, n$, and c_{K_i}, $i = 1, \ldots, n$, being any positive constants, and defining the Lyapunov function

$$V_{2a} = V_2 + \frac{1}{2}\sum_{i=1}^{n} c_{K_i} K_i\left(\hat{K}_i - \frac{1}{K_i}\right)^2, \tag{36}$$

we get

$$\begin{aligned}\dot{V}_{2a} \leq & -\frac{13}{16}\nu_1|z_1|^2 - \nu_2|z_2|^2 + 2z_2^T P_2 D^{-1}(x_1)K z_3 \\ & - (\hat{\beta} - \theta_1)\tau_1 - \frac{1}{2}\sum_{i=1}^{n}\sigma_{K_i}c_{K_i}K_i\left(\hat{K}_i - \frac{1}{K_i}\right)^2 + d_1 + \frac{1}{2}\sum_{i=1}^{n}\frac{\sigma_{K_i}c_{K_i}}{K_i}. \end{aligned} \tag{37}$$

[1] For notational convenience, we drop the arguments of functions when no confusion will result.

Step 3: Define the Lyapunov function

$$V_3 = V_{2a} + z_3^T P_3 z_3 \tag{38}$$

with P_3 being any $n \times n$ symmetric positive definite matrix. From (11), (16), and (26), we note that x_3^* can be considered a function of $(x_1, x_2, \hat{\beta}, \hat{K}, q_r, \dot{q}_r, \ddot{q}_r)$. Hence, from (29) and (38), we can write

$$\begin{aligned}\dot{V}_3 \leq & -\frac{13}{16}v_1|z_1|^2 - v_2|z_2|^2 + 2z_2^T P_2 D^{-1}(x_1)Kz_3 \\ & -(\hat{\beta}-\theta_1)\tau_1 - \frac{1}{2}\sum_{i=1}^{n}\sigma_{K_i}c_{K_i}K_i\left(\hat{K}_i - \frac{1}{K_i}\right)^2 + d_1 + \frac{1}{2}\sum_{i=1}^{n}\frac{\sigma_{K_i}c_{K_i}}{K_i} \\ & + 2z_3^T P_3\left\{x_4 - \frac{\partial x_3^*}{\partial x_1}x_2 + \frac{\partial x_3^*}{\partial x_2}D^{-1}(x_1)[\tau_{n1} + K(x_1 - x_3)]\right. \\ & \left. - \frac{\partial x_3^*}{\partial \hat{\beta}}\dot{\hat{\beta}} - \sum_{i=1}^{n}\frac{\partial x_3^*}{\partial \hat{K}_i}\dot{\hat{K}}_i - \sum_{i=0}^{2}\frac{\partial x_3^*}{\partial q_r^{(i)}}q_r^{(i+1)}\right\}.\end{aligned} \tag{39}$$

Using (26), (27), and (29), the term $(x_1 - x_3)$ can be written in the form

$$x_1 - x_3 = \zeta_1(x_1, \hat{K})z_1 + \zeta_2(x_1, x_2, q_r, \dot{q}_r, \hat{\beta}, \hat{K})z_2 - z_3 - \hat{K}D(x_1)\ddot{q}_r \tag{40}$$

where ζ_1 and ζ_2 denote

$$\zeta_1(x_1, \hat{K}) = \hat{K}D(x_1)P_2^{-1}P_1 \tag{41}$$

$$\begin{aligned}\zeta_2(x_1, x_2, q_r, \dot{q}_r, \hat{\beta}, \hat{K}) = & \hat{K}D(x_1)\left[G_2 + \frac{1}{2}c_2P_2^{-1} + G_1 + \frac{8}{v_1}G_1^4P_2\right] \\ & + \frac{1}{2}\hat{\beta}\mu_1(x_1, q_r, x_2 + G_1x_1 - G_1q_r - \dot{q}_r)\hat{K}D^{-1}(x_1)P_2.\end{aligned} \tag{42}$$

Defining $\theta_1^* = \max\{\theta_1, |K|^2\}$, we have

$$\begin{aligned}\left|2z_3^TP_3\frac{\partial x_3^*}{\partial x_2}D^{-1}(x_1)[\tau_{n1} + K(x_1 - x_3)]\right| \leq & \frac{3}{16}v_1|z_1|^2 + \frac{1}{4}v_2|z_2|^2 + c_3|z_3|^2 \\ & + \frac{1}{16}v_1|q_r|^2 + |\dot{q}_r|^2 + c_q|\ddot{q}_r|^2 + \kappa_1 \\ & + \theta_1^*\mu_2z_3^TP_3\frac{\partial x_3^*}{\partial x_2}D^{-2}(x_1)\left(\frac{\partial x_3^*}{\partial x_2}\right)^TP_3z_3\end{aligned} \tag{43}$$

where c_3 and c_q are any positive constants and μ_2 denotes the function of $(x_1, x_2, q_r, \dot{q}_r, \hat{\beta}, \hat{K})$ given by

$$\begin{aligned}\mu_2 = & \frac{32}{v_1}\rho_{11}(|x_1|^2) + \rho_{12}(|x_2|^2)\left[\frac{8}{v_2} + \frac{16}{v_1}|G_1|^2 + 1\right] + 1 + \frac{16}{v_1}|\zeta_1(x_1, \hat{K})|^2 \\ & + \frac{8}{v_2}|\zeta_2(x_1, x_2, q_r, \dot{q}_r, \hat{\beta}, \hat{K})|^2 + \frac{1}{c_3} + \frac{1}{c_q}|\hat{K}D(x_1)|^2.\end{aligned} \tag{44}$$

Also,

$$|2z_2^TP_2D^{-1}(x_1)Kz_3| \leq \frac{1}{8}v_2|z_2|^2 + \frac{8}{v_2}\theta^*|P_2D^{-1}(x_1)|^2|z_3|^2. \tag{45}$$

The virtual control law for x_4 is picked as

$$x_4^* = -G_3 z_3 - c_3 P_3^{-1} z_3 - \frac{4}{v_2}\hat{\beta}|P_2 D^{-1}(x_1)|^2 P_3^{-1} z_3 + \frac{\partial x_3^*}{\partial x_1} x_2$$
$$-\frac{1}{2}\hat{\beta}\mu_2 \frac{\partial x_3^*}{\partial x_2} D^{-2}(x_1)\left(\frac{\partial x_3^*}{\partial x_2}\right)^T P_3 z_3$$
$$+\frac{\partial x_3^*}{\partial \hat{\beta}}\left[-\sigma_\beta \hat{\beta} + \frac{1}{c_\beta}\tau_2\right] + \sum_{i=1}^{n} \frac{\partial x_3^*}{\partial \hat{K}_i}\dot{\hat{K}}_i + \sum_{i=0}^{2} \frac{\partial x_3^*}{\partial q_r^{(i)}} q_r^{(i+1)} \tag{46}$$

where c_β and σ_β are any positive constants, G_3 is an $n \times n$ matrix, and τ_2 denotes

$$\tau_2 = \tau_1 + \mu_2 z_3^T P_3 \frac{\partial x_3^*}{\partial x_2} D^{-2}(x_1)\left(\frac{\partial x_3^*}{\partial x_2}\right)^T P_3 z_3 + \frac{8}{v_2}|P_2 D^{-1}(x_1)|^2 z_3^T z_3. \tag{47}$$

In (46), $\dot{\hat{K}}_i$ denotes the right hand side (i.e., the designed dynamics for $\hat{K}_i$) of (35). G_3 is picked such that

$$P_3 G_3 + G_3^T P_3 \geq v_3 I_n \tag{48}$$

with v_3 being a positive constant.

Defining

$$z_4 = x_4 - x_4^* \tag{49}$$

and using (39), (43), (45), and (46), we get

$$\dot{V}_3 \leq -\frac{10}{16}v_1|z_1|^2 - \frac{5}{8}v_2|z_2|^2 - v_3|z_3|^2 - c_3|z_3|^2 + 2z_3^T P_3 z_4$$
$$-(\hat{\beta} - \tilde{\theta}_1)\tau_2 - \frac{1}{2}\sum_{i=1}^{n}\sigma_{K_i}c_{K_i}K_i\left(\hat{K}_i - \frac{1}{K_i}\right)^2 + d_2 + \frac{1}{2}\sum_{i=1}^{n}\frac{\sigma_{K_i}c_{K_i}}{K_i}$$
$$+\xi_1\left[-\sigma_\beta\hat{\beta} + \frac{1}{c_\beta}\tau_2 - \dot{\hat{\beta}}\right] \tag{50}$$

where $\tilde{\theta}_1 = \max\{\theta_1, \theta_1^*\}$ and

$$\xi_1 = 2z_3^T P_3 \frac{\partial x_3^*}{\partial \hat{\beta}} \tag{51}$$

$$d_2 = d_1 + \frac{1}{16}v_1|q_r|^2 + |\dot{q}_r|^2 + c_q|\ddot{q}_r|^2 + \kappa_1. \tag{52}$$

Step 4: Define the Lyapunov function

$$V_4 = V_3 + z_4^T P_4 z_4 \tag{53}$$

with P_4 being any $n \times n$ symmetric positive definite matrix. From (11), (16), (26), (29), and (46), we note that x_4^* can be considered a function of the variables $(x_1, x_2, x_3, \hat{\beta}, \hat{K}, q_r, \dot{q}_r, \ddot{q}_r, \dddot{q}_r)$. Hence, from (49), (50), and (53), we can write

$$\dot{V}_4 \leq -\frac{10}{16}v_1|z_1|^2 - \frac{5}{8}v_2|z_2|^2 - v_3|z_3|^2 - c_3|z_3|^2 + 2z_3^T P_3 z_4$$

$$-(\hat{\beta} - \tilde{\theta}_1)\tau_2 - \frac{1}{2}\sum_{i=1}^{n}\sigma_{K_i}c_{K_i}K_i\left(\hat{K}_i - \frac{1}{K_i}\right)^2 + d_2 + \frac{1}{2}\sum_{i=1}^{n}\frac{\sigma_{K_i}c_{K_i}}{K_i}$$

$$+\xi_1\left[-\sigma_\beta\hat{\beta} + \frac{1}{c_\beta}\tau_2 - \dot{\hat{\beta}}\right]$$

$$+2z_4^T P_4\left\{J^{-1}\{K(x_1 - x_3) + \tau_{l2} + u\} - \frac{\partial x_4^*}{\partial x_1}x_2\right.$$

$$+\frac{\partial x_4^*}{\partial x_2}D^{-1}(x_1)[\tau_{n1} + K(x_1 - x_3)] - \frac{\partial x_4^*}{\partial x_3}x_4$$

$$\left.-\frac{\partial x_4^*}{\partial \hat{\beta}}\dot{\hat{\beta}} - \sum_{i=1}^{n}\frac{\partial x_4^*}{\partial \hat{K}_i}\dot{\hat{K}}_i - \sum_{i=0}^{3}\frac{\partial x_4^*}{\partial q_r^{(i)}}q_r^{(i+1)}\right\}. \tag{54}$$

From (46) and (49), an upper bound on x_4 can be written as

$$|x_4| \leq |z_4| + \chi_1|z_1| + \chi_2|z_2| + \chi_3|z_3| + \chi_\beta|\hat{\beta}| + \sum_{i=1}^{n}\chi_{Ki}|\hat{K}_i| + \sum_{i=1}^{3}\chi_{q,i}|q_r^{(i)}| \tag{55}$$

where

$$\chi_1 = \left|\frac{\partial x_3^*}{\partial x_1}G_1\right| \tag{56}$$

$$\chi_2 = \left\|\left[\frac{1}{c_\beta}\frac{\partial x_3^*}{\partial \hat{\beta}}\mu_1(x_1, q_r, z_2)z_2^T P_2 D^{-2}(x_1)P_2 + \frac{\partial x_3^*}{\partial x_1}\right]\right\|$$

$$+\sum_{i=1}^{n}\frac{1}{c_{K_i}}\left[2\left|\frac{\partial x_3^*}{\partial \hat{K}_i}e_i^T\right\{D(x_1)\alpha_1(z_1, z_2, \ddot{q}_r)\right.$$

$$\left.+\frac{1}{2}\hat{\beta}\mu_1(x_1, q_r, z_2)D^{-1}(x_1)P_2 z_2\right\}\left||P_2 D^{-1}(x_1)e_i|\right] \tag{57}$$

$$\chi_3 = \left\|\left[G_3 + c_3P_3^{-1} + \frac{4}{v_2}\hat{\beta}|P_2D^{-1}(x_1)|^2P_3^{-1} + \frac{1}{2}\hat{\beta}\mu_2\frac{\partial x_3^*}{\partial x_2}D^{-2}(x_1)\left(\frac{\partial x_3^*}{\partial x_2}\right)^T P_3\right.\right.$$

$$\left.\left.-\frac{1}{c_\beta}\frac{\partial x_3^*}{\partial \hat{\beta}}\left(\mu_2 z_3^T P_3\frac{\partial x_3^*}{\partial x_2}D^{-2}(x_1)\left(\frac{\partial x_3^*}{\partial x_2}\right)^T P_3 + \frac{8}{v_2}|P_2D^{-1}(x_1)|^2 z_3^T\right)\right]\right\| \tag{58}$$

$$\chi_\beta = \sigma_\beta\left|\frac{\partial x_3^*}{\partial \hat{\beta}}\right| \tag{59}$$

$$\chi_{Ki} = \sigma_{Ki}\left|\frac{\partial x_3^*}{\partial \hat{K}_i}\right|, i = 1, \ldots, n \tag{60}$$

$$\chi_{q,1} = \left|\frac{\partial x_3^*}{\partial q_r} + \frac{\partial x_3^*}{\partial x_1}\right|, \chi_{q,2} = \left|\frac{\partial x_3^*}{\partial \dot{q}_r}\right|, \chi_{q,3} = \left|\frac{\partial x_3^*}{\partial \ddot{q}_r}\right|. \tag{61}$$

Hence, using (8) and (55),

$$|2z_4^T P_4 J^{-1} \tau_{l2}| \leq \frac{3}{16}v_1|z_1|^2 + \frac{3}{16}v_2|z_2|^2 + \frac{1}{4}v_3|z_3|^2 + c_4|z_4|^2 + 2c_q \sum_{i=1}^{3} |q_r^{(i)}|^2$$
$$+ \frac{c_\beta \sigma_\beta}{8} |\hat{\beta}|^2 + \frac{1}{8} \sum_{i=1}^{n} \sigma_{K_i} c_{K_i} K_i |\hat{K}_i|^2 + \kappa_2 + \tilde{\theta}_2 \mu_3 z_4^T P_4^2 z_4 \tag{62}$$

where c_4 is any positive constant, $\tilde{\theta}_2$ is defined as the uncertain non-negative constant

$$\tilde{\theta}_2 = \theta_2 \max\left\{1, |J^{-1}|^2, \frac{|J^{-1}|^2}{\min\{K_1, \ldots, K_n\}}\right\}, \tag{63}$$

and μ_3 denotes

$$\mu_3 = \frac{16}{v_1}\Big[|G_1|^2 \rho_{23}(|x_2|^2) + 2\rho_{21}(|x_3 - x_1|^2)|\zeta_1|^2 + \rho_{22}(|x_4|^2)\chi_1^2$$
$$+ 2\rho_{24}(|x_3|^2)|I_n - \zeta_1|^2\Big] + \frac{16}{v_2}\Big[\rho_{23}(|x_2|^2) + 2\rho_{21}(|x_3 - x_1|^2)|\zeta_2|^2$$
$$+ 2\rho_{24}(|x_3|^2)|\zeta_2|^2 + \rho_{22}(|x_4|^2)\chi_2^2\Big]$$
$$+ \frac{8}{v_3}\Big[2\rho_{21}(|x_3 - x_1|^2) + 2\rho_{24}(|x_3|^2) + \rho_{22}(|x_4|^2)\chi_3^2\Big]$$
$$+ 1 + \frac{1}{c_q}\Big[\rho_{23}(|x_2|^2) + \rho_{22}(|x_4|^2)\sum_{i=1}^{3}\chi_{q,i}^2\Big]$$
$$+ \frac{2}{c_q}\Big[\rho_{21}(|x_3 - x_1|^2) + \rho_{24}(|x_3|^2)\Big]|\hat{K} D(x_1)|^2$$
$$+ \frac{1}{c_4}\rho_{22}(|x_4|^2) + \frac{8}{c_\beta \sigma_\beta}\rho_{22}(|x_4|^2)\chi_\beta^2 + \sum_{i=1}^{n}\frac{8}{c_{K_i}\sigma_{K_i}}\rho_{22}(|x_4|^2)\chi_{K_i}^2. \tag{64}$$

Using (7) and (40), we have

$$\left|2z_4^T P_4 \frac{\partial x_4^*}{\partial x_2} D^{-1}(x_1)[\tau_{n1} + K(x_1 - x_3)]\right| \leq \frac{3}{16}v_1|z_1|^2 + \frac{1}{4}v_2|z_2|^2 + c_3|z_3|^2$$
$$+ \frac{1}{16}v_1|q_r|^2 + |\dot{q}_r|^2 + c_q|\ddot{q}_r|^2 + \kappa_1$$
$$+ \theta_1^* \mu_2 z_4^T P_4 \frac{\partial x_4^*}{\partial x_2} D^{-2}(x_1)\left(\frac{\partial x_4^*}{\partial x_2}\right)^T P_4 z_4 \tag{65}$$

where μ_2 is as given in (44). Also, using (40),

$$|2z_4^T P_4 J^{-1} K(x_1 - x_3)| \leq \frac{1}{16}v_1|z_1|^2 + \frac{1}{8}v_2|z_2|^2 + \frac{1}{4}v_3|z_3|^2 + c_q|\ddot{q}_r|^2$$
$$+ \theta_2^* \mu_4 z_4^T P_4^2 z_4 \tag{66}$$

where $\theta_2^* = |J^{-1}K|^2$ and μ_4 denotes

$$\mu_4 = \frac{16}{v_1}|\zeta_1|^2 + \frac{8}{v_2}|\zeta_2|^2 + \frac{4}{v_3} + \frac{1}{c_q}|\hat{K}D(x_1)|^2. \tag{67}$$

Using (62), (65), and (66), (54) can be rewritten as

$$\begin{aligned}
\dot{V}_4 \leq & -\frac{3}{16}v_1|z_1|^2 - \frac{1}{16}v_2|z_2|^2 - \frac{1}{2}v_3|z_3|^2 + 2z_3^T P_3 z_4 \\
& - (\hat{\beta} - \tilde{\theta}_1)\tau_2 - \frac{1}{2}\sum_{i=1}^{n}\sigma_{K_i}c_{K_i}K_i\left(\hat{K}_i - \frac{1}{K_i}\right)^2 + d_3 + \frac{1}{2}\sum_{i=1}^{n}\frac{\sigma_{K_i}c_{K_i}}{K_i} \\
& + \frac{c_\beta\sigma_\beta}{8}|\hat{\beta}|^2 + \frac{1}{8}\sum_{i=1}^{n}\sigma_{K_i}c_{K_i}K_i|\hat{K}_i|^2 + \xi_1\left[-\sigma_\beta\hat{\beta} + \frac{1}{c_\beta}\tau_2 - \dot{\hat{\beta}}\right] \\
& + 2z_4^T P_4\left\{J^{-1}u - \frac{\partial x_4^*}{\partial x_1}x_2 - \frac{\partial x_4^*}{\partial x_3}x_4\right. \\
& \left. - \frac{\partial x_4^*}{\partial \hat{\beta}}\dot{\hat{\beta}} - \sum_{i=1}^{n}\frac{\partial x_4^*}{\partial \hat{K}_i}\dot{\hat{K}}_i - \sum_{i=0}^{3}\frac{\partial x_4^*}{\partial q_r^{(i)}}q_r^{(i+1)}\right\} + c_4|z_4|^2 \\
& + \beta^*(\mu_3 + \mu_4)z_4^T P_4^2 z_4 + \beta^*\mu_2 z_4^T P_4\frac{\partial x_4^*}{\partial x_2}D^{-2}(x_1)\left(\frac{\partial x_4^*}{\partial x_2}\right)^T P_4 z_4
\end{aligned} \tag{68}$$

where $\beta^* = \max\{\tilde{\theta}_1, \tilde{\theta}_2, \theta_2^*\}$ and

$$d_3 = d_2 + \frac{1}{16}v_1|q_r|^2 + |\dot{q}_r|^2 + 2c_q|\ddot{q}_r|^2 + \kappa_1 + 2c_q\sum_{i=1}^{3}|q_r^{(i)}|^2 + \kappa_2. \tag{69}$$

From (68), the control input u is designed as

$$u = -\hat{J}\alpha_2 - \frac{1}{2}\hat{\beta}\hat{J}\chi_A P_4 z_4 \tag{70}$$

where

$$\begin{aligned}
\alpha_2 = & G_4 z_4 + \frac{1}{2}c_4 P_4^{-1}z_4 + P_4^{-1}P_3 z_3 - \frac{1}{c_\beta}\xi_1\chi_A P_4 z_4 - \frac{\partial x_4^*}{\partial x_1}x_2 - \frac{\partial x_4^*}{\partial x_3}x_4 \\
& - \frac{\partial x_4^*}{\partial \hat{\beta}}\dot{\hat{\beta}} - \sum_{i=1}^{n}\frac{\partial x_4^*}{\partial \hat{K}_i}\dot{\hat{K}}_i - \sum_{i=0}^{3}\frac{\partial x_4^*}{\partial q_r^{(i)}}q_r^{(i+1)}
\end{aligned} \tag{71}$$

$$\chi_A = (\mu_3 + \mu_4)I_n + \mu_2\frac{\partial x_4^*}{\partial x_2}D^{-2}(x_1)\left(\frac{\partial x_4^*}{\partial x_2}\right)^T, \tag{72}$$

G_4 is an $n \times n$ matrix, and $\hat{J}$ is an $n \times n$ diagonal matrix. The matrix $\hat{J} = \text{diag}\{\hat{J}_1, \ldots, \hat{J}_n\}$ is an adaptation estimate for the uncertain diagonal matrix J. In (71), $\dot{\hat{\beta}}$ denotes the right hand side of (76) (i.e., the designed dynamics of $\hat{\beta}$) and $\dot{\hat{K}}_i$ denotes the right hand side of (35) (i.e., the designed dynamics of $\hat{K}_i$).

G_4 is picked such that

$$P_4 G_4 + G_4^T P_4 \geq v_4 I_n \tag{73}$$

with v_4 being a positive constant.

Defining

$$\tau_3 = \tau_2 + z_4^T P_4 \chi_A P_4 z_4 \tag{74}$$

and using (70), (71), and (72), (68) reduces to

$$\begin{aligned}
\dot{V}_4 \leq & -\frac{3}{16}v_1|z_1|^2 - \frac{1}{16}v_2|z_2|^2 - \frac{1}{2}v_3|z_3|^2 - v_4|z_4|^2 \\
& -(\hat{\beta} - \beta^*)\tau_3 - \frac{1}{2}\sum_{i=1}^{n}\sigma_{K_i}c_{K_i}K_i\left(\hat{K}_i - \frac{1}{K_i}\right)^2 + d_3 + \frac{1}{2}\sum_{i=1}^{n}\frac{\sigma_{K_i}c_{K_i}}{K_i} \\
& + \frac{c_\beta\sigma_\beta}{8}|\hat{\beta}|^2 + \frac{1}{8}\sum_{i=1}^{n}\sigma_{K_i}c_{K_i}K_i|\hat{K}_i|^2 + \xi_1\left[-\sigma_\beta\hat{\beta} + \frac{1}{c_\beta}\tau_3 - \dot{\hat{\beta}}\right] \\
& - 2z_4^T P_4 J^{-1}(\hat{J} - J)\left[\alpha_2 + \frac{1}{2}\hat{\beta}\chi_A P_4 z_4\right].
\end{aligned} \tag{75}$$

The adaptation dynamics for $\hat{\beta}$ is designed as

$$\dot{\hat{\beta}} = -\sigma_\beta\hat{\beta} + \frac{1}{c_\beta}\tau_3. \tag{76}$$

Since J and $\hat{J}$ are diagonal matrices, we can write

$$2z_4^T P_4 J^{-1}(\hat{J} - J)\left[\alpha_2 + \frac{1}{2}\hat{\beta}\chi_A P_4 z_4\right] = \sum_{i=1}^{n}\frac{1}{J_i}(\hat{J}_i - J_i)\varpi_i \tag{77}$$

with

$$\varpi_i = e_i^T\tilde{\varpi}e_i,\ i = 1, \ldots, n \tag{78}$$

where, as in (33), $e_i \in \mathcal{R}^n$ is the ith unit vector, i.e., with i^{th} element being 1 and zeros elsewhere. $\tilde{\varpi}$ denotes

$$\tilde{\varpi} = 2\left[\alpha_2 + \frac{1}{2}\hat{\beta}\chi_A P_4 z_4\right]z_4^T P_4. \tag{79}$$

Designing the adaptation dynamics for $\hat{J}_i$, $i = 1, \ldots, n$, as

$$\dot{\hat{J}}_i = -\sigma_{J_i}\hat{J}_i + \frac{1}{c_{J_i}}\varpi_i,\ i = 1, \ldots, n \tag{80}$$

with σ_{J_i}, $i = 1, \ldots, n$ and c_{J_i}, $i = 1, \ldots, n$, being any positive constants, and defining the Lyapunov function

$$V = V_4 + \frac{1}{2}\sum_{i=1}^{n}c_{J_i}\frac{1}{J_i}(\hat{J}_i - J_i)^2 + \frac{1}{2}c_\beta(\hat{\beta} - \beta^*)^2, \tag{81}$$

(75) yields

$$\dot{V} \leq -\frac{3}{16}v_1|z_1|^2 - \frac{1}{16}v_2|z_2|^2 - \frac{1}{2}v_3|z_3|^2 - v_4|z_4|^2 - \frac{c_\beta\sigma_\beta}{4}(\hat{\beta} - \beta^*)^2$$

$$-\sum_{i=1}^{n}\frac{c_{J_i}\sigma_{J_i}}{2J_i}(\hat{J}_i-J_i)^2-\frac{1}{4}\sum_{i=1}^{n}\sigma_{K_i}c_{K_i}K_i\left(\hat{K}_i-\frac{1}{K_i}\right)^2+d_4 \tag{82}$$

where

$$d_4=d_3+\frac{3}{4}\sum_{i=1}^{n}\frac{\sigma_{K_i}c_{K_i}}{K_i}+\sum_{i=1}^{n}\frac{c_{J_i}\sigma_{J_i}}{2}J_i+\frac{3c_\beta\sigma_\beta}{4}(\beta^*)^2. \tag{83}$$

From (82), we see that

$$\dot{V}\leq-\Upsilon(\chi)+d_4 \tag{84}$$

where $\chi=(z_1,z_2,z_3,z_4,\hat{K}_1-\frac{1}{K_1},\ldots,\hat{K}_n-\frac{1}{K_n},\hat{J}_1-J_1,\ldots,\hat{J}_n-J_n,\hat{\beta}-\beta^*)$, and

$$\Upsilon(\chi)=\frac{3}{16}v_1|z_1|^2+\frac{1}{16}v_2|z_2|^2+\frac{1}{2}v_3|z_3|^2+v_4|z_4|^2+\frac{c_\beta\sigma_\beta}{4}(\hat{\beta}-\beta^*)^2$$

$$+\sum_{i=1}^{n}\frac{c_{J_i}\sigma_{J_i}}{2J_i}(\hat{J}_i-J_i)^2+\frac{1}{4}\sum_{i=1}^{n}\sigma_{K_i}c_{K_i}K_i\left(\hat{K}_i-\frac{1}{K_i}\right)^2. \tag{85}$$

From (14), (19), (36), (38), (53), and (81), it follows that $\Upsilon(\chi)\geq\Pi\,V$ where

$$\Pi=\min\left\{\frac{3v_1}{16\lambda_{max}(P_1)},\frac{v_2}{16\lambda_{max}(P_2)},\frac{v_3}{2\lambda_{max}(P_3)},\frac{v_4}{\lambda_{max}(P_4)},\right.$$

$$\left.\frac{\sigma_{K_1}}{2},\ldots,\frac{\sigma_{K_n}}{2},\sigma_{J_1},\ldots,\sigma_{J_n},\frac{\sigma_\beta}{2}\right\}. \tag{86}$$

Thus,

$$\dot{V}\leq-\Pi V+d_4. \tag{87}$$

Applying the Comparison Lemma (Khalil 1996),

$$V(t)\leq V(0)e^{-\Pi t}+\int_0^t e^{-\Pi(t-s)}d_4(s)ds. \tag{88}$$

Since the reference signal $\theta_{ref}(t)$ and its first three derivatives are bounded, $d_4(t)$ is a bounded time-varying signal. V is a smooth, positive definite, and radially unbounded function of χ. Thus, from (88), V and hence χ are bounded along the trajectories of the system. By routine signal chasing, it is seen that boundedness of χ implies boundedness of all closed-loop signals. Furthermore, the bound in (88) implies that solutions tend to the compact set in which

$$V\leq\Pi^{-1}\bar{d}_4 \tag{89}$$

where $\bar{d}_4=\limsup_{t\to\infty}d_4(t)$ is a finite positive constant since the signal $d_4(t)$ is uniformly bounded. In the set defined by (89), the tracking error $z_1=(q_1-q_r)$ satisfies the inequality

$$z_1^T P_1 z_1\leq\Pi^{-1}\bar{d}_4. \tag{90}$$

From (90), it is seen that for any given constant $\epsilon_e>0$, the controller parameter P_1 can be picked large enough to regulate $|z_1|$ to $[-\epsilon_e,\epsilon_e]$. Hence, the designed dynamic controller achieves practical trajectory tracking of the joint angles q_1 to a reference trajectory q_r.

SIMULATION STUDIES

The dynamic controller designed in Section **Control Design** is simulated here for a two-link manipulator with flexibility in both joints. The manipulator parameters are specified as:

- Masses: 1.1 kg for link 1 and 0.7 kg for link 2
- Link lengths: 0.7 m for link 1 and 0.65 m for link 2; center of gravity for each link at the center of the corresponding link
- Joint stiffness: 80 Nm/rad for joint 1 and 150 Nm/rad for joint 2
- Motor inertias: 1 kg.m^2 for each actuator
- $\tau_{l1} = \left[0.1\left(q_{1,1} + \frac{\pi}{2}\right) + 0.2\sin(0.05t),\ 0.2q_{1,2}^3 + 0.3\right]^T$; $\tau_{l2} = -0.6x_4$ where $q_1 = [\dot{q}_{1,1}, \dot{q}_{1,2}]^T$
- Orientation of the links relative to gravity specified such that $q_1 = \left[-\frac{\pi}{2}, 0\right]^T$ corresponds to the downward position of the links.

The controller parameters are picked as

- $G_1 = \text{diag}(7, 7)$; $G_2 = \text{diag}(1, 1)$; $G_3 = \text{diag}(30, 30)$; $G_4 = \text{diag}(75, 75)$
- $P_1 = \text{diag}(40, 40)$; $P_2 = \text{diag}(0.3, 0.3)$; $P_3 = \text{diag}(1, 1)$; $P_4 = \text{diag}(1, 1)$
- $c_2 = 0.5$; $c_3 = 0.5$; $c_4 = 1$; $c_q = 1$
- $\sigma_\beta = 2$; $c_\beta = 50$; $\sigma_{K_1} = \sigma_{K_2} = 0.1$; $c_{K_1} = c_{K_2} = 10$; $\sigma_{J_1} = \sigma_{J_2} = 0.1$; $c_{J_1} = c_{J_2} = 1$.

The reference trajectories are specified to be smooth-start sinusoids given by $-\frac{\pi}{2} + 0.5(1 - e^{-0.25t^2})\sin(0.5t)$ for the first joint and $0.2(1 - e^{-0.25t^2})\sin(0.3t)$ for the second joint. The robotic manipulator is initially in the downward position given by $q_1 = q_2 = [-\frac{\pi}{2}, 0]^T$ and $\dot{q}_1 = \dot{q}_2 = [0, 0]^T$. The initial conditions for the controller adaptation state variables are specified as $\hat{\beta} = 0.001$, $\hat{K}_1 = 0.01$, $\hat{K}_2 = 0.005$, $\hat{J}_1 = 1.5$, and $\hat{J}_2 = 1.5$. The simulation results are shown in Figs. 1 and 2. The joint angles $q_1 = [q_{1,1}, q_{1,2}]^T$, joint velocities $\dot{q}_1 = [\dot{q}_{1,1}, \dot{q}_{1,2}]^T$, actuator angles $q_2 = [q_{2,1}, q_{2,2}]^T$, and actuator velocities $\dot{q}_2 = [\dot{q}_{2,1}, \dot{q}_{2,2}]^T$ are shown in Fig. 1. The tracking errors ($q_1 - q_r = [e_{(1,1)}, e_{(1,2)}]^T$), the adaptation state variables ($\hat{\beta}$, $\hat{K}_1$, $\hat{K}_2$, $\hat{J}_1$, $\hat{J}_2$), and the control inputs ($u = [u_1, u_2]^T$) are shown in Fig. 2. It is seen that the designed dynamic controller achieves tracking of the specified reference trajectories with tracking errors regulated to around a degree. The tracking errors can be further reduced by increasing the controller gains. Note that the controller only requires the knowledge of $D(x_1)$, and hence therefore of the corresponding combination of link masses and link lengths that determines the inertia matrix $D(x_1)$, but does not require any knowledge of other parameters such as joint stiffness, motor inertias, load torques, etc. To illustrate the robustness of the controller, the joint stiffness parameters were changed to 60 Nm/rad for joint 1 and 170 Nm/rad for joint 2, and the motor inertia parameters were changed to 0.8 kg.m^2 for the actuator on joint 1 and 1.2 kg.m^2 for the actuator on joint 2. Keeping the controller parameters as given above, the simulations with these perturbed system parameters are shown in Figs. 3 and 4. It is seen that the trajectory tracking performance is very similar to the

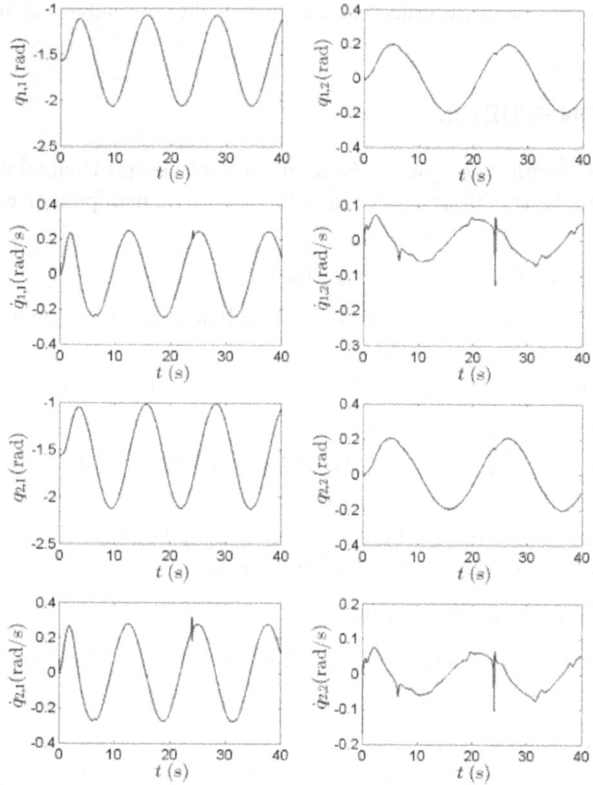

Figure 1. Simulation results for a two-link flexible joint manipulator. Joint angles $q_1 = [q_{1,1}, q_{1,2}]^T$, joint velocities $\dot{q}_1 = [\dot{q}_{1,1}, \dot{q}_{1,2}]^T$, actuator angles $q_2 = [q_{2,1}, q_{2,2}]^T$, and actuator velocities $\dot{q}_2 = [\dot{q}_{2,1}, \dot{q}_{2,2}]^T$.

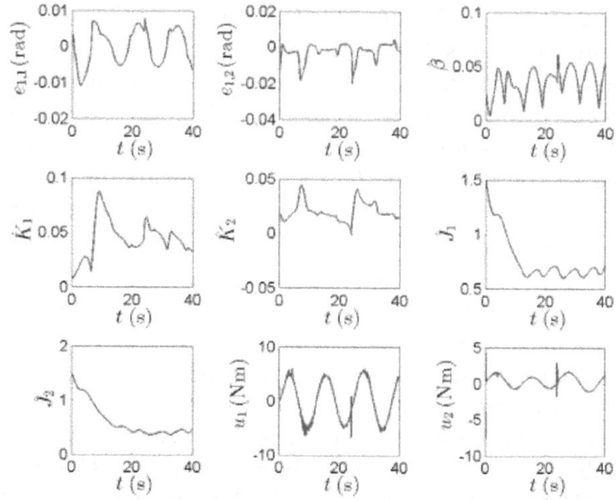

Figure 2. Simulation results for a two-link flexible joint manipulator. Tracking errors ($q_1 - q_r = [e_{(1,1)}, e_{(1,2)}]^T$), adaptation state variables ($\hat{\beta}, \hat{K}_1, \hat{K}_2, \hat{J}_1, \hat{J}_2$), and control inputs ($u = [u_1, u_2]^T$).

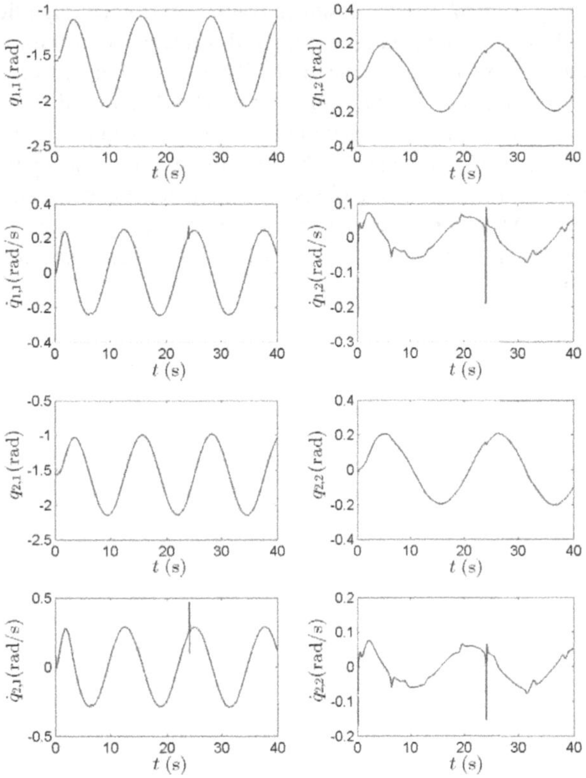

Figure 3. Simulation results for a two-link flexible joint manipulator with perturbations in system parameters K and J. Joint angles $q_1 = [q_{1,1}, q_{1,2}]^T$, joint velocities $\dot{q}_1 = [\dot{q}_{1,1}, \dot{q}_{1,2}]^T$, actuator angles $q_2 = [q_{2,1}, q_{2,2}]^T$ and actuator velocities $\dot{q}_2 = [\dot{q}_{2,1}, \dot{q}_{2,2}]^T$.

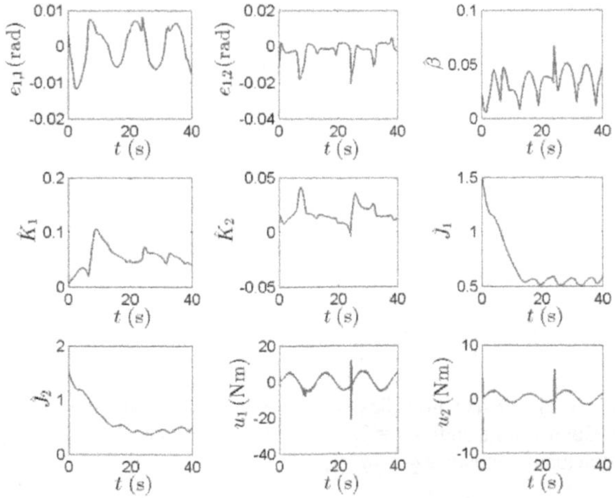

Figure 4. Simulation results for a two-link flexible joint manipulator with perturbations in system parameters K and J. Tracking errors ($q_1 - q_r = [e_{(1,1)}, e_{(1,2)}]^T$), adaptation state variables ($\hat{\beta}$, $\hat{K}_1$, $\hat{K}_2$, $\hat{J}_1$, $\hat{J}_2$), and control inputs ($u = [u_1, u_2]^T$).

simulations in Figs. 1 and 2, i.e., the designed controller provides significant robustness to variations in the motor parameters. To further illustrate the robustness of the controller to time-varying system parameters, the joint stiffness parameters were changed to (80 −20 sin(t)) Nm/rad for joint 1 and (150 + 20 cos(0.9t)) Nm/rad for joint 2, and the motor inertia parameters were changed to (1− 0.2 sin(0.8t)) kg.m^2 for the actuator on joint 1 and (1 + 0.2 cos(0.7t)) kg.m^2 for the actuator on joint 2. Keeping the controller parameters as given above, the simulations with these perturbed system parameters are shown in Figs. 5 and 6 and it is seen that the designed dynamic controller still provides similar trajectory tracking performance even with the time-varying perturbations in the system parameters.

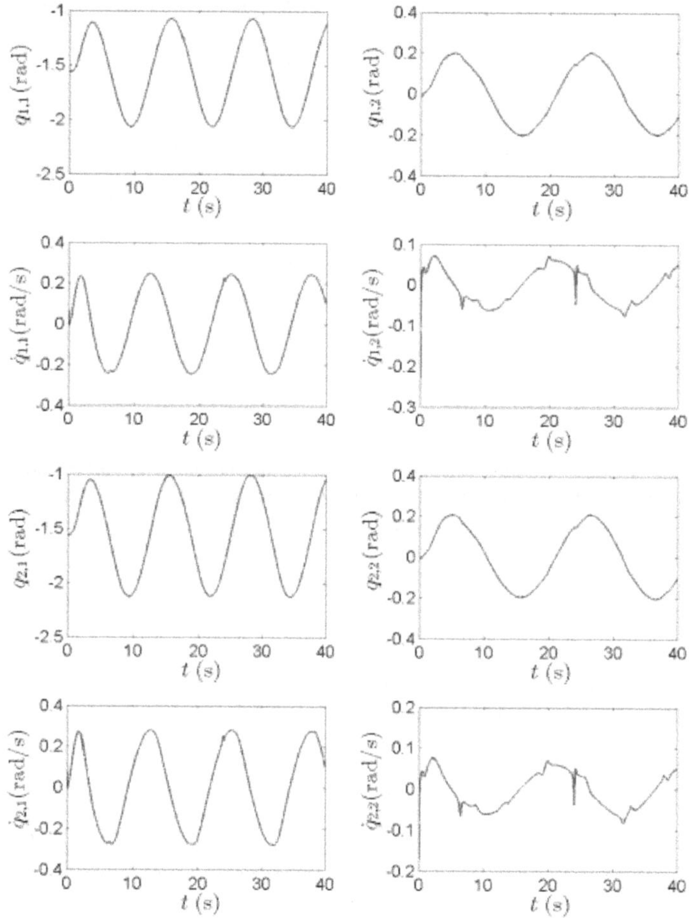

Figure 5. Simulation results for a two-link flexible joint manipulator with time-varying perturbations in system parameters K and J. Joint angles $q_1 = [q_{1,1}, q_{1,2}]^T$, joint velocities $\dot{q}_1 = [\dot{q}_{1,1}, \dot{q}_{1,2}]^T$, actuator angles $q_2 = [q_{2,1}, q_{2,2}]^T$, and actuator velocities $\dot{q}_2 = [\dot{q}_{2,1}, \dot{q}_{2,2}]^T$.

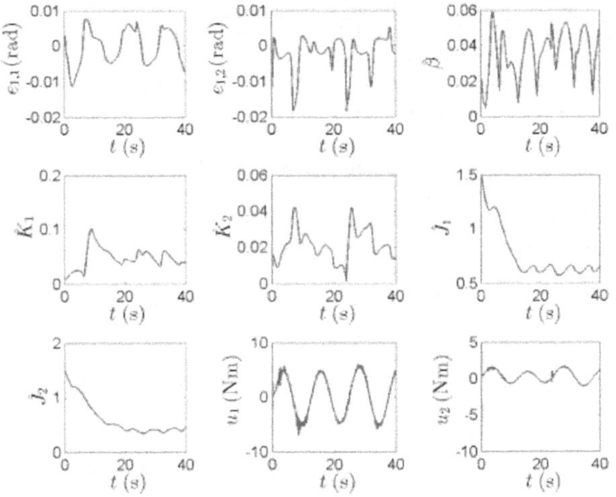

Figure 6. Simulation results for a two-link flexible joint manipulator with time-varying perturbations in system parameters K and J. Tracking errors ($q_1 - q_r = [e_{(1,1)}, e_{(1,2)}]^T$), adaptation state variables ($\hat{\beta}$, $\hat{K}_1$, $\hat{K}_2$, $\hat{J}_1$, $\hat{J}_2$), and control inputs ($u = [u_1, u_2]^T$).

CONCLUSION

A robust adaptive backstepping based controller was developed considering a general uncertain dynamic model of a multi-link robotic manipulator. The designed dynamic controller only requires knowledge of the inertia matrix of the rigid links of the manipulator and is robust to uncertainties in joint stiffness, Coriolis and centrifugal terms, friction, gravity, load torques and disturbances, and actuator inertia. The controller achieves global stabilization and practical tracking of reference trajectories specified for the joint variables. The efficacy of the designed dynamic controller was demonstrated through simulation studies for a two-link manipulator with flexible joints.

REFERENCES

Chien, M. -C. and A. -C. Huang. 2007. Adaptive control for flexible-joint electrically driven robot with time-varying uncertainties. IEEE Transactions on Industrial Electronics, 54: 1032–1038, April 2007.

Dawson, D. M., Z. Qu, M. M. Bridges and J. J. Carrol. 1991. Robust tracking of rigid-link flexible-joint electrically-driven robots. pp. 1409–1412. *In*: Proceedings of the 30th Conference on Decision and Control (Brighton, England), Dec. 1991.

Isidori, A. 1999. Nonlinear Control Systems II. London, UK: Springer, 1999.

Jain, S. and F. Khorrami. 1998. Robust adaptive control of flexible joint manipulators. Automatica, 34: 609–615, May 1998.

Jankowski, K. P. and H. V. Brussel. 1993. Inverse dynamics task control of flexible joint robots i: Continuous-time approach. Mechanism and Machine Theory, 28(6): 741–749, Nov. 1993.

Khalil, H. 1996. Nonlinear Systems. Upper Saddle River, New Jersey: Prentice Hall Inc., Second ed., 1996.

Khorasani, K. and P. V. Kokotović. 1985. Feedback linearization for a flexible manipulator near its rigid body manifold. Systems and Control Letters, 6: 187–192, Aug. 1985.

Khorrami, F., P. Krishnamurthy and H. Melkote. 2003. Modeling and Adaptive Nonlinear Control of Electric Motors. New York, USA: Springer, 2003.

Krishnamurthy, P. and F. Khorrami. 2003. Robust adaptive control for non-linear systems in generalized output-feedback canonical form. International Journal of Adaptive Control and Signal Processing, 17: 285–311, May 2003.

Krishnamurthy, P. and F. Khorrami. 2008. Dual high-gain-based adaptive output-feedback control for a class of nonlinear systems. International Journal of Adaptive Control and Signal Processing, 22(1): 23–42, Feb. 2008.

Krstić, M., I. Kanellakopoulos and P. V. Kokotović. 1995. Nonlinear and adaptive control design. New York, USA: John Wiley and Sons, 1995.

Lozano, R. and B. Brogliato. 1992. Adaptive control of robot manipulators with flexible joints. IEEE Transactions on Automatic Control, 37: 174–181, Feb. 1992.

Luca, A. D., A. Isidori and F. Nicolò. 1985. Control of a robot arm with elastic joints via nonlinear dynamic feedback. pp. 1671–1679. *In*: Proceedings of the 24th IEEE Conference on Decision and Control (Ft. Lauderdale), Dec. 1985.

Nicosia, S. and P. Tomėi. 1992. A method to design adaptive controllers for flexible joint robots. pp. 701–706. *In*: Proceedings of the 1992 International Conference on Robotics and Automation (Nice, France), May 1992.

Park, C. -W. and Y. -W. Cho. 2003. Adaptive tracking control of flexible joint manipulator based on fuzzy model reference approach. IEE Proceedings on Control Theory and Applications, 150: 198–204, March 2003.

Qu, Z. H. 1995. Input-output robust tracking control design for flexible joint robots. IEEE Transactions on Automatic Control, 40: 78–83, Jan. 1995.

Readman, M. C. 1994. Flexible Joint Robots. Boca Raton, FL, USA: CRC Press, 1994.

Sankaranarayanan, S. and F. Khorrami. 1997. Friction compensation via variable structure control: regulation and low-velocity tracking. pp. 645–650. *In*: Proceedings of the IEEE International Conference on Control Applications (Hartford, CT), Oct. 1997.

Sepulchre, R., M. Janković and P. Kokotović. 1997. Constructive Nonlinear Control. London, UK: Springer, 1997.

Spong, M. W. 1987. Modeling and control of elastic joint robots. Transactions of the ASME: Journal of Dynamical Systems, Measurement and Control, 109: 310–319, Dec. 1987.

Spong, M. W. 1989. Adaptive control of flexible joint manipulators. Systems and Control Letters, 13: 15–21, July 1989.

Spong, M. W., K. Khorasani and P. V. Kokotović. 1987. An integral manifold approach to the feedback control of flexible joint robots. IEEE Journal of Robotics and Automation, Vol. RA–3, pp. 291–300, Aug. 1987.

Wang, Z. and F. Khorrami. 2000. Robust trajectory tracking for manipulators with joint flexibility via backstepping. pp. 2849–2853. *In*: Proceedings of the American Control Conference (Chicago, IL), June 2000.

14

Adaptive Switching Iterative Learning Control of Robot Manipulator

P. R. Ouyang[1,*] and *W. J. Zhang*[2]

ABSTRACT

In this chapter, a new adaptive switching learning control approach, called adaptive switching learning PD control (ASL-PD), is proposed for trajectory tracking of robot manipulators in an iterative operation mode. The ASL-PD control method is a combination of the feedback PD control law with a gain switching technique and the feedforward learning control law with the input torque profile from the previous iteration. The ASL-PD method achieves the asymptotical convergence based on the Lyapunov's method, and it possesses both adaptive and learning capabilities with a simple control structure. Also, a new ILC called switching gain PD-PD (SPD-PD) type ILC is presented for trajectory tracking control of time-varying nonlinear systems with uncertainty and disturbance. In the SPD-PD ILC, a PD feedback control with switching gains in the iteration domain and a PD-type ILC based on the previous iteration combine together into one updating law. The proposed SPD-PD ILC can be viewed as online-offline ILC. It is theoretically proven that the boundednesses of the state error and the final tracking error are guaranteed in the presence of uncertainty, disturbance, and initialization error of the nonlinear systems. Simulation and experiments are conducted for trajectory tracking control of robot manipulators.

[1] Department of Aerospace Engineering, Ryerson University, 350 Victoria St. Toronto, ON M5B 2K3, Canada.
E-mail: pouyang@ryerson.ca

[2] Department of Mechanical Engineering, University of Saskatchewan, 57 Campus Drive, Saskatoon, SK S7N 5A9, Canada.
E-mail: wjz485@mail.usask.ca

* Corresponding author

Keywords: Adaptive control, iterative learning, switching gain control, PD control, trajectory tracking, robot manipulator, online-offline learning

INTRODUCTION

Trajectory tracking control of robot manipulators has attracted many attentions due to its wide applications in industrial systems. Basically, the control methods can be classified into the following three types. The first type is the traditional feedback control (proportional-integral-derivative (PID) control or proportional-derivative (PD) control (Chen et al. 2001, Craig 1986, Kerry 1997, Qu 1995) where the errors between the desired and the actual performance are treated in certain ways (proportional, derivative, and integral), multiplied by control gains, and fed back as the "correct" input torque. The second type is the adaptive control (Choi and Lee 2000, Craig 1988, Li et al. 1996, Li et al. 1998, Slotine and Li 1987, Tomei 1991) where the controller modifies its behaviour in response to the changes in the dynamics of the robot manipulator and the characteristics of the disturbances received by the manipulator system. The third type is the iterative learning control (ILC) (Arimoto et al. 1984, Chen and Moore 2002, Kawamura et al. 1988, Kue et al. 1991, Tayebi 2003, Yan et al. 2001) where the previous torque profile is added to the current torque in a certain manner. Some other control methods, including the robust control, model based control, switching control, and sliding mode control can be in one or another way reviewed either as specialization and/or combination of the three basic types, or are simply different names due to different emphases when the basic types are examined.

The use of traditional PD/PID control is very popular not only because of its simple structure and easy implementation but also its acceptable performance for industrial applications. It is known that the PD control can be used for trajectory tracking with the asymptotic stability if the control gains are carefully selected (Chen et al. 2001, Kerry 1997, Qu 1995). However, the PD control is not satisfactory for applications which require high tracking accuracy. This limitation with the PD control is simply due to the inherent "mismatch" between the nonlinear dynamics behaviour of a robot manipulator and the linear regulating behaviour of the PD controller. Such a limitation is also true for PID control.

The adaptive control can cope with parameter uncertainties, such as the link length, mass, inertia, and frictional nonlinearity, with a self-organizing capability. Having such a capability, however, requires extensive computation and thus compromises its application for real-time control problems (especially in high-speed operations). In addition, since the adaptive control generally does not guarantee that the estimated parameters of robot manipulators converge to their true values (Sun and Mills 1999), the tracking errors would repeatedly be brought into the system as the manipulators repeat their tasks.

Another active area of research in the control theory is the switching control (Fu and Barmish 1986, Martensson 1985, Middleton et al. 1988, Nussbaum 1983). In the switching control technique, the control of a given plant can be switched among several controllers, and each controller is designed for a specific "nominal model" of the plant. A switching control scheme usually consists of an inner loop (where a candidate controller is connected in closed-loop with the system) and an outer loop (where a supervisor decides which controller to be used and when to switch to a different one).

As such, the switching of controllers is taken place in the time domain. This underlying philosophy may be modified to perform such a switching with respect to the iteration of learning.

Robot manipulators are usually used for repetitive tasks. In this case, the reference trajectory is repeated over a given operation time. This repetitive nature makes it possible to apply ILC for improving tracking performance from iteration to iteration. It should be noted that ILC can be further classified into two kinds: off-line learning and on-line learning. In the case of off-line learning control, information in the controlled torque in the current iteration does not come from the current iteration but from the previous one. Philosophically, the learning in this case is shifted to the off-line mode. This then releases a part of the control workload at real-time, which implies the improvement of real-time trajectory tracking performance. In the case of the on-line learning control, the feedback control decision incorporates ILC at real-time.

ILC (Arimoto et al. 1984) based on the system's repetitive operations has drawn increasing attention because of its simple control structure and good tracking performances. The basic principle behind ILC is to use information collected from previous executions of the same task repetitively to form the control action for the current operation in order to improve tracking performances from iteration to iteration. Examples of systems that operate in a repetitive manner include robot manipulators, assembly line tasks, chemical batch processes, reliability testing rigs, and so on. In each of these tasks the system is required to perform the same action over and over again with high precision. In general, ILC improves the tracking performance by a self-tuning process without using a system model. The key purpose of using ILC is to achieve high performance after a few iterations. One advantage of ILC is that there is no requirement for dynamic model of the controlled system.

The classic type of ILC algorithms is a feed forward control system that is theoretically capable of reducing the tracking error to zero as the number of iterations increases towards infinity (Ahn et al. 2007, Bristow et al. 2006, Moore et al. 1992). According to the learning action type, ILC can be classified as P-type (Longman 2000, Xu et al. 2004), D-type (Arimoto et al. 1984, Yan et al. 2001, Song et al. 2005), PD-type (Chen and Moore 2002, Zhang et al. 2006), and PID-type (Madady 2008, Park et al. 1999). A PD-type averaged iterative learning control is proposed for linear systems in (Park 2005). Some detailed surveys for ILC can be found in (Ahn et al. 2007, Bristow et al. 2006, Moore et al. 1992).

Another direction in the development of ILC is using the current iteration tracking errors to form the control signal that can be viewed as online ILC (Amann et al. 1995, Cai et al. 2008, Ouyang et al. 2006, Ouyang et al. 2007, Owens and Munde 1998, Qu et al. 1991, Sun and Wang 2002, Xu et al. 1995). P-type online ILC systems based on the use of current tracking errors were proposed (Owens and Munde 1998, Sun and Wang 2002, Xu et al. 1995). A similar control algorithm was used for discrete time systems (Amann et al. 1995). D-type online ILC (Sun and Wang 2002, Yan et al. 2001), PD-type online ILC (Park 2005, Ouyang et al. 2006, Ouyang et al. 2007, Xu et al. 1995), and PID-type online ILC (Cai et al. 2008, Qu et al. 1991) were developed and applied into different systems control. As demonstrated in these papers, the iterative learning control with current tracking error signal can achieve fast convergence rates by selecting high feedback control gains.

It is well known that PD/PID feedback control is widely used in industrial applications such as robot systems and process control systems to achieve good performance. But PD/PID feedback control cannot achieve exactly the desired tracking performance because a non-zero error signal is required to activate the feedback control. Therefore, PD/PID control alone is not adequate for achieving a perfect trajectory tracking performance, especially when the system has nonlinearities and uncertainties. The combination of feedback control and iterative learning control is a promising technique to achieve good tracking performance and speed up the convergence process. P-type feedback plus P-type ILC control algorithms (P-P type) for continuous and discrete time-varying systems were proposed (Doh et al. 1999, Norrlöf and Gunnarsson 2005, Pi et al. 2000). A P-P ILC algorithm was developed for kinematic path-tracking of mobile robots in (Kang et al. 2005). High order D-D ILCs (Chen et al. 1996a, Xu et al. 2004) and PD-PD ILCs (Chen et al. 1996b, Ouyang 2009) were developed. A P-D ILC was proposed in (Yu et al. 2002) for discrete linear time-invariant systems.

In this chapter, we present two new control methods. The basic concept of these new control methods is to combine several control methods by taking advantage of each of them into a hybrid one. The architecture of this hybrid control method is as follows: (1) the control is a learning process through several iterations of off-line operations of a manipulator, (2) the control structure consists of two parts: a PD feedback part and a feedforward learning part using the torque profile obtained from the previous iteration, and (3) the gains in the PD feedback law are adapted according to the gain switching strategy with respect to the iteration. The first new control method is called the adaptive switching learning PD (ASL-PD) control method (Ouyang et al. 2006), and the second one is called switching gain PD-PD (SPD-PD) iterative learning control method (Ouyang et al. 2011).

The remainder of this chapter is organized as follows. In Section **Adaptive Switching Iterative Learning PD Control**, the ASL-PD control method is described, and its features are discussed, followed by the analysis of the asymptotic convergence of the ASL-PD control method using the Lyapunov's method. In Section **Switching Gain PD-PD Iterative Learning Control**, the SPD-PD ILC is introduced to control a general nonlinear system, and the convergence analysis is conducted. In Section **Simulation Study**, simulation studies are conducted for trajectory tracking control of robot manipulators where comparison studies are presented. Section **Experimental Verifications of ASL-PD Control** presents some experimental verification results to demonstrate the effectiveness of the proposed adaptive switching learning control methods. Conclusions are given in Section **Conclusion**.

ADAPTIVE SWITCHING ITERATIVE LEARNING PD CONTROL

Dynamic Model of A Robot Manipulator

Consider a robot manipulator with n joints running in repetitive operations. Its dynamics can be described by a set of nonlinear differential equations in the following form (Craig 1986)

$$D(q^j(t))\ddot{q}^j(t) + C(q^j(t),\dot{q}^j(t))\dot{q}^j(t) + G(q^j(t),\dot{q}^j(t)) + T_a(t) = T^j(t) \tag{1}$$

where $t \in [0, t_f]$ denotes the time and $j \in \mathbb{N}$ denotes the operation or iteration number. $q^j(t) \in \Re^n$, $\dot{q}^j(t) \in \Re^n$, and $\ddot{q}^j(t) \in \Re^n$ are the joint position, joint velocity, and joint acceleration vectors, respectively. $D(q^j(t)) \in \Re^{n\times n}$ is the inertia matrix, $C(q^j(t), \dot{q}^j(t))\dot{q}^j(t) \in \Re^n$ denotes the vector containing the Coriolis and centrifugal terms, $G(q^j(t), \dot{q}^j(t)) \in \Re^n$ is the gravitational plus frictional force, $T_a(t) \in \Re^n$ is the repetitive unknown disturbance, and $T^j(t) \in \Re^n$ is the input torque vector.

It is common knowledge that robot manipulators have the following properties (Craig 1986):

P1) $D(q^j(t))$ is a symmetric, bounded, and positive definite matrix;

P2) The matrix $\dot{D}(q^j(t)) - 2C(q^j(t), \dot{q}^j(t))$ is skew symmetric. Therefore,

$x^T(\dot{D}(q^j(t)) - 2C(q^j(t), \dot{q}^j(t)))\,x = 0, \forall\, x \in \Re^n$

Assume that all parameters of the robot are unknown and that:

A1) The desired trajectory $q_d(t)$ is of the third-order continuity *for* $t \in [0, t_f]$.

A2) For each iteration, the same initial conditions are satisfied, which are

$q_d(0) - q^j(0) = 0, \dot{q}_d(0) - \dot{q}^j(0) = 0, \forall\, j \in \mathbb{N}.$

ASL-PD Controller Design

The ASL-PD control method has two operational modes: the single operational mode and the iterative operational mode. In the single operational mode, the PD control feedback with the gain switching is used, where information from the present operation is utilized. In the iterative operational mode, a simple iterative learning control is applied as feedforward where information from previous operations is used. Together with these two operational modes, all information from the current and previous operations is utilized. Specially, the ASL-PD control method can be described as follows:

Consider the *j*th iterative operation for system in Eq. (1) with properties (P1 and P2) and assumptions (A1 and A2) under the following control law

$$T^j(t) = \underbrace{K_p^j e^j(t) + K_d^j \dot{e}^j(t)}_{feedback} + \underbrace{T^{j-1}(t)}_{feedforward} \qquad j = 0,1,\cdots,N \tag{2}$$

with the following gain switching rule

$$\begin{cases} K_p^j = \beta(j) K_p^0 \\ K_d^j = \beta(j) K_d^0 \qquad j = 1,2,\cdots,N \\ \beta(j+1) > \beta(j) \end{cases} \tag{3}$$

where $T^{-1}(t) = 0$, $e^j(t) = q_d(t) - q^j(t)$, $\dot{e}^j(t) = \dot{q}_d(t) - \dot{q}^j(t)$ and K_p^0 and K_d^0 are the initial PD control gain matrices that are diagonal positive definite. The matrices K_p^0 and K_d^0 are called the initial proportional and derivative control gains, while matrices K_p^j and K_d^j are the control gains of the *j*th iteration. $\beta(j)$ is the gain switching factor where $\beta(j) > 1$ *for* $j = 1, 2,\cdots, N$, and it is a function of the iteration number.

The gain switching law in Eq. (3) is used to adjust the PD gains from iteration to iteration. Such a switching in the ASL-PD control method acts not in the time domain but in the iteration domain. This is the main difference between the ASL-PD control method and the traditional switching control method (where switching occurs in the time domain). Therefore, the transient process of the switched system, which must be carefully treated in the case of the traditional switching control method, does not occur in the ASL-PD control method.

From Eqs. (2) and (3) it can be seen that the ASL-PD control law is a combination of feedback (with the switching gain in each iteration) and feedforward (with the learning scheme). The ASL-PD control method possesses an adaptive ability, which is demonstrated by the adoption of different control gains in different iterations; see Eq. (3). Such a switching takes place at the beginning of each iteration. Therefore, a rapid convergence speed for the trajectory tracking can be expected.

Furthermore, in the ASL-PD control law, the learning occurs due to the memorization of the torque profiles generated by the previous iterations that include information about the dynamics of a controlled system. It should be noted that such learning is direct in the sense that it generates the controlled torque profile directly from the existing torque profile in the previous iteration without any modification.

Because of the introduction of the learning strategy in the iteration, the state of the controlled object changes from iteration to iteration. This requires an adaptive control to deal with those changes, and the ASL-PD has such an adaptive capability.

In the next section, the proof of the asymptotic convergence of the ASL-PD control method for both position tracking and velocity tracking will be given.

Asymptotic Convergence Analysis of ASL-PD Method

Equation (1) can be linearized along the desired trajectory $(q_d(t), \dot{q}_d(t), \ddot{q}_d(t))$ in the following way

$$D(t)\ddot{e}^j(t)+[C(t)+C_1(t)]\dot{e}^j(t)+F(t)e^j(t)+n(\ddot{e}^j,\dot{e}^j,e^j,t)-T_a(t)=H(t)-T^j(t) \quad (4)$$

where $D(t) = D(q_d(t))$

$$C(t)=C(q_d(t),\dot{q}_d(t))$$

$$C_1(t)=\frac{\partial C}{\partial \dot{q}}\bigg|_{q_d(t),\dot{q}_d(t)}\dot{q}_d(t)+\frac{\partial G}{\partial \dot{q}}\bigg|_{q_d(t),\dot{q}_d(t)}$$

$$F(t)=\frac{\partial D}{\partial q}\bigg|_{q_d(t)}\ddot{q}_d(t)+\frac{\partial C}{\partial q}\bigg|_{q_d(t),\dot{q}_d(t)}\dot{q}_d(t)+\frac{\partial G}{\partial q}\bigg|_{q_d(t)}$$

$$H(t)=D(q_d(t))\ddot{q}_d(t)+C(q_d(t),\dot{q}_d(t))\dot{q}_d(t)+G(q_d(t))$$

The term $n(\ddot{e}^j, \dot{e}^j, e^j, t)$ contains the higher order terms $\ddot{e}^j(t)$, $\dot{e}^j(t)$, and $e^j(t)$, and it can be negligible. Therefore, for the jth and $j + 1$th iterations, Eq. (4) can be rewritten, respectively, as follows

$$D(t)\ddot{e}^j(t)+[C(t)+C_1(t)]\dot{e}^j(t)+F(t)e^j(t)-T_a(t)=H(t)-T^j(t) \quad (5)$$

$$D(t)\ddot{e}^{j+1}(t)+[C(t)+C_1(t)]\dot{e}^{j+1}(t)+F(t)e^{j+1}(t)-T_a(t)=H(t)-T^{j+1}(t) \quad (6)$$

For the simplicity of analysis, let $K_p^0 = \Lambda K_d^0$ for the initial iteration, and define the following parameter

$$y^i(t) = \dot{e}^j(t) + \Lambda e^j(t) \tag{7}$$

The following theorem can be proved.

Theorem 1. *Suppose robot system Eq. (1) satisfies properties (P1, P2) and assumptions (A1, A2). Consider the robot manipulator performing repetitive tasks under the ASL-PD control method Eq. (2) with the gain switching rule Eq. (3). The following should hold for all* $t \in [0, t_f]$

$$q^j(t) \xrightarrow{j\to\infty} q_d(t)$$

$$\dot{q}^j(t) \xrightarrow{j\to\infty} \dot{q}_d(t)$$

provided that the control gains are selected so that the following relationships hold

$$l_p = \lambda_{\min}(K_d^0 + 2C_1 - 2\Lambda D) > 0 \tag{8}$$

$$l_r = \lambda_{\min}(K_d^0 + 2C + 2F/\Lambda - 2\dot{C}_1/\Lambda) > 0 \tag{9}$$

$$l_p l_r \geq ||F/\Lambda - (C + C_1 - \Lambda D||^2_{\max} \tag{10}$$

where $\lambda_{min}(A)$ is the minimum eigenvalue of matrix A, and $||M||_{\max} = \max ||M(t)||$ for $0 \leq t \leq t_f$. Here, $||M||$ represents the Euclidean norm of M.

Proof. Define a Lyapunov function candidate as

$$V^j = \int_0^t e^{-\rho\tau} y^{jT} K_d^0 y^j d\tau \geq 0 \tag{11}$$

where $K_d^0 > 0$ is the initial derivative gain of PD control, and ρ is a positive constant.
Also, define $\delta y^j = y^{j+1} - y^j$ and $\delta e^j = e^{j+1} - e^j$. Then, from Eq. (7)

$$\delta y^j = \delta\dot{e}^j + \Lambda\delta e^j \tag{12}$$

and from Eq. (10.2)

$$T^{j+1}(t) = K_p^{j+1} e^{j+1}(t) + K_d^{j+1} \dot{e}^{j+1}(t) + T^j(t) \tag{13}$$

From Eqs. (5–7, 12–13), one can obtain the following equation

$$D\delta\dot{y}^j + (C + C_1 - \Lambda D + K_d^{j+1})\delta y^j + (F - \Lambda(C + C_1 - \Lambda D))\delta e^j = -K_d^{j+1} y^j \tag{14}$$

From the definition of V^j, for the $j + 1$th iteration, one can get

$$V^{j+1} = \int_0^t e^{-\rho\tau} y^{j+1^T} K_d^0 y^{j+1} d\tau$$

Define $\Delta V^j = V^{j+1} - V^j$. Then from Eqs. (11–12) and (14), we obtain

$$
\begin{aligned}
\Delta V^j &= \int_0^t e^{-\rho\tau}(\delta y^{jT} K_d^0 \delta y^j + 2\delta y^{jT} K_d^0 y^j) d\tau \\
&= \frac{1}{\beta(j+1)} \int_0^t e^{-\rho\tau}(\delta y^{jT} K_d^{j+1} \delta y^j + 2\delta y^{jT} K_d^{j+1} y^j) d\tau \\
&= \frac{1}{\beta(j+1)} \left\{ \int_0^t e^{-\rho\tau} \delta y^{jT} K_d^{j+1} \delta y^j d\tau - 2\int_0^t e^{-\rho\tau} \delta y^{jT} D \delta \dot{y}^j d\tau \right. \\
&\quad \left. -2\int_0^t e^{-\rho\tau} \delta y^{jT} ((C + C_1 - \Lambda D + K_d^{j+1}) \delta y^j + (F - \Lambda(C + C_1 - \Lambda D)) \delta e^j) d\tau \right\}
\end{aligned}
$$

Applying the partial integration and from (A2), we have

$$
\begin{aligned}
\int_0^t e^{-\rho\tau}\, \delta y^{jT} D \delta \dot{y}^j\, d\tau &= e^{-\rho\tau}\, \delta y^{jT} D \delta y^j \Big|_0^t - \int_0^t (e^{-\rho\tau}\, \delta y^{jT} D)'\, \delta y^j\, d\tau \\
&= e^{-\rho\tau}\, \delta y^{jT}(t)\, D(t) \delta y^j(t) + \rho \int_0^t e^{-\rho\tau}\, \delta y^{jT} D \delta y^j\, d\tau \\
&\quad - \int_0^t e^{-\rho\tau}\, \delta y^{jT} D \delta \dot{y}^j\, d\tau - \int_0^t e^{-\rho\tau}\, \delta y^{jT} \dot{D} \delta y^j\, d\tau
\end{aligned}
$$

From (P1), one can get

$$
\int_0^t \delta y^{jT} \dot{D} \delta y^j d\tau = 2\int_0^t \delta y^{jT} C \delta y^j d\tau
$$

Then

$$
\begin{aligned}
\Delta V^j = \frac{1}{\beta(j+1)} &\left\{ -e^{-\rho t}\, \delta y^{jT}(t) D(t) \delta y^j(t) - \rho \int_0^t e^{-\rho\tau} \delta y^{jT} D \delta y^j d\tau \right. \\
&- 2\int_0^t e^{-\rho\tau} \delta y^{jT} (F - \Lambda(C + C_1 - \Lambda D)) \delta e^j d\tau \\
&\left. - \int_0^t e^{-\rho\tau} \delta y^{jT} (K_d^{j+1} + 2C_1 - 2\Lambda D) \delta y^j d\tau \right\}
\end{aligned}
\tag{15}
$$

From Eq. (3), we have

$$
\begin{aligned}
\int_0^t e^{-\rho\tau} \delta y^{jT} K_d^{j+1} \delta y^j d\tau &= \beta(j+1) \int_0^t e^{-\rho\tau} \delta y^{jT} K_d^0 \delta y^j d\tau \\
&\geq \int_0^t e^{-\rho\tau} \delta y^{jT} K_d^0 \delta y^j d\tau
\end{aligned}
\tag{16}
$$

Substituting Eq. (12) into Eq. (15) and noticing Eq. (16), we obtain

$$
\begin{aligned}
\Delta V^j \leq \frac{1}{\beta(j+1)} &\left\{ -e^{-\rho t}\, \delta y^{jT}(t) D(t) \delta y^j(t) - \rho \int_0^t e^{-\rho\tau} \delta y^{jT} D \delta y^j d\tau \right. \\
&- \int_0^t e^{-\rho\tau} \delta \dot{e}^{jT} (K_d^0 + 2C_1 - 2\Lambda D) \delta \dot{e}^j d\tau - 2\Lambda \int_0^t e^{-\rho\tau} \delta e^{jT} (K_d^0 + 2C_1 - 2\Lambda D) \delta \dot{e}^j d\tau \\
&- 2\int_0^t e^{-\rho\tau} \delta \dot{e}^{jT} (F - \Lambda(C + C_1 - \Lambda D)) \delta e^j d\tau - \Lambda^2 \int_0^t e^{-\rho\tau} \delta e^{jT} (K_d^0 + 2C_1 - 2\Lambda D) \delta e^j d\tau \\
&\left. - 2\Lambda \int_0^t e^{-\rho\tau} \delta e^{jT} (F - \Lambda(C + C_1 - \Lambda D)) \delta e^j d\tau \right\}
\end{aligned}
$$

Applying the partial integration again gives

$$\int_0^t e^{-\rho\tau}\delta e^{jT}(K_d^0+2C_1-2\Lambda D)\delta\dot{e}^j d\tau$$
$$=e^{-\rho\tau}\delta e^{jT}(K_d^0+2C_1-2\Lambda D)\delta e^j\Big|_0^t+\rho\int_0^t e^{-\rho\tau}\delta e^{jT}(K_d^0+2C_1-2\Lambda D)\delta e^j d\tau$$
$$-\int_0^t e^{-\rho\tau}\delta\dot{e}^{jT}(K_d^0+2C_1-2\Lambda D)\delta e^j d\tau+2\int_0^t e^{-\rho\tau}\delta e^{jT}(\Lambda\dot{D}-\dot{C}_1)\delta e^j d\tau$$

Therefore,

$$\Delta V^j\le\frac{1}{\beta(j+1)}\Bigg\{-e^{-\rho t}\,\delta y^{jT}D\delta y^j-\rho\int_0^t e^{-\rho\tau}\delta y^{jT}D\delta y^j d\tau-\Lambda e^{-\rho t}\,\delta e^{jT}(K_d^0+2C_1-2\Lambda D)\delta e^j$$
$$-\rho\Lambda\int_0^t e^{-\rho\tau}\,\delta e^{jT}(K_d^0+2C_1-2\Lambda D)\delta e^j\,d\tau-\int_0^t e^{-\rho\tau}w\,d\tau\Bigg\} \tag{17}$$
$$\le\frac{1}{\beta(j+1)}\Bigg\{-e^{-\rho t}\,\delta y^{jT}D\delta y^j-\Lambda e^{-\rho\tau}\,\delta e^{jT}\,l_p\delta e^j-\rho\int_0^t e^{-\rho\tau}\delta y^{jT}D\delta y^j d\tau$$
$$-\rho\Lambda\int_0^t e^{-\rho\tau}\,\delta e^{jT}l_p\delta e^j\,d\tau-\int_0^t e^{-\rho\tau}w\,d\tau\Bigg\}$$

where $w=\delta\dot{e}^{jT}(K_d^0+2C_1-2\Lambda D)\delta\dot{e}^j+2\Lambda\delta\dot{e}^{jT}(F/\Lambda-(C+C_1-\Lambda D)\delta e^j$
$+\Lambda^2\delta e^{jT}(K_d^0+2C+2F/\Lambda-2\dot{C}_1/\Lambda)\delta e^j$

Let $Q=F/\Lambda-(C+C_1-\Lambda D)$. Then from Eqs. (8–9), we obtain

$$w\ge l_p\|\delta\dot{e}\|^2+2\Lambda\,\delta\dot{e}^TQ\delta e+\Lambda^2 l_r\|\delta e\|^2$$

Applying the Cauchy-Schwartz inequality gives

$$\delta\dot{e}^TQ\delta e\ge-\|\delta\dot{e}\|\,\|Q\|_{\max}\,\|\delta e\|$$

From Eqs. (8–10)

$$w\ge l_p\|\delta\dot{e}\|^2-2\Lambda\,\|\delta\dot{e}\|\,\|Q\|_{\max}\,\|\delta e\|+\Lambda^2 l_r\|\delta e\|^2 \tag{18}$$
$$=l_p(\|\delta\dot{e}\|-\frac{\Lambda}{l_p}\|Q\|_{\max}\|\delta e\|)^2+\Lambda^2(l_r-\frac{1}{l_p}\|Q\|_{\max}^2)\,\|\delta e\|^2\ge 0$$

From (P1) and Eq. (8), based on Eq. (17), it can be ensured that $\Delta V^j\le 0$. Therefore,

$$V^{j+1}\le V^j \tag{19}$$

From the definition, K_d^0 is a positive definite matrix. From the definition of V^j, $V^j>0$, and V^j is bounded. As a result, $y^j(t)\to 0$ when $j\to\infty$. Because $e^j(t)$ and $\dot{e}^j(t)$ are two independent variables, and Λ is a positive constant. Thus, if $j\to\infty$, then $e^j(t)\to 0$ and $\dot{e}^j(t)\to 0$ for $t\in[0,t_f]$.

Finally, the following conclusions hold

$$\begin{cases} q^j(t)\xrightarrow{j\to\infty}q_d(t) \\ \dot{q}^j(t)\xrightarrow{j\to\infty}\dot{q}_d(t)\end{cases}\quad\text{for } t\in[0,t_f] \tag{20}$$

From the above analysis it can be seen that the ASL-PD control method can guarantee that the tracking errors converge arbitrarily close to zero as the number of iterations increases. The following case studies based on simulation will demonstrate this conclusion.

SWITCHING GAIN PD-PD ITERATIVE LEARNING CONTROL

Dynamic Model and SPD-PD Control Law

In this Section, the considered problem is a nonlinear time-varying system with non-repetitive uncertainty and disturbance as follows:

$$\begin{cases} \dot{x}_k(t) = f(x_k(t),t) + B(t)u_k(t) + \eta_k(t) \\ y_k(t) = C(t)x_k(t) + \xi_k(t) \end{cases} \tag{21}$$

where k denotes the iteration index. $x \in \Re^n$, $u \in \Re^r$, and $y \in \Re^m$ are the state, control input and output of the system, respectively. $f(x(t),t) \in \Re^n$ is the system function, $B(t) \in \Re^{n\times r}$ the input matrix, $C(t) \in \Re^{m\times n}$ the output matrix, $\eta_k(t) \in \Re^n$ the uncertainty of the system, and $\xi_k(t) \in \Re^m$ the disturbance, respectively.

In this section, to evaluate tracking convergence, the following notational conventions for norms are adopted:

$$\|f\| = \max_{1\le i\le n} |f_i|$$

$$\|M(t)\| = \max_{1\le i\le m}\left(\sum_{j=1}^{n} |m_{i,j}| \right)$$

$$\|h(t)\|_\lambda = \sup_{t\in[0,T]} e^{-\lambda t}\|h(t)\|, \quad \lambda > 0$$

where $f = [f_1,\ldots,f_n]^T$ is a vector, $M = [m_{i,j}] \in \Re^{m\times n}$ is a matrix, and $h(t)$ ($t \in [0, T]$) is a real function where T is the time period of a repetitive task.

To restrict the discussion, the following assumptions are made for the system.

A3) The desired trajectory $y_d(t)$ is first-order continuous for $t \in [0, T]$.

A4) The control input matrix $C(t)$ is first-order continuous for $t \in [0, T]$.

A5) The function $f(x(t), t)$ is globally, uniformly Lipschitz in x for $t \in [0, T]$. That means $\|f(x_{k+1}(t), t) - f(x_k(t), t)\| \le c_f \|x_{k+1}(t) - x_k(t)\|$ where k is the iteration index and c_f (> 0) is the Lipschitz constant.

A6) Uncertainty and disturbance terms $\eta_k(t)$ and $\xi_k(t)$ are bounded as follows: $\forall t \in [0, T]$ and $\forall k$, we have $\|\eta_k(t)\| \le b_\eta$, $\|\xi_i(t)\| \le b_\xi$, and $\|\dot{\xi}_i(t)\| \le b_{\dot{\xi}}$.

To control the nonlinear system stated in Eq. (21), we propose the following SPD-PD ILC law

$$\begin{aligned} u_{k+1}(t) &= u_k(t) + K_{p1}(k+1)e_{k+1}(t) + K_{d1}(k+1)\dot{e}_{k+1}(t) \\ &\quad + K_{p2}(t)e_k(t) + K_{d2}(t)\dot{e}_k(t) \end{aligned} \tag{22}$$

with the feedback control gains given by:

$$\begin{cases} K_{p1}(k) = s(k)K_{p1}(0) \\ K_{d1}(k) = s(k)K_{d1}(0) \end{cases} \tag{23}$$
$$with \quad s(k+1) > s(k)$$

where $e_{k+1}(t) = y_d(t) - y_{k+1}(t)$, $e_k(t) = y_d(t) - y_k(t)$, $\dot{e}_{k+1}(t) = \dot{y}_d(t) - \dot{y}_{k+1}(t)$ and $\dot{e}_k(t) = \dot{y}_d(t) - \dot{y}_k(t)$ are position errors and velocity errors of the output vector for the $k + 1^{th}$ iteration and the kth iteration, respectively. $K_{pi} \in \Re^{m\times r}$ and $K_{di} \in \Re^{m\times r}$ are the proportional and derivative gain matrices, respectively. Equation (23) represents the switching gains of the feedback control, and $s(k) > 1$ is a monotonically increasing function of the iteration index.

From Eq. (22), we can see that the proposed SPD-PD ILC law consists of two control loops. The first loop includes a PD feedback controller with switching gains in the iteration domain, and the second loop is a standard PD-type ILC. Therefore, SPD-PD ILC is effectively a hybrid control that aims to take the advantages offered by both feedback control and ILC. The key purpose of introducing switching gains in the feedback loop is to expedite the convergence of the iterative operations and avoid vibration of systems. Also, one can see that the proposed control algorithm is an extension of the ASL-PD in (Ouyang et al. 2006).

We assume that, in each iteration, the repeatability of the initial state setting is satisfied within the following admissible deviation level:

$$\|x_k(0) - x_0(0)\| \le \varepsilon_x \text{ for } k = 1, 2,... \tag{24}$$

where ε_x is a small positive constant that represents the acceptable accuracy of the designed state vector and $x_0(0)$ represents the desired initial state value.

For briefness of the convergence analysis, some notations are introduced first and used in the following sections:

$$B_{d1} = \max_{t\in[0,T]} \left\|K_{d1}(0)C(t)\right\|, \quad B_{d2} = \max_{t\in[0,T]} \left\|K_{d2}C(t)\right\|$$

$$B_{pd1} = \max_{t\in[0,T]} \left\|K_{p1}(0)C + K_{d1}(0)\dot{C}\right\|, B_{pd2} = \max_{t\in[0,T]} \left\|K_{p2}C + K_{d2}\dot{C}\right\|$$

$$B_{Kp1} = \max_{t\in[0,T]} \left\|K_{p1}(0)\right\|, \quad B_{Kp2} = \max_{t\in[0,T]} \left\|K_{p2}\right\|$$

$$B_{Kd1} = \max_{t\in[0,T]} \left\|K_{d1}(0)\right\|, \quad B_{Kd2} = \max_{t\in[0,T]} \left\|K_{d2}\right\|$$

$$B_B = \max_{t\in[0,T]} \left\|B(t)\right\|, \ B_C = \max_{t\in[0,T]} \left\|C(t)\right\|, \ B_s = \max \left\|s(k)\right\|$$

$$K_{B1} = \frac{B_{d1} + c_f B_{pd1}}{\lambda - c_f}, K_{B2} = \frac{B_{d2} + c_f B_{pd2}}{\lambda - c_f}$$

$$\rho_1 = \max_{t\in[0,T]} \left\|(I + K_{d1}(0)CB)^{-1}\right\|, \ \rho_2 = \max_{t\in[0,T]} \left\|I_m - K_{d2}CB\right\|$$

$$\rho = \frac{\rho_1}{1 - \rho_1 B_s B_B K_{B1}}, \ \beta = \rho_2 + B_B K_{B2}$$

SPD-PD ILC Law and Convergence Analysis

Theorem 2. For the nonlinear time-varying system Eq. (21), if the SPD-PD type iterative learning control law Eq. (22) is applied and the switching gain algorithm Eq. (23) is adopted, then the final state error and the output tracking error are bounded and the boundednesses are given by

$$\begin{cases} \lim_{k\to\infty}\|\delta x_k(t)\|_\lambda \le \dfrac{1}{\lambda - c_f}\left(\dfrac{\rho\Phi}{1-\rho\beta} + Tb_\eta + \varepsilon_x\right) \\ \lim_{k\to\infty}\|e_k(t)\|_\lambda \le \dfrac{B_C}{\lambda - c_f}\left(\dfrac{\rho\Phi}{1-\rho\beta} + Tb_\eta + \varepsilon_x\right) + b_\xi \end{cases} \tag{25}$$

where $\Phi = (B_s K_{B1} + K_{B2} + B_s B_{d1} + B_{d2}) b_\eta + (B_s B_{Kp1} + B_{Kp2}) b_\xi + (B_s B_{Kd1} + B_{Kd2}) b_\xi + (B_s K_{B1} + K_{B2})\varepsilon_x$.

Provided the control gain $k_{d1}(0)$ and the learning gain $K_{d2}(t)$ are selected such that $I_m + k_{d1}(0)\, B(t)C(t)$ is non-singular, and

$$\begin{cases} \max_{t\in[0,T]} \left\|(I + K_{d1}(0)B(t)C(t))^{-1}\right\| = \rho_1 < 1 \\ \max_{t\in[0,T]} \|I - K_{d2}(t)B(t)C(t)\| = \rho_2 < 1 \end{cases} \tag{26}$$

Also we propose the following initial state learning algorithm:

$$x_{k+1}(0) = (1 + B(0)K_{d1}(0)C(0))^{-1}\{x_k(0) + B(0)K_{d1}(0)y_d(0) + B(0)K_{d2}(0)$$
$$(y_d(0) - y_k(0))\} \tag{27}$$

In this Section, the λ–norm is used to examine the convergence of the tracking error for the proposed SPD-PD ILC algorithm. First of all, a relation between norm and λ–norm is represented by Lemma 1.

Lemma 1. Suppose that $x(t) = [x_1(t), x_2(t),..., x_n(t)]^T$ is defined in $t \in [0, T]$. Then

$$\left(\int_0^t \|x(\tau)\| d\tau\right) e^{-\lambda t} \le \frac{1}{\lambda}\|x(t)\|_\lambda \quad \text{for } \lambda > 0 \tag{28}$$

Proof.

$$\begin{aligned} \left(\int_0^t \|x(\tau)\| d\tau\right) e^{-\lambda t} &= \int_0^t \|x(\tau)\| e^{-\lambda\tau} e^{-\lambda(t-\tau)} d\tau \\ &\le \sup_{t\in[0,T]}\left\{\|x(t)\|e^{-\lambda t}\right\} \int_0^t e^{-\lambda(t-\tau)} d\tau \\ &\le \sup_{t\in[0,T]}\left\{\|x(t)\|e^{-\lambda t}\right\} \frac{1-e^{-\lambda t}}{\lambda} \\ &\le \frac{1}{\lambda}\|x(t)\|_\lambda \end{aligned}$$

[End of proof].

We define the following four variables. $\delta x_k \triangleq x_d - x_k$, $e_k \triangleq y_d - y_k$, $\delta u_k \triangleq u_d - u_k$ and $\delta f_k = f_d - f(x_k)$.

From Eq. (21), we can calculate the tracking errors as:

$$\begin{cases} e_k = C(t)\delta x_k - \xi_k \\ e_{k+1} = C(t)\delta x_{k+1} - \xi_{k+1} \end{cases} \tag{29}$$

The derivative of the tracking errors can be represented as

$$\begin{cases} \dot{e}_k = C(t)\delta \dot{x}_k + \dot{C}(t)\delta x_k - \dot{\xi}_k \\ \dot{e}_{k+1} = C(t)\delta \dot{x}_{k+1} + \dot{C}(t)\delta x_{k+1} - \dot{\xi}_{k+1} \end{cases} \tag{30}$$

From Eq. (22) we have:

$$\delta u_{k+1} = \delta u_k - s(k+1)K_{p1}(0)e_{k+1}(t) - s(k+1)K_{d1}(0)\dot{e}_{k+1}(t) - K_{p2}(t)e_k(t) - K_{d2}(t)\dot{e}_k(t) \tag{31}$$

Submitting Eqs. (29) and (30) into Eq. (31) gets:

$$\begin{aligned} \delta u_{k+1} = \delta u_k - K_{p2}\{C\delta x_k - \xi_k\} - K_{d2}\{C\delta\dot{x}_k + \dot{C}\delta x_k - \dot{\xi}_k\} - s(k+1)K_{p1}(0)\{C\delta x_{k+1} - \xi_{k+1}\} \\ -s(k+1)K_{d1}(0)\{C\delta\dot{x}_{k+1} + \dot{C}\delta x_{k+1} - \dot{\xi}_{k+1}\} \end{aligned} \tag{32}$$

Also from Eq. (21), we can get the following equations:

$$\begin{cases} \delta\dot{x}_k = \delta f_k + B\delta u_k - \eta_k \\ \delta\dot{x}_{k+1} = \delta f_{k+1} + B\delta u_{k+1} - \eta_{k+1} \end{cases} \tag{33}$$

Submitting Eq. (32) into Eq. (33) and reorganizing gets:

$$\begin{aligned} (I + s(k+1)K_{d1}(0)CB)\delta u_{k+1} = (I - K_{d2}CB)\delta u_k - s(k+1)(K_{p1}(0)C + K_{d1}(0)\dot{C})\delta x_{k+1} \\ -(K_{p2}C + K_{d2}\dot{C})\delta x_k - s(k+1)K_{d1}(0)(C\delta f_{k+1} - C\eta_{k+1} - \dot{\xi}_{k+1}) \\ -K_{d2}(C\delta f_k - C\eta_k - \dot{\xi}_k) + s(k+1)K_{p1}(0)\xi_{k+1} + K_{p2}\xi_k \end{aligned} \tag{34}$$

From A5 we have $\|\delta f_k\| \le c_f \|\delta x_k\|$, and $\|\delta f_{k+1}\| \le c_f \|\delta x_{k+1}\|$. As $s(k+1) > 1$, to choose a proper control gain $K_{p1}(0)$ and from Eq. (26) we can ensure:

$$\left\|(I + s(k+1)K_{d1}(0)CB)^{-1}\right\| < \left\|(I + K_{d1}(0)CB)^{-1}\right\| = \rho_1 < 1 \tag{35}$$

Applying Eq. (35), Eq. (34) can be rewritten in the λ-norm and simplified as:

$$\begin{aligned} \|\delta u_{k+1}\|_\lambda \le \rho_1\{\rho_2\|\delta u_k\|_\lambda + B_s(B_{pd1} + B_{d1}c_f)\|\delta x_{k+1}\|_\lambda + (B_{pd2} + B_{d2}c_f)\|\delta x_k\|_\lambda + (B_sB_{d1} + B_{d2})b_\eta \\ + (B_sB_{Kp1} + B_{Kp2})b_\xi + (B_sB_{Kd1} + B_{Kd2})b_{\dot{\xi}}\} \end{aligned} \tag{36}$$

For the kth iteration, the state vector can be written as:

$$x_k(t) = x_k(0) + \int_0^t \big(f(x_k(\tau),\tau) + B(\tau)u_k(\tau)\big)d\tau + \int_0^t \eta_k(t)d\tau \tag{37}$$

From Eq. (21), we also have:

$$x_d(t) = x_d(0) + \int_0^t \big(f(x_d(\tau),\tau) + B(\tau)u_d(\tau)\big)d\tau \tag{38}$$

From Eqs. (37) and (38), we get:

$$\delta x_k = \int_0^t \big(f(x_d(\tau),\tau) - f(x(\tau),\tau)\big)d\tau + \int_0^t \big(B(t)\,\delta u_k - \eta_k\big)d\tau + \delta x_k(0) \tag{39}$$

Applying A5 to Eq. (39), we get

$$\delta x_k \le \int_0^t c_f \delta x_k d\tau + \int_0^t \big(B(t)\,\delta u_k - \eta_k\big)d\tau + \delta x_k(0) \tag{40}$$

Equation (40) can be written in the norm form as:

$$\|\delta x_k\| \le \int_0^t c_f \|\delta x_k\| d\tau + \int_0^t \big(\|B(t)\,\delta u_k\| + \|\eta_k\|\big)d\tau + \|\delta x_k(0)\| \tag{41}$$

According to the definition of λ-norm, for $\lambda > c_f$, applying Lemma 1 to Eq. (41) obtains:

$$\|\delta x_k\|_\lambda \le \frac{B_B}{\lambda - c_f}\|\delta u_k\|_\lambda + \frac{T}{\lambda - c_f} b_\eta + \frac{\varepsilon_x}{\lambda - c_f} \tag{42}$$

For the $k + 1^{th}$ iteration, we can get a very similar result:

$$\|\delta x_{k+1}\|_\lambda \le \frac{B_B}{\lambda - c_f}\|\delta u_{k+1}\|_\lambda + \frac{T}{\lambda - c_f} b_\eta + \frac{\varepsilon_x}{\lambda - c_f} \tag{43}$$

Submitting Eqs. (42–43) into Eq. (36) and simplifying it gets:

$$(1 - \rho_1 B_s B_B K_{B1})\|\delta u_{k+1}\|_\lambda \le \rho_1(\rho_2 + B_B K_{B2})\|\delta u_k\|_\lambda + \rho_1 \Phi \tag{44}$$

Equation (44) can be simplified as

$$\|\delta u_{k+1}\|_\lambda \le \rho\beta\|\delta u_k\|_\lambda + \rho\Phi \tag{45}$$

From Eq. (26) we have $\rho_1\rho_2 < 1$. If we choose:

$$\lambda > \frac{\rho_1 B_B\big(c_f(B_s B_{pd1} + B_{pd2}) + B_s B_{d1} + B_{d2}\big)}{1 - \rho_1\rho_2} + c_f$$

Then we can guarantee $\rho\beta < 1$.

From Eq. (45) we can get:

$$\lim_{k\to\infty}\left\|\delta u_k(t)\right\|_\lambda = \frac{\rho\Phi}{1-\rho\beta} \tag{46}$$

From Eq. (46), we can see that the control input is bounded and is close to the desired control input.

Submitting Eq. (46) into Eq. (42), we can get:

$$\lim_{k\to\infty}\left\|\delta x_k(t)\right\|_\lambda = \frac{1}{\lambda - c_f}\left(\frac{\rho\Phi}{1-\rho\beta} + Tb_\eta + \varepsilon_x\right) \tag{47}$$

Equation (47) proves that the state error is bounded.

Finally, from Eq. (29) we can get:

$$\lim_{k\to\infty}\left\|e_k(t)\right\|_\lambda = \frac{B_C}{\lambda - c_f}\left(\frac{\rho\Phi}{1-\rho\beta} + Tb_\eta + \varepsilon_x\right) + b_\xi \tag{48}$$

Equation (48) demonstrates that the output error is bounded.

Remark 1. If the initial state updating law Eq. (27) is used, we will ensure $\lim_{k\to\infty} x_k(0) = x_0(0)$. In this manner, we can get $\lim_{k\to\infty} \varepsilon_x = 0$. Therefore,

$$\lim_{k\to\infty}\left\|e_k(t)\right\|_\lambda = \frac{B_C}{\lambda - c_f}\left(\frac{\rho\Phi}{1-\rho\beta} + Tb_\eta\right) + b_\xi \tag{49}$$

Remark 2. If there is no uncertainty and disturbance in Eq. (21), then the final tracking error bound becomes:

$$\lim_{k\to\infty}\left\|e_k(t)\right\|_\lambda = \frac{B_C\varepsilon_x}{\lambda - c_f} \tag{50}$$

Remark 3. If the initial state updating law Eq. (27) is applied, and there is no uncertainty and disturbance, then the final tracking error is $\lim_{k\to\infty} \|e_k(t)\|_\lambda = 0$. Such a conclusion can be derived directly from Remark 1 and Remark 2.

Remark 4. The convergence condition Eq. (26) doesn't include the proportional gains. Therefore, it provides some extra freedom for the choices of K_{p1} and K_{p2} in the proposed control law Eq. (22).

SIMULATION STUDY

ASL-PD Results

In order to have some idea about how effective the proposed ASL-PD control method would be, we conducted a simulation study; specifically we simulated two robot manipulators. The first one was a serial robot manipulator with parameters taken directly from (Choi and Lee 2000) for the purpose of comparing the ASL-PD method with the method proposed in Ref. (Choi and Lee 2000) called the adaptive ILC. It is noted that the result for the serial manipulator may not be applicable to the parallel manipulator. Therefore, the second one is a parallel robot manipulator for which we show the effectiveness of the ASL-PD control method both in the trajectory tracking error and the required torque in the motor.

A. Trajectory Tracking of a Serial Robot Manipulator

A two degrees of freedom (DOF) serial robot is shown in Fig. 1, which was discussed in (Choi and Lee 2000) with an adaptive ILC method.

The physical parameters and desired trajectories are the same as in (Choi and Lee 2000) and listed as follows.

Physical parameters:

$m_1 = 10$ kg, $m_2 = 5$ kg, $l_1 = 1$ m, $l_2 = 0.5$ m, $l_{c_1} = 0.5$ m, $l_{c_2} = 0.25$ m, $I_1 = 0.83$ kgm^2 and $I_2 = 0.3$ kgm^2.

Desired trajectories and the repetitive disturbances:

$$q_1 = \sin 3t, \quad q_2 = \cos 3t \quad \text{for } t \in [0,5]$$

$$d_1(t) = 0.3a\sin t, \; d_2(t) = 0.1a(1 - e^{-t}) \quad \text{for } t \in [0,5]$$

where a is a constant used to examine the capability of the ASL-PD control to deal with the repetitive disturbances.

The control gains were also set to be the same as in (Choi and Lee 2000).

$$K_p^0 = K_d^0 = \text{diag}\{20,10\}$$

In the ASL-PD control method, the control gains were switched from iteration to iteration based on the following rule

$$K_p^j = 2jK_p^0, \; K_d^j = 2jK_d^0 \quad j = 1, 2, \cdots, N$$

First, consider $a = 1$. In that way, the repetitive disturbances were the same as in (Choi and Lee 2000). Figure 2a shows the tracking performance for the initial iteration, where only the PD control with small control gains was used, and no feedforward is

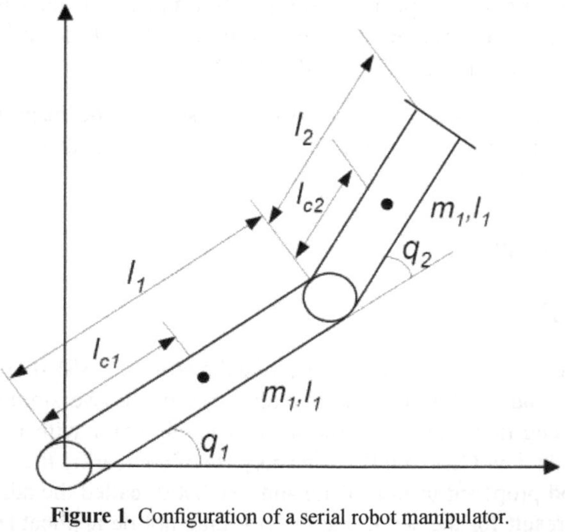

Figure 1. Configuration of a serial robot manipulator.

used. It can be seen that the tracking performance was not acceptable because the errors were too large for both joints. However, at the sixth iteration where the ASL-PD control method was applied, the tracking performance was improved dramatically as shown in Fig. 2b. At the eighth iteration, the performance was very good (Fig. 2c).

The velocity tracking performance is shown in Fig. 3. From it one can see that the velocity errors reduced from 1.96 (rad/s) at the initial iteration to 0.0657 (rad/s) at the sixth iteration, and further to 0.0385 (rad/s) at the eighth iteration for joint 1. The similar decreasing trend can be found for joint 2. From Figs. 2 and 3 it can be seen that the tracking performances were improved incrementally with the increase of the iteration number.

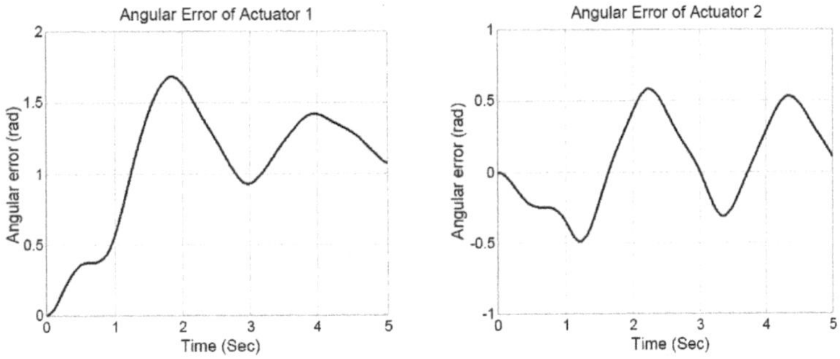

(a) Angular errors for two joints in the initial iteration.

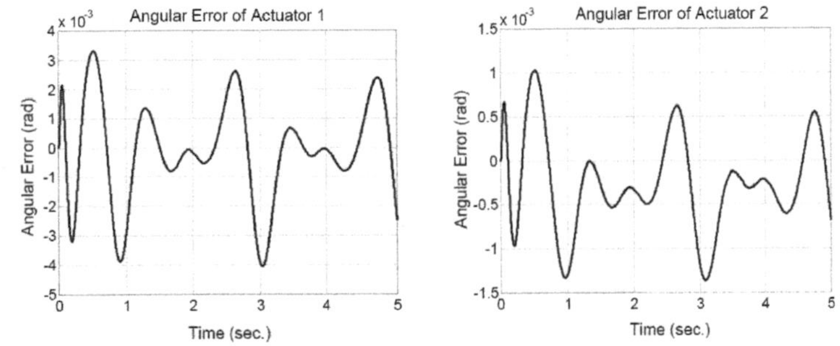

(b) Angular errors for two joints at the 6th iteration.

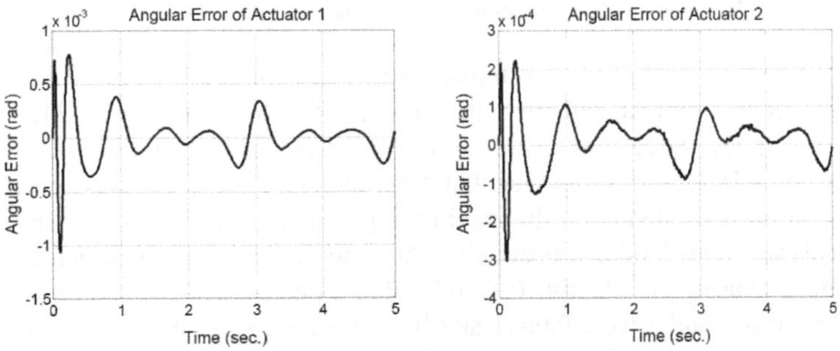

(c) Angular errors for two joints at the 8th iteration.

Figure 2. Position tracking errors for different iterations under ALS-PD control.

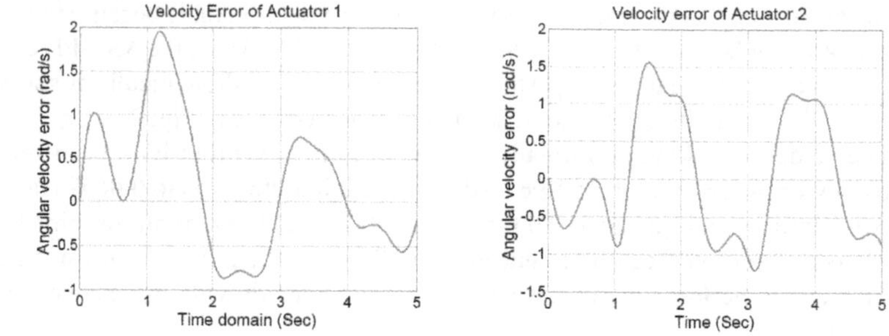

(a) Velocity errors for two joints in the initial iteration.

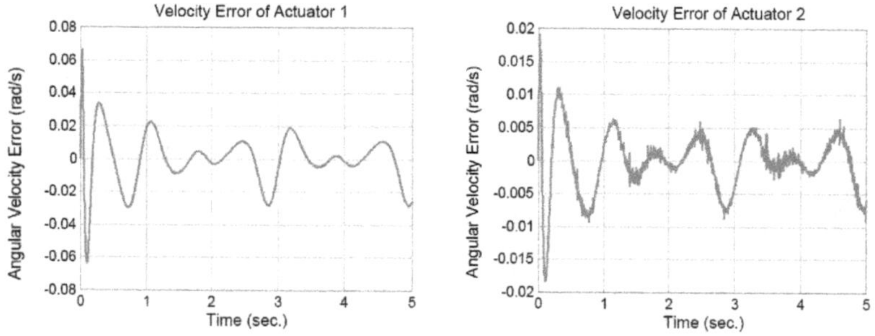

(b) Velocity errors for two joints at the 6th iteration.

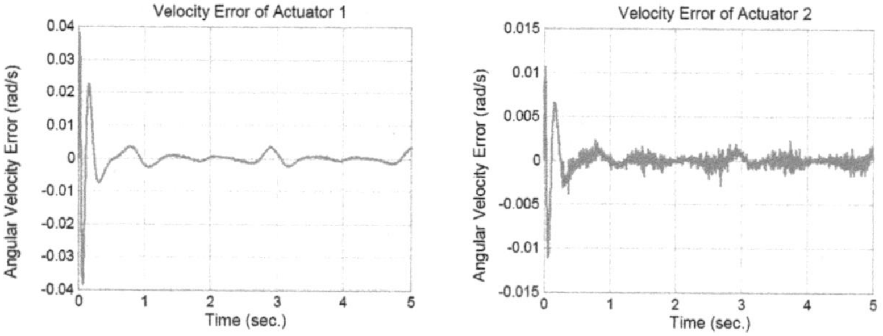

(c) Velocity errors for two joints at the 8th iteration.

Figure 3. Velocity tracking errors for different iterations under ASL-PD control.

As the gain switching rule was introduced at each iteration, the convergence rate increased greatly compared with the control method developed in (Choi and Lee 2000). Table 1 shows the trajectory tracking errors from the initial iteration to the eighth iteration. From Table 1 it can be seen that the tracking performance was considerably improved at the sixth iteration. The maximum position errors for joints 1 and 2 were 0.0041 rads and 0.0014 rads, respectively, while the similar results were achieved after thirty iterations using the adaptive ILC in Choi and Lee 2000 (The maximum position errors for joints 1 and 2 were 0.0041 and 0.0046 (rad), respectively). Therefore, the comparison of their method and our method demonstrates a fast convergence rate with the ASL-PD control method. It should be noted that the comparison of the velocity errors was not done as such information was not presented in Ref. Choi and Lee 2000.

Table 1. Trajectory tracking errors from iteration to iteration.

Iteration	0	2	4	6	8
$\max\lvert e_1^j \rvert$ (rad)	1.6837	0.4493	0.0433	0.0041	0.0011
$\max\lvert e_2^j \rvert$ (rad)	0.5833	0.1075	0.0122	0.0014	0.0003
$\max\lvert \dot{e}_1^j \rvert$ (rad/s)	1.9596	0.7835	0.1902	0.0657	0.0385
$\max\lvert \dot{e}_2^j \rvert$ (rad/s)	1.5646	0.2534	0.0523	0.0191	0.0111

It is further noted that there were repetitive disturbances at each iteration in the simulation. To examine the capacity of the ASL-PD under the repetitive disturbance condition, different levels of the repetitive disturbances were applied in the simulation. Figure 4 shows the maximum tracking errors from iteration to iteration for different repetitive disturbances which are expressed by a constant a. A larger constant a means a more disturbance. It should be noted that in Fig. 4 $a = 0$ means there is no repetitive disturbance in the simulation, and $a = 100$ means a large repetitive disturbance included in the simulation (specifically, the disturbance acted in joint 1 was about 20% of the required torque and the disturbance acted in joint 2 was about 40% of the required torque). From this figure, one can see that although the tracking errors for the initial iteration increased with the increase of the disturbance level, the final tracking errors of both the position and the velocity were the same for the different repetitive disturbance levels at the final 2 iterations. Therefore, we conclude that the ASL-PD control method has an excellent capability in terms of both rejecting the repetitive disturbance and robustness with respect to the disturbance level.

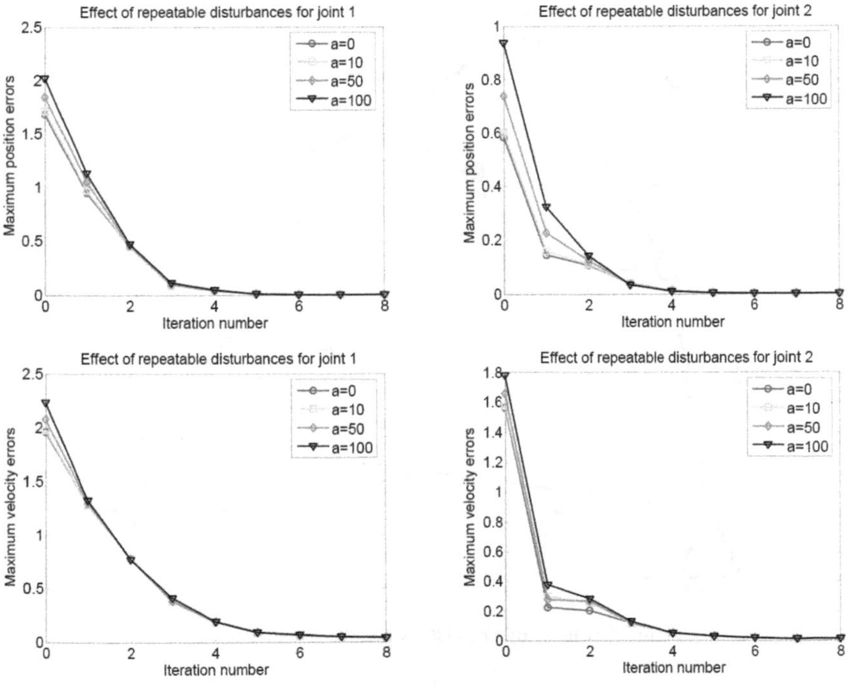

Figure 4. Effect of the repetitive disturbance on tracking errors.

B. Trajectory Tracking of a Parallel Robot Manipulator

A two DOFs parallel robot manipulator is shown in Fig. 5. Table 2 lists its physical parameters. The robot system can be viewed as two serial robotic systems with some constraints; that is, the two end-effectors of these two serial robotic systems reach the same position. Because of this constraint, the dynamics is more complex than that of its serial counterpart. The details about the dynamics of the parallel robot manipulator can be founded in Ouyang et al. 2004.

The end-effector of the robot was required to move from point A (0.7, 0.3), to point B (0.6, 0.4), and to point C(0.5,0.5). The time duration between two nearby points was 0.25 seconds. The control was carried out at the joint level where the inverse kinematics was used to calculate the joint position and velocity associated with the specific path of the end-effector. The path was designed to pass through these three points with the objective of meeting the positions, velocities, and accelerations at these three points using the motion planning method by Ouyang and Zhang 2005.

In this example, the control gains were selected as follows

$$K_p^0 = \text{diag}\{20, 20\}, K_d^0 = \text{diag}\{12, 12\}$$

The gain switching rule was set to be

$$K_p^j = 2jK_p^0, K_d^j = 2jK_d^0 \text{ for } j = 1, 2, \cdots, N$$

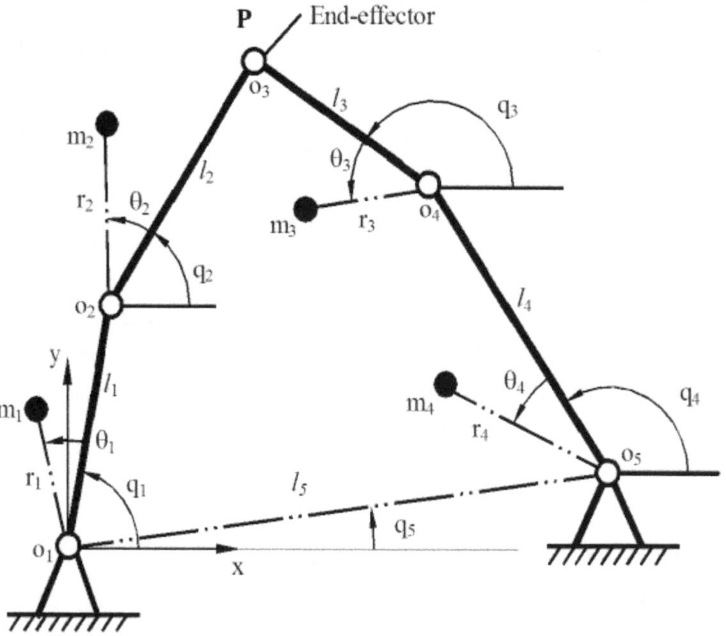

Figure 5. Scheme of a two DOFs parallel robot manipulator.

Table 2. Physical parameters of the parallel robotic manipulator.

Link	m_i(kg)	l_i(m)	r_i(m)	I_i(kgm^2)	θ_i(rad)
1	1	0.4	0.2	0.5	0
2	1.25	0.6	0.3	1	0
3	1.5	0.8	0.3	1	0
4	1	0.6	0.2	0.5	0
5	–	0.6	–	–	–

Figure 6 shows the position tracking performance improvement for the two actuators from iteration to iteration. From it one can see that, at the initial iteration, the maximum position errors were about 0.11 and 0.38 rad; only after four iterations, the maximum position errors were reduced to 0.08 and 0.05 rad. Finally, after eight iterations, the maximum errors were reduced to 0.0003 and 0.0008 rad. Figure 7 shows the velocity tracking performance improvement for the two actuators. At the initial iteration, the maximum velocity errors were about 1.17 and 2.68 rad/s in the two actuators, respectively. But after four iterations, the maximum values were reduced to 0.15 and 0.14 rad/s. After eight iterations, the maximum errors in the two actuators became 0.0046 and 0.0102 rad/s for velocity, respectively.

It should be noted that, while the tracking performance was improved from iteration to iteration, the torques required to drive the two actuators were nearly the same from iteration to iteration after a few iterations. This can be seen from Fig. 8, especially from the fifth iteration to the eighth iteration. It can be seen also from Fig. 8 that the profiles of the required torques were very smooth even as the control gains become larger as

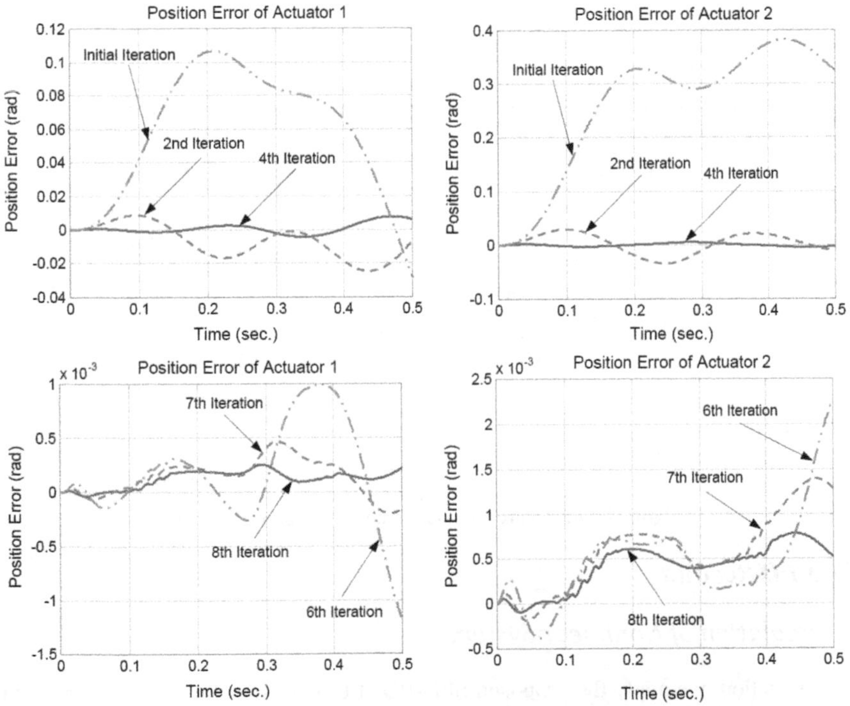

Figure 6. Position tracking performance improvement from iteration to iteration.

the iteration number is increased. Such a property is very useful for the safe use of the actuators and the attenuation of vibration of the controlled plant. It is noted that this property was missed in the switching technique in the time domain.

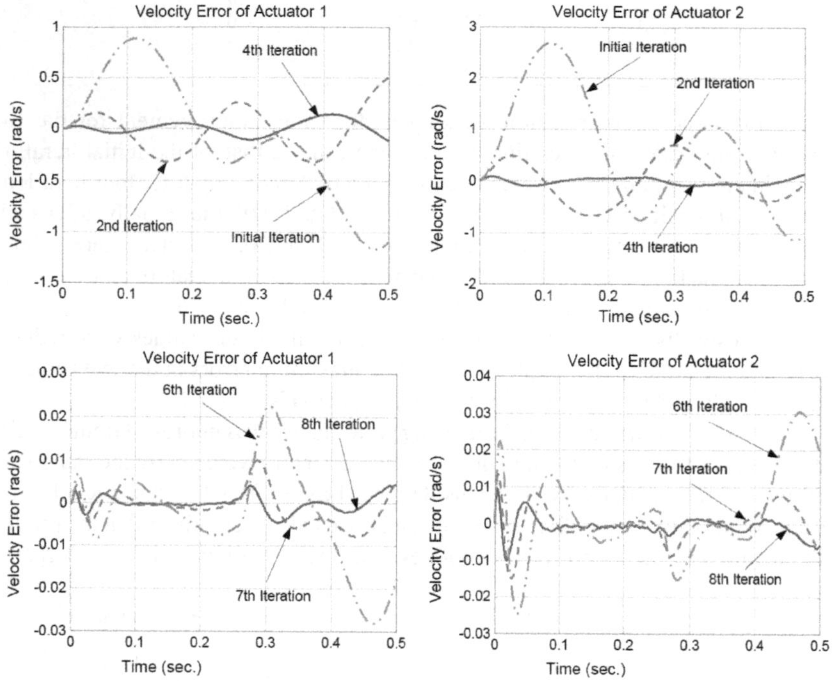

Figure 7. Velocity tracking performance improvement from iteration to iteration.

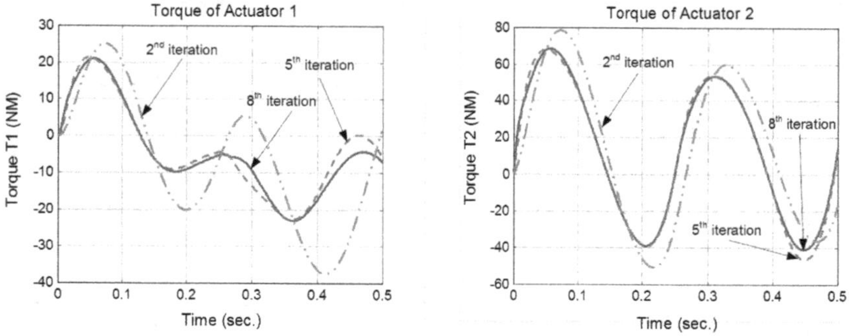

Figure 8. The required torque profiles for iteration $j = 2,5,8$.

SPD-PD Results

A. Simulation of Nonlinear System

In this section, we apply the proposed SPD-PD ILC algorithm for trajectory tracking control of a nonlinear system to improve its tracking performance through repetitive operations. The simulation example used in (Chen et al. 1996b) is adopted for the purpose of comparison. The nonlinear system is described by

$$\begin{bmatrix} \dot{x}_{1k}(t) \\ \dot{x}_{2k}(t) \end{bmatrix} = \begin{bmatrix} \sin(x_{2k}(t)) & 1+\sin(x_{1k}(t)) \\ 2-5t & -3-2t \end{bmatrix} \begin{bmatrix} x_{1k}(t) \\ x_{2k}(t) \end{bmatrix} + \begin{bmatrix} 1 & 0 \\ 0 & 2 \end{bmatrix} \begin{bmatrix} u_{1k}(t) \\ u_{2k}(t) \end{bmatrix} + (0.5+k\alpha_0) \begin{bmatrix} \cos(2\pi f_0 t) \\ 2\cos(4\pi f_0 t) \end{bmatrix}$$

$$\begin{bmatrix} y_{1k}(t) \\ y_{2k}(t) \end{bmatrix} = \begin{bmatrix} 4 & 0 \\ 0 & 1 \end{bmatrix} \begin{bmatrix} x_{1k}(t) \\ x_{2k}(t) \end{bmatrix} + (0.5+k\alpha_0) \begin{bmatrix} \sin(2\pi f_0 t) \\ 2\sin(4\pi f_0 t) \end{bmatrix}$$

With f = 5 Hz. The desired tracking trajectories are set as

$$y_{1d}(t) = y_{2d}(t) = 12t^2(1-t) \qquad for\ t \in [0,1]$$

To test the robustness of the proposed SPD-PD ILC algorithm, several simulation experiments are conducted using different control gains and learning gains. For all the cases, the initial states are set as $x_1(0) = 0.3$, and $x_2(0) = -0.3$. That means there are some initial state errors. Also, we assume the matrix B in the initial state learning schedule Eq. (27) is not accurate (The estimated value is 0.4B). In the following sections, the classic PD ILC is obtained from control law Eq. (22) by setting $K_{p1}(0) = K_{d1}(0) = 0$.

Example 1. *Repetitive Uncertainty and Disturbance*

In the first simulation experiment, we set $\alpha_0 = 0$ which means that the uncertainty and disturbance is repetitive from iteration to iteration. The following feedback control gains and learning control gains are chosen:

For classic PD ILC: $K_{p2} = diag\{1,1\}$, $K_{d2} = diag\{0.25, 0.5\}$.
For SPD-PD ILC: $K_{p1}(0) = diag\{0.5, 2.5\}$, $K_{d1}(0) = diag\{1.5\}$, $s(k) = k$
$K_{p2} = 0.6 * diag\{1,1\}$, $K_{d2} = 0.6 * diag\{0.25, 0.5\}$.

Referring to Eq. (21), we have $BC = diag\{4, 2\}$ for this nonlinear system. Therefore, the perfect learning gain is $K_{d2} = diag\{0.25, 0.5\}$ according to Eq. (26) that is used in the classic PD ILC. According to the chosen gains, from Eq. (26) we have $\rho_2 = 0$ for the classic ILC, and $\rho_1 = 0.2$ and $\rho_2 = 0.4$ for the SPD-PD ILC. That means inaccurate knowledge of matrices B and C are considered in the learning control gain design for the SPD-PD ILC.

Figure 9 shows the maximum tracking error bounds for the proposed SPD-PD ILC and the classic PD ILC from iteration to iteration. From this figure, we can see that the SPD-PD ILC algorithm can obtain a very fast convergence rate (7 iterations) and very small and monotonic decreased tracking errors. But for the classic PD ILC case, although the best learning gain K_{d2} is used (as $\rho_2 = 0$), the tracking error bounds were still in a good-bad-good mode before reaching a stable boundedness, and more iterations (18 iterations) were needed in order to obtain a relatively acceptable tracking performance. We can see that the tracking errors using PD ILC are still relatively large compared with those using the SPD-PD ILC. Similar results were shown in (Chen et al. 1996b) where more than 20 iterations are needed to achieve a stable tracking performance. This example results demonstrate that the SPD-PD ILC is more powerful in terms of reducing the tracking error and facilitating the convergence.

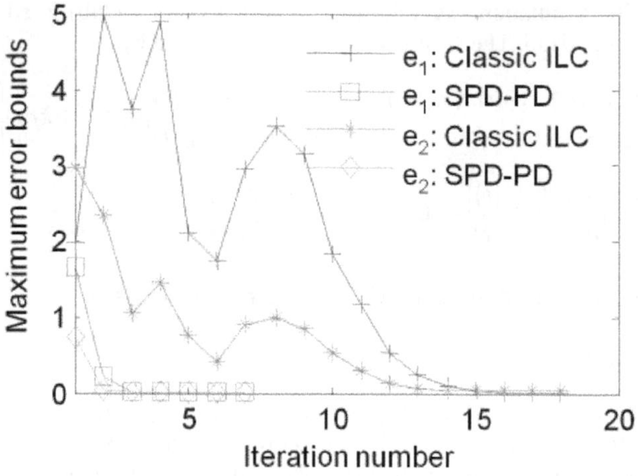

Figure 9. Maximum error bounds for Example 1.

Example 2. *Varying Uncertainty and Disturbance*

In this example, we consider a more general situation where the uncertainty and disturbance are increased from iteration to iteration by set $\alpha_0 = 0.05$. All the feedback control gains and iterative learning gains are set the same as in Example 1. Figure 10 shows the simulation results.

From Fig. 10 we can see that a good tracking performance can be achieved using the SPD-PD ILC algorithm, even in the situations where the uncertainty and disturbance are varied from iteration to iteration. Such a feature can be attributed to the switching gain feedback control which can compensate the disturbance by increasing the control gains from iteration to iteration. But for the classic PD ILC algorithm, there are large stable tracking error bounds (around 0.1 for e_1 and 0.2 for e_2). That is because the classic PD ILC cannot compensate the current iteration disturbance due to the limitation of the offline learning strategy.

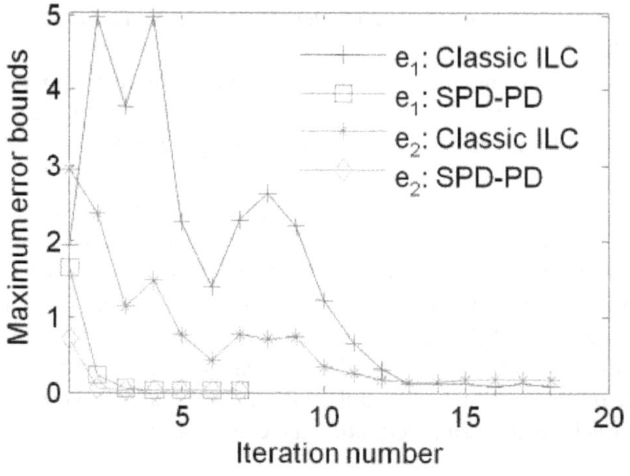

Figure 10. Maximum error bounds for Example 2.

Comparison on Different Feedback Control Gains

To adjust the final tracking error bounds, different feedback control gains are used for the SPD-PD ILC algorithm. In this example, the iterative learning control gains are set the same as the previous two examples. The following feedback control gains are used in the simulation experiments. For all three different gain cases, the switching function is set as $s(k) = k$. The middle gains are chosen the same as in Example 1.

High gains: $K_{p1}(0) = 1.5diag\{0.5,2.5\}$, $K_{d1}(0) = 1.5diag\{1,5\}$

Middle gains: $K_{p1}(0) = diag\{0.5,2.5\}$, $K_{d1}(0) = diag\{1,5\}$

Low gains: $K_{p1}(0) = 0.5diag\{0.5,2.5\}$, $K_{d1}(0) = 0.5diag\{1,5\}$

Table 3 lists the maximum final tracking error bounds under different feedback control gains in the SPD-PD ILC. From this table, we can see that, with the increase of the feedback control gains, the final tracking errors become smaller, and the convergence rate becomes faster. It demonstrates that the control parameter ρ_1 has significant effect to the convergence rate.

Table 4 shows the final tracking error bounds (stable boundary) for SPD-PD ILC compared to a fixed gain PD-PD ILC ($s(k) = 1$) using the same learning gains. From this table, it is clearly shown that the SPD-PD ILC can obtain much better tracking performance (at least 5 times) even with the existence of the non-repetitive uncertainty and disturbance (Example 2). Figure 11 shows the tracking error bound results from iteration to iteration based on SPD-PD ILC and the fixed gain PD-PD ILC.

From Table 4 and Fig. 11 we can see that the tracking performances can be improved if high feedback control gains are used in the developed SPD-PD ILC algorithm. It also clearly shows that the tracking error bounds monotonically decrease from iteration to iteration.

The Effect of Proportional Gains

From the convergence analysis conducted in the previous section, we can see that the proportional feedback gain K_{p1} and iterative learning gain K_{p2} are not factors for guaranteeing convergence as they are not included in the convergence condition

Table 3. Comparison of maximum error bounds under different control gains.

Control	Example 1			Example 2		
	Iteration	$\max\lvert e_1\rvert$	$\max\lvert e_2\rvert$	Iteration	$\max\lvert e_1\rvert$	$\max\lvert e_2\rvert$
Low gains	9	0.0005	0.0007	10	0.0058	0.0034
Mid. gains	7	0.0003	0.0007	7	0.0039	0.0022
High gains	6	0.0003	0.0006	6	0.0032	0.0017

Table 4. Comparison maximum error bounds for SPD-PD and PD-PD learning control.

Control	Example 1			Example 2		
	Iteration	$\max\lvert e_1\rvert$	$\max\lvert e_2\rvert$	Iteration	$\max\lvert e_1\rvert$	$\max\lvert e_2\rvert$
PD-PD	7	0.0023	0.0038	7	0.0241	0.0125
SPD-PD	7	0.0003	0.0007	7	0.0039	0.0022

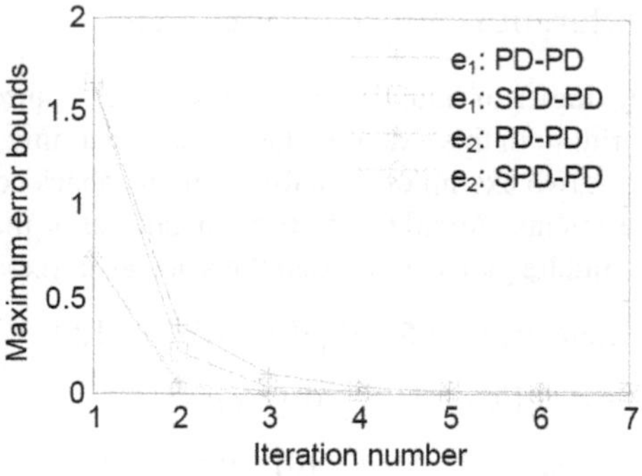

(a) Tracking error bounds for Example 1.

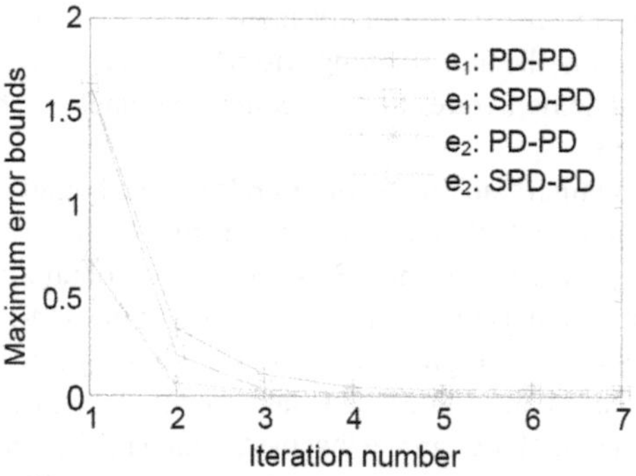

(b) Tracking error bounds for Example 2.

Figure 11. Comparison of SPD-PD and PD-PD learning control.

Eq. (26). Therefore, it is a good point to examine the effect of proportional gains on the tracking performance improvement. Figure 12 shows one simulation result of the tracking error bounds from iteration to iteration using SPD-PD ILC, SD-PD ILC (K_{p1} = 0), SPD-D ILC (K_{p2} = 0), and SD-D ILC (K_{p1} = K_{p2} = 0), respectively. It is clearly shown that K_p gains have effect only on the first few (3 in this simulation) iterations. After that, K_p gains make little contribution to the convergence rate of the system. The simulation results show that the final tracking error bounds using these four learning control laws are almost the same.

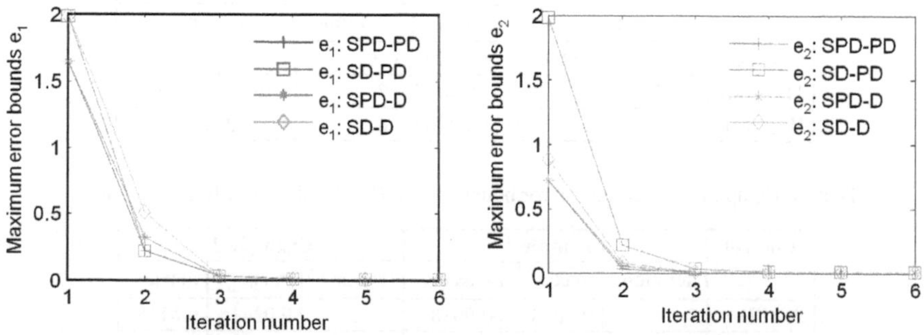

Figure 12. Effect of proportional control gains on convergence rate.

B. Simulation of Trajectory Tracking for a Planar Robotic System

It's well known that the dynamic model of a robotic system can be expressed as a second-order differential equation. If we introduce some state variables, the second-order differential equation can be easily transformed to a first-order nonlinear system expressed in (1). Therefore, the proposed SPD-PD ILC can be applied to tracking control of the robotic system. In this chapter, a 2-DOFs revolute planar robot discussed in (Choi and Lee 2001) is simulated for the trajectory tracking control, and the structural parameters of the robotic system can be found in Table 1.

The desired trajectories and the disturbances for both links are chosen as:

$$\begin{bmatrix} q_{1d} \\ q_{2d} \end{bmatrix} = \begin{bmatrix} \sin 3t \\ \cos 3t \end{bmatrix}, \quad \begin{bmatrix} \eta_1 \\ \eta_2 \end{bmatrix} = \alpha \begin{bmatrix} 0.3\sin t \\ 0.1\left(1-e^{-t}\right) \end{bmatrix} \quad t \in [0,5]$$

Tracking Performance Improvement from Iteration to Iteration

In this simulation experiment, the feedback control gains and iterative learning gains are set as:

$$K_{p1}(0) = K_{d1}(0) = diag\{20,10\} \text{ and } K_{p2} = K_{d2} = diag\{10,5\}.$$

The switching function is set as $s(k) = 2k$, and the disturbance factor is $\alpha = 10$.

Figures 13 and 14 show the trajectory tracking performance improvements and the required control torques from iteration to iteration for joint 1 and joint 2, respectively. At the first iteration, there are very large tracking errors for both joints, and the required control torques are irregular. After 3 iterations, the tracking performances improved dramatically. After 5 iterations, we can see that the controlled joint motions are very close to the desired ones, while the required torques show the desired periodic features. Figure 15 shows the maximum tracking error bounds from iteration to iteration. From Fig. 15 we can see that the tracking errors decrease monotonically from iteration to iteration. Simulation results show that only 14 iterations are needed to achieve a very good tracking performance (maximum errors are 1.75E-4 and 4.75E-5 for joints 1 and 2 from Table 5, respectively), while the maximum error bounds are 0.0041 and 0.0046 for joints 1 and 2 after 30 iterations using the control law developed in Choi and Lee 2000. Therefore, the SPD-PD ILC has a fast convergence rate, small tracking error bounds, and no vibration in the actuators. These figures demonstrate the effectiveness of the SPD-PD ILC for the trajectory tracking control of robotic systems.

Robustness for Rejecting Disturbance

To verify the effectiveness and robustness of the SPD-PD algorithm for the rejection of disturbances, different levels of disturbances are introduced in the dynamic model by adjusting the parameter α. Table 5 lists the maximum tracking error bounds for three different levels of disturbances. It shows that all the finally tracking error bounds after a few iterations (9 iterations in this example) are almost the same. From this table, we can conclude that the proposed SPD-PD ILC algorithm is robust to the disturbances.

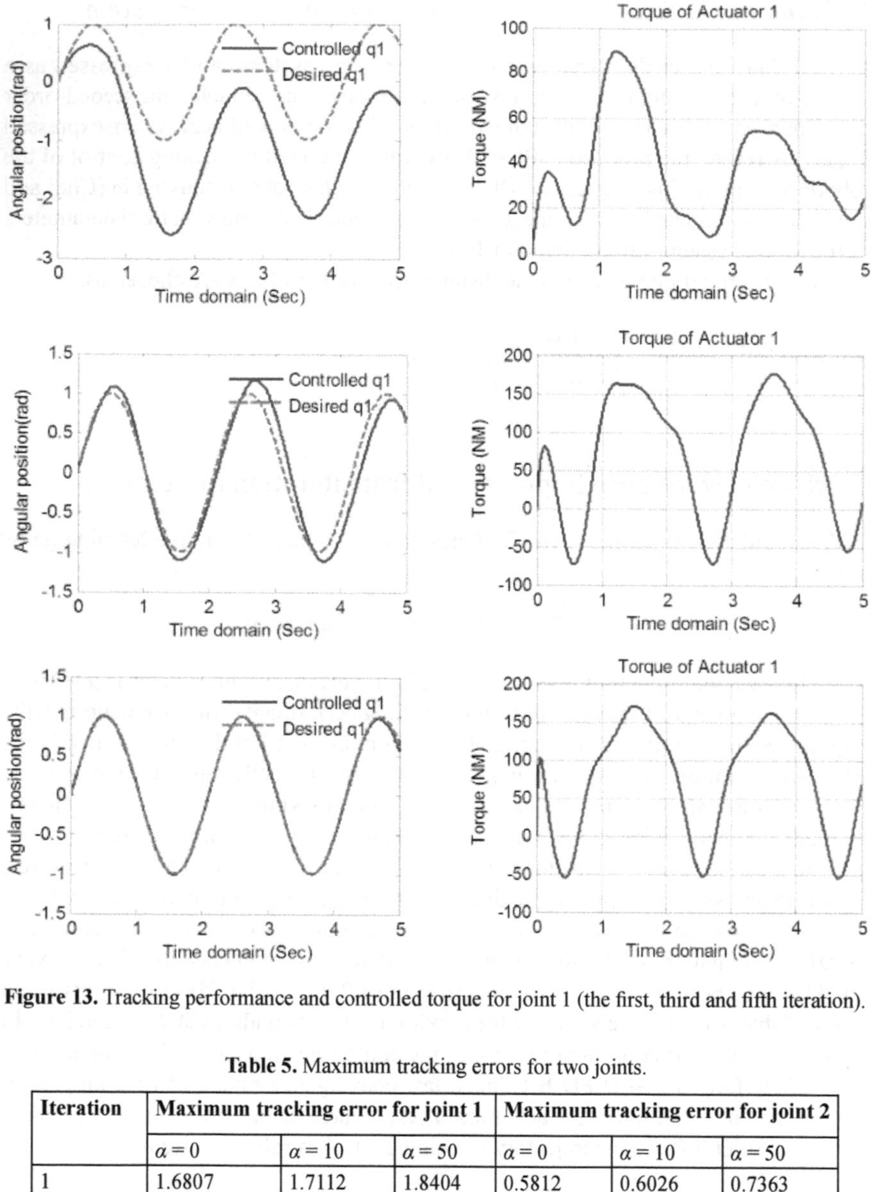

Figure 13. Tracking performance and controlled torque for joint 1 (the first, third and fifth iteration).

Table 5. Maximum tracking errors for two joints.

Iteration	Maximum tracking error for joint 1			Maximum tracking error for joint 2		
	$\alpha = 0$	$\alpha = 10$	$\alpha = 50$	$\alpha = 0$	$\alpha = 10$	$\alpha = 50$
1	1.6807	1.7112	1.8404	0.5812	0.6026	0.7363
3	0.3579	0.3578	0.3565	0.1096	0.1122	0.1213
6	0.0137	0.0136	0.0135	0.0034	0.0033	0.0034
9	0.001	0.001	0.001	2.93E-4	2.92E-4	2.93E-4
12	3.13E-4	3.14E-4	3.04E-4	8.81E-5	8.90E-5	8.68E-5
14	1.76E-4	1.74E-4	1.82E-4	4.75E-5	4.75E-5	5.17E-5

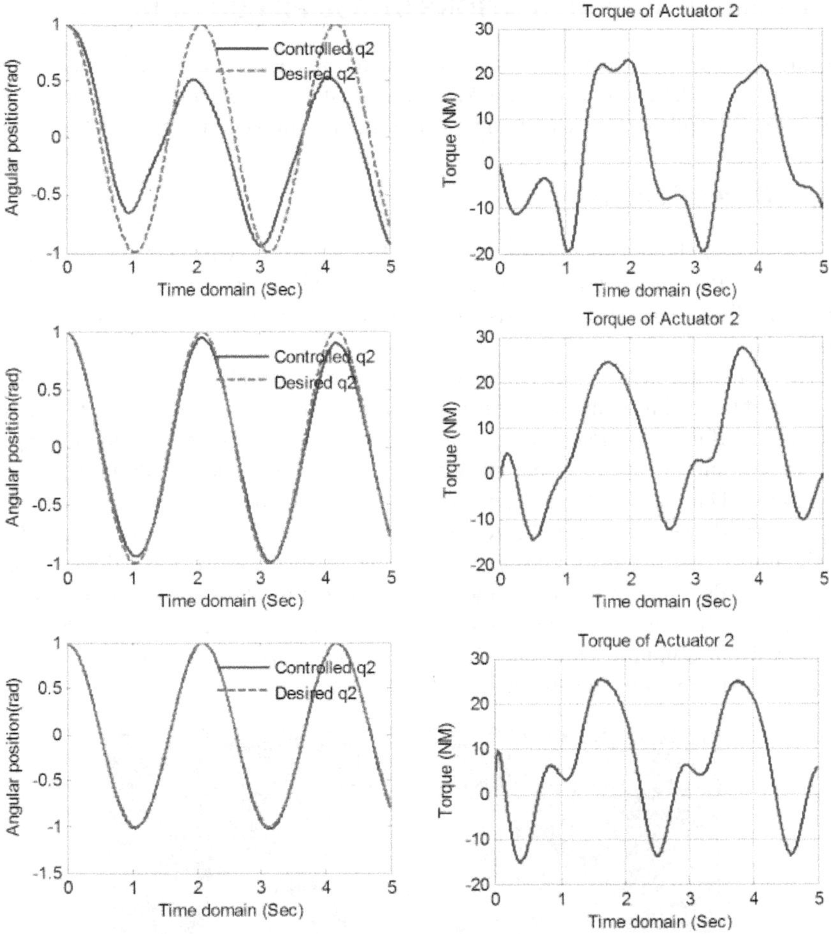

Figure 14. Tracking performance and controlled torque for joint 2 (the first, third and fifth iteration).

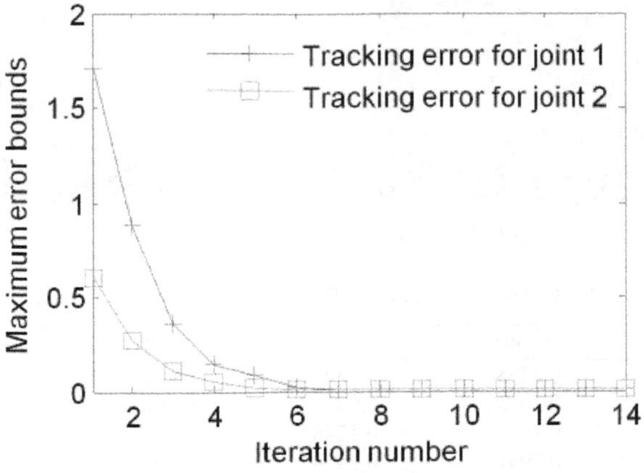

Figure 15. Tracking error bounds from iteration to iteration.

EXPERIMENTAL VERIFICATIONS OF ASL-PD CONTROL

In this section, some experiments are conducted to verify the effectiveness of the proposed ASL-PD for trajectory tracking control of a parallel robot manipulator, and comparison results are presented under different operation conditions.

Experimental Setup

A 2 DOF parallel robot manipulator shown in Fig. 16 was used to test the effectiveness of the ASL-PD control for the trajectory tracking where two servomotors drive the two input links directly. Table 6 lists the mechanical properties of the parallel robot manipulator.

The distance between the axes of servomotors 1 and 2, denoted by Link 5, is adjustable. In this study, it length is set to be 0.211 *m* in order to ensure the full rotatability of the robot manipulator. The sensors used to measure the rotation angles of the servomotors are incremental encoders with a resolution of 8000 pulses per revolution. The measured angular position is fed back and compared to the desired input. The difference between the desired position and the measured one is termed the position error. The velocity information is obtained from the measured positions by using the backwards finite-difference method. Based on the position error and the velocity information, the computer generates a voltage signal by using a given control law. This voltage signal is in turn sent to the Galil™ motion analyzer, the servo amplifiers, and then to the servomotors to generate the required torques.

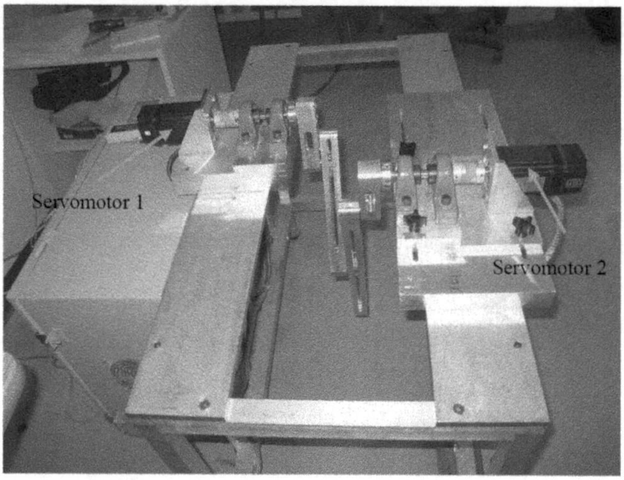

Figure 16. Parallel robot manipulator test bed.

Table 6. Parameters of the 2 DOF parallel robot manipulator.

	m_i(m)	r_i(m)	L_i(m)	I_i(m^4)
Link 1	0.2258	0.0342	0.0894	2.17×10^{-4}
Link 2	0.2623	0.0351	0.0943	3.14×10^{-4}
Link 3	0.4651	0.1243	0.2552	3.11×10^{-3}
Link 4	0.7931	0.2454	0.2352	1.67×10^{-2}
Link 5	0.2110	0	0	0

Trajectory Planning for Servomotors

The trajectory tracking of a robot manipulator can be performed in either Cartesian space (i.e., trajectory of the end-effector) or joint space (i.e., trajectory of the motors). However, regardless of the trajectory tracking approach, since the control of a robot is performed at the joint level, the conversion of the end-effector coordinates to joint coordinates is essentially required. This section is concerned with the joint or actuator trajectory tracking. For the purposes of this study, the desired trajectories of the two servomotors are planed as functions of time. In particular, this function is a Hermite polynomial of the fifth degree with continuous bounded conditions for position, velocity, and acceleration, which is adopted from (Ouyang 2005). The mathematical expression of the trajectories for two servomotors are given by, respectively,

$$q_1^d(t) = q_{10}^d + (6\frac{t^5}{t_f^5} - 15\frac{t^4}{t_f^4} + 10\frac{t^3}{t_f^3})(q_{1f}^d - q_{10}^d)$$

$$q_2^d(t) = q_{20}^d + (6\frac{t^5}{t_f^5} - 15\frac{t^4}{t_f^4} + 10\frac{t^3}{t_f^3})(q_{2f}^d - q_{20}^d)$$

where $q_1^d(t)$ and $q_2^d(t)$ are the desired trajectories of the two servomotors, q_{10}^d, q_{1f}^d, q_{20}^d, and q_{2f}^d are the desired initial and final positions of the servomotors, respectively, and t_f represents the time period required for the servomotors to reach the desired final position. In this study, the initial positions for the two servomotors is set $q_{10}^d = 0$ and $q_{20}^d = 0$ for simplification; and two cases with different operating speeds are investigated experimentally. In the first case, the robot manipulator is operated at a lower speed with

$$t_f = 4\text{sec} \quad q_{1f}^d = \frac{\pi}{2}, \quad \text{and } q_{1f}^d = \frac{\pi}{2}$$

Thus, the motor speed is $\omega = 60/(4 \times 4) = 3.75$ rpm for this case. In the second case, the robot manipulator is operated at a higher speed with

$$t_f = 2\text{sec} \quad q_{1f}^d = \pi, \quad \text{and } q_{2f}^d = \pi$$

Thus, the motor speed is $\omega = 60/(2 \times 2) = 15$ rpm for this case. For convenience, the first case is referred to as the low speed case and the second one as the high speed case in the rest of this section.

ASL-PD Control Law Results

In the experiments for the ASL-PD control, the initial PD control gains and the switching factor β are selected as follows:

For the low speed case:

Servomotor 1: $K_{p1} = 0.000022$, $K_{d1} = 0.000002$, $\beta = 2.5$
Servomotor 2: $K_{p1} = 0.000036$, $K_{d1} = 0.000004$, $\beta = 2.5$

For the high speed case:

Servomotor 1: $K_{p1} = 0.000052$, $K_{d1} = 0.0000052$, $\beta = 2$
Servomotor 2: $K_{p1} = 0.000048$, $K_{d1} = 0.0000062$, $\beta = 2$

It can be seen that the initial PD control gains of the ASL-PD control are selected to be the same as that of the adaptive NPD-LC control. The reason for this is to be able to compare the performance improvement and the convergence rate of different ILC control techniques. Figure 17 shows the tracking errors from iteration to iteration for the low speed case. From this figure, one can see that the tracking performances are improved from iteration to iteration under the ASL-PD control. Tables 7 and 8 show the experimental results for the maximum absolute tracking errors and required torques at low and high speed cases from iteration to iteration.

From these results it is clear how the position tracking errors are reduced from one iteration to the next. At low speeds, for the first iteration, the maximum tracking errors were 0.324 rad and 0.278rad for servomotors 1 and 2, respectively. After 6 iterations, the maximum tracking error was about 0.009 rad for both servomotors. The tests at

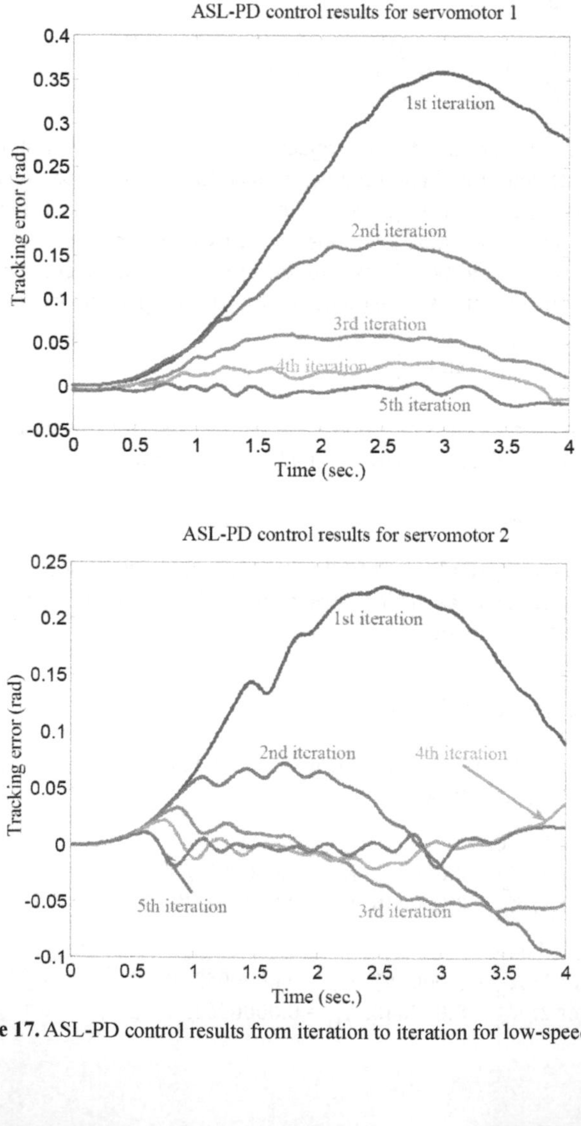

Figure 17. ASL-PD control results from iteration to iteration for low-speed case.

high speeds also demonstrated similar conclusions regarding the reduction in errors using the ASL-PD control method. In this case, for the first iteration, the maximum tracking errors were 0.135 rad and 0.110 rad for actuators 1 and 2, respectively. After 4 iterations, the maximum tracking errors were 0.017 rad and 0.016 rad for joints 1 and 2, respectively. The above figures and discussion provide a clear presentation of the effectiveness of the ASL-PD control law at both high and low speeds.

Table 7. Performance improvement with the ASL-PD control at low-speed case.

Generation	max abs(e_1)	max abs(e_2)	max abs(T_1)	max abs(T_2)
Iteration 1	0.324	0.278	0.023	0.027
Iteration 2	0.107	0.076	0.021	0.025
Iteration 3	0.045	0.055	0.022	0.022
Iteration 4	0.029	0.012	0.020	0.019
Iteration 5	0.019	0.010	0.021	0.020
Iteration 6	0.009	0.009	0.022	0.020

Table 8. Performance improvement with the ASL-PD control at high-speed case.

Generation	max abs(e_1)	max abs(e_2)	max abs(T_1)	max abs(T_2)
Iteration 1	0.135	0.110	0.054	0.051
Iteration 2	0.034	0.047	0.052	0.052
Iteration 3	0.020	0.032	0.049	0.048
Iteration 4	0.017	0.016	0.050	0.051

Comparison of the ASL-PD and the Adaptive NPD-LC

As shown in Ouyang 2005, the adaptive NPD-LC and the ASL-PD control could be used to improve the tracking performance of the parallel robotic manipulator. It should be noted that the convergence rate was different for these two learning control laws. Generally speaking, the ASL-PD control has a faster convergence rate than the adaptive NPD-LC. Figure 18 shows the comparison results obtained from the experiments. From this figure, one can see that the tracking errors after 5 iterations using the ASL-PD control were smaller than the tracking errors after 7 iterations using the adaptive NPD-LC. Such a conclusion agrees with the theoretical analysis and the simulation results as presented in Ouyang 2005. Therefore, through experiment studies for a parallel robot manipulator, the effectiveness of the ASL-PD control is demonstrated for trajectory tracking control under different operation conditions.

CONCLUSIONS

In this chapter, a new adaptive switching learning PD (ASL-PD) control method is proposed. This control method is a simple combination of a traditional PD control with a gain-switching strategy as feedback and an iterative learning control using the input torque profile obtained from the previous iteration as feedforward. The ASL-PD control incorporates both adaptive and learning capabilities; therefore, it can provide an incrementally improved tracking performance with the increase of the iteration number. The ASL-PD control method achieves the asymptotic convergence based on the Lyapunov's method. The position and velocity tracking errors monotonically decrease with the increase of the iteration number. The concept of integrating the switching

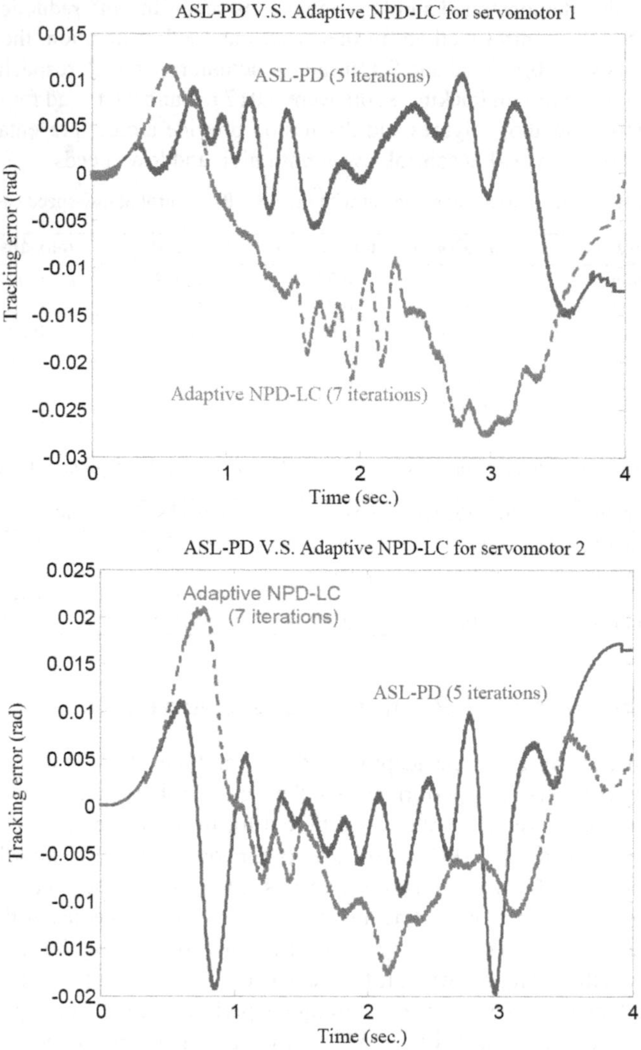

Figure 18. Comparing results of the adaptive NPD-LC and the ASL-PD.

technique and the iterative learning scheme works very well. The simulation study has demonstrated the effectiveness of the ASL-PD control method. Its distinct features are the simple structure, easy implementation, fast convergence, and excellent performance.

In addition, a new iterative learning control called SPD-PD ILC is also proposed that is a combination of a PD feedback control with switching control gains in the iteration domain and a PD-type ILC based on the previous iteration information in the updating control law. The proposed SPD-PD ILC takes the advantages offered by feedback control and iterative learning control. The proposed control law uses the current and previous system information to enhance the stability characteristics and quickly drive the tracking trajectories to the desired ones within bounds. The new SPD-PD ILC achieves tracking accuracy with very fast convergence rate and is robust against unpredictable disturbances and uncertainties by adjusting the feedback control gains. In addition, the SPD-PD ILC can provide extra degrees of freedom for the choices of the learning gains. The final tracking error bounds and the convergence rate can be adjusted by the switching gain of the PD feedback control that makes this control scheme more promising from a practical viewpoint. Nonlinear systems and a planar robotic system are used as examples to test the effectiveness of the proposed SPD-PD ILC.

Finally, the control strategies have been implemented and tested on an experimental environment of a 2 DOF parallel robot manipulator. The performances of the developed control strategies have been evaluated in terms of tracking performances. The experimental results show that the ASL-PD control results in the best performance in terms of tracking errors and iteration generations.

ACKNOWLEDGMENT

This work was supported by the Natural Sciences and Engineering Research Council of Canada (NSERC) to the authors.

REFERENCES

Ahn, H. S., Y. Q. Chen and K. L. Moore. 2007. Iterative learning control: brief survey and categorization. IEEE Transactions on Systems, Man and Cybernetics Part C: Applications and Reviews, 37(6): 1099–1121.

Amann, N., D. H. Owens and E. Rogers. 1995. Iterative learning control for discrete time systems using optimal feedback and feedforward actions. Proc. of the 34th Conference on Decision & Control, pp. 1696–1701.

Arimoto, S., S. Kawamura and F. Miyasaki. 1984. Bettering operation of robots by learning. Journal of Robotic Systems, 1(2): 123–140.

Bristow, D. A., M. Tharayil and A. G. Alleyne. 2006. A survey of iterative learning control: a learning-based method for high-performance tracking control. IEEE Control Systems Magazine, 26(3): 96–114.

Cai, Z., C. T. Freeman, P. L. Lewin and E. Rogers. 2008. Experimental comparison of stochastic iterative learning control. Proceedings of the American Control Conference, pp. 4548–4553.

Chen, Q. J., H. T. Chen, Y. J. Wang and P. Y. Woo. 2001. Global stability analysis for some trajectory tracking control schemes of robotic manipulators. Journal of Robotic Systems, 18(2): 69–75.

Chen, Y. Q. and K. L. Moore. 2002. An optimal design of PD-type iterative learning control with monotonic convergence. Proceedings of IEEE International Symposium on Intelligent Control, pp. 55–60.

Chen, Y. Q. and K. L. Moore. 2002. PI-type iterative learning control revisited. pp. 2138–2143. *In*: Proceedings of the American Control Conference, Anchorage, AK, USA.

Chen, Y. Q., J. X. Xu and T. H. Lee. 1996a. Current iterative tracking error assisted iterative learning control of uncertain discrete-time systems. Proceedings of the IEEE Conference on Decision and Control, pp. 2355–3592.

Chen, Y. Q., J. X. Xu and T. H. Lee. 1996b. An iterative learning controller using current iteration tracking error information and initial state learning. Proc. of 35th IEEE Conf. Decision and Control, pp. 3064–3069.

Choi, J. Y. and J. S. Lee. 2000. Adaptive iterative learning control of uncertain robotic systems. IEE Proc. Control Theory Application, 147(2): 217–223.

Craig, J. J. 1986. Introduction to robotics: mechanics and control. Reading, Addison-Wesley, MA.

Craig, J. J. 1988. Adaptive control of mechanical manipulators, Addison-Wesley.

Doh, T. Y., J. H. Moon, K. B. Jin and M. J. Chung. 1999. Robust iterative learning control with current feedback for uncertain linear systems. International Journal of Systems Science, 30(1): 39–47.

Fu, M. and B. R. Barmish. 1986. Adaptive stabilization of linear systems via switching control. IEEE Trans. on Automatic Control, 31(6): 1097–1103.

Kang, M. K., J. S. Lee and K. L. Han. 2005. Kinematic path-tracking of mobile robot using iterative learning control. Journal of Robotic Systems, 22(2): 111–121.
Kawamura, S., F. Miyazaki and S. Arimoto. 1988. Realization of robot motion based on a learning method. IEEE Trans. on Systems, Man, and Cybernetics, 18(1): 126–123.
Kerry, R. 1997. PD control with desired gravity compensation of robotic manipulators: a review. The International Journal of Robotics Research, 16(5): 660–672.
Kuc, T. Y., K. Nam and J. S. Lee. 1991. An iterative learning control of robot manipulators. IEEE Trans. on Robotics and Automation, 7(6): 835–842.
Li, Q., A. N. Poo, C. L. Teo and C. M. Lim. 1996. Developing a neuro-compensator for the adaptive control of robots. IEE Proc. Cont. Theory and Appl., 142(6): 562–568.
Li, Q., S. K. Tso and W. J. Zhang. 1998. Trajectory tracking control of robot manipulators using a neural-network-based torque-compensator. Proceedings of I. Mech. E, Part I, J. of System and Control, 212(5): 361–372.
Longman, R. W. 2000. Iterative learning control and repetitive control for engineering practice. International Journal of Control, 73(10): 930–954.
Madady, A. 2008. PID type iterative learning control with optimal gains. International Journal of Control, Automation, and Systems, 6(2): 194–203.
Martensson, B. 1985. The order of any stabilizing regulator is sufficient a priori information for adaptive stabilizing. Systems & Control Letters, 6(2): 87–91.
Middleton, R. H., G. C. Goodwin, D. J. Hill and D. Q. Mayne. 1988. Design issues in adaptive control. IEEE Trans. on Automatic Control, 33(1): 50–58.
Moore, K. L., M. Dahleh and S. P. Bhattacharyya. 1992. Iterative learning control: a survey and new results. Journal of Robotic Systems, 9(5): 563–594.
Norrlöf, M. and S. Gunnarsson. 2005. A note on causal and CITE iterative learning control algorithms. Automatica, 41(2): 345–350.
Nussbaum, R. D. 1983. Some remarks on a conjecture in parameter adaptive control. Systems & Control Letters, 3: 243–246.
Ouyang, P. R. 2005. Integrated design, modeling, and control of hybrid systems, Ph.D. Dissertation, University of Saskatchewan, Saskatoon, Canada.
Ouyang, P. R. 2009. PD-PD type learning control for uncertain nonlinear systems. The 2009 ASME International Design Engineering Technical Conferences and Computers and Information in Engineering Conference, pp. 699–707.
Ouyang, P. R., Q. Li, W. J. Zhang and L. S. Guo. 2004. Design, modelling and control of a hybrid machine system. Mechatronics, 14(10): 1197–1217.
Ouyang, P. R., B. A. Petz and F. F. Xi. 2011. Iterative learning control with switching gain feedback for nonlinear systems. ASME Journal of Computational and Nonlinear Dynamics, 6(1): 011020–7.
Ouyang, P. R. and P. Pipatpaibul. 2010. Iterative learning control: a comparison study. ASME 2010 International Mechanical Engineering Congress and Exposition, 8: 939–945.
Ouyang, P. R. and W. J. Zhang. 2005. Force balancing of robotic mechanisms based on adjustment of kinematic parameters. ASME Journal of Mechanical Design, 127(3): 433–440.
Ouyang, P. R., W. J. Zhang and M. M. Gupta. 2006. An adaptive switching learning control method for trajectory tracking of robot manipulators. Mechatronics, 16(1): 51–61.
Ouyang, P. R., W. J. Zhang and M. M. Gupta. 2007. PD-type on-line learning control for nonlinear system with uncertainty and disturbance. Control and Intelligent Systems, 35(6): 351–358.
Owens, D. H. and G. S. Munde. 1998. Universal adaptive iterative learning control. Proc. of the 37th IEEE Conference on Decision & Control, pp. 181–185.
Park, K. H. 2005. An average operator-based PD-type iterative learning control for variable initial state error. IEEE Trans. on Automatic Control, 50(6): 865–869.
Park, K. H., Z. Bien and D. H. Hwang. 1999. A study on the robustness of a PID-type iterative learning controller against initial state error. International Journal of Systems Science, 30(1): 49–59.
Pi, D. Y., S. Seborg, J. Shou, Y. Sun and Q. Lin. 2000. Analysis of current cycle error assisted iterative learning control for discrete nonlinear time-varying systems. IEEE International Conference on Systems, Man, and Cybernetics, 5: 3508–3513.
Qu, Z. H. 1995. Global stability of trajectory tracking of robot under PD control. Dynamics and Control, 5(1): 59–71.
Qu, Z. H., J. Dorsey, M. Dawson and R. W. Johnson. 1991. A new learning control scheme for robots. Proceedings of the 1991 IEEE ICRA, pp. 1463–1468.

Slotine, J. J. and W. Li. 1987. On the adaptive control of robot manipulators. The International Journal of Robotics Research, 6(3): 49–59.

Song, Z. Q., J. Q. Mao and S. W. Dai. 2005. First-order D-type iterative learning control for nonlinear systems with unknown relative degree. ACTA Automatica Sinca, 31(4): 555–561.

Sun, D. and J. K. Mills. 1999. Performance improvement of industrial robot trajectory tracking using adaptive-learning scheme. Trans. of the ASME, Journal of Dynamic Systems, Measurement and Control, 121(2): 285–292.

Sun, M. and D. Wang. 2002. Closed-loop iterative learning control for nonlinear systems with initial shifts. International Journal of Adaptive Control and Signal Processing, 16(7): 515–538.

Tayebi, A. 2003. Adaptive iterative learning control for robot manipulators. pp. 4518–4523. *In*: Proceedings of the American Control Conference, Denver, CO, USA.

Tomei, P. 1991. Adaptive PD controller for robot manipulators. IEEE Trans. on Robot and Automation, 7(4): 565–570.

Xu, J. X., T. H. Lee and H. W. Zhang. 2004. Analysis and comparison of iterative learning control schemes. Engineering Applications of Artificial Intelligence, 17(6): 675–686.

Xu, J. X., X. W. Wang and L. T. Heng. 1995. Analysis of continuous iterative learning control systems using current cycle feedback. Proceedings of the American Control Conference, pp. 4221–4225.

Xu, J. X., L. Sun, T. Chai and D. Tan. 2004. High-order open and closed loop iterative learning control scheme with initial state learning. 8th International Conference on Control, Automation, Robotics and Vision, pp. 637–641.

Yan, X. G., I. M. Chen and J. Lam. 2001. D-type learning control for nonlinear time-varying systems with unknown initial states and inputs. Trans. of the Institute of Measurement and Control, 23(2): 69–82.

Yu, S. J., J. H. Wu and X. W. Yan. 2002. A PD-type open-closed-loop iterative learning control and its convergence for discrete systems. Proceedings of the First International Conference on Machine Learning and Cybernetics, pp. 659–662.

Zhang, B., G. Tang and S. Zheng. 2006. PD-type iterative learning control for nonlinear time-delay system with external disturbance. Journal of Systems Engineering and Electronics, 17(3): 600–605.

15

Adaptive Robust Control Design for Robot Manipulators Based on Online Estimation of the Lumped Time-Varying Model Uncertainties

Meysar Zeinali[1,*] and *Ebrahim Esmailzadeh*[2]

ABSTRACT

This chapter presents a systematic method to design an adaptive robust control based on online estimation of the lumped time-varying model uncertainties for tracking control of the robot manipulators. The proposed adaptive robust control methodology is based on approaches developed in recent publications, which allows eliminating the chattering while retaining the system robustness in the presence of bounded time-varying uncertainties. By online estimation of the lumped model uncertainties, main drawbacks of conventional robust sliding mode control, and adaptive control of manipulators namely: the chattering phenomenon; the requirement for a priori knowledge of the bounds of the uncertainties, and the requirement for linearly parameterized dynamic model of the manipulator with known constant bound of the parameters are addressed. The global stability and robustness of the proposed controller are established in the presence of time-varying uncertainties using Lyapunov's approach and

[1] School of Engineering, Mechanical and Mchatronics Program, Laurentian University, Sudbury, Ontario, P3E 2C6, Canada.
E-mail: mzeinali@laurentian.ca

[2] Department of Automotive, Mechanical and Manufacturing Engineering, Faculty of Engineering and Applied Science.
E-mail: Ebrahim.Esmailzadeh@uoit.ca

* Corresponding author

fundamentals of sliding mode theory. Based on the simulations and experimental results, the proposed controller performs remarkably well in comparison to adaptive control in terms of the tracking error convergence and robustness against un-modeled dynamics, external disturbances, and time-varying parameters and its high performance and simplicity makes this method attractive for industrial applications.

Keywords: Adaptive robust control, chattering-free sliding mode control, robot manipulator, nonlinear uncertain system, estimation of uncertainties

INTRODUCTION

Many dynamic systems such as robot manipulators to be controlled have unknown or time-varying uncertain parameters. For instance, it is impossible or very difficult to obtain an exact dynamic model of the robot manipulator, due to the presence of large flexibility, Coulomb friction, backlash, unknown disturbances, large dynamic coupling between different links, and time-varying parameters such as parameters relating to the robot age (tear and wear) which are all uncertain to some extent, or robot manipulators may carry large payload with unknown inertial parameters or joint friction dependent on the operating condition or various adjustment (Slotine and Li 1991). To deal with uncertainties, in the last few decades, much research effort has been devoted to the design or improvement of the controller of uncertain systems (Gutman 1979, Chandrasekharan 1996, Zhou et al. 1996). Some of the main components of nonlinear robust control methods are sliding mode control (SMC), (Utkin 1977), adaptive control (AC) (Slotine and Li 1991), and adaptive sliding mode control (ASMC), (Zeinali and Notash 2010, Zeinali 2015). Generally, the basic objective of adaptive control is to maintain performance and stability of the closed-loop system in the presence of structured uncertainties (i.e., parametric uncertainties), and unstructured uncertainties (i.e., un-modeled dynamic and external disturbances). The traditional approach to achieve above objective is on-line estimation of the uncertain parameters or equivalently, the corresponding controller parameters, based on the system performance, and measured system signals and then using the estimated parameters to compute the control input (Slotine and Li 1991). This approach requires linearly parameterized model of the robot manipulator or controller, which may lead to a large number of parameters.

The adaptive control strategies proposed in Craig et al. 1987, Sadegh and Horowitz's 1987, Slotine and Li 1991, mainly addressing the difficulties arising from parametric uncertainty, and the controller proposed in (Seraji 1987) requires linearized model of the manipulator for each operating point and it is based on model reference adaptive control (MRAC). The work by Craig et al. and Sadegh and Horowitz are based on the generalization of the computed torque (inverse dynamic) approach and the objective of the adaptive controller of this class is to render the closed-loop system linear and decupled even though the parameters are unknown. The adaptive controller of Slotine and Li is conceptually different from the adaptive inverse dynamic approach in the sense that the controller does not feedback linearize the closed-loop system, while the proposed controller has the advantages of both approaches. As will be shown perfect linearization can be achieved in the presence of slowly time-varying uncertainties, and approximately linearized closed-loop can be obtained in the presence of time-varying uncertainties.

The adaptive robust controller presented in this chapter is designed using the theory of the sliding mode controller, since SMC have widely been used in many applications due to their capability to deal with uncertainties, good transient performance and their simplicity (Utkin 1977). One of the underlying assumptions in the SMC is that the control input can be switched from one value to another infinitely fast. In practice switching is not instantaneous, because finite delays are inherent in all physical actuators (Young et al. 1999). The uncertainties, imperfections in switching devices or delays, lead to high-frequency oscillations called chattering, which can cause a fast breakdown of mechanical elements in actuators, e.g., high wear. It may also excite un-modelled high-frequency dynamics, which may lead to instability. The performance of a SMC is therefore measured by its robustness and severity of chattering. Thus, it is of primary importance to reduce or eliminate the chattering, in order to make SMC to be practical. To reduce or eliminate the chattering, various approaches have been proposed by many researchers which are summarized in (Young et al. 1999, Plestan et al. 2010, Utkin and Poznyak 2013). The first and most commonly cited approach to eliminate the chattering has been the boundary layer method (Slotine and Li 1991), which requires a trade-off between performance and chattering and it is also very difficult to determine the thickness of boundary layer correctly in most cases. In (Boiko et al. 2007) it is proven that the chattering-free property associated with the continuous SMC algorithms proposed in references (Man et al. 1994) and (Yu 1998) is a misconception, and the algorithms can lead to a periodic motions (chattering).

Another issue limiting the application of the sliding mode control is that, a priori knowledge of the upper and lower bounds of the uncertainties (i.e., worst case scenario) is required to obtain robustness and convergence. This issue recently tackled in (Lee and Utkin 2007, Plestan et al. 2010) by proposing online gain adaptation algorithm. However, in both of these approaches tuning of parameter ε is a hard task and controller is discontinuous. According to Young et al. 1999, if uncertainties are sufficiently compensated there is no need to use discontinuous control law to achieve sliding mode, thus the chattering can be reduced or eliminated if the online estimation of the uncertainties is properly designed. Therefore, an alternative method to reduce or eliminate the chattering and to overcome the drawbacks of using the bounds of uncertainties is to use estimated uncertainties. The above approach has been studied and discussed in Elmali and Olgac 1992, Kim et al. 1998, Curk and Jezernik 2001, Saengdeejing and Qu 2002, Lee and Utkin 2007, Dong et al. 2006, Li and Xu 2010. However, an adaptive and chattering-free SMC based on estimated uncertainties for robot manipulators are not fully developed and due to a lack of a unified systematic approach is remained mainly unsolved.

In uncertainty estimation-based approach, the control law consists of a continuous model-based component and a component constructed based on the estimated uncertainties for compensation of the uncertainties. In this chapter an adaptive robust controller is systematically constructed based on an effective on-line estimation of the lumped time-varying uncertainties using the dynamic behaviour of a sliding function which is defined later. The formulation of the proposed uncertainty estimation is derived from the systematic stability analysis of the closed-loop system using Lyapunov stability method in Section **The Proposed Adaptive Robust Controller Design Methodology** and **Stability Analysis of Closed-Loop System in the Presence of Time-Varying Uncertainties**. The estimated uncertainty term is used to replace the discontinuous component, also called the switching component, of the standard sliding mode control.

This, in turn, eliminates the chattering and the requirements for the bounds of the uncertainties. In this method, trajectories are forced to converge to sliding surface in finite time by suitably adjusting the estimated uncertainty term of the controller.

The remainder of this chapter is organized as follows. Section **Problem Formulation and Definition** presents the system description and problem formulation. Section **The Proposed Adaptive Robust Controller Design Methodology** describes the adaptive robust control design methodology. Stability of the closed-loop system in the presence of the time-varying uncertainties is proven in Section **Stability Analysis of Closed-Loop System in the Presence of Time-Varying Uncertainties.** The Simulation results are presented in Section **Simulation Results.** Section **Experimental Results and Application to a Two-DOF Flexible Link Robot** describes the experimental results and application of the proposed method to a two-DOF Flexible Link robot. Section **Conclusion** concludes the chapter.

PROBLEM FORMULATION AND DEFINITION

Robot Manipulator Dynamics

The dynamic model (equation of motion) of a rigid robot manipulator can generally be described by the following second-order differential equation, in active joint space Codourey 1998, Angeles 2003.

$$\tau = M(q)\ddot{q} + C(q, \dot{q})\dot{q} + F_f(q, \dot{q}) + g(q) + d(t) \tag{1}$$

where $\tau = [\tau_1, \tau_2, ..., \tau_n]^T$ is the vector of joints generalized forces (torques/forces), n denotes the number of generalized coordinates of a robot manipulator, for instance, the number of DOF for serial robot manipulators or the number of actuated joints (active joints) for parallel robot manipulator. The generalized coordinates of a mechanical system (e.g., a robot manipulator) are all those displacement variables, whether rotational or translational, that determines uniquely a configuration of the system. Vectors $q = [q_1, q_2, ..., q_n]^T$, $\dot{q} = [\dot{q}_1, \dot{q}_2, ..., \dot{q}_n]^T$ and $\ddot{q} = [\ddot{q}_1, \ddot{q}_2, ..., \ddot{q}_n]^T$ denote the vectors of joints displacements, velocities and accelerations respectively, $M(q)$ is an $n \times n$ robot manipulator inertia matrix (which is symmetric positive definite), $C(q, \dot{q})$ is an $n \times n$ matrix of centripetal and Coriolis terms, $g(q)$ is an $n \times 1$ vector of gravity terms, F_f is an $n \times 1$ vector denoting the torque due to viscous and Coulomb friction, and $d(t)$ is an $n \times 1$ vector denoting the un-modeled dynamics and external disturbances.

For parallel robot manipulators, because of the closed-loop kinematic chains and existence of passive joints (not all joints of closed-loop kinematic chains are actuated) a set of constraint equations derived from kinematics of a robot manipulator is required to derive the equations of motion. Based on the literature Ghorbel et al. 2000, to derive the dynamic model for general n-DOF parallel robot manipulator, first the equations of motion are formulated in terms of, for instance $n'(n' > n)$, dependent generalized coordinates. Following this $n' - n$ holonomic (i.e., geometric) constraints are eliminated to end up with n independent differential equations with n independent generalized coordinates corresponding to the number of the *actuated joints* of the parallel robot manipulator. According to Angeles (Angeles 2003, page 473), by proper definition of the inertia matrix and quadratic terms of inertia force (matrix of centripetal and Coriolis terms), the dynamic model of a parallel robot manipulator also can be expressed in the form of Eq. (1). For detailed information the reader is referred to Angeles 2003.

Properties of Robot Manipulator Dynamics

The dynamic model presented in Eq. (1) contains two important structural properties, which are useful for developing control laws as well as for the identification of the parameters of the dynamic model (e.g., link mass and moment of inertia). The two important properties are reviewed in order to clarify the discussions in the following sections and use them for design and analysis purposes of the proposed controller throughout this chapter. The properties are as follows.

***Skew-symmetry property*:** this property refers to an important relationship between the inertia matrix $M(q)$ and the matrix $C(q, \dot{q})$ appearing in Eq. (1). The property states that matrix $N(q,\dot{q}) = \dot{M}(q) - 2C(q,\dot{q})$ is a skew-symmetric matrix, i.e., the components satisfy $n_{ij} = -n_{ji}$. The proof of this property for serial, parallel and redundantly actuated parallel robot manipulators can be found in Ghorbel et al. 2000. The skew-symmetry property is employed in the stability and robustness analysis of the proposed control methodology in Section **The Proposed Adaptive Robust Controller Design Methodology** of this chapter.

***Linearity-in-the-parameters property*:** the dynamic model presented in Eq. (1) can be defined in terms of certain parameters, such as the link mass and moment of inertia, friction coefficient, etc. Usually a priori knowledge about these parameters is not complete and therefore, they must be determined in order to modify the dynamic model or tune a model-based controller. Fortunately, it can be shown that the nonlinear dynamic model presented in Eq. (1) is linear in terms of above mentioned parameters in the following sense (Slotine and Li 1991). There exists an $n \times p$ matrix function $Y(q, \dot{q}, \ddot{q})$, which is called the regressor, and a $p \times 1$ vector Φ such that the dynamic model can be written as

$$\tau = M(q)\ddot{q} + C(q, \dot{q})\dot{q} + F_f(q, \dot{q}) + g(q) = Y(q, \dot{q}, \ddot{q})\,\Phi \tag{2}$$

$$\tau = Y(q, \dot{q}, \ddot{q})\,\Phi$$

where Φ is a $p \times 1$ vector of dynamic parameters and p is the number of parameters. The linearity in the dynamic parameters property is a base for the traditional adaptive control design and on-line parameter estimation method. This property is only used to explain the existing control methodologies and clarify the discussions, and it is not used for the design and analysis of the proposed controller in this chapter.

After discussing the dynamic model of the manipulator and its properties, it should be noted that the dynamic model of the system is an approximation of the real system, due to the presence of complex phenomena such as large flexibility, Coulomb friction, backlash, payload variation, external disturbances and time-varying parameters, which are all uncertain to some extent and cannot be modeled mathematically. Therefore, in the presence of uncertainties, the components of the dynamic model are of the form

$$\begin{aligned} M(q) &= \hat{M}(q) + \Delta M(q) \\ C(q, \dot{q}) &= \hat{C}(q, \dot{q}) + \Delta C(q, \dot{q}) \\ g(q) &= \hat{g}(q) + \Delta g(q) \end{aligned} \tag{3}$$

where $\hat{M}(q)$, $\hat{C}(q, \dot{q})$ and $\hat{g}(q)$ are the known (estimated) parts and $\Delta M(q)$, $\Delta C(q, \dot{q})$ and $\Delta g(q)$ are the unknown parts. In this work, without loss of generality, it is assumed that the term denoting the viscous and Coulomb friction effects is completely unknown. Therefore, in the presence of the parameter uncertainty, un-modelled dynamics and external disturbances, Eq. (1) can be written as follows

$$\tau = \hat{M}\ddot{q} + \hat{C}(q, \dot{q}, t)\dot{q} + \hat{g}(q) + \delta(q, \dot{q}, \ddot{q}, t) \tag{4}$$

$$\delta(q, \dot{q}, \ddot{q}, t) = \Delta M\ddot{q} + \Delta C\dot{q} + \Delta g + F_f(q, \dot{q}) + d(t) \tag{5}$$

where $\delta(q, \dot{q}, \ddot{q}, t)$ is an unknown bounded function that *lumps together various uncertain terms* (i.e., un-modelled dynamics, parameter uncertainty, external disturbances). It is worth noting that inclusion of the complexities in the vector of the lumped uncertainty has two advantages:

1. It reduces the order of the dynamic model in the sense that higher order un-modelled dynamics terms can be included in uncertainty term (i.e., additive representation) and also the number of unknown parameters, which are needed to be estimated and identified.
2. As a consequence of reducing the number of unknown parameters, the computation burden is also reduced in real-time estimation. Since, only one uncertain parameter per joint needs to be estimated this, in turn, simplifies the controller design to some extent in the sense that only one adaptation gain per joint is required to be tunned.

Along this line, it should be noted that the potential vibration modes of flexible link robots and cables in the cable-based parallel robot manipulators and other types of uncertainties might be lumped into the uncertainty vector and the additive representation of uncertainties is selected to describe the nonlinear uncertain system (i.e., Eq. (4)). According to Zhou and Doyle 1996 and Sanchez-Pena and Sznaier 1998, the above additive representations of uncertainty is well suited to represent the modelling errors caused by flexibility, by neglected high-frequency dynamics such as measurement noise, and by parametric uncertainties, if they are not included in the dynamic model. Therefore, in order to develop the controller and analyze the stability of closed-loop system, the "nominal" dynamic model of the system—that is, the system without uncertainty ($\delta(q, \dot{q}, \ddot{q}, t) = 0$)-is built based on the rigid body mode and all uncertain terms lumped into one uncertainty term which results in the general Euler-Lagrange dynamic model of the robot manipulators, Eq. (4), and is used throughout this chapter.

At this point, the following assumption is made in order to design the controller and analyze the stability of the closed loop system.

Assumption 1. The uncertainty $\delta(q, \dot{q}, \ddot{q}, t)$ is bounded in Euclidian norm and so are its partial derivatives:

$$\|\delta(q, \dot{q}, \ddot{q}, t)\| \le \rho(q, \dot{q}, \ddot{q}, t)$$

where $\rho(q, \dot{q}, \ddot{q}, t)$ is the unknown bounding function of the uncertainties, $\|\cdot\|$ is Euclidian norm.

THE PROPOSED ADAPTIVE ROBUST CONTROLLER DESIGN METHODOLOGY

The standard robust sliding mode control law for a nonlinear uncertain system described by Eqs. (4) and (5), which guarantees the stability and convergence is a discontinuous and has the following form:

$$u(t) = u_{eq}(t) - K_b sgn(S) \tag{6}$$

where K_b is the bounds of uncertainty vector which have to be known a priori, and $sgn(S)$ is a signum function, $u(t)$ is the control input and can be divided into two components, $u_{eq}(t)$ which is the model-based component, and $u_{sw} = -K_b sgn(S)$ which is the discontinuous term, and $S(q, t)$ is the sliding variable which is defined as (Slotine and Li 1991):

$$S(q, t) = (\frac{d}{dt} + \Lambda)^2 (\int edt) = \dot{e} + 2\Lambda e + \Lambda^2 \int edt \tag{7}$$

where Λ is an $n \times n$ symmetric positive definite diagonal constant matrix and $e = q - q_d$, $\dot{e} = \dot{q} - \dot{q}_d$ are the tracking error and the rate of error respectively. q_d and q are the desired and measured joint variables respectively. The integral of error is included to ensure zero offset error.

The control objective is to force the system state trajectory to reach sliding surface $S(q, t) = 0$ in finite time and stay on the sliding surface $S(q, t) = 0$ for all subsequent time. To clarify the subject, the ideal and real sliding mode is defined as follows.

Definition 1 (Levant 1993). The motion that takes place strictly on $S(q, t) = 0$ is called **"ideal" sliding mode** and described by (Utkin 1977, 2013)

$$\lim_{s_i \to 0^+} \dot{s}_i < 0 \qquad \text{and} \qquad \lim_{s_i \to 0^-} \dot{s}_i > 0$$

In real world applications such as robot manipulator, due to uncertainties the "ideal" sliding mode concept as defined above, cannot be stablished. Therefore, it is required to introduce the concept of "real" sliding mode.

Definition 2 (Levant 1993). Given the sliding variable $S(q, t)$, the **"real" sliding mode** is called the motion that takes place in a small neighbourhood of the sliding surface $S(q, t) = 0$, with $\varepsilon > 0$ and is defined as:

$$|S(t)| < \varepsilon, \lim_{\varepsilon \to 0} |S(t)| = 0,$$

However, the application of control law given in Eq. (6) can be limited mainly due to chattering and unknown bounds of the uncertainties. The discontinuous term $u_{sw} = K_b \text{sgn}(S)$ is the fundamental cause of chattering and also is designed conservatively based on the bounds of uncertainties. In this study, to eliminate the chattering the discontinuous term will be replaced by an adaptive continuous term which is obtained through systematic Lyapunov design procedure. The task of the continuous sliding mode control design with the above mentioned characteristics and without knowing the bounds of uncertainties consists of two phases as follows.

- For the first phase of the design, an exponentially stable error dynamic is chosen as a desired sliding surface (i.e., $S(q, t) = \dot{e} + 2\Lambda e + \Lambda^2 \int edt = 0$) to guarantee the convergence of tracking error to a small neighbourhood of sliding surface.
- The second phase is to design a control law with variable parameters such that it guaranties the existence of real sliding mode, or makes the associated Lyapunov function a decreasing function of time in the presence of time-varying uncertainties.

Unlike the traditional SMC, and AC of robot manipulators, the proposed control law that guarantees achieving the control objectives is continuous and consist of two components:

1. Model-based component which is constructed based on approximately known dynamic model of the system and desired acceleration not the measured acceleration and is derive in Section **Control Design for Nominal System**, that is:

   $$u_{eq}(t) = \tau_n = \hat{M}\ddot{q}_r + \hat{C}(q, \dot{q}, t)\dot{q}_r + \hat{g}(q) - KS \tag{8}$$

   where $\ddot{q}_r$, $\dot{q}_r$ are the *reference acceleration* and *reference velocity*, respectively, and are defined in the sequel.
2. The second component is a continuous component constructed based on estimated uncertainty and state feedback control which replaces the discontinuous component u_{sw} in Eq. (6) and is derived in section as:

   $$\tau_c = \delta_{est} \tag{9}$$

   where δ_{est} is estimate of the lumped uncertainty term $\delta(q, \dot{q}, \ddot{q}, t)$ and is obtained from systematic Lyapunov design procedure in the next subsection, and $-KS$ is the term to improve the transient performance and enhance the closed-loop stability. Diagonal matrix K is positive definite constant and is a design parameter. Based on the above discussion the proposed ARC is designed as follows.

Control Design for Nominal System

In the first step, the controller design for the "nominal" system—that is, the system without uncertainty ($\delta(q, \dot{q}, \ddot{q}, t) = 0$)—is considered. The nominal dynamic of the robot manipulator is described by

$$\tau = \hat{M}\ddot{q} + \hat{C}(q, \dot{q}, t)\dot{q} + \hat{g}(q) \tag{10}$$

In this case the control law is designed such that it satisfies a sufficient condition for the existence and reachability of the trajectories to the small neighbourhood of sliding surface for the nominal system. This control law derives the system trajectories to the small neighbourhood of the sliding surface in a finite time, if $q(t_0)$ is actually off $q_d(t_0)$ (i.e., desired value of $q(t)$). One method to design a control law that derives the system trajectories to the sliding surface from any initial condition is the Lyapunov direct method.

Theorem 1. *Given the nominal system described by (10), with the sliding variable described by (7), and the following Lyapunov function, the asymptotic stability of closed-loop system is guaranteed, and there exist a finite time* t_F *such that a real sliding mode* *is stablished for all* $t > t_F$ (i.e., *all system trajectories converge to small neighborhood of sliding surface*).

$$V = \frac{1}{2}(S^T \hat{M} S) \tag{11}$$

For the sake of simplicity, the arguments of sliding variable $S(q, t)$ are eliminated. Now, taking the derivative of Lyapunov function (11) results in

$$\dot{V} = \frac{1}{2} S^T \dot{\hat{M}} S + S^T \hat{M} \dot{S} \tag{12}$$

From Eq. (7) the derivative of $S(t)$ is determined as

$$\dot{S} = \ddot{e} + 2\Lambda\dot{e} + \Lambda^2 e$$
$$= (\ddot{q} - (\ddot{q}_d - 2\Lambda\dot{e} - \Lambda^2 e)) \tag{13}$$

where $\ddot{q}_d - 2\Lambda\dot{e} - \Lambda^2 e$ in Eq. (13) is called the *reference acceleration* and denoted by $\ddot{q}_r$. Thus, the derivative of the sliding variable can be determined as

$$\dot{S} = \ddot{q} - \ddot{q}_r \tag{14}$$

Substituting (11) into (12) and using equation (10) yields

$$\dot{V} = \frac{1}{2} S^T \dot{\hat{M}} S + S^T \hat{M}(\ddot{q} - \ddot{q}_r) \tag{15}$$

$$\dot{V} = \frac{1}{2} S^T \dot{\hat{M}} S + S^T (\tau - \hat{C}\dot{q} - g) - S^T \hat{M} \ddot{q}_r) \tag{16}$$

From equation (7) i.e., $S = \dot{e} + 2\Lambda e + \Lambda^2 \int e dt$, one has the following

$$S = \dot{q} - \dot{q}_r \tag{17}$$

where $\dot{q}_r = \dot{q}_d - 2\Lambda e - \Lambda^2 \int e dt$, using equation (17) and replacing for $\dot{q}$ in equation (16), yields

$$\dot{V} = \frac{1}{2} S^T \dot{\hat{M}} S + S^T (\tau - \hat{C}(S + \dot{q}_r) - \hat{g} - \hat{M}\ddot{q}_r) \tag{18}$$

$$\dot{V} = \frac{1}{2} S^T (\dot{\hat{M}} - 2\hat{C}) S + S^T (\tau - (\hat{M}\ddot{q}_r + \hat{C}\dot{q}_r + \hat{g})) \tag{19}$$

Considering the skew symmetry property of matrix $(\dot{\hat{M}} - 2\hat{C})$, which is one of the structural properties of the robot manipulators dynamic model, the following holds

$$S^T(\dot{\hat{M}} - 2\hat{C})S = 0 \tag{20}$$

Therefore, the skew symmetry property can be used to eliminate the term $1/2\, S^T (\dot{\hat{M}} - 2\hat{C})S$ in Eq. (19) and $\dot{V}$ is as:

$$\dot{V} = S^T (\tau - (\hat{M}\ddot{q}_r + \hat{C}\dot{q}_r + \hat{g})) \tag{21}$$

In order to guarantee the stability of the system, derivative of Lyapunov function needs to be a negative definite function. Therefore, $\dot{V}$ can be chosen as follows:

$$\dot{V} = -S^T K S \leq 0 \tag{22}$$

$$\dot{V} = -S^T KS = S^T(\tau - \hat{C}\dot{q}_r - \hat{g} - \hat{M}\ddot{q}_r) \tag{23}$$

$$S^T(\tau - \hat{C}\dot{q}_r - \hat{g} - \hat{M}\ddot{q}_r) + S^T KS = 0 \tag{24}$$

$$S^T(\tau - \hat{C}\dot{q}_r - \hat{g} - \hat{M}\ddot{q}_r + KS) = 0 \tag{25}$$

By solving the above equation for the control input (i.e., τ), the following expression is obtained for τ, which is called nominal control and denoted with τ_n in this study, that is

$$\tau_n = \hat{M}\ddot{q}_r + \hat{C}\dot{q}_r + \hat{g} - KS \tag{26}$$

The above control input τ_n makes $\dot{V} \leq 0$ for nominal system, which guarantees that the nominal closed-loop system is uniformly asymptotically stable. Using Barbalat's lemma, it can be shown that the semi negative derivative of Lyapunov function guarantees the reachability of the system trajectories to small neighborhood of sliding surface in a finite time, which means real sliding mode (i.e., Definition 2) can be stablished in finite time, if $q(t_0)$ is actually off $q_d(t_0)$ (i.e., desired value of $q(t)$).

Remark 1. The proportional term KS in equation (26), is in agreement with practical consideration which dictates that the control input should be reduced as the state trajectory converge to the sliding surface. But, if the control law is designed based on the bound of uncertainties a high control input is imposed for all $t \geq t_0$.

Compensation of Time-Varying Uncertainties

In the second step, it is assumed that the lumped time-varying uncertainty present (i.e., $\delta(q, \dot{q}, \ddot{q}, t) \neq 0$). As it is described in Section **Problem Formulation and Definition** (Assumption 1), the uncertainties are bounded, but this bound is not known, and system dynamic described by Eq. (4). To compensate for the effect of uncertainties an additional adaptive control component is required. This additional component which is denoted by τ_a in this manuscript, should be designed such that the overall control $\tau = \tau_n + \tau_a$ satisfies a sufficient condition for existence and reachability of sliding mode for the dynamic (4). Since it is assumed that the matching condition is satisfied therefore, the uncertain term $\delta(q, \dot{q}, \ddot{q}, t)$ appears in the same channel the control input. For the sake of brevity, the arguments of the functions are dropped. Therefore, if δ were known, then including the term $\tau_a = \delta$ in the control law would compensate for the effects of all the uncertainties. Since δ is not known, the control component τ_a can be designed based on the estimate of the lumped uncertainty term δ, which is denoted by δ_{est}, hereafter. Based on the above discussion, including $\tau_a = \delta_{est}$ in the control law could compensate for the effects of the uncertainties, provided that the estimated uncertainty, δ_{est}, approaches δ as time tends to infinity. Therefore, τ_a is a variable (adaptive) term which is constructed based on online estimation of the uncertainties and defined as follows

$$\tau_a = \delta_{est} \tag{27}$$

By combining the components derived in the above steps the proposed robust adaptive control law is obtained as:

$$\tau = \tau_n + \delta_{est}$$

$$\tau = \hat{M}\ddot{q}_r + \hat{C}(q, \dot{q}, t)\dot{q}_r + \hat{g}(q) - KS + \delta_{est} \tag{28}$$

In the following section stability of the closed-loop system with the above proposed control law is analyzed and proved and an appropriate estimation mechanism also called update law is derived systematically through Lyapunov redesign method to estimate δ_{est} (Khalil, page 579). It will also be shown that real sliding mode can be reached in finite time.

STABILITY ANALYSIS OF CLOSED-LOOP SYSTEM IN THE PRESENCE OF TIME-VARYING UNCERTAINTIES

Theorem 2. *Given the MIMO nonlinear uncertain system (the multi-link robot manipulator) described by Eqs. (4), (5), with the sliding variable (7) and the robust control law of (28), and the adaptive law that will be derived later in this section, the stability of closed-loop system in the presence of model uncertainties and disturbances is guaranteed, and there exist a finite time t_r such that a real sliding mode is stablished for all $t > t_r$.*

Proof: To prove the robust stability of the proposed controller, the following Lyapunov function is considered.

$$V = \frac{1}{2}(S^T MS + \tilde{\delta}^T \Gamma^{-1} \tilde{\delta}) \tag{29}$$

where Γ is a symmetric positive definite diagonal constant, and

$$\tilde{\delta} = \delta_{est} - \delta \tag{30}$$

Taking the derivative of Lyapunov function (29) results in

$$\dot{V} = \frac{1}{2} S^T \dot{M} S + S^T M \dot{S} + \tilde{\delta}^T \Gamma^{-1} \dot{\tilde{\delta}} \tag{31}$$

From Eq. (14) the derivative of $S(t)$ is as

$$\dot{S} = \ddot{q} - \ddot{q}_r \tag{32}$$

Substituting (32) into (31) yields

$$\dot{V} = \frac{1}{2} S^T \dot{M} S + S^T M(\ddot{q} - \ddot{q}_r) + \tilde{\delta}^T \Gamma^{-1} \dot{\tilde{\delta}} \tag{33}$$

Substituting $M\ddot{q}$ from Eq. (1), Eq. (29) can be written as

$$\dot{V} = \frac{1}{2} S^T \dot{M} S + S^T(\tau - (C\dot{q} + F_f + g + d(t))) - S^T M \ddot{q}_r + \tilde{\delta}^T \Gamma^{-1} \dot{\tilde{\delta}} \tag{34}$$

Similar to the reference acceleration, another term, called the reference velocity, can be derived from the definition of sliding surface as follows

$$S = \dot{e} + 2\Lambda e + \Lambda^2 \int e dt \tag{35}$$

Replacing $\dot{e} = \dot{q} - \dot{q}_d$ in Eq. (35) yields

$$S = \dot{q} - (\dot{q}_d - 2\Lambda e - \Lambda^2 \int e dt) \tag{36}$$

Substituting $\dot{q}_r = \dot{q}_d - 2\Lambda e - \Lambda^2 \int e dt$ in Eq. (36) leads to

$$S = \dot{q} - \dot{q}_r \tag{37}$$

Replacing $\dot{q}$ in terms of $\dot{q}_r$ and S in Eq. (34), yields

$$\dot{V} = \frac{1}{2} S^T \dot{M} S + S^T[\tau - (C(S + \dot{q}_r) + F_f + g + d(t))] - S^T M \ddot{q}_r + \tilde{\delta}^T \Gamma^{-1} \dot{\tilde{\delta}} \tag{38}$$

Rearranging Eq. (34) yields

$$\dot{V} = \frac{1}{2} S^T (\dot{M} - 2C)S + S^T[\tau - (M\ddot{q}_r + C\dot{q}_r + F_f + g + d(t))] + \tilde{\delta}^T \Gamma^{-1} \dot{\tilde{\delta}} \tag{39}$$

As mentioned earlier in Eq. (20), using skew-symmetry property of robot matrix $(\dot{M} - 2C)$, the term $0.5S^T (\dot{M} - 2C)S$ in Eq. (39) is equal to zero, and thus Eq. (39) can be written as

$$\dot{V} = S^T [\tau - (M\ddot{q}_r + C\dot{q}_r + F_f + g + d(t))] + \tilde{\delta}^T \Gamma^{-1} \dot{\tilde{\delta}} \tag{40}$$

Substituting for control input from Eq. (28) into (40) yields

$$\dot{V} = S^T [(\hat{M}\ddot{q}_r + \hat{C}\dot{q}_r + \hat{g}(q) - KS + \delta_{est}) - (M\ddot{q}_r + C\dot{q}_r + F_f + g(q) + d(t))] + \tilde{\delta}^T \Gamma^{-1} \dot{\tilde{\delta}}$$

$$\dot{V} = S^T [\hat{M}\ddot{q}_r + \hat{C}\dot{q}_r + \hat{g}(q) - KS + \delta_{est} - (M\ddot{q}_r + C\dot{q}_r + F_f + g(q) + d(t))] + \tilde{\delta}^T \Gamma^{-1} \dot{\tilde{\delta}} \tag{41}$$

$$\dot{V} = S^T [((\hat{M} - M)\ddot{q}_r + (\hat{C} - C)\dot{q}_r + (\hat{g} - g) - F_f(\dot{q}, t) - d(t)) - KS + \delta_{est}] + \tilde{\delta}^T \Gamma^{-1} \dot{\tilde{\delta}}$$

$$\dot{V} = S^T [-\delta - KS + \delta_{est}] + \tilde{\delta}^T \Gamma^{-1} \dot{\tilde{\delta}} \tag{42}$$

$$\dot{V} = -S^T KS + S^T \tilde{\delta} + \tilde{\delta}^T \Gamma^{-1} \dot{\tilde{\delta}}$$

From equation (30) $\dot{\tilde{\delta}} = \dot{\delta}_{est} - \dot{\delta}$, substituting this into equation (42) yields

$$\dot{V} = -S^T KS + S^T \tilde{\delta} + \tilde{\delta}^T \Gamma^{-1} (\dot{\delta}_{est} - \dot{\delta}) \tag{43}$$

The above equation implies that if the adaptive law is chosen as

$$\dot{\delta}_{est} = -\Gamma S \tag{44}$$

Equation (43) can be written as

$$\dot{V} = -S^T KS + S^T \tilde{\delta} - S^T \Gamma \Gamma^{-1} \tilde{\delta} - \tilde{\delta}^T \Gamma^{-1} \dot{\delta} \tag{45}$$

Equating $\Gamma\Gamma^{-1} = I$ (I is identity matrix of appropriate size), using the symmetric property of matrix Γ, results in

$$\dot{V} = -S^T KS - \tilde{\delta}^T \Gamma^{-1} \dot{\delta} \tag{46}$$

The second term of Eq. (46), i.e., $\tilde{\delta}^T \Gamma^{-1} \dot{\delta}$ shows the combined effects of the estimation error, estimation gain and the rate of change of the uncertainties with respect to time on $\dot{V}$. Since the complete knowledge of the above term is not available, the best possible way is worse case analysis, when $\tilde{\delta}^T \Gamma^{-1} \dot{\delta} < 0$. Therefore, Eq. (46) can be written as follows:

$$\dot{V} = -S^T KS - \tilde{\delta}^T \Gamma^{-1} \dot{\delta} \leq -S^T KS + \|\tilde{\delta}^T \Gamma^{-1} \dot{\delta}\| \tag{47}$$

As mentioned earlier K and Γ symmetric positive definite diagonal matrices and the following relationship hold

$$\lambda_{min}(K)||S||^2 \leq S^T KS \leq \lambda_{max}(K)||S||^2 \tag{48}$$

where $\lambda_{min}(K)$, and $\lambda_{max}(K)$ denote the minimum and maximum eigenvalues of the matrix K, respectively. Thus, Eq. (48) can be written as

$$\dot{V} \leq -\lambda_{min}(K)||S||^2 + ||\tilde{\delta}^T \Gamma^{-1} \dot{\delta}|| \leq -\lambda_{min}(K)||S||^2 + ||\tilde{\delta}^T \dot{\delta}||\ ||\Gamma^{-1}|| \tag{49}$$

$$\dot{V} \leq -\lambda_{min}(K)||S||^2 + \frac{||\tilde{\delta}^T \dot{\delta}||}{\lambda_{min}(\Gamma)}$$

From Eq. (45), $\dot{V} < 0$ if

$$\lambda_{min}(K)||S||^2 > \frac{||\tilde{\delta}^T \dot{\delta}||}{\lambda_{min}(\Gamma)} \tag{50}$$

or, equivalently

$$\dot{V} < 0 \text{ if } ||S|| > \left(\frac{||\tilde{\delta}^T \dot{\delta}||}{\lambda_{min}(K)\,\lambda_{min}(\Gamma)}\right)^{\frac{1}{2}} > \varepsilon \tag{51}$$

where ε is a positive scalar value and as can be seen from (51) its magnitude depends on the choice of K and Γ. The designer of the controller can make it arbitrarily small by using the design parameters, e.g., large K and Γ, and hence, the magnitude of $S(t)$ for which $\dot{V}$ is negative definite, can be reduced arbitrarily using the design parameters K and Γ in the presence of time-varying uncertainties. Equation (51) also implies that:

Case 1. If initially $||S|| > \varepsilon$, all trajectories will converge to the ε-vicinity of the sliding surface $S(t) = 0$ in finite time, because, $\dot{V}$ is negative definite for all $||S|| > \varepsilon$ and real sliding mode is stablished. Therefore, the finite time convergence of all solutions to a domain $||S|| \leq \varepsilon$ is guaranteed from any initial condition $||S(0)|| > \varepsilon$.

Case 2. If initially $||S|| < \varepsilon$, then $\dot{V}$ is sign indefinite and $||S||$ can increase over ε. Once $||S|| > \varepsilon$ situation will return to the Case 1 and $\dot{V}$ become negative definite. Therefore, closed-loop system is stable and all solutions of the closed-loop system are uniformly ultimately bounded.

The above behavior of the proposed controller is illustrated using the simulation results for two links robot, Fig. 1. As can be seen the system trajectory is converged to small vicinity of the $S(t) = 0$ and real sliding mode is stablished in finite time.

Case 3. Equation (46) implies that, for the slowly time-varying uncertainties (i.e., the variation of uncertainty with respect to time $\dot{\delta} \approx 0$), one can achieve asymptotic stability with the proposed controller and Eq. (46) can be written as:

$$\dot{V} = -S^T KS \leq 0 \tag{52}$$

Condition (52) implies that the system trajectories can reach small neighborhood of sliding surface in finite time and the real sliding mode is stablished. Furthermore, the system trajectories will reach the surface $S = 0$ as time tends to infinity. The only concern is that the system cannot get "stuck" at a stage where $\dot{V} = 0$, while q does not equal q_d. This problem can be solved readily using Barbalat's lemma. The reader is referred to Slotine and Li 1991 for more detail.

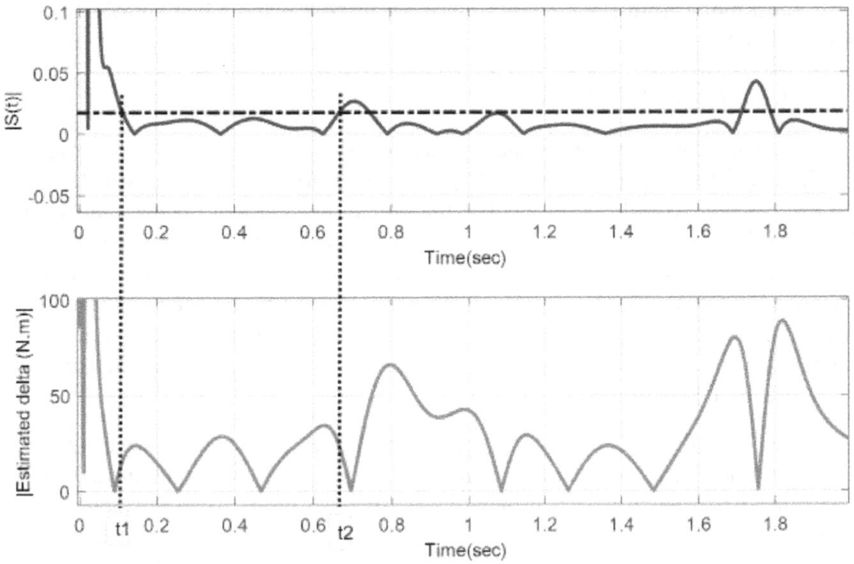

Figure 1. Behavior of sliding variable $S(t)$, and estimated uncertainty δ_{est} vs. time around ε-vicinity.

Discussion. The proposed adaptive robust control law of (28) along with adaptive law of (44), that is:

$$\begin{cases} \tau = \tau_n + \delta_{est} \\ \tau = \hat{M}\ddot{q}_r + \hat{C}(q, \dot{q}, t)\dot{q}_r + \hat{g}(q) - KS + \delta_{est} \\ \dot{\delta}_{est} = -\Gamma S \end{cases} \tag{53}$$

Works as follows:

- As can be seen from Fig. 1, there exist a time instant t_1 such that $|S(t_1)| < \varepsilon$, which means the sliding variable reached to the ε-domain of sliding surface and a real sliding mode is stablished. This implies that the estimated uncertainty term plus proportional term $\delta_{est} + KS$ is large enough to compensate for the bounded uncertainty and estimated uncertainty term will increase due to adaptive law of (44) until the sliding variable converges to $S(t) = 0$, although proportional term decreases as S(t) tends to zero.
- At this point $S(t)$ crosses the sliding surface and its sign changes. The sign change causes δ_{est} to decrease and $S(t)$ starts increasing as well as the proportional term KS, when $\|S\| > \varepsilon$, situation will return to Case 1 explained above.
- However, if the time-varying uncertainty exceeds some values that makes $|S(t_2)| > \varepsilon$ at time t_2 (see Fig. 1), then δ_{est} and KS will increase up to the value that can counteract the uncertainty and makes it converge to ε-domain again, and this is the next reaching time.

To complete the derivation of the proposed control law, the estimated lumped uncertainty can be obtained simply by integrating Eq. (44), as follows

$$\delta_{est} = -\Gamma \int S dt \tag{54}$$

Substituting from Eq. (55) into Eq. (28), the following adaptive robust control law is obtained

$$\tau = \hat{M}\ddot{q}_r + \hat{C}\dot{q}_r + \hat{g} - KS - \Gamma \int S dt \tag{55}$$

Equation (54) shows that control torque consists of two main components: (i) a model-based component, i.e., $\tau_m = \hat{M}\ddot{q}_r + \hat{C}\dot{q}_r + \hat{g}$, which can be constructed based on the available knowledge of the dynamic model of the robot manipulator, without knowing the model parameters; and (ii) a sliding function-based proportional-integral component $\tau_{PI} = -KS - \Gamma\int S dt$, which is constructed based on dynamic behavior of sliding function. Equation (54) also shows that the proposed control law can be implemented using approximately known inertia matrix, centripetal matrix and gravity terms and PI component accounts for the parametric uncertainties and non-parametric uncertainties such as external disturbances.

Remark 2. The main features of the proposed adaptive robust control is that the control law is continuous, thus chattering is eliminated, and there is no need to find a boundary layer thickness, which is very difficult to be obtained properly. In this approach the uncertainty is estimated on-line with a proper choice of the adaptation gain and using adaptation law given in Eq. (40). Therefore, a priori knowledge of the uncertainty bounds is not needed.

SIMULATION RESULTS

In this section, the performance of the proposed controller is evaluated by simulation study using analytical dynamic model of a 2-DOF serial manipulator, Fig. 2. The dynamic model of the manipulator is given below (Slotine and Li 1991, page 396). This is because a 2-DOF planar serial robot manipulator is widely used in the literature for illustration of the performance of the developed controller Slotine and Li 1991, Seraji 1987, Craig et al. 1987, and therefore, simulation results can be compared with the existing results.

The inverse dynamic model of a 2-DOF robot manipulator is given as

$$\begin{bmatrix}\tau_1\\ \tau_2\end{bmatrix} = \begin{bmatrix}M_{11} & M_{12}\\ M_{21} & M_{22}\end{bmatrix}\begin{bmatrix}\ddot{q}_1\\ \ddot{q}_2\end{bmatrix} + \begin{bmatrix}-C\dot{q}_2 & -C(\dot{q}_1+\dot{q}_2)\\ C\dot{q}_1 & 0\end{bmatrix}\begin{bmatrix}\dot{q}_1\\ \dot{q}_2\end{bmatrix} + \begin{bmatrix}F_{fr1}\\ F_{fr2}\end{bmatrix} + \begin{bmatrix}g_1(t)\\ g_2(t)\end{bmatrix} + \begin{bmatrix}d_1(t)\\ d_2(t)\end{bmatrix} \tag{56}$$

where

$$\begin{cases} M_{11} = a_1 + 2a_3\cos(q_2) - 2a_4\sin(q_2)\\ M_{12} = M_{21} = a_2 + a_3\cos(q_2) + a_4\sin(q_2)\\ M_{22} = a_2\\ \begin{bmatrix}F_{fr1}\\ F_{fr2}\end{bmatrix} = \begin{bmatrix}F_{c1} & 0\\ 0 & F_{c2}\end{bmatrix}\begin{bmatrix}\mathrm{sgn}(\dot{q}_1)\\ \mathrm{sgn}(\dot{q}_2)\end{bmatrix} + \begin{bmatrix}v_1 & 0\\ 0 & v_2\end{bmatrix}\begin{bmatrix}\dot{q}_1\\ \dot{q}_2\end{bmatrix}\\ C = a_3\sin(q_2) - a_4\cos(q_2)\\ g_1 = a_5\cos(q_1) + a_6\cos(\delta_e + q_1 + q_2)\\ g_2 = a_6\cos(\delta_e + q_1 + q_2)\end{cases} \tag{57}$$

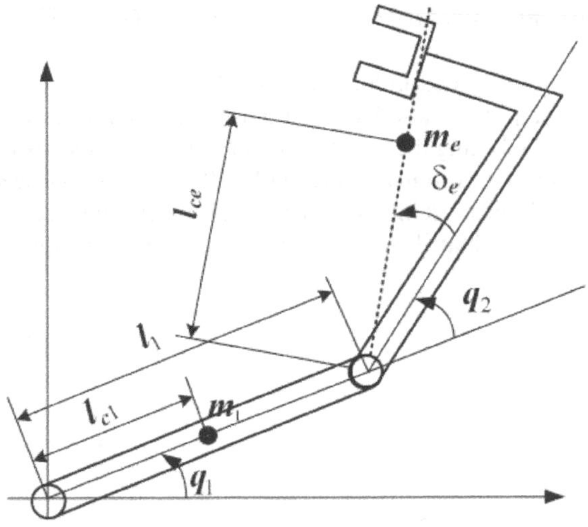

Figure 2. 2-DOF serial manipulator (Zeinali and Notash 2010).

with

$$\begin{cases} a_1 = I_1 + m_1 l_{c1}^2 + I_e + m_e l_{ce}^2 + m_e l_1^2 \\ a_2 = I_e + m_e l_{ce}^2 \\ a_3 = m_e l_{ce} l_1 \cos \delta_e \\ a_4 = m_e l_{ce} l_1 \sin \delta_e \\ a_5 = (m_1 l_{c1} + m_e l_1) g \\ a_6 = m_e l_{ce} g \end{cases} \tag{58}$$

where $[\delta_e \ v_1 \ v_2 \ F_{c1} \ F_{c2}]^T = [30^0 \ 5.5 \ 2.7 \ 5 \ 5]^T$, and the robot is modelled as two rigid links with the following parameters:

$$\begin{cases} m_1 = 1 \text{ [kg]} \ I_1 = 0.12\text{[kg.m}^2\text{]} \quad l_1 = 1.0 \text{ [m]} \quad l_{c1} = 0.5\text{[m]} \\ m_e = 2 \text{ [kg]} \ I_e = 0.25\text{[kg.m}^2\text{]} \quad l_{ce} = 0.25\text{[m]} \end{cases} \tag{59}$$

where m_1 and m_e are respectively the masses of links one and two concentrated at the center of mass of the corresponding links as shown in Fig. 5.3 (link two and end effector are considered as one rigid body), I_1 and I_e are the moments of inertia relative to the centers of mass of the corresponding link, l_1 is the length of link one, and l_{c1} and l_{ce} are the distance of centre of mass of the two links from the respective joint axes, as indicated in Fig. 2. The viscous friction coefficients ν_i and Coulomb friction forces F_{ci} are included at the two joints.

In the first part of the simulation, the performance of the controller in terms of trajectory tracking capability and eliminating the chattering is investigated using the following trajectories.

$$q_d^{\mathrm{T}} = [{}^{\pi}/_{6}(1 - \cos(2\pi t) + \sin(4\pi t)) \;\; {}^{\pi}/_{4}(1 - \cos(2\pi t) + \cos(1.5\pi t))]^{\mathrm{T}} \tag{60}$$

The results are illustrated in Fig. 3(a) and 3(b), and 4. As shown in Fig. 3, the tracking a desired trajectory described by Eq. (60) from non-zero initial condition is successfully performed using the proposed controller. The associated sliding function (i.e., $S(t)$) of joint one is illustrated in Fig. 4 (Top Row). As can be seen the chattering phenomenon which is the most important issue that limiting the practical application of the standard SMC is eliminated while the performance and robustness of the proposed controller is preserved.

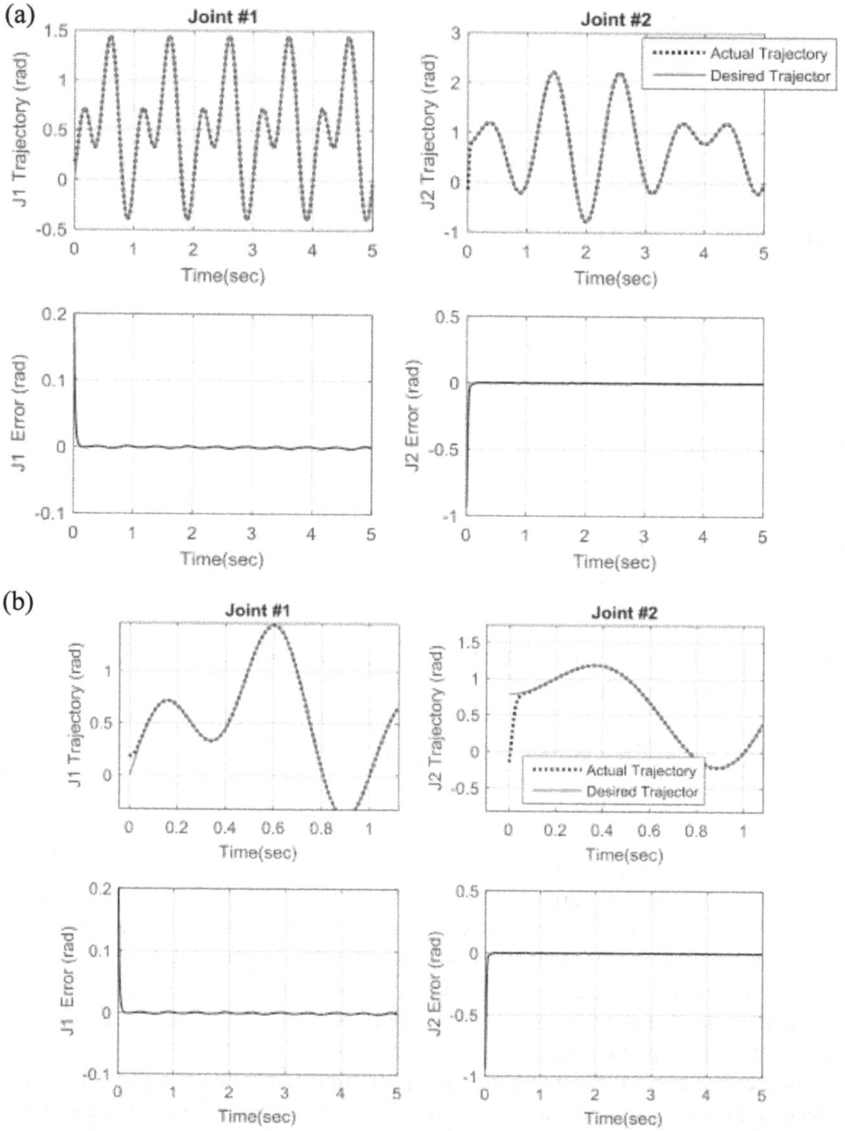

Figure 3. Trajectory tracking performance of the proposed controller from non-zero initial condition.

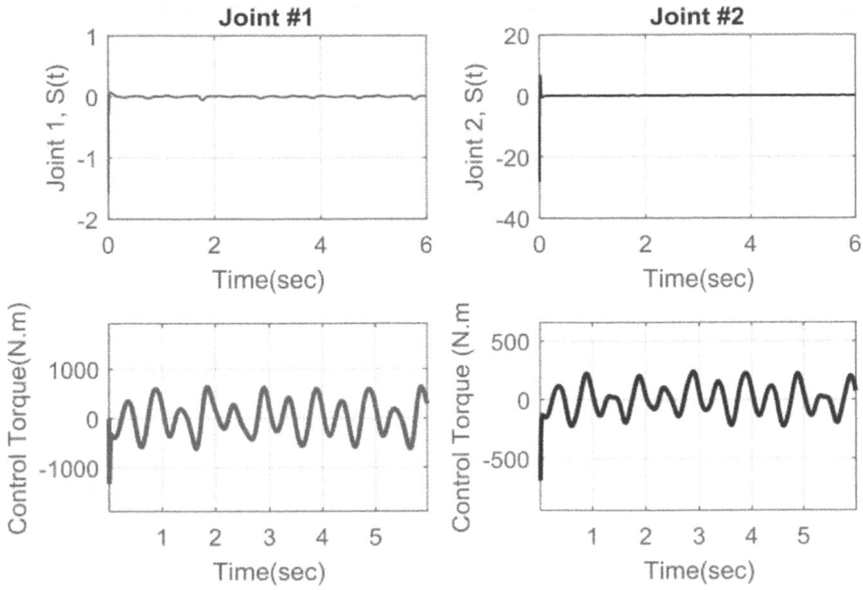

Figure 4. Convergence of system trajectories to $S(t) = 0$ (sliding surface) in finite time: (top Row) Sliding function behavior in time domain; (bottom Row) Control input in time domain.

At this point it is aimed to demonstrate the performance, adaptive capability, and robustness of the proposed controller, in the presence of the following external disturbances.

$$\begin{bmatrix} d_1(t) \\ d_2(t) \end{bmatrix} = \begin{bmatrix} 200 + 20\sin(2\pi t) \\ 100 + 15\sin(3\pi t) \end{bmatrix} \tag{61}$$

As shown in Fig. 5, the manipulator is disturbed by the external disturbance $d(t)$, at $t = 2.5$ seconds, to evaluate the stability and robustness of the closed-loop system. As can be seen from Fig. 5(a) and (b), large model uncertainties, and external disturbances are tolerated by the closed-loop system. The results also show how tracking error rapidly converges to zero after the manipulator is disturbed by the external disturbance $d(t)$, at $t = 2.5$, which is an indication of the adaptive capability of the proposed controller.

Figure 6 shows a comparison of the estimated uncertainty with the model uncertainty. As can be seen, the proposed uncertainty estimator, i.e., $\delta_{est} = -\Gamma\int Sdt$, can successfully track the time-varying complex disturbances. This implies that at the steady state, the uncertainty estimation is only subjected to a small estimation error, which can be compensated for by the proportional term, i.e., term KS in Eq. (28) of the control law.

In the second part of the simulation, trajectory tracking performance and disturbance attenuation capability of the proposed controller is compared with Slotine and Li adaptive control.

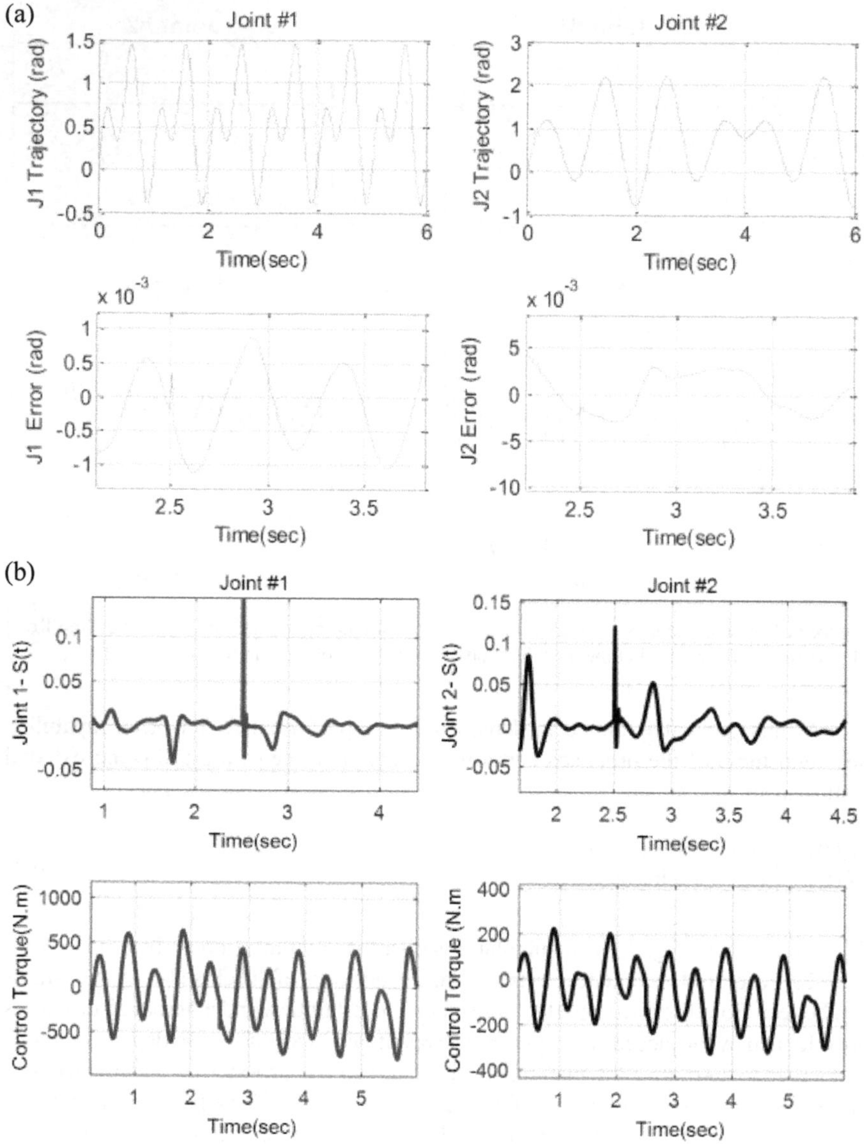

Figure 5. Trajectory tracking performance of the proposed controller in the presence of large external disturbances.

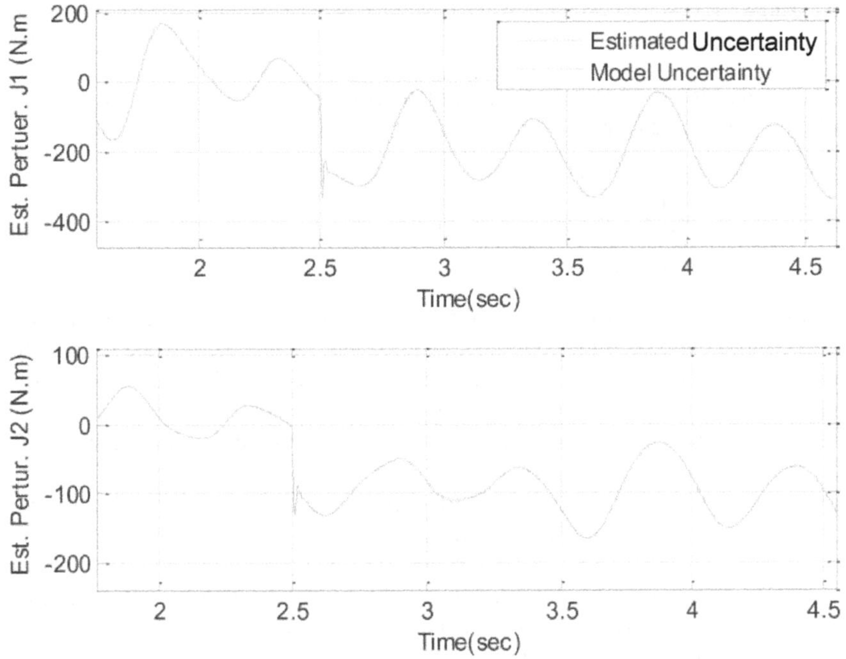

Figure 6. Comparison of model uncertainty and estimated uncertainties for joints one and two.

Comparison of the Proposed Control and Slotine and Li Adaptive Control

The control law of the SLAC method introduced by (Slotine and Li 1988) is as follows:

$$\tau = Y(q, \dot{q}, \dot{q}_r, \ddot{q}_r)\,\hat{\Phi} - KS \tag{62}$$

where matrix function $Y(q, \dot{q}, \dot{q}_r, \ddot{q}_r)$ is called the regressor, and $\hat{\Phi}$ is the estimated vector of the dynamic model parameters Φ which was defined in Eq. (2). In Slotine and Li 1991, for the 2-DOF example robot manipulator, the following parameters, which are defined in Eq. (58), are considered as the unknown parameters to be estimated.

$$\hat{\Phi} = [a_1\ a_2\ a_3\ a_4]^{\mathrm{T}} \text{ with } \hat{\Phi}\,(0) = 0 \tag{63}$$

The parameter update law of SLAC method is:

$$\dot{\hat{\Phi}} = -\Gamma_{\mathrm{SL}} Y(q, \dot{q}, \dot{q}_r, \ddot{q}_r) S \tag{64}$$

where Γ_{SL} is a symmetric positive definite matrix, and $\dot{q}_r$, $\ddot{q}_r$ and S are defined as:

$$S = \begin{bmatrix} \dot{e}_1 + \lambda_1 e_1 \\ \dot{e}_2 + \lambda_2 e_2 \end{bmatrix}, \dot{q}_r = \begin{bmatrix} \dot{q}_{\mathrm{d}1} - \lambda_1 e_1 \\ \dot{q}_{\mathrm{d}2} - \lambda_2 e_2 \end{bmatrix} \text{ and } \ddot{q}_r = \begin{bmatrix} \ddot{q}_{\mathrm{d}1} + \lambda_1 \dot{e}_1 \\ \ddot{q}_{\mathrm{d}2} + \lambda_2 \dot{e}_2 \end{bmatrix} \tag{65}$$

where $\lambda_1 = 50$; $\lambda_2 = 15$. In order to have the same basis for comparison of SLAC and the controller proposed in Eq. (28), the sliding function defined in Eq. (7) are used for both controller without the integral term of the sliding function. Furthermore, the same gain matrix K are chosen for both controllers as:

$$K = \begin{bmatrix} 300 & 0 \\ 0 & 200 \end{bmatrix} \tag{66}$$

The adaptation gains for SLAC controller are extracted from Slotine and Li (1991), and for the proposed controller are chosen as:

$$\Gamma_{SL} = \begin{bmatrix} 0.2 & 0 & 0 & 0 \\ 0 & 0.15 & 0 & 0 \\ 0 & 0 & 0.1 & 0 \\ 0 & 0 & 0 & 0.2 \end{bmatrix} \text{ and } \Gamma_{ARC} = \begin{bmatrix} 25000 & 0 \\ 0 & 25000 \end{bmatrix} \tag{67}$$

In this comparison study, the goal is to show that SLAC controller results in an offset error in the case of the consistent disturbances (undamped), while the adaptive capability of adaptive robust controller compensates for the effects of the consistent disturbances. For both control laws the simulation is conducted ***without disturbances*** and ***with disturbances*** as shown in Figs. 7 and 8. The disturbance vector defined in Eq. (64) applied at $t = 2.5$ second.

As can be seen from Fig. 7, before applying disturbance the performance of both methods in terms of the tracking errors are almost the same. But in the presence of consistent disturbance SLAC controller has considerable tracking error which depends on the amplitude of disturbance and parameter adaptation, while the tracking performance of the proposed method remains unchanged. The comparison of the two methods with disturbance is illustrated in Fig. 8.

The comparison of the two methods with disturbance is illustrated in Figure 8.

$$\begin{cases} \mathbf{d}(t) = \begin{bmatrix} d_1(t) \\ d_2(t) \end{bmatrix} = \begin{bmatrix} 0 \\ 0 \end{bmatrix} & t < 2.5 \\ \mathbf{d}(t) = \begin{bmatrix} d_1(t) \\ d_2(t) \end{bmatrix} = \begin{bmatrix} 200 + 20\sin(2\pi t) \\ 100 + 15\sin(3\pi t) \end{bmatrix} & t > 2.5 \end{cases} \tag{68}$$

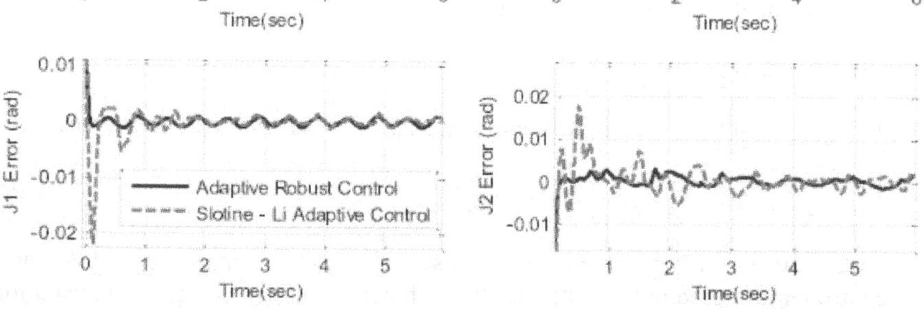

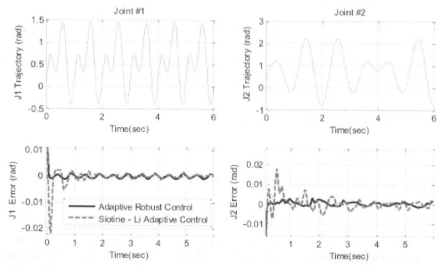

Figure 7. Comparison of the proposed Adaptive Robust Control with SLAC before applying a disturbance.

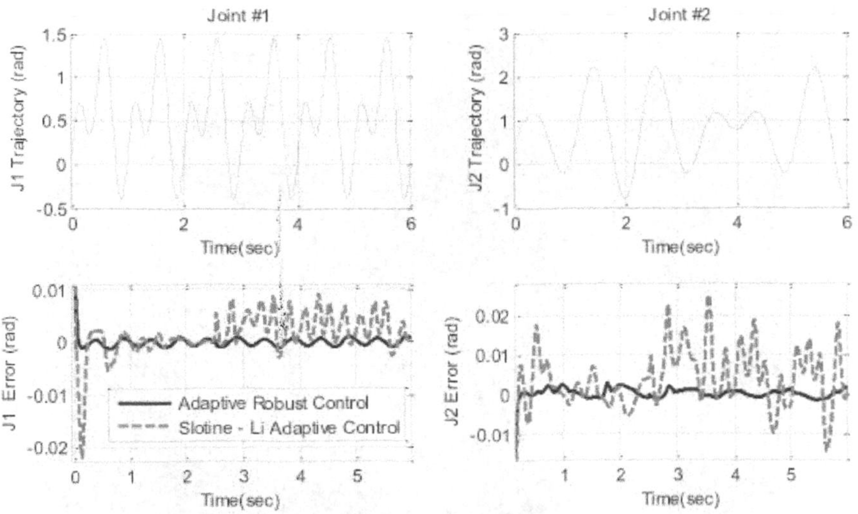

Figure 8. Comparison of proposed ARC with SLAC Adaptive Control after applying disturbance.

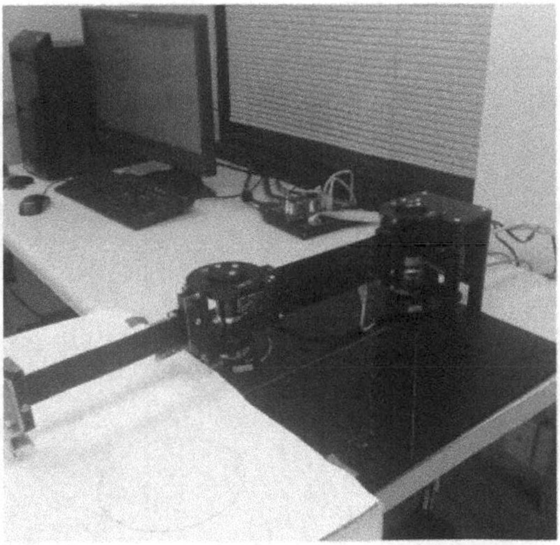

Figure 9. The 2-DOF serial flexible link (COURTESY OF QANSER Inc.).

EXPERIMENTAL RESULTS AND APPLICATION TO A TWO-DOF FLEXIBLE LINK ROBOT

The experimental test-bed used to validate the controller is an open-architecture 2-DOF serial flexible link robot (2D-SFLR), which has been designed and built by QUANSER Company. Figure 9 shows the general configuration of the 2-DOF SFLR. This experimental system consists of two DC motors each driving harmonic gearboxes (zero backlash), and two serial flexible linkages. As a result of link flexibility, vibration in the manipulator appears during the control, which in turn, makes the control more complex than rigid-link robots.

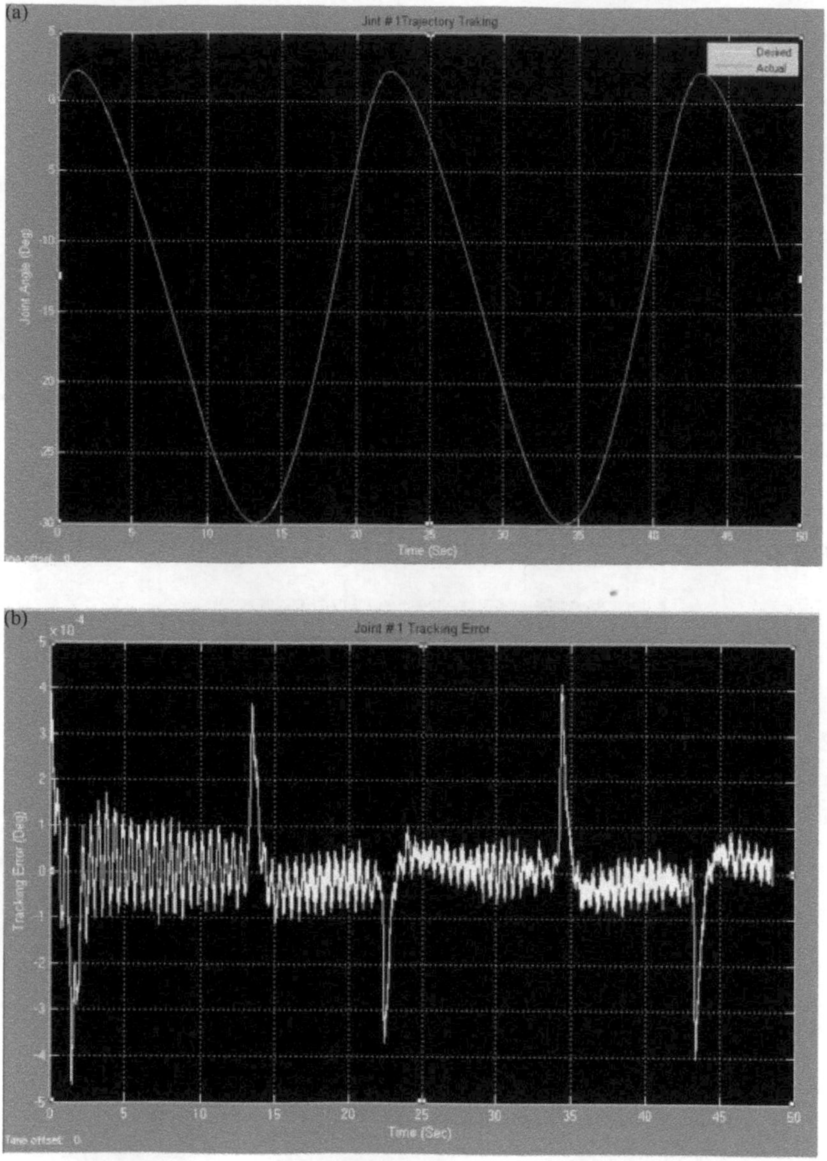

Figure 10. Trajectory Tracking Performance of the Proposed Controller: (a) Joint one angle; and (b) Joint one tracking error.

To demonstrate the performance, adaptive capability and robustness of the developed controller extensive experimental studies were performed using 2-DOF robot. In all experiments it is assumed that there exists very large parametric uncertainty (90% error in robot dynamic parameters) and an external disturbance in the form of external torque which is equal to joint torque. Figures 10(a), and (b), illustrate the tracking performance of the proposed controllers for a sinusoidal trajectory for both joint one and joint two in the presence of the model uncertainty. Figure 11 show the trajectory

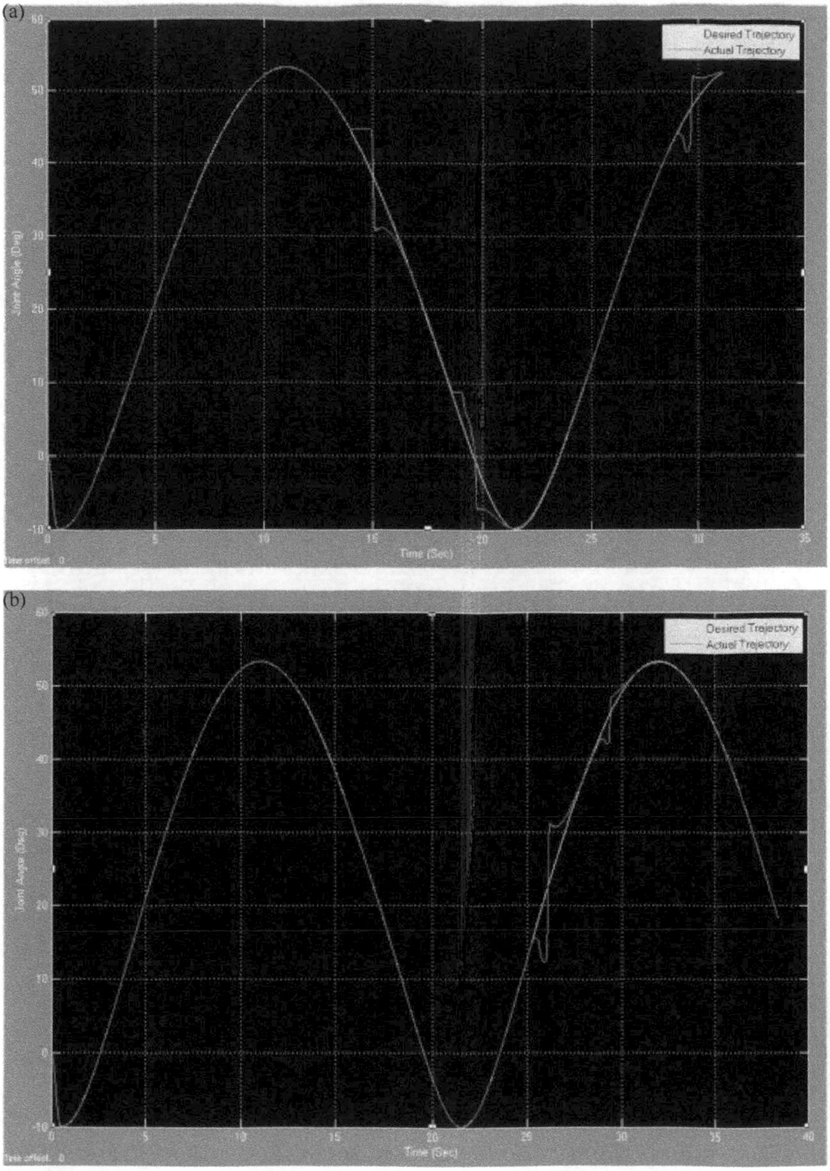

Figure 11. The Trajectory tracking performance of the proposed controller in the presence of severe external disturbances: (a) joint one; (b) Joint two.

tracking performance, and robustness of the proposed controller in the presence of the external disturbances. The results also confirmed that the proposed ARC completely eliminates the chattering, which is illustrated in Fig. 12. Unlike to the boundary layer method, in this method chattering eliminated while accuracy and performance were not compromised.

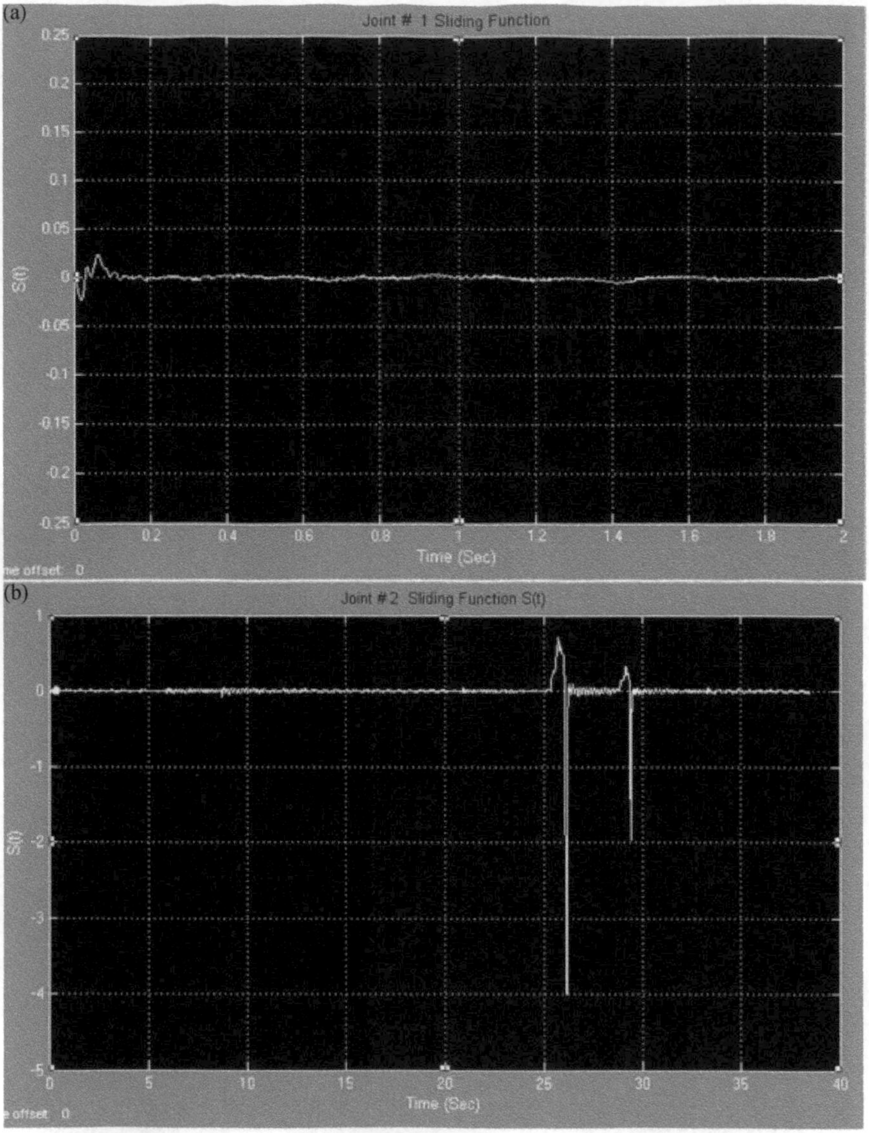

Figure 12: Convergence of sliding function $S(t)$ to zero, and staying in small neighborhood of $S(t) = 0$, also there is no chattering: (a) Joint one sliding function convergence; (b) Joint two sliding function convergence after applying disturbance two times at about $t = 26$ sec and $t = 29$ sec.

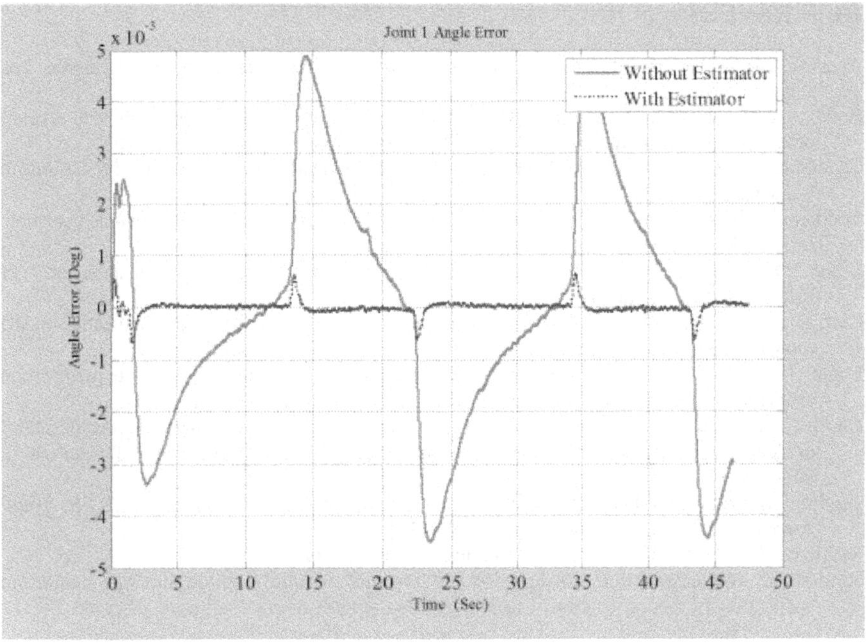

Figure 13. Comparison of tracking error with and without online estimation of the uncertainties.

CONCLUSION

In this work a systematic methodology to design an adaptive robust control using fundamentals of sliding mode control, adaptive control and Lyapunov redesign method for real-time applications of multi-link robot manipulator is developed and tested. The proposed method uses the advantages of sliding mode, adaptive control and PID techniques, while the disadvantages attributed to these methods are remedied by each other. The main contributions of the proposed approach are as follows. (1) It is proven that in the presence of fast time-varying uncertainties all trajectories will approaches the ε vicinity of the sliding surface $S(t) = 0$, because, $\dot{V}$ is negative definite for all $\|S\| > \varepsilon$, i.e., Eq. (51). (2) The controller is suitable for real-time applications due to simplicity in terms of implementation and computation burden. The proposed controller can be implemented with or without any knowledge of the manipulator dynamics and the size of the uncertainties. (3) The developed continuous adaptive robust control law can effectively eliminate the chattering. Simulation and experimental results confirmed that the proposed controller performs remarkably well in terms of the robustness, tracking error convergence and disturbance attenuation.

ACKNOWLEDGMENT

The author would like to acknowledge the financial support of Research, Development & Creativity Office (RDC Office) of Laurentian University.

REFERENCES

Angeles, J. 2003. Fundamental of robotic mechanical systems, theory, methods, and algorithms, 2nd Edition, Springer.

Boiko, I., L. Fridman, A. Pisano and E. Usai. 2007. Analysis of chattering in systems with second-order sliding modes. Automatic Control, IEEE Transactions on, 52(11): 2085–2102.

Chandrasekharan, P. C. 1996. Robust Control of Linear Dynamical Systems. Academic Press Limited, London.

Codourey, A. 1998. Dynamic modeling of parallel robots for computed-torque control implementation. Int. J. of Robotics Research, 17(2): 1325–1336.

Craig, J. J., P. Hsu and S. S. Sastry. 1987. Adaptive control of mechanical manipulators. Int. J. of Robotics Research, 6(2): 16–28.

Curk, B. and K. Jezernik. 2001. Sliding mode control with perturbation estimation: application on DD robot mechanism. Robotica, 19: 641–648.

Elmali, H. and N. Olgac. 1992. Theory and implementation of sliding mode control with perturbation estimation. IEEE, Int. Conf. on Robotics and Automation, 3: 2114–2119.

Ghorbel, F. H., O. Chételat, R. Gunawardana and R. Longchamp. 2000. Modeling and set point control of closed-chain mechanisms: Theory and experiment. IEEE Trans. Contr. Syst. Technol., 8(5): 801–815.

Gutman, S. 1979. Uncertain dynamical systems: a Lyapunov min-max approach. IEEE, Trans. Automatic Control AC-24, Issue 3, pp. 437–443.

Khalil, K. H. 2002. Nonlinear Systems (3rd Edition), Prentice Hall.

Kim, N., C. W. Lee and P. H. Chang. 1998. Sliding mode control with perturbation estimation: application to motion control of parallel manipulators. Control Engineering Practice, 6(11): 1321–1330.

Lee, H. and V. I. Utkin. 2007. Chattering suppression methods in sliding mode control systems. Annual Reviews in Control, 31: 179–188.

Levant, A. 1993. Sliding order and sliding accuracy in sliding mode control. Int. J. Control, 58: 1247–1263.

Man, Z., A. P. Poplinsky and H. R. Wu. 1994. A robust terminal sliding-mode control scheme for rigid robot manipulators. IEEE Trans. Autom. Control, 39(12): 2465–2469.

Sadeghn, N. and R. Horowitz. 1987. Stability analysis of an adaptive controller for robotic manipulators. Proc. IEEE Int. Conf. on Robotics and Automation, 3: 1223–1229.

Saengdeejing, A. and Z. Qu. 2002. Recursive estimation of unstructured uncertainty and robust control design. Proc. IEEE Conf. on Decision and Control, pp. 2220–2225.

Sanchez-Pena, R. and M. Sznaier. 1998. Robust systems theory and applications, John Wiley and Sons Inc.

Seraji, H. 1987. A new approach to adaptive control of manipulators. ASME, J. of Dynamic Systems, Measurement and Control, 109(3): 193–202.

Slotine, J. J. E. and W. Li. 1991. Applied Nonlinear Control, Prentice-Hall Inc.

Utkin, V. 1977. Variable structure systems with sliding modes. IEEE, Trans. Auto. Control, AC-22, No. 2, pp. 212–222.

Young, K. D., V. I. Utkin and U. Ozguner. 1999. A control engineer's guide to sliding mode control. IEEE Trans. on Cont. Systems Technology, 7(3): 328–342.

Yu, T. 1998. Terminal sliding-mode control for rigid robots. Automatica, 34(1): 51–56.

Zeinali, M. and L. Notash. 2010. Adaptive sliding mode control with uncertainty estimator for robot manipulators. Journal of Mechanism and Machine Theory, 45(1): 80–90.

Zeinali, M. 2015, Adaptive chattering-free sliding mode control design for robot manipulators based on online estimation of uncertainties and its experimental verification. Journal of Mechatronics, 3(2): 85–97.

Zhou, K., C. J. Doyle and K. Glover. 1996. Robust and Optimal Control, Prentice-Hall Inc., New Jersey.

Zhou, K., C. J. Doyle and K. Glover. 1996. Robust Optimal Control, Prentice-Hall, Inc.

16

Evaluation of Microgenetic and Microimmune Algorithms for Solving Inverse Kinematics of Hyper-redundant Robotic Manipulators On-line

D. Dužanec,[1] *S. Glumac,*[2] *Z. Kovačić*[3,*] and *M. Pavčević*[4]

ABSTRACT

This chapter concerns with evaluation of the micro-genetic and micro-immune optimization algorithms for solving inverse kinematics of a redundant robotic arm as a prerequisite for adaptive trajectory planning in various manipulation tasks. To make sure the quickest search of the optimal solution, the structure and parameters of genetic algorithms might be optimized. The advantage of the micro-genetic algorithms is that they work with smaller genetic populations than standard genetic algorithms. Describe the method of determining the optimal (best possible) mutation interval for micro-genetic algorithms based on the analysis of their behaviour. Then define a multi-objective criterion to compare performances of micro-immune and micro-genetic algorithms. Use multi-objective optimization to determine a set of best algorithms, and the micro-immune algorithm proved superior in solving a selected optimization

[1] S Tempera d.o.o., Construction and Development Department, Zagreb, Croatia.

[2] AVL d.o.o., Zagreb, Croatia.

[3] University of Zagreb, Faculty of EE&C, Department of Control and Computer Engineering, Zagreb, Croatia.

[4] University of Zagreb, Faculty of EE&C, Department of Applied Mathematics, Zagreb, Croatia.

* Corresponding author: zdenko.kovacic@fer.hr

problem. The results obtained analytically and numerically for the chosen vector optimization problem (the search for the inverse kinematics solution of a hyper-redundant robot) demonstrate the practical value of these methods. Make assessment of the time needed for algorithms to solve the inverse kinematics problem on-line based on the tests carried out on a low-cost micro-controller.

Keywords: Optimization algorithm, microgenetic algorithm, microimmune algorithm, inverse kinematics, hyper-redundant robots, real time execution

INTRODUCTION

The construction of hyper-redundant robotic manipulators is characterized by a large number of links and joints which significantly exceed the minimum number of links and joints needed for executing tasks. Typical forms of hyper-redundant robots are the forms of snakes, tentacles and elephant trunks. It is a well known fact that even the simplest robot configurations like a two degrees of freedom (DOF) planar rotational robot can be kinematically redundant. In Fig. 1 we can see a 4 DOF planar robot executing a rectangular path. Addition of two DOF has increased the redundancy of the robot and allowed the execution of a given robot task in four possible ways. Further increasing the number of joints will eventually turn a simple two DOF robot into a hyper-redundant planar robot.

By definition, a kinematically redundant robot is one whose forward kinematics is a non-bijective function. According to this definition, robots with less than 6 degrees of freedom can be considered redundant as long as there are two or more joint configurations that give the same end effector pose. This definition also means that a unique inverse kinematics function for such robot does not exist. Instead, there is a set of possible inverse kinematic solutions which satisfy the given robot pose. Obviously,

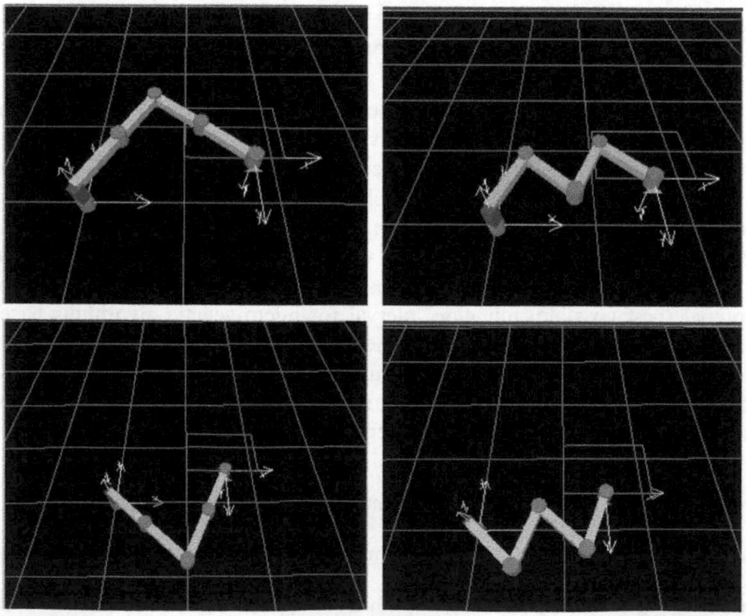

Figure 1. Four-way execution of a rectangular trajectory by a redundant 4 DOF planar robot.

such set for a 4 DOF redundant planar robot shown in Fig. 1 contains four concurrent candidate solutions.

The excessive number of redundant DOF becomes very beneficial in applications like inspection and maintenance in nuclear power plants, manufacturing and assembly in very confined spaces or medical diagnosis of the human intestine. Hyper-redundant robots are able to avoid obstacles in the workspace in much easier way than standard robots and also, they can adapt to variations of workspace if such changes occur during task execution.

Depending on the type of a hyper-redundant robot, different locomotion patterns can be targeted to resemble the locomotion patterns seen in nature (Chirikjian and Burdick 1995). This imposes a handful of design requirements, from solving the inverse kinematics problem to generating the trajectories which are both feasible and close to optimal. Real time trajectory planning for hyper-redundant robots designed to operate in industry-driven scenarios is a computationally demanding task. Usually complex mechanical constructions of hyper-redundant robots disallow easy assessment of robot's dynamical constraints affecting the trajectory planning.

Prompt adaptation of planned trajectories to static and dynamic changes in the workspace depends on the timeliness of an inverse kinematics solution for a new situation. This has become a driving force for research described herein—to find computationally effective ways of solving inverse kinematics in real-time or at least in the timeframe as close as possible to the real time.

The concept adopted here is that an inverse kinematics solution can be found using optimization algorithms (Davidor 1991, Aydin and Kocaoglan 1995, Huang et al. 2012, Wampler 1986). If the least square criterion is defined and the forward kinematics is a smooth function, gradient techniques can be employed. In the case of hyper-redundant robots, calculating a forward kinematics Jacobian matrix becomes computationally demanding and scales quadratically with the number of joints. This approach is also prone to singularity issues. In order to handle them even more computational effort is required. Providing this solution is computationally tractable it can be difficult to deal with constraints such as workspace obstacles.

The ultimate goal is to develop a real time inverse kinematics algorithm capable of handling nonlinear constraints (Glumac and Kovačić 2013). Computational complexity of calculating the forward kinematics function scales linearly with the number of joints. Since genetic algorithms (GA) only require evaluation of candidate solutions, it is assumed that this algorithm tends to be faster than gradient-based algorithms (Goldberg 1989, Michlewicz 1992, Michlewicz 2005, Alvarez 2002, Mezura et al. 2006). This approach eliminates problems due to mechanical and algorithmic singularities. In order to avoid permanent attraction to local optima, GA use techniques such as crossover, mutation, rejection of the least fit and selection of the fittest (elitism) trying to imitate the natural process of evolution.

Bearing in mind the optimization problem, GA designs may vary depending on the type of structure, the GA operators and parameters (Kelly Senecal 2000, Tian et al. 2005, Simões and Costa 2003, Zamanifar and Koorangi 2007, Derigs et al. 1997, Elmihoub et al. 2004, Beaty et al. 1990). Understanding what happens to it when an operator changes or a parameter changes is a prerequisite for a successful GA design. Defining the Schema Theorem, Holland tried to make a theoretical analysis of GA performance (Holland 1975, Srivastava and Goldberg 2001). Liepins and Vose in

(1991) and (Vose 1993) have modeled the genetic search directly, instead of looking at the scheme. They considered infinite populations and observed them as a probability distribution, and then studied the way in which such distribution is changing under the influence of genetic operators. Prügel-Bennett and Shapiro analyzed GA using a statistical mechanics theory (Prügel-Benett and Shapiro 1974).

Although there is a number of contributions to a better understanding of the performance of GA (Alvarez 2002, Schmitt and Rothlauf 2001, Zhang and Mühlenbein 2004, Hartl 1990), some practical GA design criteria for determining the optimal GA parameters for a particular optimization problem are not still there. Usually, only simple GA structures are analyzed and parameters for the optimal behavior of the algorithm are obtained heuristically. In other words, the "optimal" GA parameters are regularly determined by practical experiments.

The so-called microgenetic algorithms (μGA) appear promising for use in online inverse kinematics solvers as they work with small populations (Tian et al. 2005, Krishnakumar 1989, Toscano and Coello 2001, Fuentes and Coello 2007). They use a mutation algorithm that periodically saves only the fittest solution in the population, while other solutions are replaced with new randomly selected entities (Kelly Senecal 2000, Tian et al. 2005). The rate of mutation affects the speed of convergence of μGA but useful criteria for selecting the optimal interval for a particular mutation optimization problem do not exist. Here we demonstrate a method for determining the "optimal" (best possible) μGA mutation interval based on a theoretical analysis of μGA. The theoretical analysis is based on the Markov chain theory (Stewart 2009, Ching et al. 2007), which has been used for analysis of GA by other authors (Poli et al. 2001).

Further improvement can be searched through the consideration of other evolutionary computational approaches such as optimization using the clonal selection principle (Nunes de Castro and Von Zuben 2002), and microimmune algorithms (μIA). These algorithms do not use crossover operation, but also work with small populations (Nunes de Castro and Timmis 2002, Dasgupta 2006, Wong and Lau 2009, Zhang 2007, Chen et al. 2010, Lin and Chen 2011). This makes them rational candidates for use in on-line inverse kinematics solvers.

ROBOT KINEMATICS

A hyper-redundant robot is represented as a rigid body serial chain composed of N joints represented by a joint variable vector $\mathbf{q} = [q_1\, q_2 \ldots q_N]^T$. A kinematic model of such chain is described with homogeneous transformation matrices.

The robot model used for testing optimization algorithms is presented Fig. 2. Such robot has rotational joints which can be described with the following homogeneous matrices (Glumac and Kovačić 2013):

$$\mathbf{T}_{2k}^{2k+1} = \begin{bmatrix} \cos(q_{2k+1}) & 0 & -\sin(q_{2k+1}) & 0 \\ \sin(q_{2k+1}) & 0 & \cos(q_{2k+1}) & 0 \\ 0 & -1 & 0 & 0 \\ 0 & 0 & 0 & 1 \end{bmatrix} \tag{1}$$

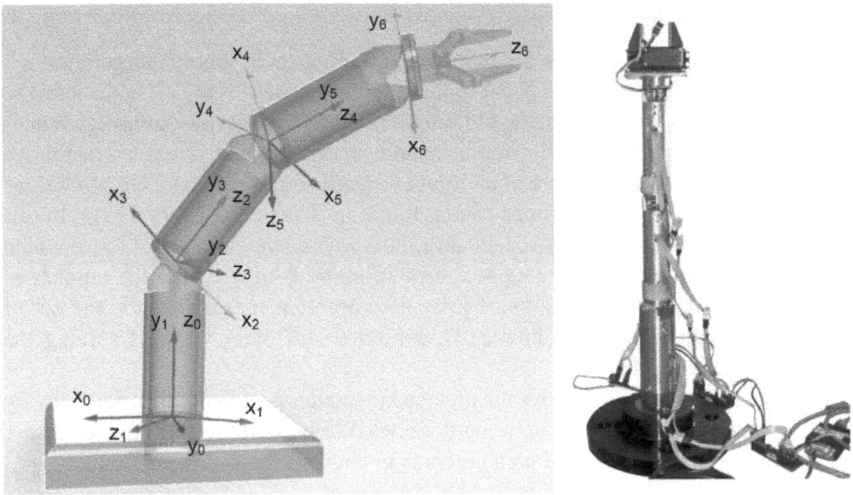

Figure 2. A kinematic model of a robotic arm with six revolute joints used for testing inverse kinematics algorithms presented in this chapter.

$$\mathbf{T}_{2k+1}^{2k+2} = \begin{bmatrix} \cos(q_{2k+1}) & 0 & \sin(q_{2k+1}) & 0 \\ \sin(q_{2k+1}) & 0 & -\cos(q_{2k+1}) & 0 \\ 0 & 1 & 0 & l \\ 0 & 0 & 0 & 1 \end{bmatrix} \tag{2}$$

where $l = 12.5$ cm is a link length, $k \in \{0; 1; 2; \ldots\}$ is a link index and q_i is the angle of rotation for joint $i \in \{1; 2; \ldots; N\}$.

Robot configuration $\mathbf{q} \in S$ is influenced by mechanical limitations of each joint q_i; $q_{2k+1} \in [-\pi \text{ rad}, \pi \text{ rad}]$, $q_{2k+2} \in [-\pi/4 \text{ rad}, \pi/4 \text{ rad}]$. These limitations are due to mechanical construction and electrical wiring of the robot. From the algorithmic point of view, a set of possible robot configurations is a bounded box $S \subset \mathbb{R}^n$.

Homogeneous transformation which transforms the end effector coordinate system into the base coordinate system is presented with the following matrix:

$$\mathbf{T}_0^n = \mathbf{T}_0^1 \mathbf{T}_1^2 \ldots \mathbf{T}_{N-1}^N = \begin{bmatrix} x_1 & y_1 & z_1 & p_1 \\ x_2 & y_2 & z_2 & p_2 \\ x_3 & y_3 & z_3 & p_3 \\ 0 & 0 & 0 & 1 \end{bmatrix} \tag{3}$$

Forward kinematics is the problem of determining the end effector pose $\mathbf{w}$ (position $\mathbf{p}$ and orientation ϕ) for a given joint configuration $\mathbf{q}$. We assume that the end effector orientation is represented by an approach vector $\mathbf{a} = [z_1\ z_2\ z_3]^{\mathrm{T}}$. Such representation is good enough for many standard applications where robot grippers rotate around the approach vector. The main reason for using this representation is a simple way of calculating and visualizing an orientation error.

Let W denote a set of feasible end effector poses. Then a forward kinematics function $f: S \rightarrow W$ calculates the end effector pose $\mathbf{w}$ for given joint parameters $\mathbf{q}$:

$$\mathbf{w} = f(\mathbf{q}) = \begin{bmatrix}\mathbf{p}\\ \mathbf{a}\end{bmatrix} = [p_1\, p_2\, p_3\, z_1\, z_2\, z_3] \tag{4}$$

Solving inverse kinematics is a problem of finding some robot configuration $\mathbf{q}_{opt}$ in which the end effector pose is equal to the reference pose $\mathbf{w} = \mathbf{w}_{ref}$. It is desirable to find a closed form solution which would be extracted from a set of solution candidates after satisfying some additional criteria (spatial distribution of obstacles in robot's workspace, shortest displacement of all joints for a given pose change, etc.) Solutions for a large number of configurations can be found in (Crane and Duffy 1998). In this chapter we are focused on hyper-redundant robots with a large number of joints which prevent elegant formulation of a closed form solution. By increasing the number of joints and having finer resolutions of joint displacement measurements, the set of possible solutions increases dramatically, asking so for other ways of solving the inverse kinematics problem.

A frequently used way to solve the inverse kinematics problem is by using velocity based control. The relationship between the rate of change of the end effector's pose and the rate of change of N joint parameters is given with the following relation:

$$\dot{\mathbf{w}} = \mathbf{J}\dot{\mathbf{q}} \tag{5}$$

where $\mathbf{J} = \frac{\partial \mathbf{w}}{\partial \mathbf{q}}$ is the Fréchet derivative of (4). Dimensions of Jacobian $\mathbf{J}$ are $6 \times N$ and the solution of inverse kinematics problem using velocity control may attain the following form:

$$\dot{\mathbf{q}} = \mathbf{J}^{*}\dot{\mathbf{w}} \tag{6}$$

where $\mathbf{J}^{*}$ is the pseudo inverse of Jacobian $\mathbf{J}$. This approach is prone to singularity issues which may be solved by introducing regularization or by augmenting the end effector pose with energy or manipulability constraints (Galicki 2007, Brock et al. 2008).

This approach has an issue of large computational complexity with respect to the number of joints. The forward kinematics of the robot used for case study in this chapter can be calculated using (3). This calculation is linear $o(N)$ in the number of joints as it requires multiplication of joint transformation matrices. The Jacobian matrix of (4) needs calculation of the rate of change of the end effector coordinate system with respect to the change of a robot joint displacement:

$$\frac{\partial \mathbf{T}_0^N}{\partial q_k} = \mathbf{T}_0^1 \mathbf{T}_1^2 \ldots \frac{\partial \mathbf{T}_{k-1}^k}{\partial q_k} \ldots \mathbf{T}_{N-1}^N \tag{7}$$

which makes that its calculation is quadratic $o(N^2)$ in the number of joints. After calculation of the Jacobian matrix $\mathbf{J}$ follows additional calculation of pseudoinverse $\mathbf{J}^{*}$ or its regularized form which further increases the computational complexity. For robots with a large number of joints this may be a serious limitation in using this approach for solving inverse kinematics.

Optimization Approach

When dealing with hyper-redundant robots different inverse kinematics solutions may need to be employed. In other words the inverse kinematics problem is determined as a problem of minimizing a suitable penalty function f_p:

$$\mathbf{q}_{opt} = \arg\min_{\mathbf{q}\in S} f_p(\mathbf{q}) \tag{8}$$

As stated in (8), the solution $\mathbf{q}_{opt}$ minimizes a penalty function $f_p : S \rightarrow R$ over the space of possible robot configurations *S*. A penalty function f_p needs to be defined so that its minima are robot configurations which achieve the reference end effector pose $\mathbf{w}_{ref}$. The penalty function is defined as follows (Glumac and Kovačić 2013):

$$f_p(\boldsymbol{q}) = \left\|\boldsymbol{p}_{ref} - \boldsymbol{p}\right\|_1 + l\left\|\boldsymbol{a}_{ref} - \boldsymbol{a}\right\|_1 \tag{9}$$

where $||.||_1$ is a vector norm defined as $||\mathbf{x}||_1 = \Sigma_{i=0}^{n}|x_i|$ and *l* is a parameter used to level impacts of two deviation terms.

Graphical presentation of the terms used in this criterion is given in Fig. 3.

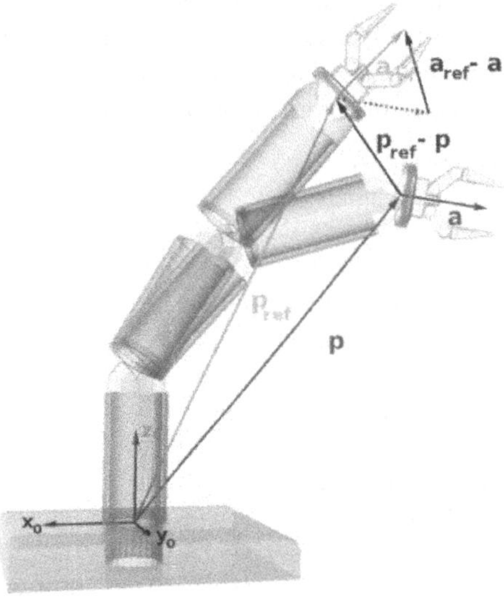

Figure 3. A reference position $\mathbf{p}_{ref}$ and approach vector $\mathbf{a}_{ref}$ are presented in light gray and a candidate solution (position **p** and approach vector **a**) are in black. Differences in respective vectors are colored black. © [2013] IEEE. Reprinted, with permission, from Glumac, S. and Z. Kovačić. 2013. Microimmune algorithm for solving inverse kinematics of redundant robots. Proceedings of the 2013 IEEE International Conference on Systems, Man, and Cybernetics, pp. 201–207, Manchester, CPS (Conference Publishing Services), 2013.

COMPARISON OF MICROGENETIC AND MICROIMMUNE OPTIMIZATION ALGORITHMS

The µGA and µIA optimization algorithms presented in this section are used in solving the problem (8) over a bounded box $S \subset \mathbb{R}^n$. Without loss of generality it can be assumed that the search space *S* is the Cartesian product of *n* unit intervals $S = [0, 1]^n$. Any bounded box can be reduced to *S* using translation and scaling.

µGA and µIA considered in this analysis are described respectively in pseudocode blocks Algorithm 1 and Algorithm 2. µGA is taken from (Dužanec and Kovačić 2009) while µIA was developed as a simplification of µGA (Glumac and Kovačić 2013).

While searching for an optimal solution both algorithms use the contraction operator (Dužanec and Kovačić 2009). This operator contracts the search space around the latest best solution found. At the beginning of the search initial search space is set to $S^0 = S$. Let $\mathbf{q}^v = [q_1^v\ q_2^v \ldots q_n^v]^T$ denote a best solution found after *v* contractions of search space. Search space after the *v*-th contraction is given with the following expression:

$$S^v = \left\{ \mathbf{q} \in \mathbb{R}^n : \mathbf{q} \in \prod_{i=1}^{n} \left[q_i^v - \frac{\lambda^v}{2}, q_i^v + \frac{\lambda^v}{2} \right] \text{and } \mathbf{q} \in S \right\} \tag{10}$$

where λ is a contraction factor $\lambda \in [0, 1)$ and Π is a Cartesian product of sets. In μGA contraction of search space is used after every i_s steps. A step is one iteration of algorithm's main loop. In μIA contraction operator is alternating with the mutation operator.

The value λ^v is named the diameter of the search space S^v. The set S^v has the form of a ball with the radius length defined with $\|.\|_\infty$ norm and λ^v as its diameter if the constraint on S in (10) is ignored. When the diameter of the search space becomes smaller than an arbitrary value $d^* \in [0, 1)$, it is assumed that the search space has become so small that no better solution can be found. This eventually leads to stopping condition used in the experimental evaluation of optimization algorithms.

Algorithm 1. Pseudocode of μGA[1]
initialize *Population* of *L* individuals
while $\lambda^v > d^*$ **do**
    select two *Parents* and *Bad individual* from *Population*
    create a *Child* using **crossover** from *Parents*
    remove *Bad* individual from *Population*
    insert *Child* into *Population*
        if i_m iterations have passed **then**
            mutate *Population* with elitism
            if i_s iterations have passed **then**
                contract *Search space*
            end if
        end if
end while

Algorithm 2. Pseudocode of μIA[2]
initialize *Population* of *L* individuals
while $\lambda^v > d^*$ **do**
    mutate *Population* with elitism
    contract *Search space*
end while

In the case the minimum value of penalty function is known, the value of the penalty function can be used to terminate the search. Since the value of (9) is known to be 0 for the optimal solution, the algorithm can terminate the search when the best found solution assesses better than some threshold parameter f_p^*. Parameter f_p^* should be a small value close to 0. This stopping condition is used in the algorithm executed on the microcontroller platform.

Population initialization is generation of L candidate solutions. Candidate solutions are sampled from the uniform distribution over the initial search space S. This step is the same for μIA and μGA.

Mutation reinitializes population with preserving elitism. This saves the best solution and samples new $L-1$ candidate solutions from the uniform distribution over the contracted search space S^v. μGA mutates population after every i_m crossovers, while μIA mutates population along with every search space contraction.

Another difference between μIA and μGA is the use of crossover. Like most genetic algorithms μGA uses crossover, while μIA does not. Since artificial immune system algorithms do not use crossover, the microimmune algorithm is named after this property. μGA uses a 3-tournament selection to select three individuals from the population. Of three selected individuals, two more fitting individuals are used as parents (the mother and the father) while the third is removed from the population. Let $\mathbf{q}_f = [q_{f1}\, q_{f2} \dots q_{fn}]^T$ denote the father and $\mathbf{q}_m = [q_{m1}\, q_{m2} \dots q_{mn}]^T$ denote the mother. The child $\mathbf{q}_c$ is generated using the uniform distribution in the set (bounded box determined by its parents):

$$S^c = \left\{ \mathbf{q} \in \mathbb{R}^n : \mathbf{q} \in \prod_{i=1}^{n} [\min(q_i^f,\, q_i^m), \max(q_i^f,\, q_i^m)] \right\} \tag{11}$$

and inserted in the population.

The crossover operation employed here creates new children (solutions) that reside in the portion of S bounded by their parents. This creates the possibility that both parents possess the same genes at the same positions in their genetic codes, and crossover would not be able to change them. The effect is that only a part of S is searched and an algorithm can easily get stuck in a local optimum. In order to prevent that, the mutation operator is applied. The rate of mutation must be chosen carefully; take too large, the crossover effect will be weakened by predominantly stochastic nature of GA. Take too small, the rate of mutation would increase the probability of missing the global optimum.

Memory requirements of μIA are independent of its parameters. Amount of memory used by μIA depends only on the number of dimensions in the penalty function domain, i.e., the number of robot joints N. During search, μIA needs space to store the best solution and values of search space limits, generate a new candidate solution and calculate a penalty function. The algorithm adds almost no memory requirements than those required to calculate the penalty function. μGA is also a memory efficient algorithm, but as opposed to μIA it needs to store an entire population of candidate solutions. Since the property of μGA is that number of individuals are small, this is not a big increase in memory requirements.

Both μIA and μGA algorithms are not guaranteed to find an optimal solution in search space S. A final solution of an algorithm run may not be even close to optimal. In order to accommodate for this, an algorithm parameters need to be tuned. When properly tuned, μIA and μGA are expected to produce high quality solutions.

Mutation optimization in μGA

Bearing in mind the task of solving inverse kinematics of a hyper-redundant robot with N DOF the optimal solution vector $\mathbf{q}_{opt}$ we are looking for lies in the space of all possible solutions S. In order to find $\mathbf{q}_{opt}$, we use the penalty function $f_p(\mathbf{q})$ defined in (9).

Theoretical analysis of mutation effects in μGA

Generally, the solution space S of the N-dimensional optimization problem is the N-dimensional solid. Let us divide each dimension of the solution space into three subgroups:

$$\begin{aligned} A &= \left\{ x \middle| x \in \left[x_1, x_{opt} - \Delta_{opt} \right) \right\} \\ B &= \left\{ x \middle| x \in \left[x_{opt} - \Delta_{opt}, x_{opt} + \Delta_{opt} \right] \right\} \\ C &= \left\{ x \middle| x \in \left(x_{opt} + \Delta_{opt}, x_2 \right] \right\} \end{aligned} \tag{12}$$

Then the N-dimensional solution space is divided into 3^N subgroups. Figure 4 shows the solution space subgroups for $N = 1$ and $N = 2$.

Subgroup B is the region around the optimal solution given by the required precision of optimization problem solution. In the case of the N-dimensional optimization problem the chromosomes that have all the vector elements in the subgroup B represent a global solution.

The idea of behavioral analysis of μGA is to calculate the probability that the optimization problem will be solved near the global optimum. To get a better insight

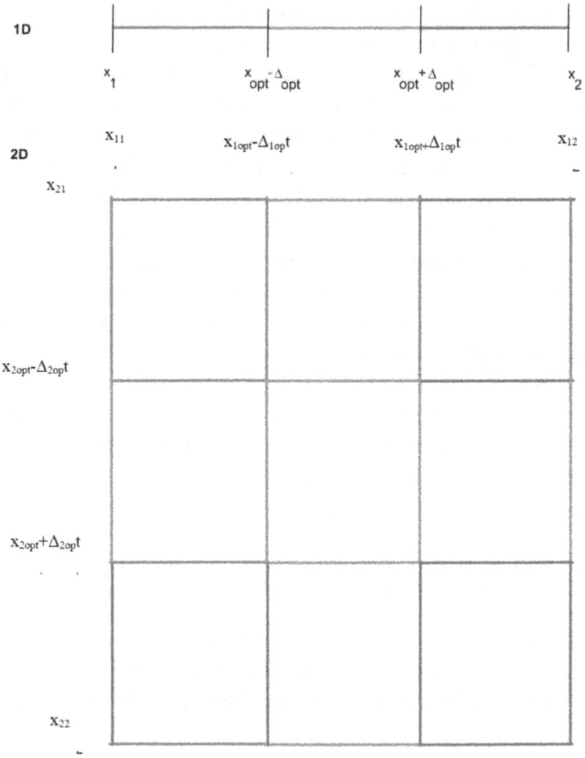

Figure 4. Subgroups for solution space dimensions $N = 1, 2$.

into the matter, first one-dimensional solution space is analyzed. Figure 5 shows an example of one-dimensional penalty function.

The probability of randomly generated number x from S to belong to subgroups A, B or C can be determined using geometric probability:

$$p_A = \frac{(x_{opt} - \Delta_{opt}) - x_1}{x_2 - x_1} \qquad p_B = \frac{2 \cdot \Delta_{opt}}{x_2 - x_1} \qquad p_C = \frac{x_2 - (x_{opt} + \Delta_{opt})}{x_2 - x_1} \tag{13}$$

As described in (11), crossover operation results with a new chromosome (child) lying in the section of S bounded by both parents, which can be rewritten in the following way:

$$q_{id} = \min(q_{ip1}, q_{ip2}) + \text{rand}(|q_{ip1} - q_{ip2}|), \forall i \in [1, N] \tag{14}$$

where q_{id}, q_{ip1} and q_{ip2} represent the ith elements of the child and parent chromosomes (vectors), respectively.

In the one-dimensional optimization problem, this means that the crossover operation can generate a child in subgroup B (find a solution) if one parent is picked from the subgroup A, and the other from the subgroup B. To make sure that μGA will find a solution, the chromosomes of the population should not be harvested from the same subgroup.

Let us index subgroups A, B and C as follows: 0 (A), 1 (B) and 2 (C). To determine all the possible states of the population, the *association matrix* must be defined, whose elements contain data that indicate whether the possible solutions are belonging to a subgroup A, B and C. Using the above introduced indexation the association matrix can be defined as follows:

$$\mathbf{Pr} = [pr_{ij}] = mod\left(floor\left(\frac{i}{3^j}\right), 3\right), i \in [0, 3^N - 1], j \in [0, N - 1] \tag{15}$$

Now, the probability of generating a chromosome in a solution space subgroup must be defined. For a one-dimensional optimization problem the elements of the probability vector $\mathbf{d}_s$ are defined with (13):

$$\mathbf{d}_s = [p_A \, p_B \, p_C]^T \tag{16}$$

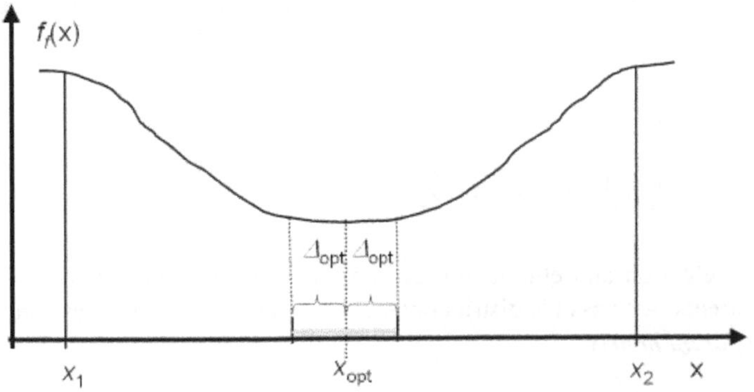

Figure 5. Example of a penalty function of a one-dimensional optimization problem.

The probability is actually proportional to the "size" of a subgroup in relation to the whole space of solutions, so the probability vector $\mathbf{d}_s$ can also be called the size vector. Using (15) and (16) the size vector for the N-dimensional space of solutions can be defined as follows:

$$\mathbf{d}_s = [d_{sv(i)}], \in [0, 3^N - 1]$$
$$d_{sv(i)} = \prod_{j=0}^{N-1} d_{s(Pr_{ji})}, \ \forall i \in [0, 3^N] \tag{17}$$

In Eq. (17) we assume that the solution space of each vector element is the same. Generally, this is not true, however, the probabilities p_A, p_B and p_C really depend on the relations between ($x_2 - x_1$, x_{opt} and Δ_{opt}), and not actual values. Therefore, in order to simplify the calculation the reasonable assumption is that these probabilities are the same for each vector element.

L solutions of the population can be distributed to 3^N subgroups in $\bar{C}_{3^N}^{(L)}$ different ways. The *combination matrix* $\mathbf{T}_{vp}$ that contains all possible distributions of L chromosomes in 3^N subgroups is defined by the expression:

$$\mathbf{T}_{vp} = [\mathbf{t}^T{}_{vpi}] = [t_{vpij}], 1 \le i \le \bar{C}_{3^N}^{(L)}, 1 \le j \le 3^n$$
$$\mathbf{t}_{vpi}^T = [t_{pi1} \ \cdots \ t_{pi3^N}], 0 \le t_{pij} \le \bar{C}_{3^N}^{(L)} L, \sum_{j=1}^{3^N} t_{pij} = L, \forall i \in \left[1, \bar{C}_{3^N}^{(L)}\right] \tag{18}$$

The probability of generating the initial population for each distribution of solutions as defined in the $\mathbf{T}_{vp}$ can be defined using the size vector:

$$\mathbf{p}_{v0} = [p_{v0i}], \ i \in \left[1, \overline{C_L^{3^N}}\right]$$
$$p_{v0i} = L! \prod_{j=1}^{3^N-1} \frac{1}{t_{vpij}} \left[\frac{d_{vsj}}{(x_2 - x_1)^N}\right]^{t_{vpij}} \tag{19}$$

The next step of the genetic algorithm is selection. In our case, the three-chromosome tournament selection operator is used. Therefore, all possible distributions of three selected chromosomes in 3^N subgroups must be defined. Therefore the *selection state matrix* is defined as follows:

$$\mathbf{T}_{vs} = \left[\mathbf{t}_{vsi}^T\right] = \left[t_{vsij}\right], \ 1 \le i \le \overline{C}_{3^N}^{(3)}, \ 1 \le j \le 3^N$$
$$\mathbf{t}_{vsi}^T = \left[t_{si1} \ \ldots \ t_{si3^N}\right], \ 0 \le t_{sij} \le 3, \ \sum_{j=1}^{3^N} t_{sij} = 3, \forall i \in \left[1, \overline{C}_{3^N}^{(3)}\right] \tag{20}$$

After selection and elimination of one chromosome, other two chromosomes become parents. All possible distributions of two parent chromosomes can be defined using the *parent matrix*:

$$\mathbf{T}_{vc} = \left[\mathbf{t}_{vci}^T\right] = \left[t_{vcij}\right],\ 1 \le i \le \overline{C}_{3^N}^{(2)},\ 1 \le j \le 3^N$$
$$\mathbf{t}_{vci}^T = \left[t_{ci1} \ \ldots \ t_{ci3^N}\right],\ 0 \le t_{cij} \le 2,\ \sum_{j=1}^{3^N} t_{cij} = 2, \forall i \in \left[1, \overline{C}_{3^N}^{(2)}\right] \tag{21}$$

The *selection result matrix* ($\overline{C}_L^{3^N} \times \overline{C}_3^{3^N}$) defines the probability to jump from a state defined by the matrix $\mathbf{T}_{vp}$ to a state defined by the matrix $\mathbf{T}_{vs}$. Assuming that the selection probability for each chromosome is the same, regardless of position in the space of solutions, the elements of the selection result matrix **O** can be defined by:

$$o_{vij} = \begin{cases} \dfrac{\prod_{k=1}^{\overline{C}_{3^N}^{(3)}} \overline{C}_{t_{pik}}^{(t_{sjk})}}{\overline{C}_L^{(3)}} & t_{pil} \ge t_{sjl} \forall l \in \left[1, 3^N\right] \\ 0 & \text{else} \end{cases} \tag{22}$$
$$\forall i \in \left[1, \text{rows}(\mathbf{T}_{vp})\right], \forall j \in \left[1, \left[1, \text{rows}(\mathbf{T}_{vs})\right]\right]$$

After selection, the worst chromosome is eliminated. This means that the algorithm jumps from the state defined by the matrix $\mathbf{T}_{vs}$ to the state of the matrix $\mathbf{T}_{vc}$. Because the worst chromosome is deleted, it is not possible to delete a global solution (chromosomes that have all the vector elements in the subgroup *B*). Under this assumption the elements of the *elimination matrix* **E** are defined as follows:

$$e_{vij} = \begin{cases} \dfrac{\prod_{k=1}^{\overline{C}_{3^N}^{(2)}} \overline{C}_{t_{sik}}^{(t_{cjk})}}{3 - t_{cjb}} & t_{sib} \le t_{cjb} \\ 1 & t_{cjb} = 3 \wedge t_{sib} = 2 \\ 0 & \text{else} \end{cases} \tag{23}$$
$$\forall i \in \left[1, \text{rows}(\mathbf{T}_{vs})\right], \forall j \in \left[1, \left[1, \text{rows}(\mathbf{T}_{vc})\right]\right]$$

where *b* is the column of the association matrix describing a global solution:

$$\text{Pr}_{bi} = 1 \forall i \in \left[1, \text{rows}(\text{Pr})\right] \tag{24}$$

Selection and elimination are followed by a crossover operation defined in expression (14). Positions of children generated within the solution space depend on the positions of their parents. The probability of generating a child within a specific subgroup of *S* depends on two conditions: subgroups to which parents belong and parents' positions therein. However, the probabilities of these two conditions are unknown. Therefore, the crossover result probability is calculated using mathematical expected values within all subgroups of *S*:

$$\begin{aligned} x_{p<} &= \int_{x_1}^{x_{opt}-\Delta_{opt}} \frac{x}{x_{opt}-\Delta_{opt}-x_1} dx = \frac{x_{opt}-\Delta_{opt}+x_1}{2} \\ x_{p=} &= \int_{x_{opt}-\Delta_{opt}}^{x_{opt}+\Delta_{opt}} \frac{x}{(x_{opt}+\Delta_{opt})-(x_{opt}-\Delta_{opt})} dx = x_{opt} \\ x_{p>} &= \int_{x_{opt}+\Delta_{opt}}^{x_2} \frac{x}{x_2-(x_{opt}+\Delta_{opt})} dx = \frac{x_2+x_{opt}+\Delta_{opt}}{2} \end{aligned} \tag{25}$$

Let us define the affiliation matrix of a 2-dimensional problem in the following form:

$$\mathbf{Pr} = \begin{bmatrix} 0 & 1 & 2 & 0 & 1 & 2 & 0 & 1 & 2 \\ 0 & 0 & 0 & 1 & 1 & 1 & 2 & 2 & 2 \end{bmatrix} \tag{26}$$

A crossover probability vector based on the expected values of parents and the affiliation matrix (26), is defined by the expression:

$$\mathbf{x}_{ve}(p_1,p_2) = \begin{cases} \begin{bmatrix} 1 & 0 & 0 \end{bmatrix}^T & ,p_1=0 \wedge p_2=0 \\ \begin{bmatrix} \frac{x_{opt}-x_{p<}-\Delta_{opt}}{x_{opt}-x_{p<}} & \frac{\Delta_{opt}}{x_{opt}-x_{p<}} & 0 \end{bmatrix}^T & ,p_1=0 \wedge p_2=1 \\ \begin{bmatrix} \frac{x_{opt}-x_{p<}-\Delta_{opt}}{x_{opt}-x_{p<}} & \frac{2\Delta_{opt}}{x_{opt}-x_{p<}} & \frac{x_{p>}-x_{opt}-\Delta_{opt}}{x_{opt}-x_{p<}} \end{bmatrix}^T & ,p_1=0 \wedge p_2=2 \\ \begin{bmatrix} \frac{x_{opt}-x_{p<}-\Delta_{opt}}{x_{opt}-x_{p<}} & \frac{\Delta_{opt}}{x_{opt}-x_{p<}} & 0 \end{bmatrix}^T & ,p_1=1 \wedge p_2=0 \\ \begin{bmatrix} 0 & 1 & 0 \end{bmatrix}^T & ,p_1=1 \wedge p_2=1 \\ \begin{bmatrix} 0 & \frac{\Delta_{opt}}{x_{opt}-x_{p<}} & \frac{x_{p>}-x_{opt}-\Delta_{opt}}{x_{opt}-x_{p<}} \end{bmatrix}^T & ,p_1=1 \wedge p_2=2 \\ \begin{bmatrix} \frac{x_{opt}-x_{p<}-\Delta_{opt}}{x_{opt}-x_{p<}} & \frac{2\Delta_{opt}}{x_{opt}-x_{p<}} & \frac{x_{p>}-x_{opt}-\Delta_{opt}}{x_{opt}-x_{p<}} \end{bmatrix}^T & ,p_1=2 \wedge p_2=0 \\ \begin{bmatrix} 0 & \frac{\Delta_{opt}}{x_{opt}-x_{p<}} & \frac{x_{p>}-x_{opt}-\Delta_{opt}}{x_{opt}-x_{p<}} \end{bmatrix}^T & ,p_1=2 \wedge p_2=1 \\ \begin{bmatrix} 0 & 0 & 1 \end{bmatrix}^T & ,p_1=2 \wedge p_2=2 \end{cases} \tag{27}$$

The crossover probability vector defines the probability that an element of the child chromosome will belong to a specific subgroup of *S*, if affiliations of respective parent elements are known. To determine the affiliation of the whole solution all vector elements must be taken into account. The total probability is the product of probabilities of all elements of the vector:

$$\begin{aligned}
&\mathbf{x}_{vt}(s_1,s_2)=\left[\mathbf{x}_{vti}(s_1,s_2)\right]\\
&\mathbf{x}_{vti}(s_1,s_2)=\mathbf{x}_{ve}(pr_{i,s_1},pr_{i,s_2}),1\le i\le N\\
&\mathbf{x}_{v}(s_1,s_2)=\left[x_{vi}\right],1\le i\le 3^N\\
&x_{vi}=\prod_{k=1}^{N}x_{vt(k,pr_{k,i})}
\end{aligned} \tag{28}$$

A *crossover probability matrix* defines the probabilities that the child chromosome belongs to a subgroup of S for every possible distribution of parents' chromosomes. It is defined as follows:

$$\begin{aligned}
&\mathbf{B}_{\mathbf{v}}=\left[b_{vij}\right],0\le i\le \overline{C_{3^N}^{(2)}},0\le j\le C_{3^N}^{(2)}\\
&\mathbf{b}_{vi}=\mathbf{x}_v(t_{vci,par1},t_{vci,par_2})^T
\end{aligned} \tag{29}$$

At this point all probabilities, starting from generating the initial population, selection, elimination and crossover are defined. The result is that the probability of generating a new child chromosome within defined subgroups of S can be determined. Now the analysis can be made to determine how this affects the complete population. By elimination of the worst chromosome and generating a new one by the crossover operator the algorithm actually changes from one state of the selection state matrix $\mathbf{T}_{vs}$ to another. The probabilities of jumping from one state of the selection state matrix to another are defined with the *transition probability matrix* **K**:

$$\begin{aligned}
&k_{i,\mathrm{g}(\mathbf{T}_s,\mathrm{inc}(\mathbf{T}_{c(i)},j))}=b_{ij}\\
&\forall i\in\left[1,\mathrm{rows}(\mathbf{B})\right],\forall j\in\left[1,\mathrm{cols}(\mathbf{B})\right]
\end{aligned} \tag{30}$$

where g is the function that returns the index of the matrix row containing a set of values:

$$\mathrm{g}(\mathbf{M},\mathbf{x})=i\to\left\{\mathbf{M}_i=\mathbf{x}^T\right\} \tag{31}$$

while inc is the function that increments the ith element of the vector:

$$\mathrm{inc}(\mathbf{x},i)\to x_i=x_i+1 \tag{32}$$

The objective of the analysis is to determine the expected state of the overall population. The probability of jumping from one state of the combination matrix $\mathbf{T}_{vp}$ to another state of the selection state matrix $\mathbf{T}_{vs}$ is defined with the selection result matrix (22). The product of the elimination matrix **E** and the transition probability matrix **K** defines the probability of jumping from one state of $\mathbf{T}_{vs}$ to another by using a crossover operator. All possible changes caused by the crossover operator can be determined by the *change matrix* $\mathbf{T}_d$, which is defined for a 1-dimensional optimization problem as follows:

$$\mathbf{T}_d = \begin{bmatrix} 0 & 0 & 0 \\ -1 & 1 & 0 \\ -1 & 0 & 1 \\ 1 & -1 & 0 \\ 0 & -1 & 1 \\ 1 & 0 & -1 \\ 0 & 1 & -1 \end{bmatrix} \quad \text{(columns: } d_<,\ d_=,\ d_> \text{)} \tag{33}$$

The first row of $\mathbf{T}_d$ describes a state where no change in the population distribution has occurred. This means that elimination of the worst chromosome and generation of a new chromosome affected the same subgroup (*A*, *B* or *C*). The second row indicates that elimination took place in *A* and generation happened in *B*. For the *N*-dimensional problem the dimension of the change matrix $\mathbf{T}_d$ becomes equal to $3^N \times \left[1 + \frac{(3^N)!}{(3^N-2)!}\right]$.

The *change probability matrix* **D** describes the probability of population distribution changes defined by $\mathbf{T}_d$:

$$d_{i,\delta_{ij}} = \begin{cases} (\mathbf{E}\cdot\mathbf{K})_{ij} & \delta_{ij} \geq 0 \\ 0 & \text{else} \end{cases} \quad \forall i \in [1, \text{rows}(\mathbf{E}\cdot\mathbf{K})], j \in [1, \text{cols}(\mathbf{E}\cdot\mathbf{K})] \tag{34}$$

where

$$\delta_{ij} = \text{g}\left[\mathbf{T}_d, (\mathbf{T}_{sj} - \mathbf{T}_{si})^T\right] \tag{35}$$

The probability of jumping from one state of the $\mathbf{T}_{vp}$ can now be defined with the *transition matrix* **C**:

$$c_{i,\varepsilon_{ij}} = \begin{cases} (\mathbf{O}\cdot\mathbf{D})_{ij} & \varepsilon_{ij} \geq 0 \\ 0 & \text{else} \end{cases} \quad \forall i \in [1, \text{rows}(\mathbf{O}\cdot\mathbf{D})], j \in [1, \text{cols}(\mathbf{O}\cdot\mathbf{D})] \tag{36}$$

where

$$\varepsilon_{ij} = \text{g}\left[\mathbf{T}_p, (\mathbf{T}_{pi} + \text{T}_{dj})^{\text{T}}\right] \tag{37}$$

The state of population in the ith iteration can now be determined as follows:

$$\mathbf{P}(i) = \mathbf{P}^T{}_0 \mathbf{C}^i \tag{38}$$

The probability of finding the optimal solution is equal to the probability that the population has reached the state where all the affiliation matrix elements are equal to 1:

$$p_{vopt}(i) = \mathbf{p}_v(i)_k, \text{ where } pr_{l,k} = 1, \forall l \in [1, N] \tag{39}$$

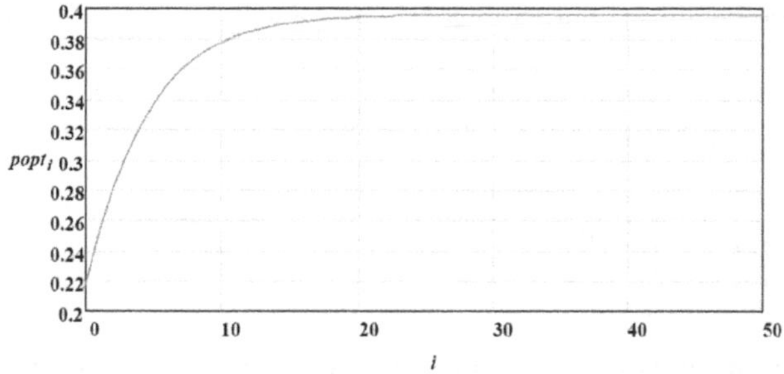

Figure 6. Probability of finding the optimal solution for a 2-dimensional optimization problem by using only the crossover operation for $x_1 = 0$, $x_2 = 255$, $x_{opt} = 127$, $\Delta_{opt} = 1$ and $L = 5$.

The probability of finding the optimal solution by using only the crossover operation as defined in (36) is shown in Fig. 6. The whole process starts with a limited genetic material brought in by a randomly generated initial population, which does not guarantee that the optimal solution will be found. For example, the crossover of two chromosomes from the same subgroup of S will result with a new chromosome in the same subgroup, not necessarily the optimal one. Mutation operator inserts new genetic material into the population. In the microgenetic algorithms, mutation is performed periodically after every i_m iterations. All chromosomes except the best one are deleted and replaced with new randomly generated chromosomes.

The problem is how to determine the number of iterations i_m for the optimal mutation interval. According to the graph shown in Fig. 6, the probability of finding the optimal solution saturates approximately after 33 crossovers (the whole genetic material becomes exhausted), which clearly indicates that this number of iterations corresponds to the optimal mutation interval i_m.

To analyze the influence of mutation on the behavior of the genetic algorithm, the *mutation matrix*, which contains all the possible combinations of generating L-1 new chromosomes, must be defined:

$$\mathbf{T}_{vm} = \left[\mathbf{t}_{vmi}^T\right] = \left[t_{vmij}\right],\ 1 \le i \le \overline{C}_{3^N}^{(L-1)},\ 1 \le j \le 3^N$$
$$\mathbf{t}_{vmi}^T = \left[t_{vmi1}\ \ \cdots\ \ t_{vmi3^N}\right],\ 0 \le t_{vmij} \le \overline{C}_{3^N}^{(L-1)} L,\ \sum_{j=1}^{3^N} t_{vmij} = L-1,\ \forall i \in \left[1, \overline{C}_{3^N}^{(L-1)}\right] \tag{40}$$

Bearing in mind that only the best fitting chromosome will not be deleted during mutation, an assumption is made that all chromosomes except the one around x_{opt} have the same probability to survive. This can be described with the *best probabilities matrix* **U** defined as follows:

$$u_{ij} = \begin{cases} \dfrac{t_{mpij}}{L} & t_{pi1} = 0 \\ 1 & j = 1 \wedge t_{mpi1} > 0 \\ 0 & else \end{cases} \tag{41}$$

$$\forall i \in \left[1, \text{rows}(\mathbf{T}_{mp})\right], j \in \left[1, \text{cols}(\mathbf{T}_{mp})\right]$$

Now the probabilities of generating L-1 chromosomes in the subgroups of S must be defined:

$$\mathbf{p}_{vm} = \left[p_{vmi}\right],\ i \in \left[1, \overline{C_{L-1}^{3^N}}\right]$$
$$p_{vmi} = (L-1)! \prod_{j=1}^{3^N-1} \frac{1}{t_{vmij}} \left[\frac{d_{vsj}}{(x_2 - x_1)^N}\right]^{t_{vmij}} \tag{42}$$

From the probabilities of generating new L-1 chromosomes and the probability of the best fitting chromosome, the probability of a new generated population state can be defined with the *mutation transfer matrix* **M**:

$$m_{i,g\left[\mathbf{T}_p, \text{inc}(\mathbf{T}_{vmj}, k)\right]} = \sum_{j=1}^{\text{rows}(\mathbf{T}_m)} p_{vMj} \cdot u_{i1}$$
$$\forall i \in \left[1, \text{rows}(\mathbf{T}_{vp})\right], \forall k \in \left[1, \text{rows}(\mathbf{T}_{vm})\right] \tag{43}$$

The population state in the ith iteration can now be determined as follows:

$$\mathbf{P}(i) = \mathbf{P}_0^{\ T}\left[\left(\mathbf{C}^{i_m}\mathbf{M}\right)^{\text{round}(\frac{i}{i_m})} \mathbf{C}^{\text{mod}(i, i_m)}\right] \tag{44}$$

Figure 7 shows the probability of finding the optimal solution for a 2-dimensional optimization problem by using crossover and mutation for $x_1 = 0$, $x_2 = 256$, $x_{opt} = 128$, $\Delta_{opt} = 1$, $L = 5$ and $i_m = 33$. It can be seen that every mutation gradually increases the probability of finding the optimal solution and this probability finally settles in 1, thus guaranteeing that the solution will be found after a given number of iterations.

μGA Optimization - Numerical Experiments

The described microgenetic algorithm was tested in the case of solving the inverse kinematics problem of a hyper-redundant robot arm shown in Fig. 2, which required

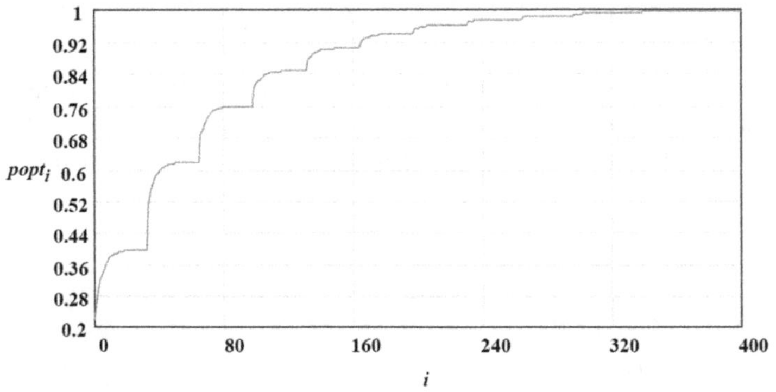

Figure 7. Probability of finding the optimal solution for a 2-dimensional optimization problem by using crossover and mutation for $x_1 = 0$, $x_2 = 256$, $x_{opt} = 128$, $\Delta_{opt} = 1$, $L = 5$ and $i_m = 33$.

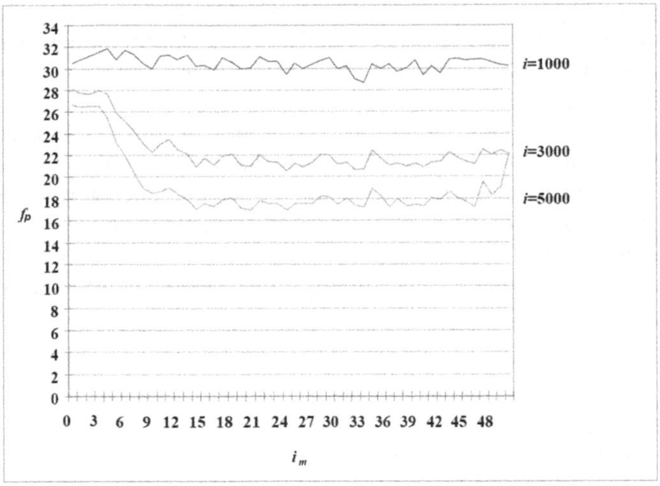

Figure 8. Penalty function vs. mutation interval after 1000, 3000 and 5000 μGA iterations (Dužanec and Kovačić 2009).

solving a vector optimization problem (Dužanec and Kovačić 2009). To reduce the randomization effects in initial genetic populations, the algorithm was started 100 times with the same set of parameters, while the value of the mutation interval i_m was changed in the range of 1–50.

Results were used to calculate the average penalty function based on the best solutions found. Figure 8 shows the penalty function as a function of mutation interval after 1000, 3000 and 5000 iterations. We know that the optimal value of the mutation interval lies where the penalty function value is the lowest. As shown in Fig. 8, it can be seen that the numerically determined values of the best possible mutation intervals lie in the wide range of $i_m \approx 15–40$, which is a good match with the behavior of μGA determined analytically ($i_m \approx 33$) (Dužanec and Kovačić 2009) and practically (Dužanec and Kovačić 2009).

MULTIOBJECTIVE OPTIMIZATION OF ALGORITHMS

Selection and crossover change the way an algorithm seeks an optimal solution. Depending on the choice of penalty function their use can slow down the convergence of an algorithm without increasing the quality of the final solution (Glumac and Kovačić 2013). From the previous statement two criteria for evaluating the algorithm can be identified; the quality of solutions and the execution time. For inverse kinematics problem and algorithms being presented, these criteria are in conflict as experimentally shown in this section. There is no single fastest, most accurate algorithm and the trade-off between these two criteria is needed. Multiobjective optimization is used to determine which algorithms provide the best trade-off.

Let X denote a set of algorithms that are evaluated, $f_1 : X \rightarrow \mathrm{R}$ a criterion representing the quality of the final solutions and $f_2 : X \rightarrow \mathrm{N}$ the criterion representing the execution time. Algorithm $x \in X$ is dominated by algorithm $y \in X$ if y is strictly better than x by one criterion, and better than or equal by the other criterion. The goal of multiobjective optimization is to find the Pareto frontier; the subset of algorithms such that no algorithm in this set is dominated by some other algorithm. An algorithm which is not in the Pareto frontier is strictly inferior in performance than at least one algorithm in the Pareto frontier.

In order to find the Pareto frontier of X, criteria f_1 and f_2 need to be defined. Let $\mathbf{q}_f$ denote the stochastic variable which models the final algorithm solution. In that case $f_p(\mathbf{q}_f)$ is the stochastic variable presenting the penalty function value. Since $f_p(\mathbf{q}_f)$ is a stochastic variable, its expected value is used as a criterion $f_1 : S \rightarrow \mathrm{R}$. In order to approximate the value of f_1 the arithmetic mean for a final number of optimization runs is used. Samples from the distribution of $f_p(\mathbf{q}_f)$ are generated in the manner described in Algorithm 3.

Algorithm 3. Generation of random criterion[3]
uniformly sample a robot configuration $\mathbf{q} \in S$
calculate a pose $\mathbf{w}$ for the sampled configuration $\mathbf{q}$
generate a penalty function (9) using $\mathbf{w}_{ref} = \mathbf{w}$

The execution time must be measured in quantity which is independent of processor and implementation details. The number of penalty function calls m is a property of the algorithm itself and the penalty function calculation is considered computationally the most demanding task of one algorithm iteration. This makes m ideal for evaluating execution time. Since m is almost proportional to execution time measured in seconds, approximate time for algorithm run can be calculated using this property. The number of penalty function calls is a deterministic value and can be calculated for given parameters of μIA (Glumac and Kovačić 2013):

$$m = L + (L - 1)\log_{\lambda} d^* \tag{45}$$

The multiobjective optimization is performed by brute force search using the experimental data described in the next section. A final result of this optimization is the comparison of Pareto frontiers of a μIA set and a μGA set presented in the next section.

NUMERICAL EXPERIMENTS WITH μGA AND μIA

Let $\mu\mathrm{IA}_{(L,\lambda)}$ denote the instance of μIA with a population size L and a contraction factor λ. Let $\mu\mathrm{GA}_{(L,\lambda,im,is)}$ denote the instance of μGA with a population size L, a contraction factor λ, a mutation step i_m and a contraction step i_s. Experiments are done on the set of $\mu\mathrm{IA}_{(L,\lambda)}$ for each combination of $L \in \{3, 4, \ldots, 29\}$, $\lambda \in \{0.4, 0.41, \ldots, 0.99\}$ and the set of $\mu\mathrm{GA}_{(L,\lambda,im,is)}$ for each combination of $\lambda \in \{0.55, 0.65, \ldots, 0.95\}$, $i_m \in \{1, 2, \ldots, 29\}$, $i_s \in \{ i_m, 2\, i_m, \ldots\}$. In these experiments the minimal diameter of search space is set to be $d^* = 0.001$.

In order to evaluate the quality of the final solution each instance of algorithm is run 500 times. Samples of criterion $f_p(\mathbf{q}_f)$ are generated as in Algorithm 3 for each of those runs. The criterion f_1 is plotted in Fig. 9. Since the minimum value of this criterion is known to be 0, that plot can be used to assess how good solution can be reached using

[3] © [2013] IEEE. Reprinted, with permission, from Glumac, S. and Z. Kovačić. 2013. Micro-immune algorithm for solving inverse kinematics of redundant robots. Proceedings of the 2013 IEEE International Conference on Systems, Man, and Cybernetics, pp. 201–207, Manchester, CPS (Conference Publishing Services), 2013.

μIA with some combination of parameters. It can be seen that both for bigger L and bigger λ the quality of the solution increases (Glumac and Kovačić 2013).

As described in the previous section, the execution time of each μIA instance is evaluated in algorithm with the number of penalty function calls (45). This criterion is plotted in Fig. 10. From the analysis of (45) it can be concluded that the number of penalty function calls increases as L or λ increases.

In order to perform multiobjective optimization both μIA set and μGA set need to be presented in the objective space. The set of μIA and the set of μGA are presented respectively in Fig. 11 and Fig. 12. The comparison of their Pareto frontiers is presented in Fig. 13. It can be observed that the Pareto frontier of the μIA set tends to dominate the Pareto frontier of the μGA set. The high percentage (88%) of algorithms in the Pareto frontier of μGA set have a small mutation step $i_m \in \{1, 2\}$. This can suggest that best algorithms in the set of μGA tend to decrease crossover frequency in favor of mutation (Glumac and Kovačić 2013).

To test how the quality of μIA solution scales with the number of robot joints, the quality of the solution and the execution time of μIA set was measured for the robot model with $N = 12$ joints. Results in the objective space are presented in Fig. 14 and the comparison of Pareto frontiers of this set and the μIA set for the robot model with 6 joints is presented in Fig. 15.

ALGORITHM PERFORMANCE TESTING ON THE MICROCONTROLLER PLATFORM

A modified μIA was tested on the Microchip dsPIC30F4011 microcontroller platform (Glumac and Kovačić 2013). This microcontroller has 32 kB of program memory and 4 kB of working memory. While performing experiments its instruction frequency

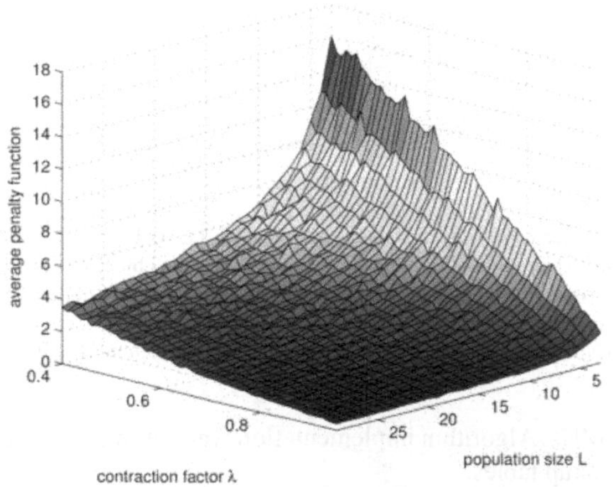

Figure 9. The mean value of μIA solution penalty function in μIA set for each combination of $L \in \{3, 4, \ldots, 29\}$, $\lambda \in \{0.4, 0.41, \ldots, 0.95\}$. © [2013] IEEE. Reprinted, with permission, from Glumac, S. and Z. Kovačić. 2013. Micro-immune algorithm for solving inverse kinematics of redundant robots. Proceedings of the 2013 IEEE International Conference on Systems, Man, and Cybernetics, pp. 201–207, Manchester, CPS (Conference Publishing Services), 2013.

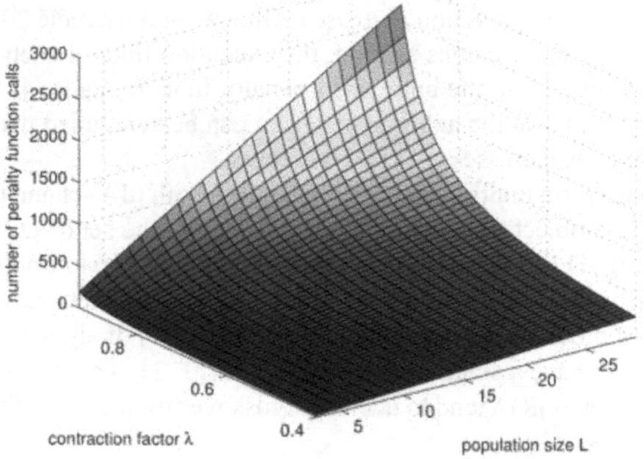

Figure 10. The number of penalty function calls in µIA set for each combination of $L \in \{3, 4, \ldots, 29\}$, $\lambda \in \{0.4, 0.41, \ldots, 0.93\}$. © [2013] IEEE. Reprinted, with permission, from Glumac, S. and Z. Kovačić. 2013. Micro-immune algorithm for solving inverse kinematics of redundant robots. Proceedings of the 2013 IEEE International Conference on Systems, Man, and Cybernetics, pp. 201–207, Manchester, CPS (Conference Publishing Services), 2013.

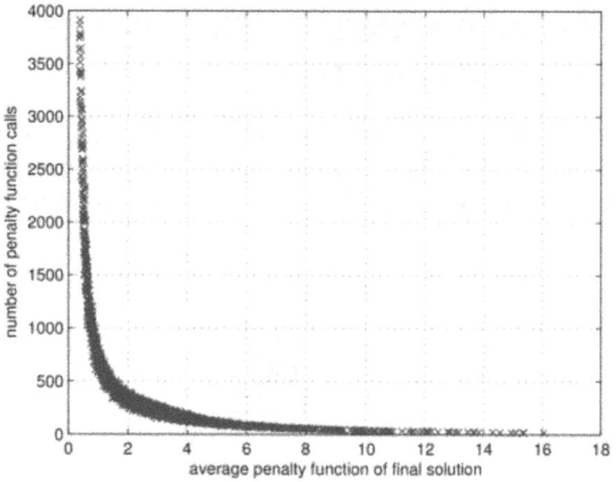

Figure 11. µIA set for each combination of $L \in \{3, 4, \ldots, 29\}$, $\lambda \in \{0.4, 0.41, \ldots, 0.99\}$ used in solving the inverse kinematics of a robot with $N = 6$ joints (presented in the objective space). © [2013] IEEE. Reprinted, with permission, from Glumac, S. and Z. Kovačić. 2013. Micro-immune algorithm for solving inverse kinematics of redundant robots. Proceedings of the 2013 IEEE International Conference on Systems, Man, and Cybernetics, pp. 201–207, Manchester, CPS (Conference Publishing Services), 2013.

was set to 29.49 MHz. Algorithm implementation was optimized using the fixed point arithmetic and lookup tables.

µIA implemented on the microcontroller is presented in the pseudocode block Algorithm 4. Two instances of µIA were tested; $\mu IA_{(7,0.93)}$ and $\mu IA_{(5,0.91)}$. Each instance was run 100 times on a randomly generated penalty function as in Algorithm 3. A reference pose $\mathbf{w}_{ref}$ was generated on the personal computer and sent to the microcontroller over CAN bus. The microcontroller ran the algorithm and returned a final solution and a

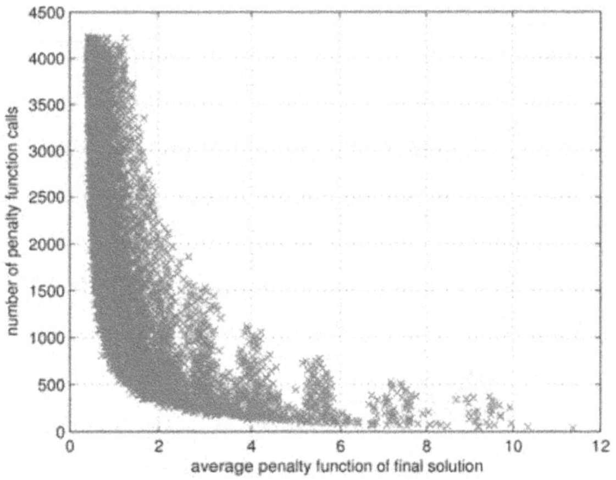

Figure 12. μGA set for each combination of $L = 5, \lambda \in \{0.55, 0.65, \ldots, 0.95\}$, $i_m \in \{1, 2, \ldots, 29\}$, $i_s \in \{i_m, 2\,i_m, \ldots\}$ used in solving the inverse kinematics of a robot with $N = 6$ joints (presented in the objective space). © [2013] IEEE. Reprinted, with permission, from Glumac, S. and Z. Kovačić. 2013. Micro-immune algorithm for solving inverse kinematics of redundant robots. Proceedings of the 2013 IEEE International Conference on Systems, Man, and Cybernetics, pp. 201–207, Manchester, CPS (Conference Publishing Services), 2013.

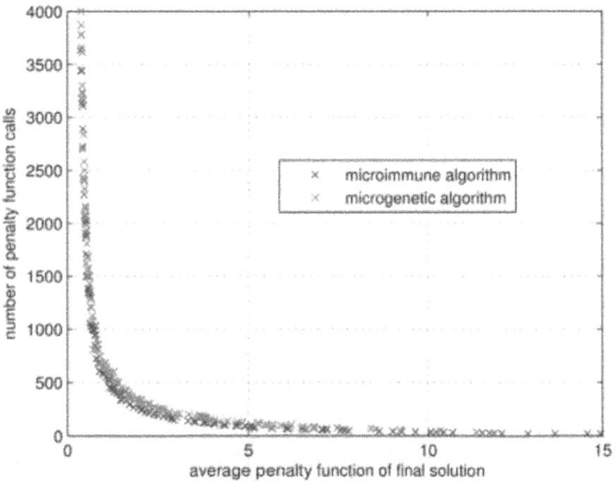

Figure 13. Pareto frontiers for the μIA set from Fig. 11 (black) and the μGA set from Fig. 12 gray. © [2013] IEEE. Reprinted, with permission, from Glumac, S. and Z. Kovačić. 2013. Micro-immune algorithm for solving inverse kinematics of redundant robots. Proceedings of the 2013 IEEE International Conference on Systems, Man, and Cybernetics, pp. 201–207, Manchester, CPS (Conference Publishing Services), 2013.

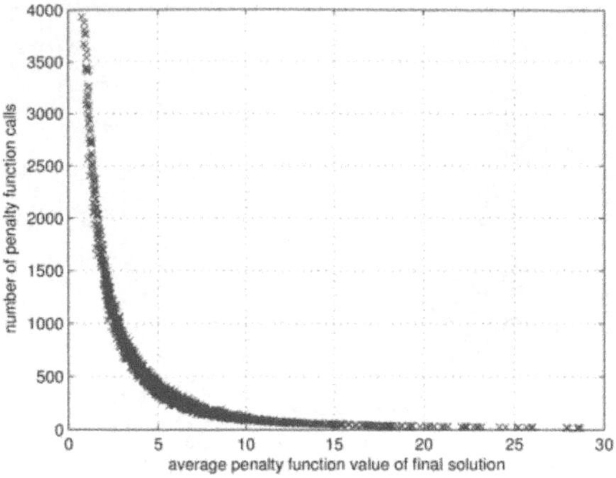

Figure 14. μIA set for each combination of $L \in \{3, 4, \ldots, 29\}$, $\lambda \in \{0.4, 0.41, \ldots, 0.99\}$ used in solving the inverse kinematics of a robot with $N = 12$ joints (presented in the objective space). © [2013] IEEE. Reprinted, with permission, from Glumac, S. and Z. Kovačić. 2013. Micro-immune algorithm for solving inverse kinematics of redundant robots. Proceedings of the 2013 IEEE International Conference on Systems, Man, and Cybernetics, pp. 201–207, Manchester, CPS (Conference Publishing Services), 2013.

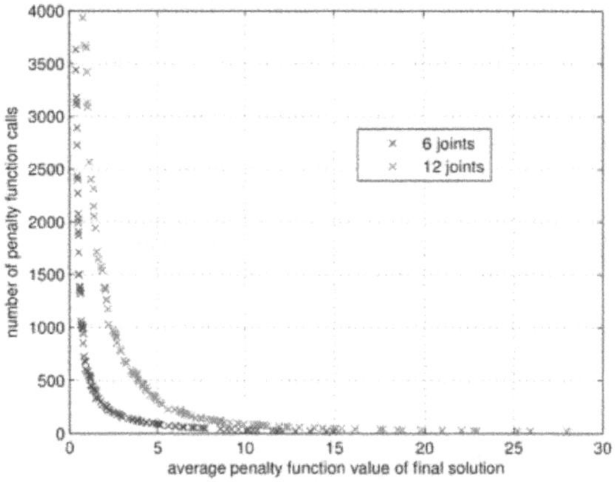

Figure 15. Pareto frontiers for the μIA set from Fig. 11 (black) and the μIA set from Fig. 14 (gray). © [2013] IEEE. Reprinted, with permission, from Glumac, S. and Z. Kovačić. 2013. Micro-immune algorithm for solving inverse kinematics of redundant robots. Proceedings of the 2013 IEEE International Conference on Systems, Man, and Cybernetics, pp. 201–207, Manchester, CPS (Conference Publishing Services), 2013.

measured execution time to the personal computer. Execution times for each algorithm run are presented in Fig. 16 and Fig. 17. The maximum execution times for $\mu IA_{(5,0.91)}$ and $\mu IA_{(7,0.93)}$ are respectively 55 ms and 98 ms. These execution times appear short enough for on-line inverse kinematics solvers in some robotic applications.

Algorithm 4 Pseudocode of μ1A[4]
initialize Population
while $J_w(\mathbf{q}) > J^*$ **and** $d < d^*$ **do**
mutate Population
contract Search space
end while

Validity of using the number of penalty function calls as a measure for execution time relies on the assumption that the maximum execution time in seconds is proportional to the maximum number of penalty function calls. This assumption appears correct for algorithms tested in this section and allows the maximum execution time of any μIA on dsPIC4011 to be assessed.

Since algorithm used on the microcontroller is given a wanted penalty function value f_p^*, the percentage of algorithm runs able to achieve that value is presented in Fig. 18. This figure presents the fact that μIA offers no guarantee to find an optimal solution. Despite this fact, μIA can find high quality solutions (Glumac and Kovačić 2013). Since the value of (9) cannot be used to determine position and orientation errors of the algorithm's final solution, these quantities are presented respectively in Fig. 19 and Fig. 20. Figure 19 shows the Euclidean distance deviation of the final solution from a reference end-effector position $||\mathbf{p} - \mathbf{p}_{ref}||_2$. Likewise, Fig. 20 shows the angular deviation of the approach vector obtained from a reference approach vector of end-effector $\cos^{-1}(\mathbf{a}^T \cdot \mathbf{a}_{ref})$. The quality of these solutions needs to be evaluated in relation to the actual size of the robotic arm. The robot used for evaluation (see Fig. 2) is a three-link arm with each link of length equal to 12.5 cm. In the hard-home position the distance between the end effector and the arm base is 37.5 cm.

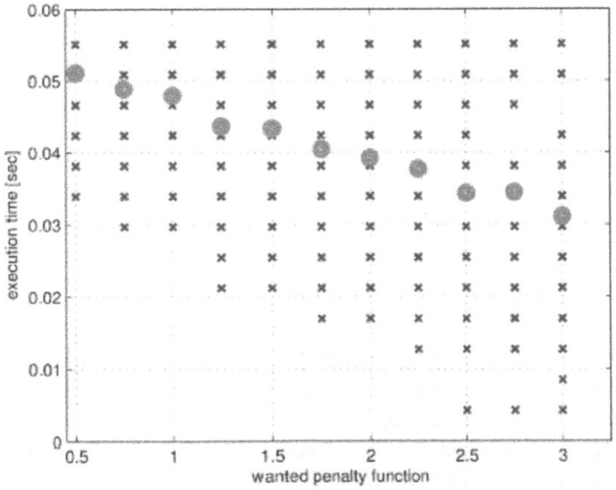

Figure 16. Run times in seconds for $\mu IA_{(5,0.91)}$ executed on PIC microcontroller, circles are the average case. © [2013] IEEE. Reprinted, with permission, from Glumac, S. and Z. Kovačić. 2013. Microimmune algorithm for solving inverse kinematics of redundant robots. Proceedings of the 2013 IEEE International Conference on Systems, Man, and Cybernetics, pp. 201–207, Manchester, CPS (Conference Publishing Services), 2013.

[4] © [2013] IEEE. Reprinted, with permission, from Glumac, S. and Z. Kovačić. 2013. Microimmune algorithm for solving inverse kinematics of redundant robots. Proceedings of the 2013 IEEE International Conference on Systems, Man, and Cybernetics, pp. 201–207, Manchester, CPS (Conference Publishing Services), 2013.

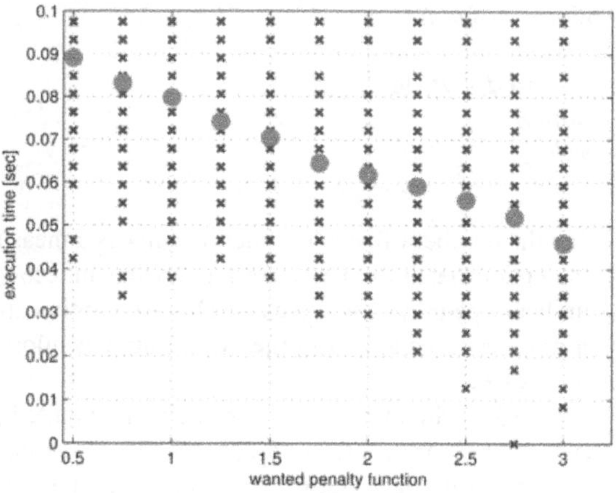

Figure 17. Run times in seconds for $\mu IA_{(7,0.93)}$ executed on PIC microcontroller with instruction, circles are the average case. © [2013] IEEE. Reprinted, with permission, from Glumac, S. and Z. Kovačić. 2013. Micro-immune algorithm for solving inverse kinematics of redundant robots. Proceedings of the 2013 IEEE International Conference on Systems, Man, and Cybernetics, pp. 201–207, Manchester, CPS (Conference Publishing Services), 2013.

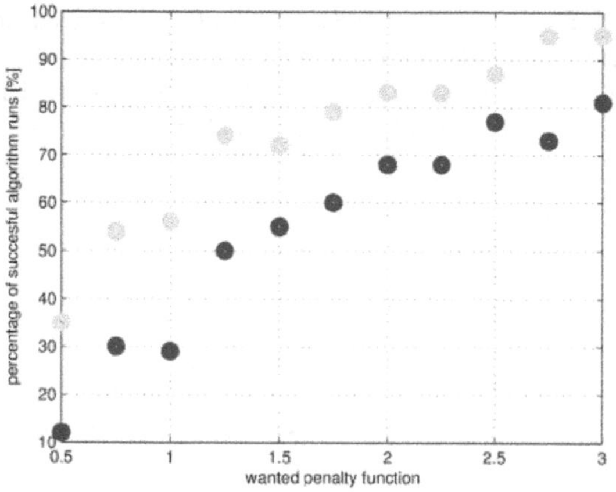

Figure 18. Ratio of runs for $\mu IA_{(5,0.91)}$ (black) and $\mu IA_{(7,0.93)}$ (light gray) that have found the solution with the wanted value of the penalty function. © [2013] IEEE. Reprinted, with permission, from Glumac, S. and Z. Kovačić. 2013. Micro-immune algorithm for solving inverse kinematics of redundant robots. Proceedings of the 2013 IEEE International Conference on Systems, Man, and Cybernetics, pp. 201–207, Manchester, CPS (Conference Publishing Services), 2013.

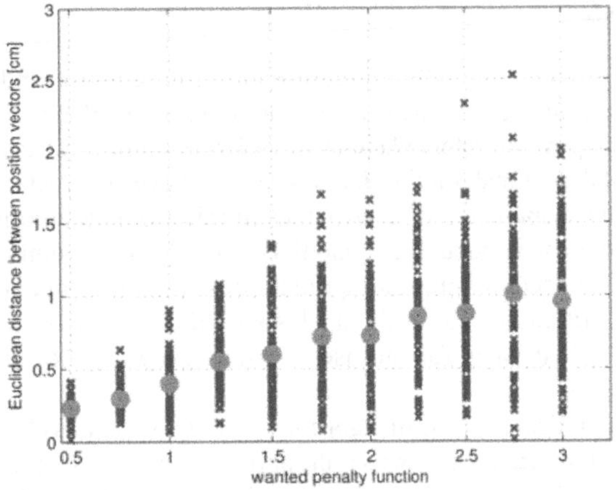

Figure 19. Position error for $\mu IA_{(7,0.93)}$ measured as the Euclidean distance between a reference and a calculated position of end-effector in cm. © [2013] IEEE. Reprinted, with permission, from Glumac, S. and Z. Kovačić. 2013. Micro-immune algorithm for solving inverse kinematics of redundant robots. Proceedings of the 2013 IEEE International Conference on Systems, Man, and Cybernetics, pp. 201–207, Manchester, CPS (Conference Publishing Services), 2013.

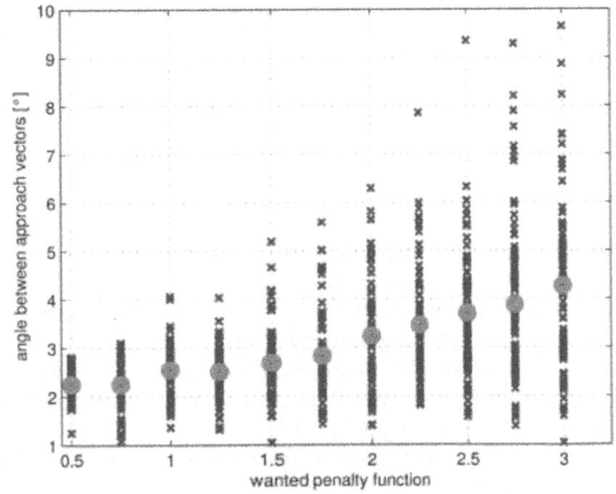

Figure 20. Orientation error for $\mu IA_{(7,0.93)}$ measured as the angle between a reference and a calculated approach vector of end-effector in degrees. © [2013] IEEE. Reprinted, with permission, from Glumac, S. and Z. Kovačić. 2013. Micro-immune algorithm for solving inverse kinematics of redundant robots. Proceedings of the 2013 IEEE International Conference on Systems, Man, and Cybernetics, pp. 201–207, Manchester, CPS (Conference Publishing Services), 2013.

CONCLUSIONS

This chapter is concerned with evaluation of microgenetic and microimmune optimization algorithms for solving inverse kinematics of a redundant robotic arm as a prerequisite for adaptive trajectory planning in various manipulation tasks. To ensure the quickest search of the optimal solution, it is desirable to have the structure and parameters of genetic algorithms optimized. The advantage of microgenetic algorithms is that they work with smaller genetic populations and thus allow for faster optimization runs.

The method of determining the optimal (best possible) mutation interval for microgenetic algorithms based on the analysis of their behavior is described. This provides a theoretical basis for the use of microgenetic algorithms in concrete optimization problems.

Since the theoretical analysis of algorithms is not always possible, multiobjective optimization of optimization algorithms is the approach being proposed and presented in this chapter. This kind of metaoptimization provides a straightforward way to determine a good optimization algorithm and its parameters without the need for making a deep theoretical analysis. Additional benefit of this procedure is that it provides the means to find out whether the considered type of optimization algorithm is a feasible candidate for solving a particular optimization problem.

As shown first by numerical experiments with μGA and μIA, and then by performance testing of μIA on a standard low-cost microcontroller platform, μIA enables the solution for inverse kinematics problem of a redundant robot with a small expected error. This shows that simplicity of μGA/μIA and low memory consumption are features that determine their very good potential for use in robust robotic systems trying to adapt their motion to slower and faster changes of their working environment.

REFERENCES

Alvarez, G. 2002. Can we make genetic algorithms work in high–dimensionality problems? Stanford Exploration Project (SEP) report, 112, 2002.

Aydin, K. K. and E. Kocaoglan. 1995. Genetic algorithm based redundancy resolution of robot manipulators. 3rd International Symposium on Uncertainty Modelling and Analysis, p. 322.

Beaty, S., D. Whitley and G. Johnson. 1990. Motivation and framework for using genetic algorithms for microcode compaction. Proceedings of the 23rd Annual Workshop in Micro-programming and Micro-architecture (MICRO-23), pp. 117–124.

Brock, O., J. Kuffner and J. Xiao. 2008. Motion for manipulation tasks. Springer Handbook of Robotics, pp. 615–645, Springer, 2008.

Chen, J., Q. Lin and Z. Ji. 2010. A hybrid immune multiobjective optimization algorithm. European Journal of Operational Research (Elsevier), 204(2): 294–302. 16 July 2010.

Ching, W., S. Zhang and M. K. Ng. 2007. On multi-dimensional Markov chain models. Pacific Journal of Optimization, 3: 235–243.

Chirikjian, G. S. and J. W. Burdick. 1991. The kinematics of hyper-redundant robot locomotion. IEEE Transactions on Robotics and Automation, 11(6): 781–793, Dec. 1995.

Crane, C. D. and J. Duffy. 1998. Kinematic Analysis of Robot Manipulators, Cambridge University Press, 1998.

Dasgupta, D. 2006. Advances in artificial immune systems. Computational Intelligence Magazine, IEEE, 1(4): 40–49, Nov. 2006.

Davidor, Y. 1991. Genetic algorithms and robotics. World Scientific, 1991.

Derigs, U., M. Kabath and M. Zils. 1997. Adaptive genetic algorithms: A methodology for dynamic autoconfiguration of genetic search algorithms, MIC97, Sophia-Antipolis, France, pp. 231–248, 1997.

Dužanec, D. and Z. Kovačić. 2009. Performance analysis-based GA parameter selection and increase of μGA accuracy by gradual contraction of solution space. IEEE Conference on Industrial Technology ICIT 2009, 10–13 February 2009, Gippsland, Victoria, Australia.

Dužanec, D. and Z. Kovačić. 2009. Determination of optimal mutation interval for μGA based on the performance analysis of GA and μGA. Proceedings of the European Control Conference 2009, Budapest, pp. 3214–3220.

Elmihoub, T., A. A. Hopgood, L. Nolle and A. Battersby. 2004. Performance of hybrid genetic algorithms incorporating local search. Proceedings 18th European Simulation Multi-conference, Graham Horton (c) SCS Europe, pp. 154–160.

Fuentes, J. C. and C. A. Coello. 2007. Handling constraints in particle swarm optimization using a small population size. Lecture Notes in Computer Science, MICAI 2007: Advances in Artificial Intelligence, Vol. 4827, Springer, 2007.

Galicki, M. 2007. Generalized kinematic control of redundant manipulators. Lecture on Information Sciences, 2007, pp. 219–226.

Glumac, S. and Z. Kovačić. 2013. Micro-immune algorithm for solving inverse kinematics of redundant robots. Proceedings of the 2013 IEEE International Conference on Systems, Man, and Cybernetics, pp. 201–207, Manchester, CPS (Conference Publishing Services), 2013.

Goldberg, D. E. 1989. Genetic Algorithms in Search, Optimization and Machine Learning, Addison Wesley, Reading, MA, 1989.

Hartl, R. F. 1990. A Global Convergence Proof for a Class of Genetic Algorithms, Univ. of Technology, Vienna, 1990.

Holland, J. H. 1975. Adaptation in Natural and Artificial Systems, University of Michigan Press, 1975.

Huang, H. C., C. P. Chen and P. R. Wang. 2012. Particle swarm optimization for solving the inverse kinematics of 7-DOF robotic manipulators. Systems, Man and Cybernetics (SMC), 2012 IEEE International Conference on, 2012.

Kelly Senecal, P. 2000. Numerical optimization using then GEN4 micro-genetic algorithm code, Engine research center, University of Wisconsin-Madison, Draft Manuscript, August 2000.

Krishnakumar, K. 1989. Micro-genetic algorithms for stationary and non-stationary function optimization. pp. 289–296. *In*: SPIE Proceedings: Intelligent Control and Adaptive Systems, 1989.

Lin, Q. and J. Chen. 2011. A novel micro-population immune multiobjective optimization algorithm. Computers & Operations Research (Elsevier), on-line, November 2011.

Mezura, E., J. Velázquez and C. A. Coello. 2006. A comparative study of differential evolution variants for global optimization. pp. 485–492. *In*: ACM, GECCO 2006.

Michlewicz, Z. 1992. Genetic Algorithms + Data Structures = Evolution Programs, Springer Verlag, 1992.

Michlewicz, Z. 2005. Evolutionary Computation for Modelling and Optimization, Springer, 2005.

Nunes de Castro, L. and F. J. Von Zuben. 2002. Learning and optimization using the clonal selection principle. IEEE Trans. Evol. Comput., 6(3): 239–251, June 2002.

Nunes de Castro, L. and J. Timmis. 2002. Artificial Immune Systems: A new Computational Intelligence Approach, Springer, 2002.

Poli, R., J. E. Rowe and N. F. McPhee. 2001. Markov chain models for GP and variable-length GAs with homologous crossover, Proceedings of the Genetic and Evolutionary Computation Conference GECCO-2001, pp. 112–119.

Prügel-Benett, A. and J. L. Shapiro. 1974. An analysis of genetic algorithms using statistical mechanics. Phys. Rev. Lett., 72: 75–114.

Schmitt, F. and F. Rothlauf. 2001. On the mean of the second largest eigenvalue on the convergence rate of genetic algorithms. Working Paper 1/2001, Morgan Kaufmann Publishers, San Francisco, California.

Simões, A. and E. Costa. 2003. Improving the genetic algorithm's performance when using transformation, International Conference on Neural Networks and Genetic Algorithms (ICANNGA'03), pp. 175–181, Roanne, France, 23–25 April, Springer, 2003.

Srivastava, P. and David E. Goldberg. 2001. Verification of the Theory of Genetic and Evolutionary Continuation. IlliGAL Report No. 2001007, January, 2001.

Stewart, W. J. 2009. Probability, Markov chains, queues and simulation: the mathematical basis of performance modelling, Princeton University Press, 2009.

Tian, Y., J. Qian and F. Meng. 2005. Optimal design of matched load by immune micro genetic algorithm. Progress in Electromagnetics Research Symposium, Hangzhou, China, August 22–26, pp. 101–104.

Toscano, G. and Carlos A. Coello. 2001. A micro-genetic algorithm for multiobjective optimization. *In*: First International Conference on Evolutionary Multi-criterion Optimization, Lecture Notes in Computer Science, Vol. 1993, Springer, 2001, pp. 126–140.

Vose, M. D. and G. E. Lipiens. 1991. Punctuated equilibria in genetic search. Complex Systems, 5: 31–44.

Vose, M. D. 1993. A Critical Examination of the Schema Theorem, Technical Report ut-cs-93-212, University of Tennessee, Knoxville, 1993.

Wampler, C. W. 1986. Manipulator inverse kinematic solutions based on vector formulations and damped least-squares methods. Systems, Man and Cybernetics, IEEE Transactions on, 1986.

Wong, E. Y. C. and H. Y. K. Lau. 2009. Advancement in the twentieth century in artificial immune systems for optimization: Review and future outlook. Systems, Man and Cybernetics, 2009. SMC 2009. IEEE International Conference on, 2009.

Zamanifar, K. and M. Koorangi. 2007. Designing optimal binary search tree using parallel genetic algorithms. IJCSNS International Journal of Computer Science and Network Security, 7(1): 138–146, January 2007.

Zhang, Q. and H. Mühlenbein. 2004. On the convergence of a class of estimation of distribution algorithms. IEEE Transactions on Evolutionary Computation, 8(2): 127–136, April 2004.

Zhang, Z. 2007. Immune optimization algorithm for constrained nonlinear multiobjective optimization problems. Applied Soft Computing (Elsevier), 7(3): 840–857, June 2007.

Index

Q

R

S

T

U

Milton Keynes UK
Ingram Content Group UK Ltd.
UKHW021839071024
449327UK00021B/1520

9 780367 782610